The Triumph of Consciousness

Overcoming False Environmentalism, Lapdog Media and Global Government

Christopher James Clark

ProgRESS**ive**

Progressive Press

The TRIUMPH of CONSCIOUSNESS

Overcoming False Environmentalism, Lapdog Media and Global Government

Published April 16th, 2010 by Progressive Press, http://ProgressivePress.com

Library of Congress Cataloging-in-Publication Data applied for.
BISAC Subject area codes:
POL033000—POLITICAL SCIENCE/Globalization
SCI042000—SCIENCE/Earth Sciences/Meteorology & Climatology
SOC052000—SOCIAL SCIENCE/Media Studies

Length: 167,000 words on 347 pages
ISBN: 1-61577-370-3. ISBN13/EAN: 978-1-61577-370-1.

On the cover: La Pasionaria Memorial in Glasgow, Scotland, a dedication to Dolores Ibárruri, a popular defender of the Spanish republic, and to the thousands of foreigners who, during the 1930s, went to Spain to defend freedom in the face of fascism. The base of the statue bears the inscription, "Better to die on your feet than live forever on your knees."

Cover photo by Patrick McGuire, www.paddimir.com
Cover design by Christopher Clark and Jason Haynes

The Triumph of Consciousness traces the history of globalization during the past hundred-plus years by examining the writings and public statements of various financial and political elites who are determined to implement an authoritative, non-democratic, global government. At the forefront of this movement are the Anglo-American banking establishment, the Rockefeller Foundation, the Ford Foundation, the Bilderberg Group, the Council on Foreign Relations, the Trilateral Commission, the Club of Rome, and others. By various mechanisms, these same organizations control the environmental movement and promulgate its prevailing ethos: Earth is overpopulated, and humans are causing potentially catastrophic global warming. But are these postulations based on concrete, scientific evidence? Or were they contrived to further an ulterior agenda? Why don't the mainstream corporate media investigate these questions? *The Triumph of Consciousness* addresses these issues, while reminding readers that societal change begins when individuals overcome their fear and apathy and embrace their inherent strength and wisdom.

TABLE OF CONTENTS

For Linda, Kevin, Jeff and Dan.
Thanks for everything.

INTRODUCTION

"Power at its best is love implementing the demands of justice, and justice at its best is power correcting everything that stands against love."[1] — Martin Luther King, Jr.

They were open-minded, warmhearted, highly intelligent people. They seemed genuinely interested in peace, prosperity, and social justice. Nevertheless, something about them troubled me. It was a matter of perception. It was how they perceived negativity.

One winter 2009 afternoon, chance circumstances brought us all together for lunch at our local organic foods restaurant. After the prerequisite small talk, the conversation shifted towards the disastrous Bush administration and the fledgling Obama administration. While we all had been expecting "change," I alone saw Obama as simply the latest puppet, controlled by the same Wall Street puppeteers who controlled Bush. "Obama," I remarked, "received *outrageous* Wall Street campaign contributions.[2] Obama championed the banker bailouts despite the people's adamant opposition.[3] During his campaign, Obama's foreign policy advisor was Zbigniew Brzezinski, the *founder* of the Trilateral Commission.[4] Obama has since filled his administration with Council on Foreign Relations and Trilateral Commission members. If he's really dedicated to change, why are we seeing all the same old faces? If he really wants to restore the country's dignity, why won't he repeal the Patriot Act, the Military Commissions Act, the John Warner Defense Authorization Act, and NSPD 51? Collectively, these Bush-era laws gutted the Constitution, stripped Americans of vital civil liberties, and facilitated the build-up for martial law in America."

Patrick countered, "We're living in a post-9/11 world. We need to protect ourselves from terrorists." "I couldn't agree more," I replied, "but who are the terrorists? Are they disgruntled Muslims hiding out in faraway caves? Or are they treasonous corporate, military and government insiders?" Maria asked, "Are you speaking about conspiracy theories?" "I'm speaking about conspiracy facts," I replied, "but since you mention it, have you ever wondered why the media and popular culture incessantly ridicule *conspiracy theories*? Aren't most crimes conspiracies, or at least a significant portion of them? Conspiracy simply means two or more people working secretly to perpetrate some unlawful or unethical act. Is that so uncommon? Even the government's official explanation of 9/11 is a *conspiracy theory*. They claim that nineteen Islamic extremists *conspired* to fly hijacked jets into buildings. And since this explanation has never been proven, it is, by definition, a *conspiracy theory*."

Maria replied, "Yes, but you're talking about prominent people committing *high treason*. You're implying the government is lying about 9/11." "I'm not implying," I responded, "I'm saying, definitively, they are lying. Volumes of evidence support this assertion. First of all, as reported by the BBC, *Washington Post*, *Los Angeles Times* and others, at least *seven* of the alleged nineteen hijackers are *alive*.[5] They were misidentified and falsely accused. The FBI, however, never retracted nor revised their original list. Furthermore, they never formally charged Osama bin Laden. His FBI Most Wanted profile accuses him of earlier terror attacks, yet does not condemn him for 9/11."[6] In 2006, FBI spokesperson Rex Tomb explained, "The reason why 9/11 is not mentioned on Osama bin

Laden's Most Wanted page is because the FBI has no hard evidence connecting bin Laden to 9/11."[7] Astonished, Maria looked me in the eyes and asked, "You're kidding, right?"

"I wish I was," I replied, "but this is on record. And it gets worse. For example, the government contends that airplanes and fires caused the towers to collapse. According to the laws of physics, however, such gravity-driven collapses should have taken at least 60 seconds. Instead, each tower collapsed *symmetrically* into its own footprint *at the speed of gravity*—roughly 10 seconds. We should have seen *massive* piles of rubble, the remains of two *110-story* buildings. Instead, pyroclastic flows of finely pulverized dust rapidly dispersed across lower Manhattan. Pyroclastic flows occur *only* during volcanic eruptions and controlled demolitions of buildings. The World Trade Center dust, according to multiple peer-reviewed scientific studies, had the *precise* chemical signature of thermite, a high-powered explosive.[8] The government refuses to explain why."

Patrick skeptically asked, "So you're saying explosives brought down the buildings? Does any other evidence support this contention?" "Plenty," I replied. "For example, *molten metal*, at temperatures far-exceeding those physically possible from jet fuel-based fires, was found in the rubble of *all three* towers.[9] Also Columbia University recorded significant seismographic activity *before* the collapse of each tower. A 2.1 tremor occurred just before the collapse of the South Tower, a 2.3 before the North Tower, and a 0.6 before Building 7."[10]

"Building 7? What's that?" Sarah asked. I answered, "Most people only know about the twin towers. But at 5:20 that afternoon, a 47-story building collapsed at free-fall speed—just 7 seconds—again neatly into its own footprint. Oddly enough, *no plane hit Building 7*. The government claims that fires must have weakened the building's core columns. Then, somehow, every floor, truss, and column all simultaneously failed—at *precisely* the same time. That's simply preposterous. Furthermore, it would have been the first time in architectural history that a steel-structured building collapsed due to fire. Thus it should have been completely unexpected. Yet somehow Fox, CBS, and the BBC all reported on Building 7's collapse twenty minutes *before* it actually happened! Reuters released the story—just a little too soon—and the networks reported.[11] How could they have possibly known?

Frankly, every aspect of the government's official *conspiracy theory*—the Twin Towers, Building 7, the Pentagon, and the Pennsylvania crash—is absurd and completely indefensible. Furthermore, thousands of prominent architects, engineers, scientists, pilots, professors, politicians, intelligence officers, and military officials, have rejected the official story and subsequently called for a new, independent investigation.[12] For example, check out Architects & Engineers for 9/11 Truth, an organization of over 1,100 professionals (and growing) who have put their careers and reputations on the line by declaring the government's story physically impossible.[13] Furthermore, scores of scholarly books, articles and films have documented the overwhelming evidence against the official conspiracy theory."

Sarah was shocked. "I can't believe I never heard about this! If what you're say is true, then the official story was an official cover-up." "It's true," I responded, "but it's not just me saying it. Actually, among the 10 commissioners who authored the government's *official* 9/11 Commission Report, *six* have publicly doubted its veracity. For example, commissioner Max Cleland declared,

- "[The commission] was deliberately compromised by the president of the United States."[14]
- "It should be a national scandal."[15]
- "One of these days we will have to get the full story because the 9/11 issue is so important to America. But this White House wants to cover it up."[16]

Commissioner Timothy Roemer commented,

- "Never have so few commissioners reviewed such important documents with so many restrictions."[17]
- "We were extremely frustrated with the false statements we were getting."[18]

Commissioner Bob Kerry acknowledged,

- "There are ample reasons to suspect that there may be some alternative to what we outlined in our version."[19]

Commissioner John Lehman conceded,

- "We *purposely* put together a staff that had, in a way, conflicts of interest."[20]

Thomas Kean and Lee Hamilton, the Co-Chairmen of the commission, revealed that,

- The CIA "obstructed our investigation."[21]
- The commission was "set up to fail."[22]

And finally, John Farmer, the commission's Senior Counsel, lamented,

- "I was shocked at how different the truth was from the way it was described. The [NORAD] tapes told a radically different story from what had been told to us and the public for two years. This is not spin. This is not true."[23]

Maria countered, "But why would they attack their own country? It just doesn't make sense." I explained, "In the late 1990s, future Bush administration insiders including Dick Cheney, Donald Rumsfeld, Paul Wolfowitz, Elliot Abrams, Richard Perle, John Bolton, Scooter Libby and Jeb Bush (George's brother), established a think-tank called the Project for a New American Century. In a now-infamous September 2000 document, they insisted that America must wage 'simultaneous major theater wars' in the Middle East. They acknowledged, however, that the public would never support such an agenda unless some 'catastrophic and catalyzing event—like a new Pearl Harbor' were to occur on American soil.[24] Miraculously, they got their wish.

It's called false flag terrorism—an inside job. A rogue network, operating through corporate, government and military power structures, plans and orchestrates a terror attack against the citizens they purport to represent. Then they quickly blame the attack on foreign enemies to justify an otherwise unpopular course of action. Indeed, the pretexts for most wars usually involve false flag terror. The sinking of the USS Maine, for example, justified the Spanish American War. The sinking of the Lusitania justified America's entry into World War I.[25] Pearl Harbor justified America's entry into World War II.[26] The Gulf of Tonkin Incident justified the Vietnam War.[27] And of course, 9/11 justified the wars in Afghanistan and Iraq. These were *all* false flag terror attacks.

But you asked why they would attack their own country. Well, there's another reason. The so-called Americans who planned and orchestrated 9/11 and other false flags don't consider themselves Americans. They consider themselves globalists. They swear oaths to the U.S. Constitution, yet they serve an ulterior agenda commonly known as the New World Order. While posturing as Americans, they covertly work to weaken national sovereignty while strengthening various transnational organizations. Ultimately, they intend to impose global government."

Sarah interjected, "But maybe global government would be good. Maybe we could end all wars and finally establish world peace." "It's a beautiful concept," I replied, "but far from the objective these globalists are working towards. According to their own documents, they intend to implement an authoritarian, *neo-feudal* world government run by bankers and bureaucrats. Furthermore, they intend to drastically reduce world population, down to about one billion people. Through various round table groups including the Council on Foreign Relations, the Bilderberg Group and the Trilateral Commission, they have incrementally advanced this agenda over the past hundred-plus years. They've created banking panics and depressions. They've created wars and international crises. Having hijacked the environmental movement, they're now hyping fraudulent doomsday theories regarding overpopulation and global warming. All the while, they maintain that only global government can avert future catastrophe. Only global government can secure peace and prosperity. After World War II, they established the United Nations in preparation for their planned future society. Years later, as further steppingstones, the same people established the European Union and the forthcoming North American Union. This history is all well-documented."

"But this is so negative. I prefer to focus on positive things," Sarah stated. Patrick agreed, "Yes. Negativity is like a black hole. It absorbs everything in its vicinity. There are so many positive things happening in the world, so many people trying to make the world a better place. If we direct enough positive energy towards these positive endeavors, the negativity will diminish." Maria added, "I agree. We can't solve our problems by blaming politicians and bankers. Love is the answer, only unconditional love."

The conversation subsided. We focused on our meals. I considered dropping the issue altogether. However, after what seemed like several minutes, I cautiously proceeded: "I agree, ultimately we have only ourselves to blame. Our ignorance and apathy have enabled this agenda to progress. Quite simply, most people nowadays are only semiconscious. On the other hand, those who are planning and overseeing this agenda are slightly *more* conscious. I mean they are more aware, more sophisticated, because they have studied human psychology, sociology and behaviorism. They know how to manipulate and exploit populations. Perhaps it's uncomfortable to see them as *more* conscious. They would seem to be *less* conscious. After all, they behave like animals, preying upon the weak. However, *we* have capitulated to this treatment. Along the scale of human consciousness-potential, they are indeed quite un-advanced, quite crude and rudimentary. However, collectively, we are even less sophisticated. Nevertheless, consciousness is not fixed. If we resolve to do so, we can raise our consciousness and surpass these tyrants."

"Consciousness," I continued, "is awareness. To become more aware requires inquiry. To solve any problem, one must first understand and address, without judgment, the various causes of that problem. Earthquakes, for example, are problematic. They cause major death and destruction. However, earthquakes, as a phenomenon, are neither negative nor positive. They simply are what they are. Is it negative to discuss earthquakes?

Is it negative to study them and develop strategies for minimizing their destructive impact? As most people would agree, these are both positive and desirable pursuits. Through careful analysis, engineers and architects have learned how to design and construct buildings and bridges able to withstand earthquakes. Through their work, humanity has become collectively *more conscious*.

The reward for consciousness is freedom. The more conscious we are, the more freedom we have. Yet the expansion of consciousness requires energy, of which the two primary sources are fear and love. Fear-driven consciousness is extremely limiting. Nevertheless, most people, to some degree, are driven by fear. Those who serve the New World Order are driven by persistent, extreme fear. Ordinary people, on average, are driven by intermittent, lazy fear. Extreme fear can induce slightly higher levels of consciousness than lazy fear, but love-driven consciousness triumphs over all. So I agree with you, love *is* the answer, but we must understand what love actually *is*.

Love is the energy of creation. Love is active, not passive. Love never shies away from truth, no matter how uncomfortable it may be. Love is highly intelligent. Love knows that effective solutions require intelligent action. Love is the strongest force in the universe, and perhaps that's why most people are so hesitant to embrace it. Because embracing it means you might actually have to *do* something with it. In other words, love provides magnificent energy, yet demands incredible responsibility. Why are we so reluctant to honestly and objectively analyze our societal problems and formulate practical, insightful solutions? Are we afraid of identifying, exposing and condemning our manipulators? Love is not necessarily polite. Love does whatever the situation requires.

If we want genuine peace and prosperity, we must understand our societal institutions. If ruthless psychopaths are indeed imposing global tyranny, we must acknowledge this unpleasant reality. Through education and concerted action, we can harness our inherent power. And by embracing love, authentic love, we can become more consciousness, and thus more free. The globalists fully expect to achieve their New World Order plans. They believe we are simply too ignorant and too apathetic to resist them. Now is the time to rise up in defense of sovereignty and freedom. Now is the time to *be* the change you wish to see. This is the *Triumph of Consciousness*. The first step is education. We have the keys, but with them we must unlock the minds of men." Nobody spoke. But their eyes glimmered with empowerment.

PART I: TYRANNICAL GLOBAL GOVERNMENT

"Single acts of tyranny may be ascribed to the accidental opinion of a day; but a series of oppressions, begun at a distinguished period, and pursued unalterably through every change of ministers, too plainly prove a deliberate, systematical plan of reducing us to slavery."[1]*–Thomas Jefferson*

There's nothing new about it. The populations of ancient Egypt and Rome, for example, endured similar oppression. The so-called "New World Order" is simply the latest attempt by elite control-freaks to subjugate and exploit humanity. In March 2009, New Hampshire State Representative Dan Itse insightfully observed, "There's the phrase the 'New World Order.' But do you know what the New World Order is? It's the Old World Order. The New World Order is what was established here in the new world of government *by the people* with the people on top and the government at the bottom. This gravitation towards an oligarchic or monarchic society, a hierarchical top-down government, that's the Old World Order, and we've got to stop the Old World Order from being established here in America."[2]

For decades, the ruling class has been publicly announcing and promoting its planned New World Order. On September 11, 1990, for example, George H.W. Bush, speaking before a Joint Session of Congress, declared, "The crisis in the Persian Gulf, as grave as it is, also offers a rare opportunity to move toward an historic period of cooperation. Out of these troubled times, our fifth objective—a New World Order—can emerge: a new era—freer from the threat of terror, stronger in the pursuit of justice, and more secure in the quest for peace."[3]

During his presidency, Bush frequently alluded to the New World Order. Several weeks later, for example, while speaking before the United Nations in New York, he announced that the UN could "cap an historic movement towards a New World Order and a long era of peace."[4] Then, during his State of the Union address on January 29, 1991, Bush proclaimed, "What is at stake is more than one small country; it is a big idea: a New World Order, where diverse nations are drawn together in common cause to achieve the universal aspirations of mankind—peace and security, freedom, and the rule of law."[5]

Like most elites, Bush always spoke cryptically about the New World Order, highlighting, for example, its promised peace and security. Yet never would he elaborate. He never mentioned, for example, that *scientific tyranny* also provides "peace." Although the people are slaves, they nevertheless have "security." Former Soviet Premier Mikhail Gorbachev, another prominent globalist, also frequently speaks cryptically about the New World Order. In his book *On My Country and the World*, Gorbachev declared,

"It is necessary, in other words, to find roads leading to a new civilization...we understand that building such a civilization is a long-term task (although on the scale of history a task that cannot be postponed). Few people in the world are ready for the profound, fundamental changes required for the creation of this civilization. What, then, should be done? We should not try to effect immediate all-embracing changes; rather, we should move toward such changes step by step, finding urgent solutions where they are absolutely necessary, and partial solutions where nothing else can yet be done. Solutions will gradually enlarge the field of agreement and the range of possibilities for later, more substantial measures."[6]

In the same book, Gorbachev proclaimed, "The socialist idea has not lost its significance or its historical relevance…the socialist idea is inextinguishable."[7] And furthermore, "In the final analysis, a system for collective management of worldwide processes must be created, an effective form of collaboration based on equality among nations and peoples. We must know how to combine and jointly subordinate national interests and actions for the sake of worldwide interests and actions."[8] Again, Gorbachev's "new civilization" was simply a euphemism for sophisticated, scientific tyranny. Why else should such "few people" be ready for "the profound, fundamental changes required?" Why else must national interests be subordinated?

The New World Order, properly defined, is *scientific tyranny*. On December 15, 1987, Jesse Helms, a five-term U.S. Senator and Chairman of the Senate Foreign Relations Committee from 1995 to 2000, addressed the Senate regarding such tyranny. He exclaimed,

"This campaign against the American people—against traditional American culture and values—is systematic psychological warfare. It is orchestrated by a vast array of interests comprising not only the Eastern establishment but also the radical left. Among this group we find the Department of State, the Department of Commerce, the money center banks and multinational corporations, the media, the educational establishment, the entertainment industry, and the large tax-exempt foundations. Mr. President, a careful examination of what is happening behind the scenes reveals that all of these interests are working to create what some refer to as a New World Order. Private organizations such as the Council on Foreign Relations, the Royal Institute of International Affairs, the Trilateral Commission, the Dartmouth Conference, the Aspen Institute for Humanistic Studies, the Atlantic Institute, and the Bilderberg Group serve to disseminate and to coordinate the plans for this so-called New World Order in powerful business, financial, academic, and official circles."[9]

Helms concluded, "The influence of establishment insiders over our foreign policy has become a fact of life in our time. This pervasive influence runs contrary to the real long-term national security of our nation. It is an influence which, if unchecked, could ultimately subvert our constitutional order."[10]

Since the early 1900s, the Rockefeller family has tirelessly served the New World Order vision. Through banking, government, medicine, education, eugenics, phony philanthropy, and false environmentalism, Rockefeller family members have systematically destabilized the United States in preparation for world government. David Rockefeller, the family's current patriarch, openly despises the people, traditions, and Constitution of America. In his 2002 book *Memoirs*, Rockefeller wrote,

"For more than a century ideological extremists at either end of the political spectrum have seized upon well-publicized incidents such as my encounter with Castro to attack the Rockefeller family for the inordinate influence they claim we wield over American political and economic institutions. Some even believe we are part of a secret cabal working against the best interests of the United States, characterizing my family and me as 'internationalists' and of conspiring with others around the world to build a more integrated global political and economic structure—one world, if you will. If that is the charge, I stand guilty, and I am proud of it."[11]

What sort of New World Order was Rockefeller envisioning? In 1973, upon returning from a ten-day tour of China, he wrote an article for the *New York Times* recounting his experience and impressions. Rockefeller admired and praised Mao Tse-tung's authoritarian regime, which subdued the Chinese people through fear, intimidation and propaganda, and which ultimately slaughtered at least 35 million.[12] Rockefeller reflected, "One is impressed immediately by the sense of national harmony. From the loud patriotic music at the border onward, there is very real and pervasive dedication to Chairman Mao and Maoist principles. Whatever the price of the Chinese Revolution, it has obviously succeeded not only in producing more efficient and dedicated administration, but also in fostering high morale and community of purpose." Rockefeller concluded, "The social experiment in China under Chairman Mao's leadership is one of the most important and successful in human history."[13]

The globalists believe that out of chaos, they can create order. Therefore, they regularly engineer crises to advance certain aspects of their agenda. Then they tell the unsuspecting public that such crises are actually *opportunities* for global integration. For example, on September 12, 2001, prominent Council on Foreign Relations (CFR) member Gary Hart disgracefully called upon President Bush to advance the New World Order by exploiting the previous day's tragedy. During a CFR meeting, broadcast live on C-SPAN, Hart stated, "There is a chance for the President of the United States to use this disaster to carry out a phrase his father used, I think only once, and hasn't been used since, and that is a New World Order."[14] His father, of course, used the phrase incessantly, as have scores of globalists before and after. Nevertheless, Hart was exploiting the tragedy to promote an even more nightmarish future, albeit one cloaked in feel-good rhetoric. More recently, on January 5, 2009, Henry Kissinger downplayed the latest Israeli assault on Gaza by claiming it wasn't "such a crisis." Rather, it was an opportunity for Obama to create a New World Order. CNBC interviewed Kissinger on the floor of the New York Stock Exchange where, regarding President-elect Obama, Kissinger stated, "He can give a new impetus to American foreign policy, partly because the perception of him is so extraordinary around the world. I think his task will be to develop an overall strategy for America in this period when, really a New World Order can be created. It's a great opportunity. It isn't such a crisis."[15]

Chapter 1

The New World Order Agenda

Since the early 1900s, according to former Georgetown history professor Carroll Quigley, the Anglo-American ruling class has sought "nothing less than to create a world system of financial control in private hands able to dominate the political system of each country and the economy of the world as a whole. This system was to be controlled in a feudalist fashion by the central banks of the world acting in concert, by secret agreements arrived at in frequent private meetings and conferences."[1] Many of these secret meetings took place at the Council on Foreign Relations (CFR), an elite organization of which Quigley was a member. A professor at Princeton and Harvard, and from 1941 to 1976 at Georgetown University's School of Foreign Service, Quigley had an insider's view of history. During the 1950s, he also consulted for the U.S. Department of Defense, the U.S. Navy, the Smithsonian Institution, and the House Select Committee on Astronautics and Space Exploration.[2] Quigley was also an important mentor to Bill Clinton. During his acceptance speech at the 1992 Democratic National Convention, Clinton recalled, "As a teenager, I heard John Kennedy's summons to citizenship. And then, as a student at Georgetown, I heard that call clarified by a professor name Carroll Quigley, who said to us that America was the greatest Nation in history because our people had always believed in two things—that tomorrow can be better than today and that every one of us has a personal, moral responsibility to make it so."[3]

During the mid-1960s, Quigley obtained special permission to peruse the CFR's private archives. At the time, he was conducting research for his magnum opus, a 1300-page book called *Tragedy and Hope*. Published in 1966, *Tragedy and Hope* painstakingly detailed the machinations of the Anglo-American ruling class. Although his book greatly embarrassed this ruling class, Quigley was not a whistle-blower. He explained, "I know of the operations of this network because I have studied it for twenty years and was permitted for two years, in the early 1960s, to examine its papers and secret records. I have no aversion to it or to most of its aims and have, for much of my life, been close to it and to many of its instruments. I have objected, both in the past and recently, to a few of its policies (notably to its belief that England was an Atlantic rather than a European Power and must be allied, or even federated, with the United States and must remain isolated from Europe), but in general my chief difference of opinion is that it wishes to remain unknown, and I believe its role in history is significant enough to be known."[4]

The Anglo-American Banking Establishment

According to Quigley, the agenda for global domination originated in England and America. He wrote extensively about the networks of power between London, New York and Washington. For example,

- "There does exist and has existed for a generation, an international Anglophile network which operates, to some extent, in the way the radical Right believes the Communists act. In fact, this network, which we may identify as the Round Table Groups, has no aversion to cooperating with the Communists, or any other groups, and frequently does so."[5]

- "The two ends of this English-speaking axis have sometimes been called, perhaps facetiously, the English and American Establishments. There is, however, a considerable degree of truth behind the joke, a truth which reflects a very real power structure. It is this power structure which the radical Right in the United States has been attacking for years in the belief that they are attacking the Communists. This is particularly true when these attacks are directed, as they so frequently are at 'Harvard Socialism,' or at 'Left-wing newspapers' like the *New York Times* and the *Washington Post*, or at foundations and their dependent establishments."[6]

- "There grew up in the twentieth century a power structure between London and New York which penetrated deeply into university life, the press, and the practice of foreign policy. In England the center was the Round Table Group, while in the United States it was J.P. Morgan and Company or its local branches in Boston, Philadelphia, and Cleveland."[7]

For centuries, international bankers have controlled governments by controlling the monetary system and by monetizing government debt. Former Senator Barry Goldwater observed, "International bankers make money by extending credit to governments. The greater the debt of the political state, the larger the interest returned to the lenders. The national banks of Europe are actually owned and controlled by private interests."[8] Quigley elaborated on how the scheme actually works:

"The power of investment bankers over governments rests on a number of factors, of which the most significant, perhaps, is the need of governments to issue short-term treasury bills as well as long-term government bonds. Just as businessmen go to commercial banks for current capital advances to smooth over the discrepancies between their irregular and intermittent incomes and their periodic and persistent outgoes (such as monthly rents, annual mortgage payments, and weekly wages), so a government has to go to merchant bankers (or institutions controlled by them) to tide over the shallow places caused by irregular tax receipts. As experts in government bonds, the international bankers not only handled the necessary advances but provided advice to government officials and, on many occasions, placed their own members in official posts for varied periods to deal with special problems."[9]

Such placement of bankers into official posts has today reached obscene proportions. Timothy Geithner, for example, the current Treasury Secretary, was formerly President of the *private, run-for-profit,* Federal Reserve Bank of New York. Henry Paulson, the Treasury Secretary under Bush, was previously the CEO of Goldman Sachs. Such arrangements have allowed for extraordinary abuses of power. On the advice of these men, for example, the banker bailouts of 2008-2009 funneled at least $12 *trillion*, possibly $23 *trillion* (according to Neil Barofsky, special inspector general for the Treasury's Troubled Asset Relief Program) from the citizenry back to the banks.[10] Such self-serving advice is standard operating procedure for the banking establishment. Quigley explained,

"In addition to their power over government based on government financing and personal influence, bankers could steer governments in ways they wished them to go by other pressures. Since most government officials felt ignorant of finance, they sought advice from bankers whom they considered to be experts in the field. The history of the last century shows, as we shall see later, that the advice given to governments by bankers, like the advice they gave to industrialists, was consistently good for bankers, but was often disastrous for governments, businessmen, and the people generally. Such advice could be enforced if necessary by manipulation of exchanges, gold flows, discount rates, and even levels of business activity."[11]

The international banking establishment has relentlessly imposed its private central bank model around the world. Quigley explained, "Each central bank, in the hands of men like Montagu Norman of the Bank of England, Benjamin Strong of the New York Federal Reserve Bank, Charles Rist of the Bank of France, and Hjalmar Schacht of the Reichsbank, sought to dominate its government by its ability to control Treasury loans, to manipulate foreign exchanges, to influence the level of economic activity in the country, and to influence cooperative politicians by subsequent economic rewards in the business world…They made agreements on all the major financial problems of the world, as well as on many of the economic and political problems, especially in reference to loans, payments, and the economic future of the chief areas of the globe."[12]

America's Enduring Battle Against the Banks

Despite its deceptive name, the United States Federal Reserve is not a federal institution. America's central bank is an independent, privately owned, run-for-profit corporation consisting of twelve regional Federal Reserve banks. In 1982, the United States Court of Appeals definitively ruled, "Each Federal Reserve Bank is a separate corporation owned by commercial banks in its region."[13] This, however, was old news. In the early 1930s, for example, Congressman Louis McFadden famously wrangled with the Federal Reserve. In June 1932, he declared before the House of Representatives, "Some people think the Federal Reserve banks are United States Government institutions. They are not Government institutions. They are private credit monopolies which prey upon the people of the United States for the benefit of themselves and their foreign customers, foreign and domestic speculators and swindlers, and rich and predatory money lenders."[14] McFadden also referred to the Federal Reserve as "one of the most corrupt institutions the world has ever known."[15] Indeed, the American people's struggle to control their currency and monetary policy has been an enduring and dominant theme of U.S. history:

- Thomas Jefferson, the third U.S. President (1801-09), prophetically warned, "If the American people ever allow private banks to control the issue of their money, first by inflation and then by deflation, the banks and the corporations that will grow up around them will deprive the people of their prosperity until their children will wake up homeless on the continent their fathers conquered."[16]

- Andrew Jackson, the seventh U.S. President (1829-37), famously battled, and defeated, the Second Bank of the United States. In 1834, he remarked, "The mischief and dangers which flow from a national bank far overbalance all its advantages. The bold effort the present bank has made to control the government, the distress it has wantonly produced…are but premonitions of the fate that awaits the American people should they be deluded into a perpetuation of this institution, or the establishment of another like it."[17] In 1835, Jackson became the only president ever to pay off the national debt and begin managing a *national surplus*.[18]

- Abraham Lincoln, the sixteenth U.S. President (1861-65), also sparred with the bankers. In 1862, desperately seeking war funds, he approached Wall Street banks for loans. The banks demanded exorbitant interest rates, which Lincoln refused. Instead, he authorized the government to print its *own* money, known as greenbacks.[19] In 1863, Lincoln declared, "The privilege of creating and issuing money is not only the supreme prerogative of government, but it is the government's greatest creative opportunity. By the adoption of these principles, the taxpayers will be saved immense sums of interest."[20] He added, "Money will cease to be master and become the servant of humanity." In response to Lincoln's assassination in 1865, Otto von Bismarck, who would later become Germany's first Chancellor, poignantly remarked, "I fear that foreign bankers with their craftiness and tortuous tricks will entirely control the exuberant riches of America and use it systematically to corrupt modern civilization. They will not hesitate to plunge the whole of Christendom into wars and chaos in order that the Earth should become their inheritance."[21]

- James Garfield, the twentieth U.S. President from March 1881 until his assassination six months later, realized the importance of government-controlled currency. During his service in Congress, Garfield was both Chairman of the Appropriations Committee and a member of the Banking and Currency Committee. As President, he observed, "Whoever controls the volume of money in any country is absolute master of all industry and commerce...and when you realize that the entire system is very easily controlled, one way or another, by a few powerful men at the top, you will not have to be told how periods of inflation and depression originate."[22] Several weeks after making this observation, Garfield was assassinated.

- William Jennings Bryan, the U.S. Secretary of State from 1913 to 1915, was the Democratic Party Presidential Candidate in 1896, 1900 and 1908. Although he lost each election, Bryan's consistent advocacy of government-issued currency greatly distressed the banking establishment. During his 1896 speech before the Democratic National Convention, Bryan declared, "What we need is an Andrew Jackson to stand as Jackson stood, against the encroachments of aggregated wealth. We say in our platform that we believe that the right to coin money and issue money is a function of government. We believe it. We believe it is a part of sovereignty...Those who are opposed to this proposition tell us that the issue of paper money is a function of the bank and that the government ought to go out of the banking business. I stand with Jefferson rather than with them, and tell them, as he did, that the issue of money is a function of the government and that the banks should go out of the governing business."[23]

- Theodore Roosevelt, the twenty-sixth U.S. President (1901-08), declared in 1912, "Political parties exist to secure responsible government and to execute the will of the people. From these great tasks both of the old parties have turned aside. Instead of instruments to promote the general welfare, they have become the tools of corrupt interests which use them impartially to serve their selfish purposes. Behind the ostensible government sits enthroned an invisible government owing no allegiance and acknowledging no responsibility to the people. To destroy this invisible government, to befoul the unholy alliance between corrupt business and corrupt politics is the first task of the statesman of the day."[24]

- John Hylan, Mayor of New York from 1918 to 1925, publicly condemned the "invisible government," which he identified as international bankers and the Rockefeller-Standard Oil interests. On March 26, 1922, the *New York Times* reprinted his remarks before the Knights of Columbus in Chicago:

"Some years ago, a sterling American, Theodore Roosevelt, condemned what he called 'invisible government.' He denounced as malefactors of great wealth and as enemies of the Republic those men of excessive fortune who were forever trying to grasp greater gain. The warning of Theodore Roosevelt has much timeliness today, for the real menace of our republic is this invisible government which like a giant octopus sprawls its slimy length over city, state and nation. Like the octopus of real life, it operates under cover of a self-created screen. It seizes in its long and powerful tentacles our executive officers, our legislative bodies, our schools, our courts, our newspapers and every agency created for the public protection. It squirms in the jaws of darkness and thus is the better able to clutch the reins of government, secure enactment of the legislation favorable to corrupt business, violate the law with impunity, smother the press and reach into the courts.

To depart from mere generalizations, let me say that at the head of this octopus are the Rockefeller-Standard Oil interests and a small group of powerful banking houses generally referred to as the international bankers. The little coterie of powerful international bankers virtually run the United States Government for their own selfish purposes. They practically control both parties, write political platforms, make catspaws of party leaders, use the leading men of private organizations and resort to every device to place in nomination for high public office only such candidates as will be amenable to the dictates of corrupt big business. They connive at centralization of government on the theory that a small group of hand-picked, privately controlled individuals in power can be more easily handled than a larger group among whom there will most likely be men sincerely interested in public welfare. These international bankers and Rockefeller-Standard Oil interests control the majority of newspapers and magazines in this country. They use the columns of these papers to club into submission or drive out of office public officials who refuse to do the bidding of the powerful corrupt cliques which compose the invisible government."[25]

- Franklin Delano Roosevelt, the thirty-second U.S. President (1933-45), wrote in November 1933 a letter to Colonel Edward Mandell House concerning the ever-increasing threat posed by the banks. He wrote, "The real truth of the matter is, as you and I know, that a financial element in the larger centers has owned the Government ever since the days of Andrew Jackson—and I am not wholly excepting the Administration of W.W. The country is going through a repetition of Jackson's fight with the Bank of the United States—only on a far bigger and broader basis."[26]

- Congressman Wright Patman, in 1967 while Chairman of the House Banking and Currency Committee, declared, "In the United States today we have, in effect, two governments. We have the duly constituted government, then we have an independent, uncontrolled and uncoordinated government in the Federal Reserve, operating the money powers which are reserved to Congress by the Constitution."[27]

- Senator Barry Goldwater in his 1979 book *With No Apologies* commented, "The Federal Reserve is a bank of monetary issue. It is empowered to establish a national discount rate and to authorize the printing of the currency of the United States. The accounts of the Federal Reserve System have never been audited. It operates outside the control of

Congress and through its Board of Governors manipulates the credit of the United States. Under Franklin D. Roosevelt the original term of office for governors was extended from seven to fourteen years—putting the board beyond the reach of any President."[28]

• Congressman Ron Paul, currently a ranking member of the House Banking Committee, has consistently clashed with the Federal Reserve, and frequently called for its abolishment. In September 2002, Paul declared before Congress, "Abolishing the Federal Reserve will allow Congress to reassert its constitutional authority over monetary policy. The United States Constitution grants to Congress the authority to coin money and regulate the value of the currency. The Constitution does not give Congress the authority to delegate control over monetary policy to a central bank. Furthermore, the Constitution certainly does not empower the federal government to erode the American standard of living via an inflationary monetary policy."[29]

Following the Civil War, the United States prospered for many decades without a central bank. Nevertheless, during this period, the banking establishment was actively manipulating the economy, consolidating wealth and securing political power. John Davison Rockefeller and John Piermont Morgan, for example, amassed enormous fortunes with which they gained extraordinary influence over society. Quigley explained,

"For almost fifty years, from 1880 to 1930, financial capitalism approximated a feudal structure in which two great powers, centered in New York, dominated a number of lesser powers, both in New York and in provincial cities. No description of this structure as it existed in the 1920's can be given in a brief compass, since it infiltrated all aspects of American life and especially all branches of economic life. At the center were a group of less than a dozen investment banks, which were, at the height of their powers, still unincorporated private partnerships. These included J. P. Morgan; the Rockefeller family; Kuhn, Loeb and Company; Dillon, Read and Company; Brown Brothers and Harriman; and others. Each of these was linked in organizational or personal relationships with various banks, insurance companies, railroads, utilities, and industrial firms. The result was to form a number of webs of economic power of which the more important centered in New York, while other provincial groups allied with these were to be found in Pittsburgh, Cleveland, Chicago, and Boston."[30]

Quigley also explained how this group pioneered the modern corporation as a vehicle for economic power and monopolistic control:

"On the business side, they sought to sever control from ownership of securities, believing they could hold the former and relinquish the latter. On the industrial side, they sought to advance monopoly and restrict production, thus keeping prices up and their security holdings liquid. The efforts of financiers to separate ownership from control were aided by the great capital demands of modern industry. Such demands for capital made necessary the corporation form of business organization. This inevitably brings together the capital owned by a large number of persons to create an enterprise controlled by a small number of persons. The financiers did all they could to make the former number as large as possible and the latter number as small as possible. The former was achieved by stock splitting, issuing securities of low par value, and by high-pressure security salesmanship. The latter was achieved by plural-voting stock, nonvoting stock, pyramiding of holding companies, election of directors by co-optation, and similar techniques. The result of this was that larger and larger aggregates of wealth fell into the control of smaller and smaller groups of men."[31]

By 1910, Morgan, Rockefeller and Rothschild representatives were actively planning America's next central bank, the Federal Reserve.[32] In 1913, with the help of Senator Nelson Aldrich, a Morgan business associate and father-in-law of John D. Rockefeller, Jr., they passed the Federal Reserve Act, a major coup for the banking establishment. The Federal Reserve obtained exclusive rights to monetize government debt. Congressman Charles Lindbergh, father of the famous aviator, presciently warned, "When the President signs this act, the invisible government by the money power, proven to exist by the money trust investigation, will be legalized. The money power overawes the legislative and executive forces of the nation. I have seen these forces exerted during the different stages of this bill. From now on depressions will be scientifically created."[33]

Indeed, in *Tragedy and Hope*, Quigley discussed at length how the banking establishment engineered the Great Depression to advance their New World Order agenda. For example, he commented, "The historical importance of the banker-engendered deflationary crisis of 1927-1940 can hardly be overestimated. It gave a blow to democracy and to the parliamentary system which the later triumphs of these in World War II and the postwar world were unable to repair fully. It gave an impetus to aggression by those nations where parliamentary government collapsed, and thus became a chief cause of World War II...it impelled the whole economic development of the West along the road from financial capitalism to monopoly capitalism and, shortly thereafter, toward the pluralist economy."[34]

War and Crisis Management

Besides banking, war is another indispensible mechanism used by elites to manipulate and direct populations. To launch its wars, however, the ruling class must obtain some degree of popular support. They must convince the people that a foreign enemy threatens their way of life. Where no genuine threat exists, elites have no qualms about manufacturing one. Most major wars, including the current War on Terror, the Vietnam War, and both World Wars, began following staged provocations. On May 7, 1915, nine miles off the English coast, a German submarine torpedoed the Lusitania, a British merchant ship traveling from New York to Liverpool carrying 1,257 passengers, including 197 Americans. The ship sunk. 785 passengers including 128 Americans died. Most significantly, this tragedy pushed America into World War I. The sinking of the Lusitania was not, however, an unforeseeable disaster. Powerful American and English elites desperately wanted America to enter the war. The American people, however, were decidedly against this course of action. Also President Woodrow Wilson, since the beginning of the war, had vowed to keep America neutral. The Anglo-American war enthusiasts would therefore have to manufacture a crisis large enough to sufficiently sway public opinion.

By financing French, British and American interests, J.P. Morgan greatly profited from World War I. According to Murray Rothbard, the famous Austrian School economist,

"At the moment of great financial danger for the Morgans, the advent of World War I came as a godsend. Long connected to British, including Rothschild, financial interests, the Morgans leaped into the fray, quickly securing the appointment, for J.P.

Morgan and Company, of fiscal agent for the warring British and French governments, and monopoly underwriter for their war bonds in the United States. J.P. Morgan also became the fiscal agent for the Bank of England, the powerful English central bank. Not only that: the Morgans were heavily involved in financing American munitions and other firms exporting war material to Britain and France. J.P. Morgan and Company, moreover, became the central authority organizing and channeling war purchases for the two Allied nations."[35]

In order to collect on his loans, Morgan obviously needed England and France to defeat Germany. And according to Walter Page, the U.S. Ambassador to England, the Allies were doomed unless America would enter the war. Writing to President Wilson on March 5, 1917, Page explained, "The pressure of this approaching crisis, I am certain, has gone beyond the ability of the Morgan financial agency for the British and French governments. The financial necessities of the Allies are too great and urgent for any private agency to handle, for every such agency has to encounter business rivalries and sectional antagonisms. It is not improbable that the only way of maintaining our present preeminent trade position and averting a panic is by declaring war on Germany."[36]

Despite Wilson's assurances of American neutrality, he was subservient to powers greater than him. Colonel Edward Mandel House, an ambitious southern elitist, was both Wilson's advisor and, ultimately, his controller. About House, the *New York Times* acknowledged, "He likes to mold policies and decide events; and since he knows how to do it, he has been able to follow his liking—with results."[37] Moreover, Wilson admitted, "Mr. House is my second personality. He is my independent self. His thoughts and mine are one. If I were in his place I would do just as he suggested. If any one thinks he is reflecting my opinion by whatever action he takes, they are welcome to the conclusion."[38]

House was closely connected to J.P. Morgan and America's ruling class. For example, House was involved with establishing both the Federal Reserve banking system and the Council on Foreign Relations. On the day of the attack on the Lusitania, House was in London meeting with British foreign minister Edward Grey. Grey asked House, "What will America do if the Germans sink an ocean liner with American passengers aboard?" House replied, "I believe that a flame of indignation would sweep the United States and that by itself would be sufficient to carry us into war."[39] Furthermore, according to prominent author George Sylvester Viereck,

"Ten months before the election which returned Wilson to the White House in 1916 'because he kept us out of war,' Colonel House negotiated a secret agreement with England and France on behalf of Wilson, which pledged the United States to intervene on behalf of the Allies. On March 9, 1916, Woodrow Wilson formally sanctioned the undertaking. If an inkling of the conversations between Colonel House and the leaders of England and France had reached the American people before the election, it might have caused incalculable reverberations in public opinion."[40]

Leading up to the attack on the Lusitania, the British government was taking calculated measures to lure America into the war. Quigley explained,

"On November 5, 1914, they declared the whole sea from Scotland to Iceland a 'war zone,' covered it with fields of explosive floating mines, and ordered all ships going to the Baltic, Scandinavia, or the Low Countries to go by way of the English Channel, where they were stopped, searched, and much of their cargoes seized, even when these cargoes could not be declared contraband under existing international law. In reprisal the Germans on February 18, 1915 declared the English Channel a 'war zone,' announced that their submarines would sink shipping in that area, and ordered shipping for the Baltic area to use the route north of Scotland."[41]

The United States recognized the British war zone, yet "refused to accept the German war zone, and insisted that American lives and property were under American protection even when traveling on armed belligerent ships in this war zone."[42] Furthermore, they "insisted that German submarines must obey the laws of the sea as drawn for surface vessels. These laws provided that merchant ships could be stopped by a war vessel and inspected, and could be sunk, if carrying contraband, after the passengers and the ships' papers were put in a place of safety."[43]

This arrangement was extremely problematic for Germany. Quigley explained, "It was not only difficult, or even impossible, for German submarines to meet these conditions; it was often dangerous, since British merchant ships received instructions to attack German submarines at sight, by ramming if possible. It was even dangerous for the German submarines to apply the established law of neutral vessels; for British vessels, with these aggressive orders, frequently flew neutral flags and posed as neutrals as long as possible."[44] Thomas Bailey, a Stanford history professor for 40 years, also confirmed this policy, citing an official 1915 decree from the British Admiralty to masters of British merchant ships ordering them to preemptively attack all enemy submarines. A violation of international law, the decree demanded, "steer straight for her at your utmost speed, altering course if necessary to keep her ahead."[45] Thomas observed that German submarines "operated in constant fear of being fired upon or rammed."[46] Consequently, German submarines started sinking British merchant vessels with little or no warning.

One week before the Lusitania's fateful voyage, the German Embassy attempted to place advertisements in fifty American newspapers warning potential passengers not to travel. However, the State Department intervened and blocked these transactions.[47] Nevertheless, George Sylvester Viereck managed to get them published in the *New York Times* and *New York Tribune*.[48] The advertisements read, "Travelers intending to embark on the Atlantic voyage are reminded that a state of war exists between Germany and her allies and Great Britain and her allies; that the zone of war includes the waters adjacent to the British Isles; that, in accordance with formal notice given by the Imperial German Government, vessels flying the flag of Great Britain, or any of her allies, are liable to destruction in those waters and that travelers sailing in the war zone on ships of Great Britain or her allies do so at their own risk."[49]

To further provoke the German attack, the British loaded the ship with war munitions. Winston Churchill later admitted that Lusitania's cargo included rifle ammunition and shrapnel shells weighing around 346,000 pounds.[50] Also the Lusitania had no convoy accompanying it through the war zone, a violation of British naval policy. Quoting Charles Sumner, the New York General Agent for Cunard Lines, the British Steamship Company that owned the Lusitania, the *New York Times* reported, "As for submarines, he added that he had no fear of them whatever, pointing out at the same time 'there is a general system of convoying British ships,' and adding that the British Navy was responsible for all British ships and 'especially for Cunarders.'"[51] Even more suspicious was the bizarre, amateurish behavior of Captain Turner, the Lusitania's captain. For example, on April 16, the British admiralty gave ship captains the following memorandum:

"War experience has shown that fast steamers can considerably reduce the chance of successful surprise submarine attack by zigzagging, that is to say, altering the course at short and irregular intervals, say in ten minutes to half an hour. This course is almost invariably adopted by warships when cruising in an area known to be infested with submarines. The underwater speed of a submarine is very low, and it is exceedingly difficult for her to get into position to deliver an attack unless she can preserve and predict the course of the ship [to be] attacked."[52]

Captain Turner, however, claimed he had misunderstood this message to mean that zigzagging should commence only *after* spotting a submarine.[53] During the attack, Turner also refused to lower the rescue boats, a decision resulting in hundreds of unnecessary deaths. In *The Lusitania's Last Voyage*, published by Houghton Mifflin just months after the tragedy, Charles Lauriat wrote, "Captain Turner and Captain Anderson were both calling in stentorian tones not to lower away the boats…saying that there was no danger and that the ship would float."[54]

Under normal circumstances, the ship indeed should have floated. With its hull divided into 175 watertight compartments, the Lusitania, like the Titanic, was considered unsinkable. One torpedo was certainly not enough to seal its fate. The likely culprit was a secondary explosion, which numerous survivors attested to. Bailey explained, "Vessels not even one-fifth the size of the Lusitania did not sink at all after being torpedoed only once; or sank slowly; or required a second torpedo or gunfire to complete their destruction. In any event, since we can definitely eliminate a second torpedo, the odds in favor of the exploding munitions theory are considerably increased."[55]

Dudley Field Malone, Collector of the Port of New York, recalled, "It was estimated that there were between ten and eleven tons of black powder in a particular consignment of forty-two hundred cases of Springfield cartridges listed among the small-arms ammunition on the Lusitania's manifest."[56] He suggested, "The intense heat generated by the torpedo explosion ignited the ten or eleven tons of powder in the cartridges"[57] Malone concluded, "I have always felt that the attack on the Lusitania was an unjustifiable crime, but that her quick sinking and great loss of life was contributed to by the negligence of the captain of the ship in having his port-holes open, in reducing the speed of the Lusitania from twenty-five to fifteen knots an hour in dangerous waters, and in failing to have his lifeboats swung out."[58]

The suspicious consignment of rifle cartridges originated from the prime benefactor of the attack, J.P. Morgan. Horace Cornelius Peterson, in his book *Propaganda for War*, explained,

"The major part of the material shipped on the Lusitania was munitions of war. In the cargo were included twenty-four hundred cases of rifle cartridges shipped from the Remington AUMC Company (JP Morgan and Company), and twelve hundred and fifty cases of shrapnel shipped from the Bethlehem Steel Company (Charles M. Schwab) at Bethlehem, Pennsylvania…Under the rulings of the Department of Commerce and Labor, these shrapnel and rifle bullets were not considered 'high explosives' and hence it was legally possible to ship them on a passenger vessel. On the other hand it cannot be denied that this made the Lusitania a munitions ship, which was using a cargo of noncombatants, including women and children, to protect war supplies. A number of people have pointed out that on land no American sitting on an ammunition wagon could prevent its being fired-on on its way to the front and consequently Americans traveling on a munitions boat could not be expected to give it immunity. In reply to the German contention that the Lusitania carried munitions, the American government took the position that the Germans were beclouding the issue."[59]

After the attack, Secretary of State William Jennings Bryan wrote to Wilson, "Americans who took passage on a British vessel destined to pass through the German war zone did so in a measure at their own peril, and were not entitled to the full protection of the Government."[60] He also noted, "Germany has a right to prevent contraband from going to the Allies, and a ship carrying contraband should not rely upon passengers to protect her from attack — it would be like putting women and children in front of an army."[61] Seeing that Wilson would eventually commit America to the war, Bryan resigned from his post. Wilson then told Colonel House he was "greatly relieved now that Mr. Bryan had gone."[62]

America finally entered the war in April 1917. Senator Robert La Follette, during a September 20, 1917 address to the Nonpartisan League convention in St. Paul, Minnesota, condemned America's participation. He alleged that Wilson knew the Lusitania was to be loaded with munitions, but refused to warn passengers. La Follette declared, "I say this, that the comparatively small privilege, of the right of an American citizen to ride on a munitions-loaded ship flying a foreign flag, is too small to involve this government in the loss of millions and millions of lives!"[63] He received a standing ovation.

America Dodges a Fascist Coup

Most Americans would be surprised to learn that Germany, Italy and others in Europe were not the only Western nations to grapple with fascism during the 1920s and 1930s. In 1933-34, thanks primarily to the courageous resolve of one man, America narrowly avoided a fascist coup d'état. During this time, a group of Wall Street elites, including representatives from the Morgan, DuPont, Rockefeller, Pew, and Mellon families, planned and organized the ousting of President Roosevelt and his New Deal politics.[64] They hired Gerald MacGuire, a former American Legion commander, to recruit General Smedley Butler to lead the coup. Butler was a retired United States Marine Corps Major General with thirty-three years of distinguished service, during which time he twice received the Congressional Medal of Honor.[65] Jules Archer, in his book *The Plot to Seize the White House*, explained that,

> "[Butler] was perhaps the best-known, and certainly the most popular and charismatic, military figure in the United States. He also suited the plotters' plans perfectly because he was noted for a brilliant, hard-hitting style of oratory that, they undoubtedly reasoned, could be put to the service of demagoguery in the same spell binding way Hitler and Mussolini had magnetized millions into following them. His rasping voice and fiery spirit captured audiences and held them hypnotized."[66]

MacGuire's assignment was to coax Butler into leading an ostensibly patriotic coup to put America back on the gold standard and thereby prevent Roosevelt from financially destroying the country. Although he abhorred the suggestion, Butler feigned interest in order to gather evidence against the plotters and eventually bring them to justice. He soon learned they had allocated $3 million to form a 500,000-man army of veterans for the operation. If more cash was needed, they would provide up to $300 million. By late 1934, Butler had gathered enough intelligence to terminate the coup.

On November 20, 1934, a Congressional Committee led by Representatives John McCormack and Samuel Dickstein arranged an executive session in New York City to hear Butler's story. The *New York Times*, *New York Post* and other publications ridiculed Butler and his assertions. *Time* magazine, for example, declared, "No military officer of the U.S. since the late, tempestuous George Custer has succeeded in publicly floundering in so much hot water as Smedley Darlington Butler."[67] Nevertheless, Butler had the evidence to support his claims. On February 15, 1935 the McCormack-Dickstein Committee submitted its findings to the House of Representatives. They concluded,

"In the last few weeks of the committee's official life it received evidence showing that certain persons had made an attempt to establish a fascist organization in this country. No evidence was presented and this committee had none to show a connection between this effort and any fascist activity of any European country. There is no question that these attempts were discussed, were planned, and might have been placed in execution when and if the financial backers deemed it expedient. This committee received evidence from Maj. Gen. Smedley D. Butler (retired), twice decorated by the Congress of the United States. He testified before the committee as to conversations with one Gerald C. MacGuire in which the latter is alleged to have suggested the formation of a fascist army under the leadership of General Butler. MacGuire denied these allegations under oath, but your committee was able to verify all the pertinent statements made by General Butler, with the exception of the direct statement suggesting the creation of the organization. This, however, was corroborated in the correspondence of MacGuire with his principal, Robert Sterling Clark, of New York City, while MacGuire was abroad studying the various forms of veterans' organizations of Fascist character. This committee asserts that any efforts based on lines as suggested in the foregoing and leading off to the extreme right, are just as bad as efforts which would lead to the extreme left. Armed forces for the purpose of establishing a dictatorship by means of Fascism or a dictatorship through the instrumentality of the proletariat, or a dictatorship predicated on racial and religious hatreds, have no place in this country."[68]

Butler effectively thwarted the coup, yet the conspirators remained at large. During a February 1935 radio broadcast, Butler condemned the committee for protecting the implicated parties. "Like most committees, he explained, "it has slaughtered the little and allowed the big to escape. The big shots weren't even called to testify."[69] Dickstein claimed, "We didn't have the time. We'd have taken care of the Wall Street groups if we had the time. I would have had no hesitation in going after the Morgans."[70] McCormack later recalled, "The way I figure it, we did our job in the committee by exposing the plot, and then it was up to the Department of Justice to do their job—to take it from there."[71]

The criminals were never prosecuted. Indeed, their beneficiaries retain power to this day. In 1935, Butler published a book called *War is a Racket*. He summarized, "War is a racket to protect economic interests, not our country, and our soldiers are sent to die on foreign soil to protect investments by big business."[72] In 1971, while conducting research for his landmark book, Archer interviewed McCormack. When asked about the plotters' motivation, McCormack reflected, "The plotters definitely hated the New Deal because it was for the people, not for the moneyed interests, and they were willing to spend a lot of their money to dump Mr. Roosevelt out of the White House."[73] Regarding Butler, McCormack concluded, "In peace or war he was one of the outstanding Americans in our history. I can't emphasize too strongly the very important part he played in exposing the Fascist plot in the early 1930's backed by and planned by persons possessing tremendous wealth."[74]

Round Table Groups

To gradually implement their New World Order agenda, the globalists operate through exclusive, private associations of powerful people, sometimes referred to as round table groups. These groups plan the future, often decades in advance. They also determine and write public policy, which governments then implement. Discussed throughout this book, today's most prominent round table groups include the Council on Foreign Relations, the Bilderberg Group, the Trilateral Commission, and the Club of Rome. In England, during the late 1800s and early 1900s, Cecil Rhodes and Alfred Milner organized the precursors of these modern groups. According to Quigley,

> "Rhodes inspired devoted support for his goals from others in South Africa and in England. With financial support from Lord Rothschild and Alfred Beit, he was able to monopolize the diamond mines of South Africa as De Beers Consolidated Mines and to build up a great gold mining enterprise as Consolidated Gold Fields. In the middle 1890's Rhodes had a personal income of at least a million pounds sterling a year...which was spent so freely for his mysterious purposes that he was usually overdrawn on his account. These purposes centered on his desire to federate the English-speaking peoples and to bring all the habitable portions of the world under their control. For this purpose Rhodes left part of his great fortune to found the Rhodes Scholarships at Oxford in order to spread the English ruling-class tradition throughout the English-speaking world as Ruskin had wanted."[75]

Scores of Rhodes Scholars have since ascended to the highest levels of government. Notable recipients include Dean Rusk, U.S. Secretary of State from 1961-1969, John Oakes, *New York Times* Editor from 1961-1976, Stansfield Turner, CIA Director from 1977-1981, James Woolsey, CIA Director from 1993-1995, Franklin Raines, CEO of Fannie Mae from 1999-2004, and Bill Clinton, 42nd President of the United States.

Cecil Rhodes was a virulent racist, elitist and imperialist. That the establishment, after so many years, *still* awards scholarships in his name is emblematic of their true ambitions. Rhodes' writings include the following ruminations:

- "There are various races of mankind—the Yellow, the Black, the Brown, and the White. If the test be numerical, the Yellow race comes first. But if the test be the area of the world and the power to control its destinies, the primacy of the White race is indisputable. The yellow race is massed thick on one half of a single continent: the White exclusively occupies Europe, practically occupies the Americas, is colonising Australia, and is dominating Asia. In the struggle for existence the White race had unquestionably come out on top."[76]

- "I contend that we are the first race in the world, and that the more of the world we inhabit the better it is for the human race. I contend that every acre added to our territory means the birth of more of the English race who otherwise would not be brought into existence. Added to this, the absorption of the greater portion of the world under our rule simply means the end of all wars."[77]

- "Why should we not form a secret society with but one object, the furtherance of the British Empire and the bringing of the whole uncivilised world under British rule, for the recovery of the United States, for the making of the Anglo-Saxon race but one Empire?"[78]

On February 5, 1891, "Rhodes and Stead organized a secret society of which Rhodes had been dreaming for sixteen years. In this secret society Rhodes was to be leader; Stead, Brett (Lord Esher), and Milner were to form an executive committee; Arthur (Lord) Balfour, (Sir) Harry Johnston, Lord Rothschild, Albert (Lord) Grey, and others were listed as potential members of a 'Circle of Initiates'; while there was to be an outer circle known as the 'Association of Helpers' (later organized by Milner as the Round Table organization)…This group was able to get access to Rhodes' money after his death in 1902 and also to the funds of loyal Rhodes supporters."[79]

After Rhodes' death, Milner formed a group known as Milner's Kindergarten. Its members influenced and infiltrated the British government, virtually dictating foreign policy for decades. Quigley explained, "As governor-general and high commissioner of South Africa in the period 1897-1905, Milner recruited a group of young men, chiefly from Oxford and from Toynbee Hall, to assist him in organizing his administration. Through his influence these men were able to win influential posts in government and international finance and became the dominant influence in British imperial and foreign affairs up to 1939. Under Milner in South Africa they were known as Milner's Kindergarten until 1910. In 1909-1913 they organized semi-secret groups, known as Round Table Groups, in the chief British dependencies and the United States. These still function in eight countries."[80]

After World War I, these round table groups coalesced into the Royal Institute of International Affairs in England, and the Council on Foreign Relations in the United States. Later in 1925, they established the Institute of Pacific Relations in twelve different eastern countries. Quigley explained, "At the end of the war of 1914, it became clear that the organization of this system had to be greatly extended. Once again the task was entrusted to Lionel Curtis who established, in England and each dominion, a front organization to the existing local Round Table Group. This front organization, called the Royal Institute of International Affairs, had as its nucleus in each area the existing submerged Round Table Group. In New York it was known as the Council on Foreign Relations, and was a front for J.P. Morgan and Company in association with the very small American Round Table Group. The American organizers were dominated by the large number of Morgan 'experts,' including Lamont and Beer, who had gone to the Paris Peace Conference and there became close friends with the similar group of English 'experts' which had been recruited by the Milner group. In fact, the original plans for the Royal Institute of International Affairs and the Council on Foreign Relations were drawn up at Paris."[81] In its early years, according to Quigley, the CFR "was dominated by J.P. Morgan."[82]

Rear Admiral Chester Ward, like Quigley, also had an insider's view of the CFR. Besides being a Judge Advocate General of the U.S. Navy from 1956 to 1960, he was also a 16-year CFR member.[83] Regarding these round table groups, Ward explained, "The most powerful cliques in these elitist groups have one objective in common: they want to bring about the surrender of the sovereignty and national independence of the United States.

They differ only as to the entity into which our sovereignty should be merged. Some dream of taking the United States into a one-world all-powerful global government — possibly a vastly strengthened United Nations, or possibly limited to the Atlantic community. They consider that this objective is at once so idealistic (the brotherhood of peoples or Parliament of Man concept) and so urgent (a lasting world peace can be secured only by disarming all nations down to internal police levels), that their ends justifies any means."[84]

Arnold J. Toynbee

Arnold J. Toynbee was the nephew of Arnold Toynbee (the best friend of Lord Alfred Milner).[85] Arnold Toynbee was an economic historian who despite dying young — he was only 31 in 1883 when he died — had an enormous influence on the British ruling class and their plans for world domination. Quigley, in his book *The Anglo-American Establishment*, explained that Toynbee devised the strategic, organizational methodology used by the Milner Group and, later, the Royal Institute of International Affairs, the Council on Foreign Relations, the Trilateral Commission and other such organizations. This methodology was described by Benjamin Jowett in the preface to Toynbee's posthumous book *Lectures on the Industrial Revolution*: "He would gather his friends around him; they would form an organization; they would work on quietly for a time, some at Oxford, some in London; they would prepare themselves in different parts of the subject until they were ready to strike in public."[86]

According to Quigley, Arnold J. Toynbee was "the most important" member of the Royal Institute for International Affairs since its founding in 1920.[87] In June 1931, he delivered a speech in Copenhagen, later published in *Pacific Affairs*, the official journal of the Institute of Pacific Relations. His message is shocking, both for its arrogance and its bluntness. Toynbee left no doubt that international elites are absolutely opposed to national sovereignty and have been tirelessly working towards world government. He stated,

"In the spirit of determination which, happily, animates us, we shall have no inclination to under-estimate the strength of the political force which we are striving to overcome. What is this force? If we are frank with ourselves, we shall admit that we are engaged on a deliberate and sustained and concentrated effort to impose limitations upon the sovereignty and the independence of the fifty or sixty local sovereign independent states which, at present, partition the habitable surface of the earth and divide the political allegiance of mankind. The surest sign, to my mind, that this fetish of local national sovereignty is our intended victim is the emphasis with which all our statesmen and our publicists protest with one accord, and over and over again, at every step forward which we take, that, whatever changes we may make in the international situation, the sacred principle of local sovereignty will be maintained inviolable. This, I repeat, is a sure sign that, at each of these steps forward, the principle of local sovereignty is really being encroached upon and its sphere of action reduced and its power for evil restricted. It is just because we are really attacking the principle of local sovereignty that we keep on protesting our loyalty to it so loudly. The harder we press our attack upon the idol, the more pains we take to keep its priests and devotees in a fool's paradise — wrapped in a false sense of security which will inhibit them from taking up arms in their idol's defense. Perhaps, too, when we make these protestations, we are partly concerned to deceive ourselves. For let us be honest. Even the most internationally-minded among us are votaries of this false god of local national sovereignty to some extent. It is such an old-established object of worship that it retains some hold even over the most enlightened souls."[88]

Toynbee then denounced Christianity, as he consistently did throughout his life, by portraying the nation-state as an outgrowth of skewed Christian values. He continued, calling for the creation of a global institution, something like the failed League of Nations. "In plain terms, we have to retransfer the prestige and the prerogatives of sovereignty from the fifty or sixty fragments of contemporary society to the whole of contemporary society, from the local national states by which sovereignty has been usurped, with disastrous consequences, for half a millennium, to some institution embodying our society as a whole. In the world as it is to-day, this institution can hardly be a universal church. It is more likely to be something like a League of Nations."[89]

He then admitted in dramatic, almost comical fashion that the various interlocking round table groups were cunningly scheming to end national sovereignty. If the public knew what they were planning, he admitted, the New World Order would certainly crumble. Toynbee declared,

"I will merely repeat that we are at present working, discreetly, but with all our might, to wrest this mysterious political force called sovereignty out of the clutches of the local national states of our world. And all the time we are denying with our lips what we are doing with our hands, because to impugn the sovereignty of the local national states of the world is still a heresy for which a statesman or a publicist can be perhaps not quite burnt at the stake, but certainly ostracised and discredited. The dragon of local sovereignty can still use its teeth and claws, when it is brought to bay. Nevertheless, I believe that the monster is doomed to perish by our sword. The fifty or sixty local states of the world will no doubt survive as administrative conveniences. But sooner or later sovereignty will depart from them. Sovereignty will cease, in fact if not in name, to be a local affair."[90]

Toynbee reads like a caricature of himself, like some sadistic New World Order henchman, foaming at the mouth and clutching his fists. His words, however, are utterly serious.

Like H.G. Wells, another prolific elite of the day, Toynbee's arrogant, paternalistic attitude was an extension of his deeply ingrained feelings of cultural, if not racial, superiority. For him, a global system enabling economic, political and cultural unification was inevitable. While theoretically any cultural group could spearhead and direct the New World Order, Toynbee reasoned that the cultural superiority of Europeans best qualified them for the task. He wrote, "I cannot foresee a time when the outer world will be able to dispense with European culture—with the thought and the art and the ideals which radiate out from Europe over the rest of the world. If this light that shines in Europe were to be extinguished, the rest of the world would surely sink first into twilight and ultimately into darkness. If this salt that is preserved in Europe were to lose its savour, the rest of the world would surely find itself going intellectually and aesthetically stale. Therefore we must exert ourselves to safeguard the position of Europe in the new international society— and this is not only in the interests of us poor Europeans, but in the interests of mankind at large."[91]

Chapter 2

Elite Foundations

During the 1800s and early 1900s, industrialists like John D. Rockefeller, Andrew Carnegie and Henry Ford accumulated vast fortunes, more money than they could possibly spend in *several* lifetimes. Realizing that such incredible wealth represented incredible power, they established philanthropic foundations, supposedly to give back to society. Indeed, these foundations do provide *some* assistance to the poor and sick. Nevertheless, they are far more interested in social engineering—shaping society according to elite interests. Their charity merely cloaks their primary objectives. Although they invest heavily in science, education, environmental protection, and other seemingly innocuous pursuits, these foundations are attempting to forge an entirely new society, a New World Order. René Wormser, in his book *Foundations: Their Power and Influence*, explained, "It would be difficult to find a single foundation-supported organization of any substance which has not favored the United Nations, or similar global schemes; fantastically heavy foreign aid at the burdensome expense of the taxpayer; meddling in the colonial affairs of other nations; and American military commitments over the globe."[1]

Wormser was General Counsel for the Reece Committee, a congressional committee in the early 1950s tasked with investigating tax-exempt foundations. Carroll Quigley commented, "The Eighty-third Congress in July 1953 set up a Special Committee to Investigate Tax-Exempt Foundations with Representative B. Carroll Reece, of Tennessee, as chairman. It soon became clear that people of immense wealth would be unhappy if the investigation went too far and that the 'most respected' newspapers in the country, closely allied with these men of wealth, would not get excited enough about any revelations to make the publicity worth while, in terms of votes or campaign contributions. An interesting report showing the Left-wing associations of the interlocking nexus of tax-exempt foundations was issued in 1954 rather quietly. Four years later, the Reece committee's general counsel, René A. Wormser, wrote a shocked, but not shocking, book on the subject called *Foundations: Their Power and Influence*."[2]

In 1982, G. Edward Griffin conducted an interview with Norman Dodd concerning Dodd's experience with the Reece Committee. This widely available video provides extremely revealing insight into the machinations of the New World Order, specifically the foundations' efforts to shape public perception and reeducate society. In 1953, Dodd became Director of Research for the Reese Committee. The task of the Committee, Dodd explained, "was to investigate the activities of foundations as to whether or not these activities could justifiably be labeled 'un-American'…We defined that, in our way, as being a determination to effect changes in the country by un-Constitutional means."[3] Investigating the history of the big foundations, Dodd determined that their purpose was "to get control over the content of American education," and then "orient our educational system away from support of the principles embodied in the Declaration of Independence and implemented in the Constitution."

During his investigation, Dodd met with Rowan Gaither, then President of the Ford Foundation, who stated, "Mr. Dodd, all of us who have a hand in the making of policies here have had experience either with the OSS during the war, or European economic administration after the war. We've had experience operating under directives. And these directives emanate, and did emanate, from the White House. Now, we still operate under such directives. Would you like to know what the substance of these directives is?"[4] Dodd replied affirmatively. Gaither bluntly asserted, "Mr. Dodd, we operate in response to similar directives, the substance of which is that we shall use our grant-making power so as to alter life in the United States, that it can be comfortably merged with the Soviet Union." Stunned, Dodd replied, "Of course, legally, you're entitled to make grants for this purpose. But I don't think you're entitled to withhold that information from the people of the country to which you're indebted for your tax exemption. So why don't you tell the people of the country just what you told me?" Gaither replied, "We would not think of doing any such thing."[5]

Even more revealing was Dodd's experience with the Carnegie Foundation. Joseph Johnson, a prominent Bilderberg member and CFR director from 1950-1974, was then President of the Carnegie Foundation.[6] Johnson granted Dodd access to the Carnegie library, including the minute books from the foundation's meetings since its inception. Dodd sent Kathryn Casey, an attorney from his staff, to New York for the assignment. She made startling discoveries, and recorded everything on Dictaphone belts. Regarding her findings, Dodd stated, "We are now at the year 1908, which was the year that the Carnegie Foundation began operations. And in that year, the trustees, meeting for the first time, raised a specific question, which they discussed throughout the balance of the year in a very learned fashion. And the question is this: Is there any means known more effective than war, assuming you wish to alter the life of an entire people? And they conclude that no more effective means than war, to that end, is known to humanity. So then in 1909, they raise the second question, and discuss it, namely, how do we involve the United States in a war?"[7] Dodd continued,

"They answer that question as follows: 'we must control the State Department.' And then that very naturally raises the question of 'how do we do that?' They answer it by saying: 'we must take over and control the diplomatic machinery of this country.' And finally they resolve to aim at that as an objective. Then time passes and we are eventually in a war, which would be WWI. At that time they record in their minutes a shocking report in which they dispatch to President Wilson a telegram cautioning him to see that the war does not end too quickly. And finally, of course, the war is over.

At that time, their interest shifts over to preventing what they call 'a reversion of life in the United States to what it was prior to 1914 when WWI broke out.' At that point they come to the conclusion that 'to prevent a reversion, we must control education in the United States.' And they realize that is a pretty big task, too big for them alone. So they approach the Rockefeller Foundation with the suggestion that, 'that portion of education which could be considered domestic be handled by the Rockefeller Foundation, and that portion which is international should be handled by the Endowment.' They then decide that the key to the success of these two operations lay in the alteration of the teaching of American history. So they approach four of the then most prominent teachers of American history in the country — people like Charles and Mary Byrd. And their suggestion to them is this, 'will they alter the manner in which they present this subject?' And they get turned down flat. So they then decide that it is necessary for them to, as they say, 'build our own stable of historians.'

Then they approach the Guggenheim Foundation, which specializes in fellowships, and say, 'when we find young men in the process of studying for doctorates in the field of American history, and we feel that they are the right caliber, will you grant them fellowships on our say so?' And the answer is, 'yes.' So under that condition, eventually they assemble twenty. And they take these twenty potential teachers of American history to London. And there they are briefed as to what is expected of them when as, and if, they secure appointments in keeping with the doctorates they will have earned.

That group of twenty historians ultimately becomes the nucleus of the American Historical Association. And then toward the end of the 1920's, the Endowment grants to the American Historical Association $400,000 for a study of our history in a manner which points to 'what can this country look forward to in the future?' That culminates in a seven-volume study, the last volume of which is of course, in essence, a summary of the contents of the other six. The essence of the last volume is, 'the future of this country belongs to collectivism, administered with characteristic American efficiency.'"[8]

Undermining the Constitution

The various elite foundations have funded countless studies attacking the United States Constitution. In 1963, for example, the Center for Democratic Institutions, an offshoot of the Ford Foundation's Fund for the Republic,[9] published a whitepaper entitled, "The Elite and the Electorate: Is Government by the People Possible?" Senator J. William Fulbright, Chairman of the Senate Foreign Relations Committee and Bilderberg attendee,[10] contributed to the report, writing, "The case for government by elites is irrefutable insofar as it rests on the need for expert and specialized knowledge. The average citizen is no more qualified for the detailed administration of government than the average politician is qualified to practice medicine or to split an atom…Government by the people is possible but highly improbable. The difficulties of self-government are manifest throughout the world."[11] Recognizing the report's deception, an Ohio newspaper reported, "In this discussion all advocates make an obvious effort to appear sweet and reasonable at all times. While they seem to favor an increase in presidential power, they also warn that we must not lean too heavily on a professional elite. Having said this, they cross the street to claim that the necessity of a professional elite is irrefutable and we must have it."[12]

In 1987, the Rockefeller Foundation published *The Secret Constitution and the Need for Constitutional Change* by Arthur Selwyn Miller, a pro-establishment book cloaked in anti-establishment rhetoric. In his introduction, Miller clearly established his allegiances: "This book has had a long gestation period. It was begun in the fall of 1981 when I had the honor of being a scholar in residence at the Villa Serbelloni, the Rockefeller Foundation's study center at Bellagio on Lake Como in Italy. Anticipating the ceremonies in 1987 on the 200th anniversary of the constitutional convention, I became convinced that major alterations were necessary in the constitution that emerged on September 17, 1787, from the conclave in Philadelphia."[13]

Throughout the book, Miller attacked some globalist institutions including the World Bank and IMF, and heavily criticized corporatism as "the fusion of political and economic power." He explained, "Corporatism is more than a system of interest representation; it is a shorthand label for an emerging system of governance. In such a system 'public' and 'private' have for all practical purposes become indistinguishable."[14] As for "the Establishment," Miller wrote, "those who formally rule take their signals and commands, not from the electorate as a body, but from a small group of men (plus a few women). This group will be called the Establishment. It exists, even though that existence is stoutly denied; it is one of the secrets of the American social order. A second secret is the fact that the existence of the Establishment—the ruling class—is not supposed to be discussed. A third secret is implicit in what has been said—that there is really only one political party of any consequence in the United States, one that has been called the 'Property Party.' The Republicans and the Democrats are in fact two branches of the same (secret) party."[15]

Quite simply, Arthur Selwyn Miller was a Judas goat. Shepherds (the Rockefeller Foundation) typically train one goat, the Judas goat, to live amongst the others. The Judas goat earns the trust of his brethren but ultimately betrays them by leading them to slaughter. Miller argued that "the Establishment" had irresponsibly created a constitutional crisis. "Extraordinary conditions," he declared, "demand extraordinary, even unique remedies."[16] But his "unique remedies" were thoroughly toxic. Instead of the antidote, Miller offered more poison. His suggestions parroted the New World Order agenda verbatim. For example, he argued against national sovereignty: "Nationalism is not only a retrograde step in history; it is fraught with such manifest dangers that it should be seen as a dangerous social disease...New vision is required to plan and manage the future, a global vision that will transcend national boundaries and eliminate the poison of nationalistic solutions."[17]

He also echoed his globalist masters' obsession with depopulation: "Population will have to be stabilized. This is a fact that simply cannot be ignored (or refuted)...The goal should be for an 'optimum' population, one that is not necessarily stationary, as Lester Brown suggested, but one that is in a state of shifting equilibrium with available and potential resources. The relationship between population size and availability of resources is one of the principal factors determining the types of institutions that societies establish."[18] Miller also advocated expanding the definition of national security to address "environmental degradation throughout the world." He asserted, "Unless this is checked, and soon, biological systems will collapse, and the earth as it is now known will begin to resemble that of the ancient Sumerians."[19] Finally, Miller also endorsed the globalists' dream of a completely planned society: "Ours is the age of the planned society, as any number of post-World War II developments attest. No other way is possible."[20]

After appealing to the people by acknowledging the establishment's existence and its manipulation and subjugation of society, Miller offered radical solutions designed to simply further empower this same establishment. "Constitutional Band-Aids are not sufficient to the manifest need," he explained.[21] Recognizing that Americans would not support radical constitutional reform, Miller nevertheless concluded, "After all, they really have no choice, for constitutional alteration will come whether or not it is liked or planned for. We need a new republic, one suited to the demands of the present day and the foreseeable future."[22] According to Miller, the problem is not the degenerate establishment, which has hijacked the country. The problem is the country itself, specifically the Constitution.

Preparing Teachers and Students for the Future

To advance their agenda, the globalists realized they would need to control education. In 1902, with a $1 million donation from John D. Rockefeller, Congress incorporated the General Education Board (GEB). During its first six years, Rockefeller endowed GEB with over $43 million, approximately $1 trillion in today's money.[23] Also, Rockefeller's son (JDR, Jr.) sat on GEB's board. The largest Rockefeller "gift," came with the following stipulation, as written by his son: "Thirty-two million dollars, one third to be added to the permanent endowment of the board, two-thirds to be applied to such specific purposes of the board as either he [JDR] or I may, from time to time, direct."[24] What "specific purposes" did the Rockefellers have in mind? In 1906, the GEB defined its unofficial mission statement in a document called *Occasional Letter Number One:*

"In our dreams...people yield themselves with perfect docility to our molding hands. The present educational conventions fade from our minds, and unhampered by tradition we work our own good will upon a grateful and responsive folk. We shall not try to make these people or any of their children into philosophers or men of learning or men of science. We have not to raise up from among them authors, educators, poets or men of letters. We shall not search for embryo great artists, painters, musicians, nor lawyers, doctors, preachers, politicians, statesmen, of whom we have ample supply. The task we set before ourselves is very simple...we will organize children...and teach them to do in a perfect way the things their fathers and mothers are doing in an imperfect way."[25]

Eight years later, the National Education Association officially recognized the subversive nature of the big foundations. During their 1914 annual meeting in St. Paul, Minnesota, they declared, "We view with alarm the activity of the Carnegie and Rockefeller Foundations—agencies not in any way responsible to the people—in their efforts to control the policies of our State educational institutions, to fashion after their conception and to standardize our courses of study, and to surround the institutions with conditions which menace true academic freedom and defeat the primary purpose of democracy as heretofore preserved inviolate in our common schools, normal schools, and universities."[26] In 1919, Rockefeller gave another $50 million to GEB.[27] During the 1930s GEB distributed $5 million per year of Rockefeller cash to public education, especially to medical education. Finally, in 1960, the Rockefeller Foundation officially absorbed the GEB.[28]

John Taylor Gatto was the 1989, 1990 and 1991 New York City Teacher of the Year. In 1991, while also New York State Teacher of the Year, Gatto wrote an article for the *Wall Street Journal* called, "I may be a teacher, but I'm not an educator." With this article, he also announced his resignation. Gatto explained, "I just can't do it anymore. I can't train children to wait to be told what to do; I can't train people to drop what they are doing when a bell sounds; I can't persuade children to feel some justice in their class placement when there isn't any, and I can't persuade children to believe teachers have valuable secrets they can acquire by becoming our disciples. That isn't true. Government schooling is the most radical adventure in history. It kills the family by monopolizing the best times of childhood and by teaching disrespect for home and parents. An exaggeration? Hardly. Parents aren't meant to participate in our form of schooling, rhetoric to the contrary. My orders as schoolteacher are to make children fit an animal training system, not to help each find his or her personal path."[29] Throughout his many books, including *The Underground History of American Education* and *Weapons of Mass Instruction*, Gatto has exposed the manipulation of the education system by elite interests. *The Deliberate Dumbing Down of America* by Charlotte Iserbyt is another invaluable resource. Iserbyt was Senior Policy Advisor in the Office of Educational Research and Improvement (OERI), U.S. Department of Education, during the Reagan Administration.

Government Funded Indoctrination

In 1952, Nelson Rockefeller was an advisor to President-elect Dwight Eisenhower. On Rockefeller's recommendation, the Eisenhower administration established the Department of Health, Education & Welfare (HEW), of which Rockefeller became the first Under-Secretary in 1953. Over the ensuing decades, HEW conducted scores of studies, financed by Rockefeller Foundation grants. In 1969, HEW published a study called *Behavioral Science Teacher Education Program.* The study examined "the second stage of a project to develop a model teacher education program."[30] The primary objective of the program was the "Development of a new kind of elementary school teacher who is basically well-educated, engages in teaching as clinical practice, is an effective student of the capacities and environmental characteristics of human learning, and functions as a responsible agent of social change."[31] Although this sounds reasonable and pragmatic, the report defined "responsible agents of social change" as those who would indoctrinate and prepare children to accept tyranny. According to the study, if the program "is to be in tune with present and future times, two areas must be given careful attention. First, the program must consciously consider and draw on the potentialities of a changing society...A second concern in maintaining a relevant teacher education program is the faculty who develop and implement the program."[32]

The future they envisioned was one where a scientific-technological elite would dominate humanity and "strain the democratic fabric to a ripping point." For example, they asserted, "Calculations of the future and how to modify it are no longer considered obscure academic pursuits. Instead they are the business of many who are concerned about and responsible for devising various modes of social change. Future planning is of major concern to government agencies as well as social scientists."[33] The following are various attributes of the Department of Health, Education and Welfare's planned future society:

- *"Futurism as a Social Tool and Decision-Making by an Elite:* The capability of projecting present potentialities and emerging developments into the future will be increased. The complexity of the society and rapidity of change will require that comprehensive long-range planning become the rule, in order that carefully developed plans will be ready <u>before</u> changes occur. Long-range planning and implementation of plans will be made by a technological-scientific elite. Political democracy, in the American ideological sense, will be limited to broad social policy; even there, issues, alternatives, and means will be so complex that the elite will be influential to a degree which will arouse the fear and animosity of others. This will strain the democratic fabric to a ripping point."[34]

- *"Systems Approaches to Cybernetics:* The use of the systems approach to problem solving and of cybernetics to manage automation will remold the nation. They will increase efficiency and depersonalization. Man's traditional slow speed in thinking through problems, analyzing alternatives, testing and evaluating them, and implementing them will be eliminated by computers and cybernetics. Only a few people will be able to have a major role in the processes, and they will apply the remnants of the Protestant Ethic. Most of the population will seek meaning through other means or devote themselves to pleasure seeking. The controlling elite will engage in power plays largely without the involvement of most of the people. The society will be a leisurely one. People will study, play, and travel; some will be in various stages of the drug-induced experiences."[35]

- *"A Controlling Elite*: The Protestant Ethic will atrophy as more and more enjoy varied leisure and guaranteed sustenance. Work as the means <u>and</u> end of living will diminish in importance except for a few with exceptional motivation, drive, or aspiration. No major source of a sense of worth and dignity will replace the Protestant Ethic. Most people will tend to be hedonistic, and a dominant elite will provide 'bread and circuses' to keep social dissension and disruption at a minimum. A small elite will carry society's burdens. The resulting impersonal manipulation of most people's life styles will be softened by provisions for pleasure seeking and guaranteed physical necessities. Participatory democracy in the American-ideal mold will mainly disappear. The worth and dignity of individuals will be endangered on every hand. Only exceptional individuals will be able to maintain a sense of worth and dignity."[36] To indoctrinate and prepare students for their future roles, these elites would split the educational system into a "planned education program for most," and a "highly structured program for the self- and societally-selected future elite."[37] The impact on school personnel would involve the "Reduction to technician level for the bulk of school personnel who merely manipulate largely teacher-proof programs; creation of a small group of elite scholar-practitioners who help to plan and direct, in close collaboration with the total establishment."[38]

- "*Communications Capabilities and Potentialities for Opinion Control:* The range of communications capabilities will be increased significantly. Each individual will receive at birth a multi-purpose identification number which will have, among other things, extensive communications uses. None will be out of communication with those authorized to reach him. Each will be able to receive instant updating of ideas and information on topics previously identified. Routine jobs to be done in any setting can be initiated automatically by those responsible for the task; all will be in constant communication with their employers, or other controllers, and thus exposed to direct and subliminal influence. Mass media transmission will be instantaneous to wherever people are and in forms suited to their particular needs and roles. Each individual will be saturated with ideas and information. Some will be self-selected; other kinds will be imposed overtly by those who assume responsibility for others' actions (for example, employers); still other kinds will be imposed covertly by various agencies, organizations, and enterprises. Relatively few individuals will be able to maintain control over their opinions. Most will be pawns of competing opinion molders."[39]

- "*Biological Capabilities in Controlling Inherited Characteristics and Potentialities*: Biological capabilities for controlling a child's birth and his development and reactions after birth will increase. Birth control capabilities will become perfected on a semi-permanent level. Thus, most children will be wanted and 'designed' with maximum capacities for future development and minimal hindrance to projected development. The society's capacities for quality living will be enhanced by the quality of its citizenry and the level of development which will be reached by many. However, for personal-philosophical-political reasons, birth control may not be practiced, and children may be programmed without balanced characteristics and capacities. Potentialities for conflict will be increased by various kinds of elites with different values and priorities."[40]

This is the planned New World Order society. Forty years after the publication of this chilling document, much of its prognosis has already transpired. The chicanery of the foundations and the corruption of the educational system, especially via UNESCO, are further exposed in upcoming chapters.

Chapter 3

The Fabian Society

In 1884, several British intellectuals including Beatrice and Sidney Webb, H.G. Wells and George Bernard Shaw founded the Fabian Society as a vehicle for slowly imposing socialism in England and, by extension, other western nations. Rather than hasty revolutions, the Fabians preferred gradualism to achieve their ends. John Micklethwait, Editor-in-Chief of *The Economist*, explained, "They did not believe in overthrowing society, like the Marxists. They did not particularly care about winning elections, like the Labour Party they also helped found. Indeed, they tried not to tie themselves to one particular party. They hoped that socialism would come about gradually but relentlessly — by clothing collectivism in the garb of common sense and by extending government controls over one institution after another."[1] The Fabians took their name from Roman general Fabius Maximus who defeated Hannibal (the legendary Carthaginian military commander) through patience and cunning rather than full-scale engagement. Historian Adam Lowther explained, "Fabius had long advocated refusing battle to Hannibal. Instead, he implemented a strategy, which sought to starve and harass Hannibal until he was forced to withdraw from the Italian peninsula."[2]

The Fabian Society also relied on patience and cunning. According to their original *Programme of the Fabian Society*, "The Fabian Society dates from January 4, 1884. It differs from other Socialist bodies in not trying to enlist the mass of its converts in its ranks, and in encouraging its members to join and permeate other organizations. It exists for purposes of co-operation in research, internal discussion, and external propaganda."[3] Under the heading, "Basis of the Fabian Society," the *Programme* reads,

> "The Fabian Society consists of Socialists. It therefore aims at the reorganization of society by the emancipation of land and industrial capital from individual and class ownership, and the vesting of them in the community for the general benefit. In this way only can the natural and acquired advantages of the country be equitably shared by the whole people. The Society accordingly works for the extinction of private property in land and of the consequent individual appropriation, in the form of rent, of the price paid for permission to use the earth, as well as for the advantages of superior soils and sites."[4]

To advance their ideology, four Fabian members, Beatrice and Sidney Webb, Graham Wallas, and George Bernard Shaw, founded the London School of Economics in 1895. Funding came from the Rockefeller Foundation, the Carnegie United Kingdom Trust Fund and various interests tied to J.P. Morgan & Company.[5] Such collaboration between the Fabian Society and the various elite foundations is hardly surprising considering that both sought to radically reengineer society through steady propaganda and indoctrination. Micklethwait explained, "The Fabians also helped to establish the idea that socialism was an exciting way of life, not just a political creed. They founded a network of 'groups' — the women's group, the arts group, groups for education, biology and local government."[6] This indoctrination, however, was not limited to England. "Fabianism was a template," explained Micklethwait, "for something that was much more universal. Across Europe groups of intellectuals helped to establish the idea that socialism was the wave of the future, and groups of activists helped to define socialism not just as a body of ideas but also as a community. The result of all these efforts was the 'socialist movement': an ideology that was also a fraternity; a set of beliefs that could organize people's live from the cradle to the grave; a faith that could exert a relentless pressure on moderates and extract a terrible revenge on traitors."[7]

The Fabian Society also helped establish England's Labour Party in the early 20th century. By 1918, Sidney Webb was drafting Labour's new Constitution according to Fabian ideology. A.M. McBriar, in his book *Fabian Socialism and English Politics*, explained, "The new Labour Party constitution and its accompanying manifesto *Labour and the New Social Order* must be considered amongst Sidney Webb's most skilful pieces of 'permeation.' The bricks of earlier Labour Party resolutions are cleverly put together with a little mortar of Fabian doctrine to produce a recognizably Fabian edifice. The distinctive new achievement was the writing of Socialism into the party's Constitution…It was of course explained carefully in *Labour and the New Social Order* that this Socialism would be of a moderate, constitutional, evolutionary kind."[8] The Fabian Society is still active today, claiming both the current and former Prime Minister as members. The *London Guardian* reported, "Since the 1997 general election there have been around 200 Fabian MPs in the Commons, amongst whom number nearly the entire cabinet, including Tony Blair, Gordon Brown, Robin Cook, Jack Straw, David Blunkett and Clare Short. The society has pursued its role as the new Labour government's 'critical friend,' seeking to ask challenging questions and to stimulate public debate…At the start of the 21st century the society plays as crucial a role in the political life of the country as ever"[9]

Besides covertly pushing socialism, the Fabian Society was also deeply aligned with eugenics, the practice of eliminating so-called undesirables from society.[10] Most of the Fabian's founding members were ardent supporters of eugenics. George Bernard Shaw, for example, was a eugenics extremist. In 1910, while lecturing at the Eugenics Education Society in London, Shaw recommended using "lethal chambers" for eliminating undesirables. "A part of eugenic politics," he explained, "would finally land us in an extensive use of the lethal chamber. A great many people would have to be put out of existence, simply because it wastes other people's time to look after them."[11] Later in 1933 he wrote, "Extermination must be put on a scientific basis if it is ever to be carried out humanely and apologetically as well as thoroughly."[12] Sydney and Beatrice Webb were also prominent supporters of the eugenics movement. In 1896, Sydney expressed his disgust for the "degenerate hordes of demoralized 'residuum' unfit for social life."[13] As documented below, Fabian members H.G. Wells and Bertrand Russell also called for radical eugenics and extensive social reengineering.

Bertrand Russell

History often makes heroes of the most distasteful men. For example, many people remember Bertrand Russell as a brilliant philosopher and Nobel Prize winner. The *Stanford Encyclopedia of Philosophy* observed, "Over the course of his long career, Russell made significant contributions, not just to logic and philosophy, but to a broad range of other subjects including education, history, political theory and religious studies. In addition, many of his writings on a wide variety of topics in both the sciences and the humanities have influenced generations of general readers."[14] The *Stanford Encyclopedia*, however, ignored Russell's elitism, his racism, his enthusiasm for eugenics and depopulation, and his advocacy of world government. Russell sometimes masqueraded as an anti-establishment, anti-tyranny personality. Nevertheless, he strongly supported depopulation, world government and eugenics, all hallmarks of the New World Order agenda. Also, notably, Russell was a member of the Fabian Society for several years.

Population Control Requires World Government

As documented in Chapter 10, elites have always feared losing power to ever-increasing populations of so-called commoners. Therefore, for at least 200 years, they have used propaganda to generate fear about overpopulation. In recent years, they have aggressively associated overpopulation with environmental degradation. Only world government, they suggest, can save humanity from global warming and overpopulation. In 1951, when Bertrand Russell wrote *The Impact of Science on Society*, the propaganda was more simplistic. Russell merely claimed that without world government, hungry nations would band together and attack the well-fed nations:

"There are three ways of securing a society that shall be stable as regards population. The first is that of birth control, the second that of infanticide or really destructive wars, and third that of general misery except for a powerful minority…Of these three, only birth control avoids extreme cruelty and unhappiness for the majority of human beings. Meanwhile, so long as there is not a single world government there will be competition for power among the different nations. And as increase of population brings the threat of famine, national power will become more and more obviously the only way of avoiding starvation. There will therefore be blocs in which the hungry nations band together against those that are well fed. That is the explanation of the victory of communism in China. These considerations prove that a scientific world society cannot be stable unless there is a world government."[15]

Showcasing his ruthlessness, Russell explained how the world government would control the world's food and distribute it based on fixed quotas. If a nation's population were to increase, they could either (a) eat less or (b) decrease their population:

"The population of the world is increasing, and its capacity for food production is diminishing. Such a state of affairs obviously cannot continue very long without producing a cataclysm. To deal with this problem it will be necessary to find ways of preventing an increase in world population. If this is to be done otherwise than by wars, pestilences, and famines, it will demand a powerful international authority. This authority should deal out the world's food to the various nations in proportion to their population at the time of the establishment of the authority. If any nation subsequently increased its population it should not on that account receive any more food. The motive for not increasing population would therefore be very compelling. What method of preventing an increase might be preferred should be left to each state to decide."[16]

Russell lamented that war cannot effectively neutralize population growth. He seemed enthusiastic, however, about once-per-generation plagues to help cull the herd. While this strategy might be unpleasant, according to Russell, "high-minded people" are immune to the suffering of others. Those who Russell calls "high-minded," however, are actually psychopaths devoid of empathy:

"But bad times, you may say, are exceptional, and can be dealt with by exceptional methods. This has been more or less true during the honeymoon period of industrialism, but it will not remain true unless the increase of population can be enormously diminished. At present the population of the world is increasing at about 58,000 per diem. War, so far, has had no very great effect on this increase, which continued throughout each of the world wars…but perhaps bacteriological war may prove more effective. If a Black Death could spread throughout the world once in every generation, survivors could procreate freely without making the world too full. There would be nothing in this to offend the consciences of the devout or to restrain the ambitions of nationalists. The state of affairs might be somewhat unpleasant, but what of that? Really high-minded people are indifferent to happiness, especially other people's."[17]

World Government as a Preventative for War

Besides combating the threat of overpopulation, Russell also reasoned that world government is necessary to prevent war (the perennial justification for ending national sovereignty). He explained, "As regards war, the principle of unrestricted national sovereignty, cherished by liberals in the nineteenth century and by the Kremlin in the present day, must be abandoned. Means must be found of subjecting the relations of nations to the rule of law, so that a single nation will no longer be, as at present, the judge in its own cause. If this is not done, the world will quickly return to barbarism."[18] Russell claimed that social cohesion requires an enemy, whether legitimate or manufactured. Within the framework of world government, however, he failed to foresee any possible 'enemies.' Decades later, the Club of Rome revealed that manmade global warming, or more precisely, humanity itself was the enemy. But Russell presumed that a world government would have no enemies and would therefore be tyrannical. He explained,

"There would now be no technical difficulty about a single world-wide Empire. Since war is likely to become more destructive to human life than it has been in recent centuries, unification under a single world government is probably necessary unless we are to acquiesce in either a return to barbarism or the extinction of the human race. There is, it must be confessed, a psychological difficulty about a single world government. The chief source of social cohesion in the past, I repeat, has been war: the passions that inspire a feeling of unity are hate and fear. These depend upon the existence of an enemy, actual or potential. It seems to follow that a world government could only be kept in being by force, not by the spontaneous loyalty that now inspires a nation at war."[19]

Nevertheless, he assured his readers that *eventually* humanity would consent to, and perhaps even embrace, scientific tyranny: "Now there are only two sovereign States: Russia (with satellites) and the United States (with satellites). If either becomes preponderant, either by victory or by an obvious military superiority, the preponderant Power can establish a single Authority over the whole world, and thus make future wars impossible. At first this Authority will, in certain regions, be based on force, but if the Western nations are in control, force will as soon as possible give way to consent. When that has been achieved, the most difficult of world problems will have been solved, and science can become wholly beneficent."[20]

Scientific Tyranny

How would the world government win the consent of humanity? Russell wrote extensively about scientific propaganda, indoctrination and genetic manipulation. "It is possible nowadays," he explained, "for a government to be very much more oppressive than any government could be before there was scientific technique. Propaganda makes persuasion easier for the government; public ownership of halls and paper makes counter-propaganda more difficult; and the effectiveness of modern armaments makes popular risings impossible."[21] Russell also explained how a scientific oligarchy would inevitably become totalitarian: "We have seen that scientific technique increases the importance of organisations, and therefore the extent to which authority impinges upon the life of the individual. It follows that a scientific oligarchy has more power than any oligarchy could have in pre-scientific times. There is a tendency, which is inevitable unless consciously combated, for organisations to coalesce, and so to increase in size, until, ultimately, almost all become merged in the State. A scientific oligarchy, accordingly, is bound to become what is called 'totalitarian', that is to say, all important forms of power will become a monopoly of the State."[22] On mass psychology, Russell observed,

"I think the subject which will be of most importance politically is mass psychology. Mass psychology is, scientifically speaking, not a very advanced study, and so far its professors have not been in universities: they have been advertisers, politicians, and above all, dictators. This study is immensely useful to practical men, whether they wish to become rich or to acquire the government. It is, of course, as a science, founded upon individual psychology, but hitherto it has employed rule-of-thumb methods which were based upon a kind of intuitive common sense. Its importance has been enormously increased by the growth of modern methods of propaganda. Of these the most influential is what is called 'education'. Religion plays a part, though a diminishing one; the Press, the cinema and the radio play an increasing part...It may be hoped that in time anybody will be able to persuade anybody of anything if he can catch the patient young and is provided by the State with money and equipment."[23]

On education, he remarked:

- "Although this science will be diligently studied, it will be rigidly confined to the governing class. The populace will not be allowed to know how its convictions were generated. When the technique has been perfected, every government that has been in charge of education for a generation will be able to control its subjects securely without the need of armies or policemen."[24]
- "The nations which at present increase rapidly should be encouraged to adopt the methods by which, in the West, the increase of population has been checked. Educational propaganda, with government help, could achieve this result in a generation."[25]

Russell was also interested in genetic and otherwise physiological manipulation for control purposes. He foreshadowed, for example, the current use of genetically modified foods and vaccinations to physically alter and engineer human beings:

"It is to be expected that advances in physiology and psychology will give governments much more control over individual mentality than they now have even in totalitarian countries. Fichte laid it down that education should aim at destroying free will, so that, after pupils have left school, they shall be incapable, throughout the rest of their lives, of thinking or acting otherwise than as their schoolmasters would have wished. But in his day this was an unattainable ideal…In the future such failures are not likely to occur where there is dictatorship. Diet, injections, and injunctions will combine, from a very early age, to produce the sort of character and the sort of beliefs that the authorities consider desirable, and any serious criticism of the powers that be will become psychologically impossible. Even if all are miserable, all will believe themselves happy, because the government will tell them that they are so."[26]

Finally, Russell also supported eugenics. He advocated sterilizing mentally defective people and strongly encouraged "desirable parents to have a large number of children."[27] He believed that so-called superior types should dominate inferior types. In his 1929 book *Marriage and Morals*, Russell highlighted the supposed inferiority of Africans, yet stopped short of calling for their extermination because, according to him, they make for good workers in tropical regions. He explained, "In extreme cases there can be little doubt of the superiority of one race to another. North America, Australia and New Zealand certainly contribute more to the civilization of the world than they would do if they were still peopled by aborigines. It seems on the whole fair to regard negroes as on the average inferior to white men, although for work in the tropics they are indispensable, so that their extermination (apart from questions of humanity) would be highly undesirable."[28] But Russell didn't discriminate based only on race. He reasoned that elites, through selective breeding, would effectively split the species. And just as humans justify *owning* domesticated animals, elites would soon justify *owning* lower-species humans. He explained,

"The system, one may surmise, will be something like this: except possibly in the governing aristocracy, all but 5 per cent of males and 30 per cent of females will be sterilised. The 30 per cent of females will be expected to spend the years from eighteen to forty in reproduction, in order to secure adequate cannon fodder. As a rule, artificial insemination will be preferred to the natural method. The unsterilised, if they desire the pleasures of love, will usually have to seek them with sterilised partners. Sires will be chosen for various qualities, some for muscle others for brains. All will have to be healthy, and unless they are to be the fathers of oligarchs they will have to be of a submissive and docile disposition. Children will, as in Plato's Republic, be taken from their mothers and reared by professional nurses. Gradually, by selective breeding the congenital differences between rulers and ruled will increase until they become almost different species. A revolt of the plebs would become as unthinkable as an organised insurrection of sheep against the practice of eating mutton."[29]

H.G. Wells

Known as "The Father of Science Fiction,"[30] H.G. Wells was another Fabian socialist who today is rarely remembered for his overt racism and radical elitism. Wells was a student of Thomas Henry Huxley, the grandfather of Julian and Aldous. He later became a colleague of Julian Huxley, the racist eugenicist. Together, they wrote the book *The Science of Life*. Wells was a founding member of the Fabian Society and a dutiful servant of the New World Order. In 1940, he even wrote a book called *The New World Order*. He predicted, "Countless people, from maharajas to millionaires and from pukkha sahibs to pretty ladies, will hate the new world order, be rendered unhappy by frustration of their passions and ambitions through its advent and will die protesting against it. When we attempt to estimate its promise we have to bear in mind the distress of a generation or so of malcontents, many of them quite gallant and graceful-looking people."[31]

The New World Order envisioned by Wells was a Utopia for 'superior races.' In his book *A Modern Utopia*, he reflected, "The depopulation of the Congo Free State by the Belgians, the horrible massacre of Chinese by European soldiery during the Pekin expedition, are condoned as a painful but necessary part of the civilizing process of the world."[32] Thus for Wells, inferior races were obstacles to civilization. He continued, "The politically ascendant peoples of the present phase are understood to be the superior races…all uncivilised people are represented as the inferior races, unfit to associate with the former on terms of equality, unfit to intermarry with them on any terms, unfit for any decisive voice in human affairs."[33] With his 1902 book *Anticipations*, Wells established himself as one history's most vile men, easily on par with Hitler from an ideological perspective:

- "And how will the New Republic treat the inferior races? How will it deal with the black? How will it deal with the yellow man? How will it tackle that alleged termite in the civilized woodwork, the Jew? Certainly not as races at all. It will aim to establish, and it will at last, though probably only after a second century has passed, establish a world-state with a common language and a common rule. All over the world its roads, its standards, its laws, and its apparatus of control will run. It will, I have said, make the multiplication of those who fall behind a certain standard of social efficiency unpleasant and difficult."[34]

- "The Jew will probably lose much of his particularism, intermarry with Gentiles, and cease to be a physically distinct element in human affairs in a century or so. But much of his moral tradition will, I hope, never die. And for the rest, those swarms of black, and brown, and dirty-white, and yellow people, who do not come into the new needs of efficiency? Well, the world is a world, not a charitable institution, and I take it they will have to go. The whole tenor and meaning of the world, as I see it, is that they have to go. So far as they fail to develop sound, vigorous, and distinctive personalities for the great world of the future, it is their portion to die out and disappear."[35]

Wells summarized his New World Order ideology with a definitive statement: "There is only one sane and logical thing to be done with a really inferior race, and that is to exterminate it."[36] On their website, the Fabian Society today brags, "The Fabian Society has played a central role for more than a century in the development of political ideas and public policy on the left of centre."[37] They proudly recognize George Bernard Shaw, H.G. Wells, and Beatrice and Sidney Webb as "Famous Members."[38]

Chapter 4
The Council on Foreign Relations

Although most Americans have never heard about it, although most American schools have never taught about it, the Council on Foreign Relations (CFR) has been perhaps *the most powerful* organization in the U.S. since its 1921 founding. Comprised of influential politicians, bankers, CEOs, lawyers, academics, and journalists, the CFR permeates government, business, and media, and effectively functions as a shadow government. Most disturbingly, the CFR's initiatives and policy recommendations continually support the New World Order agenda of destroying national sovereignty and implementing regional, and eventually global government. As discussed in Chapter 7, for example, they were the impetus behind the impending North American Union. Also countless CFR members have publicly called for world government, including David Rockefeller.

For decades, David Rockefeller has been the most powerful, most influential member of the Council on Foreign Relations. At ninety-four years of age, he currently serves as Honorary Chairman. He previously served as Director from 1949 to 1985, Vice-President from 1950 to 1970, and Chairman of the Board from 1970 to 1985.[1] In his 2002 book *Memoirs*, Rockefeller casually discussed the vast scope of CFR projects. He wrote, "From the early 1950s on, then, the Council's program of speakers, study groups, and publications has provided a forum where critical issues are examined and discussed. Vietnam, the opening of China, détente with the Soviet Union, balancing world population with food resources, the Arab-Israeli conflict in the Middle East, economic development in the Third World, the expansion of NATO—these and many other issues have found their place on the Council's agenda through the years."[2] He neglected to mention, however, that the creation of the United Nations and the North American Union have also "found their place on the Council's agenda."

The CFR's purpose is to gradually weaken and eventually overthrow the constitutional republic of the United States. As previously noted, David Rockefeller openly acknowledged this agenda in his own book.[3] Most other prominent CFR members are also vehemently anti-national sovereignty. CFR President Richard Haas, for example, even published an article in the *Taipei Times* called, "State sovereignty must be altered in globalized era."[4] Many CFR members work for the government and thus have taken oaths of office whereby they swear to *protect and defend* the U.S. Constitution. Nevertheless, the prevailing mentality within the CFR is decidedly anti-Constitution. In 1992, the CFR even published a book asserting, "The founders of the American state, it may be thought, ought to exercise no special hold on our outlook; the earth, as Jefferson said, belongs to the living."[5]

Funding for the CFR comes mostly from corporations and foundations. According to their 2008 Annual Report, more than $28 million of their $68 million Operating Revenue Budget came from foundation grants. Another $7 million came from corporate members, $10 million from individual members, and $8 million from sales of their journal *Foreign Affairs*.[6] The top banks, corporations, and military contractors are all CFR "Corporate Members" including Goldman Sachs, Bank of America, JPMorgan Chase & Co., Chevron, Exxon Mobil, Lockheed Martin, etc.[7] In 1954, the House Select Committee to Investigate Tax-Exempt Foundations, also known as the Reece Committee, concluded that the CFR's activities "are directed overwhelmingly at promoting the globalist concept." They also determined that CFR had become "in essence an agency of the United States government...carrying its internationalist bias with it."[8]

Government Infiltration

In 1961, the *Christian Science Monitor* observed that the CFR "has staffed almost every key position of every administration since that of FDR."[9] Forty-nine years later, nothing has changed. Although many critics accuse the CFR of *influencing* the government, in many respects, the CFR *is* the government. Since its inception, CFR members have thoroughly dominated the government's most influential posts. Consequently, the organization has exerted enormous influence, especially over U.S. foreign policy. Since 1950, 15 of 18 CIA Directors have been CFR members.[10] Since 1929, 18 of 22 Secretaries of State have been CFR members.[11] Since 1949, 20 of 25 Secretaries of War/Defense have been CFR members.[12] Since 1921, 19 of 26 Secretaries of Treasury have been CFR members.[13] From 1928 to 1972, the CFR won every presidential election.[14] Every president was a member except Lyndon Johnson, who filled his administration with CFR members.

In 1976, the Trilateral Commission, another of Rockefeller's anti-sovereignty groups, took over the presidency. In 1980, the CFR held the vice-presidency with CFR member George H.W. Bush. And although President Reagan was not a member, he filled his administration with 313 CFR members.[15] In 1988, President Bush staffed his administration with 387 CFR or Trilateral Commission members.[16] In 1992, CFR, Trilateral Commission *and* Bilderberg attendee Bill Clinton became President. Although George W. Bush was not a CFR member, his team included CFR members Dick Cheney, Richard Perle, Paul Wolfowitz, Lewis Libby, Colin Powell, Robert Zoellick, and Condoleezza Rice.[17] Barack Obama, despite his promises of change, has also stacked his administration with CFR members and members of other elite organizations including the Trilateral Commission and Bilderberg Group. They include:

> Vice President Joe Biden (CFR), Secretary of Treasury Timothy Geithner (CFR, TC, Bilderberg), Secretary of State Hilary Clinton (CFR, Bilderberg), Ambassador to the United Nations Susan Rice (TC), National Security Advisor General James L. Jones (CFR, TC, Bilderberg), Deputy National Security Advisor Thomas Donilon (CFR, TC), State Department Special Envoy Henry Kissinger (CFR, TC, Bilderberg), State Department Special Envoy George Mitchell (CFR, Bilderberg), Chairman of the Economic Recovery Committee Paul Volcker (CFR, TC, Bilderberg), Director of National Security Admiral Dennis C. Blair (CFR, TC, Bilderberg), Secretary of Defense Robert Gates (CFR, TC, Bilderberg), Deputy Secretary of State James Steinberg (CFR, TC, Bilderberg), State Department Special Envoy Richard Haass (President of CFR, TC, Bilderberg), Presidential Advisor Alan Greenspan (CFR, TC, Bilderberg), State Department Special Envoy Richard C. Holbrooke (CFR, TC, Bilderberg).[18]

How the CFR Functions

The CFR website states, "CFR takes no institutional positions on matters of policy."[19] On this point, however, they are clearly insecure, like children who, after telling a lie, then ask, "You believe me, don't you?" Many CFR members feel compelled to frequently echo this "you believe me, don't you?" mantra. For example, William Bundy, Editor of the CFR's *Foreign Affairs* from 1972-1984, stated, "From the beginning, the Council has seen its function as giving running room to individual ideas and writing, and to discussion and debate. It has never, as a body, taken any position on foreign-policy problems."[20] Also David Rockefeller wrote, "But the essential point is that the Council never takes a position—official or unofficial—on any foreign policy issue even though its members are free to do so."[21] These statements are highly deceptive and ultimately false.

The CFR doesn't need to take an "official position" because its members, by their own volition, generally represent ruling-class values and interests. Chester Ward, a CFR member for 16 years and former Judge Advocate General of the U.S. Navy, explained, "As accurately stated in the excerpt from the 1972 President's Report, CFR is an 'organization of individual members.' This point has been missed by most of those who have attempted to picture CFR's power. CFR, as such, does not write the platforms of both political parties or select their respective presidential candidates, or control U.S. defense and foreign policies. But CFR members, as individuals acting in concert with other individual CFR members, do."[22] Speaking about David Rockefeller, Ward stated, "When he, or any influential member of CFR decides to take a hand in a policy or program within the cognizance of CFR, he will not act through the organization, but as leader of a sort of floating ad hoc coalition with other influential members having similar objectives. The policy-making members use CFR as an instrument rather than as an organization. It has proved to be a tool of great value, especially for propagandizing."[23]

The CFR functions like a microcosm of the New World Order. The New World has no official documents, no official statements and no official policy. It is an unofficial coalition of likeminded elites who wish to destroy the sovereignty of every nation. These elites work *through* numerous organizations and institutions, but the organizations themselves don't *officially* endorse the New World Order agenda. These organizations are highly compartmentalized, with power and influence highly centralized. There are rings within rings within rings of knowledge and understanding. The outermost rings don't even know the innermost rings exist. The outermost rings simply conduct research and provide administrative services. They legitimize the organization, thus providing cover for the innermost rings.

The outermost rings generally consist of people who genuinely believe they are working towards the betterment of humanity. They have no idea they are being co-opted. The CFR, for example, has over 4,300 members.[24] Even Warren Beatty, Michael Douglas, and Angelina Jolie are members.[25] Nobody is accusing Angelina Jolie of plotting the destruction of the Constitution and the implementation of the North American Union—the innermost rings are doing that. Ward explained, "Although, from the inside, CFR is certainly not the monolith that some members and most nonmembers consider it, this lust to surrender the sovereignty and independence of the United States is pervasive throughout most of the membership, and particularly in the leadership of the several divergent cliques that make up what is actually a polycentric organization."[26]

Ward described the innermost ring as "A much smaller group but more powerful, with a low profile but controlling billions of dollars in the United States and elsewhere."[27] He explained,

"This faction comprises the Wall Street international bankers and their key agents. Primarily, they want the world banking monopoly from whatever power ends up in control of the global government. They would probably prefer that this be an all-powerful United Nations organization; but they are also prepared to deal with a one-world government controlled by the Soviet Communists if U.S. sovereignty is ever surrendered to them. This CFR faction is headed by the Rockefeller brothers."[28]

Ward also explained how the outer rings serve as worker bees for the inner rings: "Once the ruling members of the CFR have decided that the U.S. government should adopt a particular policy, the very substantial research facilities of the CFR are put to work to develop arguments, intellectual and emotional, to support the new policy, and to confound and discredit, intellectually and politically, any opposition. The most articulate theoreticians and ideologists prepare related articles, aided by the research, to sell the new policy and to make it appear inevitable and irresistible. By following the evolution of this propaganda in the most prestigious scholarly journal in the world, *Foreign Affairs*, anyone can determine years in advance what the future defense and foreign policies of the United States will be."[29]

Prominent Critics

Through the years, many prominent people have worked to expose the CFR. Dan Smoot, for example, a former FBI agent, wrote, "Through many interlocking organizations, the Council on Foreign Relations 'educates' the public—and brings pressures upon Congress—to support CFR policies. All organizations, in this incredible propaganda web, work in their own way toward the objective of the Council on Foreign Relations: to create a one-world socialist system and to make America a part of it."[30] Barry Goldwater, a five-term Senator and former Major General in the U.S. Air Force Reserve, was also highly critical of the CFR. Goldwater understood the New World Order mentality. He knew they were simply interested in domination and willing to work with any groups necessary to achieve their ends. He explained,

"A number of writers disturbed by the influential role this organization has played in determining foreign policy have concluded the Council on Foreign Relations and its members are an active part of the communist conspiracy for world domination...I believe the Council on Foreign Relations and its ancillary elitist groups are indifferent to communism. They have no ideological anchors. In their pursuit of a new world order they are prepared to deal without prejudice with a communist state, a socialist state, a democratic state, monarchy, oligarchy—it's all the same to them."[31]

Goldwater also recognized the CFR's disdain for national sovereignty and personal freedom and liberty:

"Their goal is to impose a benign stability on the quarreling family of nations through merger and consolidation. They see the elimination of national boundaries, the suppression of racial and ethnic loyalties as the most expeditious avenue to world peace. Their rationale rests exclusively on materialism. They believe economic competition is the root cause of international tension. This approach dismisses as insignificant the form of government or the political ideology expressed by that form. It may be that if the CFR vision of the future could be realized, there would be a reduction in wars, a lessening of poverty, a more efficient utilization of the world's resources. To my mind, this would inevitably be accompanied by a loss in personal freedom of choice and the reestablishment of the restraints which provoked the American Revolution."[32]

Curtis Dall was another prominent CFR critic. A Wall Street insider during the 1920s, Dall rose to become a partner of Fenner & Beane, the firm known today as Merrill Lynch. Dall's military service spanned 13 years, in which time he became an Air Force colonel. From 1960-1964, he was National Chairman of the Constitution Party. Dall was also the son-in-law of former President Franklin Roosevelt. His remarkable career exposed him to the upper echelons of political, economic and military power, thus enabling a unique historical perspective. Dall witnessed first hand the devious machinations of the ruling class (especially through organizations like the CFR) to destroy America and impose world government. In 1967, he wrote a fascinating memoir, *FDR: My Exploited Father-in-Law*, exposing the shadow government and its true objectives. Dall explained,

"Few members of the CFR know the long-range plans of its small top-management group. Hence, giving effect to all of the foregoing status areas, ninety per cent or more of the membership do not remotely comprehend just who 'plays the piano upstairs.' The piano is continuously played, nevertheless, and no time is lost by the CFR in teaching many of our duly elected officials to dance. Hence, this situation does not exactly constitute government by the people; it is subtle dictatorship by the few! It is an internationalist 'black tie' dictatorship, surmounted upon the base of many confused and bemused status-seekers, presenting an over-all distinguished front, little known, of course, to the unsuspecting public who must not learn of it. Doubtless, I could have secured a comfortable seat at that banquet table for myself some years ago, but the realization that this Constitutional Republic of ours is something very precious and must be protected, not exploited, for me overshadowed other considerations."[33]

He also noted,

"Obviously, the real objectives of One-World Government leaders and their ever-close bankers are most devious. They have now acquired full control of the money and credit machinery of the United States of America, via the creation and establishment of the *privately owned* Federal Reserve Bank. They now plan to uproot and to gradually destroy the Spiritual background of all peoples. Initially, Christianity is the prime target, then Judaism, then all other religions! That bleak program is absolutely necessary for them to complete, if possible, before they can reach godless power — aimed to benefit a few but to make assembly-line puppets out of us, the many. When you hear and read about the word peace, so often splashed about for political purposes by United Nations leaders, ask yourself just one question — whose peace? Every government and every individual has his own definition of what that word means. Often it is merely a vague image, erected to deftly mislead and confuse us."[34]

Larry McDonald was a U.S. Congressman from 1975 until his death aboard Korean Air Lines Flight 007, shot down on September 1, 1983. McDonald, then Chairman of the John Birch Society, was a persistent critic of the Council on Foreign Relations, the Trilateral Commission and other globalist outfits. In May of 1983, he appeared on CNN's *Crossfire* with co-hosts Pat Buchanan and CFR member Tom Braden. Despite Braden's shallow, condescending demeanor, McDonald calmly and collectively explained the globalist agenda to the hosts and the national audience. The following excerpts come from this fascinating discussion:

McDonald: "The Trilateral Commission, the Council on Foreign Relations, let's face it, they've dominated the State Department for 40 years."

Buchanan: "But what are they trying to do?"

McDonald: "Well they're objective is to try to bring about a gradual transition in our society, a dissolving of sovereignty, and moving steadily to the left on the political spectrum."

Braden: "Who are they?"

McDonald: "You are looking at a group that is trying to bring about a dissolution of national sovereignty, on the road to world government. And certainly you're familiar with local professor Carroll Quigley, who has been part of your club, in which he admitted all this. He said in his book *Tragedy and Hope*, 'the only thing I disagree with is that we've worked to keep it a secret.' And you see Arthur Schlesinger, Jr. writing way back in 1947, he said, 'yes this is the hidden policy of America, but we can't tell the American public because they're too unsophisticated to see the value.'"

Braden: "Of all the things to say about Arthur Schlesinger, that's the silliest statement I ever heard, he never said anything like that."

McDonald: "Well let me suggest you read the May/June issue of the *Partisan Review* of 1947 Tom and you can read it for yourself, it's called the 'Schlesinger Manifesto.'"

Buchanan: "Isn't there some move that occurred in the post-war era that now has been dissipated because nobody believes in the utopian ideal of world government anymore?"

McDonald: "Well I think there are those that realize that moving straight from a prototype of the United Nations into world government perhaps is tactically impossible. But phasing out increasingly national sovereignty into regional government and phasing out sovereignties into international treaties in multiple areas could be a route."

Braden: "What I ought to do is read more about conspiracy."

McDonald: "What you ought to do is go back and look at your founder Edward Mandel House because he wrote the book *Philip Dru: Administrator*. Colonel House said that what he envisioned for the world was a world government along socialist lines as envisioned by Karl Marx. Now that's your leader, Tom. So you've got to go back and read his book.[35]

Foreign Affairs

First published in 1922, *Foreign Affairs* journal is the voice of the Council on Foreign Relations. According to its editorial vision, "In pursuance of its ideals *Foreign Affairs* will not devote itself to the support of any one cause, however worthy. Like the Council on Foreign Relations from which it has sprung it will tolerate wide differences of opinion. Its articles will not represent any consensus of beliefs."[36] Nevertheless, the journal has a decidedly pro-New World Order bent. Since the 1970s, *Foreign Affairs* has printed over 210 articles including the phrase 'New World Order," and over 25 articles with 'New World Order' in the title. These of course are not articles decrying the New World Order and reaffirming the authority of the U.S. Constitution. *Foreign Affairs* is written by globalists, for globalists. Ron Paul, for example, the pro-Constitution Congressman, has never contributed.

Chester Ward observed that *Foreign Affairs*, "carries the 'CFR party line' set by the ruling cliques within the CFR. This prestigious quarterly starts to mold U.S. official foreign and defense policies some five to ten years in advance of the changes it finally brings about…under the guise of 'fighting isolationism,' *Foreign Affairs* quite obviously advocates internationalism and an end to national sovereignty, including U.S. sovereignty."[37] Kingman

Brewster's "Reflections On Our National Purpose," and Richard Gardner's "The Hard Road to World Order" are two of many articles that typify *Foreign Affairs'* content.

In his 1972 article, "Reflections on our National Purpose," Kingman Brewster, Jr. relentlessly assailed what he referred to as, "The isolationism of smug, disdainful, secure, aloof Americanism."[38] 37 years later, his article is still particularly relevant. Then, like now, many young people were protesting U.S. wars of aggression and U.S. foreign policy in general. Brewster advocated first appealing to the young, and then cleverly duping them into supporting the very tyranny against which they were protesting. Brewster recognized the frustration of America's youth due to their country's meddling abroad. He explained, "For the first time in our history a very substantial number of the future leaders of the country are disenchanted with the inherited role of the United States in the world. They badly need a vision about what that role might be."[39]

But instead of encouraging them to develop their own vision, Brewster recommended indoctrinating the youth with standard New World Order ideology, the idea that nations must inevitably relinquish their sovereignty and accept global government. Brewster explained, "It is hard to see how we will engage the young, and stand any chance of competing for the respect of mankind generally, if we continue to be hold-outs, more concerned with the sovereignty of nations than with the ultimate sovereignty of peoples."[40] Accordingly, proper indoctrination could make "the oncoming generation in the United States and in other developed countries not only willing but eager to see the creation of authorities superior to national governments."[41]

Brewster predicted that some future president would have to sell the American people on the New World Order. Presumably, this President would have to hold office during a time of great turmoil. He would have to be charismatic and have millions of blind-faith supporters. Brewster explained, "Our new situation of mutual, national dependence is inescapable. If we would face it in a creative mood, we will have to take some risks in order to invite others to pool their sovereignty with ours on matters which none of us can control alone. We shall have to abide by lawfully achieved results even when we might have wished or voted otherwise. Some day, some President must convince all the American people that this is a proud and exciting call to be faced with zest rather than with reluctance."[42] Is Obama this anticipated president?

In 1974, prominent Trilateral Commission member and lifetime globalist Richard Gardner penned a now infamous article for *Foreign Affairs* entitled, "The Hard Road to World Order." Like most globalists, Gardner believed that world government was inevitable. He only lamented the public's reluctance to embrace the idea. He explained, "We are witnessing an outbreak of shortsighted nationalism that seems oblivious to the economic, political and moral implications of interdependence."[43] Frustrated with what he termed an "unhappy state of affairs" whereby "few people retain much confidence in the more ambitious strategies for world order that had wide backing a generation ago," Gardener advocated implementation by stealth, ever so gradually. He explained, "An end run around national sovereignty, eroding it piece by piece, will accomplish much more than the old-fashioned frontal assault."[44]

As previously discussed, this is precisely the globalists' game plan. And as evidenced by Gardner, the game plan has never been secret. In one short article, he publicized the entire agenda, including ending the dollar standard, establishing an authoritative global monetary institution, destructive trade agreements, a global environmental authority and widespread depopulation. Furthermore, he made clear that attempts to maintain sovereignty would not be respected: "The hopeful aspect of the present situation is that even as nations resist appeals for 'world government' and 'the surrender of sovereignty,' technological, economic and political interests are forcing them to establish more and more far-reaching arrangements to manage their mutual interdependence."[45]

Regarding ending the dollar standard, Gardner remarked, "The non-Communist nations are embarked on a long-term negotiation for the reform of the international monetary system, aimed at developing a new system of reserves and settlements to replace the dollar standard."[46] On the establishment of an authoritative global monetary institution, he called for "a revitalization of the International Monetary Fund, which would have unprecedented powers to create new international reserves and to influence national decisions on exchange rates and on domestic monetary and fiscal policies."[47]

Gardner also foreshadowed the destructive NAFTA trade agreement with his assertion that, "Among other things, we will be seeking new rules in the General Agreement on Tariffs and Trade to cover a whole range of hitherto unregulated nontariff barriers. These will subject countries to an unprecedented degree of international surveillance over up to now sacrosanct 'domestic' policies."[48] On the environment, Gardner noted, "The next few years should see a continued strengthening of the new global and regional agencies charged with protecting the world's environment."[49] He also called for a "procedure by which at least some nations agree to have certain kinds of environmental decisions reviewed by independent scientific authorities."[50] These "authorities," while by no means independent, and rather unscientific, would later become the United Nation's Intergovernmental Panel on Climate Change (IPCC).

Finally, on the globalists' depopulation agenda, Gardner asserted, "We are entering a wholly new phase of international concern and international action on the population problem, dramatized by the holding this year of the first World Population Conference to take place at the political level. By the end of this decade, a majority of nations are likely to have explicit population policies, many of them designed to achieve zero population growth by a specific target date. These national policies and targets will be established and implemented in most cases with the help of international agencies."[51] "In short," Gardner concluded, "the 'house of world order' will have to be built from the bottom up rather than from the top down. It will look like a great 'booming, buzzing confusion.'"[52]

Arthur Schlesinger, Jr. was yet another prominent *Foreign Affairs* writer from the 1960s until his death in 2007. A former Harvard professor and speechwriter for John F. Kennedy, Schlesinger strongly supported the New World Order agenda. In 1995, for example, he penned an op-ed for the *New York Times* called, "New Isolationists Weaken America." Schlesinger asserted that, with or without the approval of the American people, the United States "will continue to accept international political, economic and military commitments unprecedented in its history."[53] He then ridiculed several politicians who opposed U.S. support for a standing United Nations army. Bob Dole, for example, had asserted, "The American people will not tolerate American casualties for irresponsible internationalism." Schlesinger disregarded such foolishness. Nevertheless, he admitted, "Dying for world order when there is no concrete threat to one's own nation is a hard argument to make."[54]

Schlesinger further acknowledged that the American people feel neither obliged nor inspired to make such sacrifices: "Nor can it be said that this recoil from collective security misrepresents popular sentiment. A public opinion survey by the Chicago Council on Foreign Relations and the Gallup Organization shows that, while Americans are still ready to endorse euphonious generalities in support of internationalism, there is a marked drop-off when it comes to committing not just words but money and lives."[55] To Schlesinger however, the will of the people, traditional American values, the Constitution, and the concept of government *for and by* the people were simply outdated tenets of neo-isolationism, and direct threats to the New World Order. With sadistic anticipation, he declared, "In the United States, neo-isolationism promises to prevent the most powerful nation on the planet from playing any role in enforcing the peace system. If we refuse a role, we cannot expect smaller, weaker and poorer nations to insure world order for us. We are not going to achieve a new world order without paying for it in blood, words and money."[56] Make no mistake, Schlesinger and his fellow would-be overlords have voluntarily provided the words. By their thinking, the people must now provide the money and blood.

Assassination Attempt on the Constitution

During a speech on February 23, 1954, Senator William Jenner observed, "Today the path to total dictatorship in the United States can be laid by strictly legal means, unseen and unheard by the Congress, the President, or the people...Outwardly we have a Constitutional government. We have operating within our Government and political system another body representing another form of government, a bureaucratic elite which believes our Constitution is outmoded and is sure that it is the winning side...all the strange developments in foreign policy agreements may be traced to this group who are going to make us over to suit their pleasure."[57] The United States of America is a Constitutional Republic with power intentionally distributed between the Executive, Legislative and Judicial branches of government. The country's founders carefully crafted the Constitution to protect against would-be tyrants. While allowing for amendments congruous with the general spirit of the Constitution, they urged future Americans to defend and maintain at all costs the separation of powers.

After his closing speech at the 1787 Constitutional Convention at Philadelphia's Independence Hall, a woman approached Benjamin Franklin and asked, "Well, Doctor, what have we got, a republic or a monarchy?" Franklin responded, "A republic, if you can keep it."[58] Having studied history and governments extensively, he knew that duplicitous men would inevitably seek to undermine the Constitution. His "if you can keep it" comment was a preemptive warning about such organizations as the Council on Foreign Relations. He was also warning about such men as James McGregor Burns, a Pulitzer Prize winning presidential biographer and longtime CFR luminary.[59]

In 1984, Burns published *The Power to Lead*. The CFR's *Foreign Affairs* reviewed the book favorably, noting, "The author favors amending the Constitution to break down the separation of Executive and Congress."[60] Burns echoed Woodrow Wilson's contention that "the Federal Government lacks strength because its powers are divided, lacks promptness because its authorities are multiplied, lacks wieldiness because its processes are roundabout, lacks efficiency because its responsibility is indistinct and its action without competent direction."[61] Burns despised the Constitution and its framers. He lamented, "Let us face reality. The Framers have simply been too shrew for us. They have outwitted us. They designed separated institutions that cannot be unified by mechanical linkages, frail bridges, tinkering." Nevertheless, he advocated direct confrontation: "If we are to 'turn the Founders upside down'—to put together what they put asunder—we must directly confront the constitutional structures they erected."[62]

The Declaration of Interdependence

The World Affairs Council (WAC) is officially an independent, non-profit, non-partisan organization "working to engage and educate Americans in international affairs and foreign policy."[63] However, throughout its history, stretching back to 1918, the WAC has functioned as a CFR satellite. Former FBI agent Dan Smoot explained, "The CFR does not have formal affiliation—and can therefore disclaim official connection with—its subsidiary propaganda agencies; but the real and effective interlock between all these groups can be shown not only by their common objective (one-world socialism) and a common source of income (the foundations), but also by the overlapping of personnel: directors and officials of the Council on Foreign Relations are also officials in the interlocking organizations."[64] Many prominent CFR members including John Foster Dulles, Averell Harriman,[65] Nelson Rockefeller,[66] George Ball,[67] and Richard N. Cooper,[68] have collaborated with the WAC.

In 1976, the World Affairs Council commissioned historian Henry Steele Commager to draft the *Declaration of Interdependence*, a transparent ploy to significantly weaken national sovereignty. The *Declaration* read, "We affirm that a world without law is a world without order, and we call upon all nations to strengthen and to sustain the United Nations and its specialized agencies, and other institutions of world order, and to broaden the jurisdiction of the World Court." Commager further urged the public to abandon "narrow notions of national sovereignty."[69] The *Syracuse Herald-American* derided both Commager and his *Declaration*, stating, "As a historian, the professor doubtless has learned a vast deal of the world as it was; he manifests a pathetic naiveté about the world as it is."[70] Nevertheless, the WAC duped scores of Congressional leaders. The *New York Times* reported, "128 members of Congress have so far agreed to sign a 'Declaration of Interdependence' drafted by Henry Steele Commager, the historian. Durable though the nation-state has proved to be, interdependence has caught up with most of the difficulties that used to be regarded in the United States as primarily 'domestic.'"[71]

Representative Marjorie Holt, however, strongly opposed this subterfuge. An Ohio newspaper reported, "Rep. Holt strongly opposed the signing of this declaration. It called for surrender of our national sovereignty to international organizations. It desires that our economy should be regulated by international authorities. It proposes that we enter a 'new world order' that would redistribute the wealth created by the American people."[72] After originally supporting and signing the *Declaration*, Representative Goodloe Byron ordered the WAC to delete his signature. According to the *Associated Press*, Byron discovered "provisions in the Declaration of Interdependence I do not agree with. These provisions would restrict the sovereignty of the United States. Therefore, I did not want to continue to associate my name with this document."[73]

Recent High Treason

In 2006, CFR President Richard Haass boldly penned an ultra-provocative article for the *Taipei Times* entitled, "State sovereignty must be altered in globalized era." Haass declared, "For 350 years, sovereignty—the notion that states are the central actors on the world stage and that governments are essentially free to do what they want within their own territory but not within the territory of other states—has provided the organizing principle of international relations. The time has come to rethink this notion."[74] His diatribe against national sovereignty strongly challenged the CFR's repeated assertions that they never take a position regarding foreign policy. Here was the *president* of the CFR stating publicly, "Globalization thus implies that sovereignty is not only becoming weaker in reality, but that it needs to become weaker. States would be wise to weaken sovereignty in order to protect themselves, because they cannot insulate themselves from what goes on elsewhere. Sovereignty is no longer a sanctuary."[75] Haass further declared, "Moreover, states must be prepared to cede some sovereignty to world bodies if the international system is to function...The goal should be to redefine sovereignty for the era of globalization, to find a balance between a world of fully sovereign states and an international system of either world government or anarchy."[76]

On May 1, 2008, the CFR launched the International Institutions and Global Governance Program, a five-year program "to explore the institutional requirements for world order in the twenty-first century."[77] They targeted the usual suspects—terrorism, climate change and weapons of mass destruction—to justify stronger institutions of "global governance," a euphemism for global government first used by the United Nations in the 1990s. The CFR considered numerous models for global governance including adapting existing institutions like the UN:

> "The program will consider whether the most promising framework for governance is a formal organization with universal membership (e.g., the United Nations); a regional or sub-regional organization; a narrower, informal coalition of like-minded countries; or some combination of all three. Building on these issue-area investigations, the program will also consider the potential to adapt major bedrock institutions (e.g., the UN, G8, NATO, IMF) to meet today's challenges, as well as the feasibility of creating new frameworks. It will also address the participation of non-state actors."[78]

According to the CFR, America's love of freedom and sovereignty is the biggest obstacle to global governance. They explained, "Few countries have been as sensitive as the United States to restrictions on their freedom of action or as jealous in guarding their sovereign prerogatives."[79] They also noted that the U.S. Constitution "complicates" new international obligations: "the separation of powers enshrined in the U.S. Constitution, which gives Congress a critical voice in the ratification of treaties and endorsement of global institutions, complicates U.S. assumptions of new international obligations."[80]

The United Nations: A CFR Global Government Prototype

In the wake of the September 11th attacks, George Bush announced that America's enemies hate the U.S. Constitution: "They hate our freedoms, our freedom of religion, our freedom of speech, our freedom to vote and assemble and disagree with each other."[81] His statement, though steeped in irony, was completely accurate. Bush portrayed America's enemies as ragtag, Muslim cave dwellers. The real enemies, however, typically wear $10,000 suits, control vast amounts of wealth, and occupy leadership positions within corporations, governments, intelligence agencies, think-tanks, and financial institutions. By their own admissions, these people created the United Nations to gradually weaken and eventually destroy the U.S. Constitution. In 2002, Bob Barr, a former federal prosecutor and former Congressman, wrote about the ever-pervasive threat posed by the UN. He explained,

> "Efforts to destroy the American experiment persist, and our Constitution continues to come under attack from those abroad. Sadly, most Americans are not even aware of these threats. Though not international terrorists, those threatening our freedoms do share a

common disdain for America's freedom and strength. I speak of those individuals and governments that have hijacked the United Nations apparatus, in an effort to impose a one-world government at odds with our democratic system."[82]

Prominent globalist Mikhail Gorbachev, in his book *On My Country and the World*, confirmed this invasive United Nations agenda. He approvingly wrote, "Obviously, the goals of global management cannot be achieved all at once, in a single leap…Thus it is necessary to approach this goal step by step, to try to enhance the role of existing institutions and encourage the coordination of efforts of various governments. Above all, we are thinking about the United Nations."[83]

On their website, the United Nations poses the question, "Is the United Nations a world government?" to which they answer, "The UN is not, and was never intended to be, a world government."[84] Nevertheless, despite these assurances, plenty of documentary evidence suggests the opposite. Under the heading, "History of the United Nations," they state, "The forerunner of the United Nations was the League of Nations, an organization conceived in similar circumstances during the First World War, and established in 1919 under the Treaty of Versailles 'to promote international cooperation and to achieve peace and security.'"[85] From here, their historical account jumps to the 1945 San Francisco Conference where representatives from fifty nations met to draft the UN Charter. Therefore, they imply that nothing significant happened between 1919 and 1945. They don't mention, for example, how the Council on Foreign Relations and the Rockefeller Foundation started planning the UN in 1939. Why has the UN expunged this essential history? Clearly it would challenge their assertion that the United Nations "was never intended to be a world government."

During the early 1950s, René Wormser was General Counsel for the Reece Committee, a congressional committee investigating tax-exempt foundations. In his 1958 book *Foundations: Their Power and Influence*, Wormser commented on this insidious Rockefeller-CFR collaboration: "The Council on Foreign Relations, another member of the international complex, financed by the Rockefeller and Carnegie Foundations, overwhelmingly propagandizes the globalist concept. This organization became virtually an agency of the government when World War II broke out. The Rockefeller Foundation had started and financed certain studies known as the *War and Peace Studies*, manned largely by associates of the Council; the State Department, in due course, took these studies over, retaining the major personnel which the Council on Foreign Relations had supplied."[86] Although their planning was ultra-secretive, the CFR openly propagandized the public, priming them to accept and embrace the forthcoming United Nations organization.

In October 1939, John Foster Dulles, a founding member of the CFR and a Rockefeller Foundation trustee since 1935,[87] announced that the U.S. must prepare to make sacrifices for a coming "world politico-economic order." These sacrifices, Dulles explained, would involve the dilution of national sovereignty. The *New York Times* reported, "John Foster Dulles of New York told the National Council of the Y.M.C.A. today that 'some dilution or leveling-off of the sovereignty system' as it prevails in the world today must take place to assure a constructively peaceful world order and prevent future wars…Furthermore, he said the United States as well as other sovereign powers must be prepared to make sacrifices afterward in setting up a world politico-economic order which would level off inequalities of economic opportunity with respect to nations."[88] His remarks came more than two years *before* Japanese bombs fell upon Pearl Harbor. Although America was not even at war, Dulles was already telling Americans to prepare for postwar sacrifices. And the Council on Foreign Relations was already mapping out the coming "world politico-economic order."

In 1939, with $350,000 from the Rockefeller Foundation,[89] the Council on Foreign Relations began drafting a business plan for world government. Isaiah Bowman (the CFR's founding Director), Allan Dulles (future CIA Director and Warren Commission member), and Whitney Shepardson (former Director of John D. Rockefeller's General Education Board), were among the directors of the *War and Peace Studies*. In the CFR's officially sanctioned history, *Continuing the Inquiry*, CFR member Peter Grose admitted, "The men from the Council proposed a discreet venture reminiscent of the Inquiry: a program of independent analysis and study that would guide American foreign policy in the coming years of war and the challenging new world that would emerge after."[90]

The "Inquiry" was a group of CFR founding members who drafted significant portions of the 1919 Treaty of Versailles, including the League of Nations Charter. The Inquiry is significant because it demonstrates a pattern of treason emanating from the CFR. The CFR formed in 1921 largely *because* the Inquiry failed. Thus they had nearly two-decades to plan and prepare the League of Nations' predecessor, the United Nations, and to ensure its successful implementation. Curtis Dall, the son-in-law of Franklin Roosevelt, detailed this chain-of-events in his book *Franklin Delano Roosevelt: My Exploited Father-in-Law*:

> "The plan to launch a United Nations supra-government on us and others began in Paris, at the Peace Conference in 1919, when the League of Nations folded. The real long-range objectives of the UN are cleverly concealed, and always have been concealed, behind clouds of One World Revolutionary Socialist Propaganda — made possible by the continued application of the factor of deception. The concept of a global United Nations apparatus did not suddenly emerge out of thin air! Obviously, no one could have suddenly provided the vast sums of money needed to successfully promote it. Hence, the long-planned-for UN operation for self-serving wars, a slick undertaking, was worth a great deal of money to a few people here and abroad, ambitious for more power and more wealth. Who? In 1919 at the Paris Peace Conference, held at Versailles, when 'Peace' was featured by its noticeable absence, the *League of Nations* was unfolded. It was carefully planned by a foreign clique to be sold to President Wilson and then sold by him to this Country. World Bankers were its leading advocates and sponsors. The League of Nations failed partly because its plan was published in advance but chiefly because of the vigorous resistance to it in the U.S. Senate and from many alert citizens here who became *aware* of the secluded dangers involved in it. Not to be daunted by the failure of the League of Nations, this same high level, money clique decided to keep the 'One-World' concept alive for self-serving reasons, and promptly planned for a new vehicle, aided by plenty of spade-work in advance, that looked ahead twenty-five years or so to successfully trap the American people. In order to make certain that there would be no possible slip this time, an organization was created by them called The Council On Foreign Relations (C.F.R.), to carefully train men for various areas of operation and to suitably mold their ideological, political, financial, military, and educational objectives. In London, England, there was also created a Counterpart called The Royal Institute of International Affairs. The World phase of these two groups is known as The Bilderbergers, headed by H.R.H. Prince Bernhard of the Netherlands. Its headquarters are often in Holland."[91]

Regarding the *War and Peace Studies,* Isaiah Bowman acknowledged, "The matter is strictly confidential, because the whole plan would be 'ditched' if it became generally known that the State Department is working in collaboration with any outside group."[92] This admission certainly challenge David Rockefeller's claim that "the Council never takes a position—official or unofficial—on any foreign policy issue."[93] In his book *American Empire,* Neil Smith analyzed the scope of the *War and Peace Studies.* He explained, "Absorbed and reinvented inside the State Department, the *War and Peace Studies* project effectively ended by February 1942. It contributed some 670 reports on geographical, political, and economic issues, held 361 meetings, involved over 100 individuals, and spent almost $300,000 of Rockefeller Foundation funds in the process. The CFR's early dream of power and influence now came to fruition, and in the following years several council members including Bowman moved toward the crown of State Department power."[94]

In 1942, the State Department assumed "official" control of the project with the establishment of the Advisory Committee on Postwar Foreign Policy. The CFR, however, retained considerable influence. In his book *The Council on Foreign Relations and American Foreign Policy in the Early Cold War,* Michael Wala explained, "Secretary of State Hull directed the Committee and some Council members were invited to participate. This group and its subcommittees concentrated on the United Nations organization, the successor to the League of Nations, a subject that had always received keen attention at Council meetings."[95] The Advisory Committee effectively drafted the Moscow Declaration of November 1, 1943, signed by delegates of the United States, Great Britain, China and the Soviet Union,[96] which paved the way for the forthcoming United Nations Charter.[97] CFR member Grayson Kirk later bragged, "The arrangement between the State Department and the CFR made it possible for [an] intellectual breeze from the outside to blow through the State Department in a way, that the State Department, acting officially, itself would have been entirely too cautious to permit."[98]

Towards the conclusion of the war, Rockefeller interests began heavily propagandizing Latin American countries with news stories friendly to U.S. and Rockefeller interests. Nelson Rockefeller was intent on coercing these countries into joining an unofficial US-led voting bloc. According to John Luftus, a former Justice Department prosecutor and Army Intelligence officer who was trusted enough to hold some of the highest security clearances in the world,[99] Rockefeller strong-armed these countries into backing the U.S. during the aforementioned 1945 San Francisco Conference. Although the war was nearly over, Rockefeller and the U.S. government even compelled Argentina to officially declare war on Germany and Japan, just so they could participate in the upcoming voting.[100] William Engdahl explained, "Rockefeller's strategy was to use his bloc of Latin American nations to 'buy' the majority vote at the UN. The Latin American bloc represented nineteen votes to Europe's nine. As a result, Washington and the powerful international banking business interests shaping its postwar agenda, ended up with decisive control of the IMF, the World Bank and a dominant role in the United Nations. The Rockefeller family, generous to a fault, even donated the land for the headquarters of the new United Nations in New York."[101] On October 24, 1945, the United Nations officially came into existence with the ratification of the UN Charter.

A Conduit of Globalization

Since its unsavory beginnings at the Council on Foreign Relations, the United Nations has always been a New World Order conduit. Nevertheless, the contention that the UN "was never intended to be a world government" has some elements of truth. The UN was intended to be an *experiment*, a sort of multi-generational war-game. The Rockefeller-CFR coalition did not intend to take the experiment live, at least not immediately. The New World Order agenda was nowhere near maturity during the 1940s. More time was required to slowly merge the various world cultures and collectively guide them towards a one-world ideology. Furthermore, the modern fake environmental movement, critical to the long-term agenda, was barely getting underway. In the interim, the CFR has closely monitored this grandiose UN experiment.

One now-public document from 1961 clearly reveals what they've been planning for. Dean Rusk, a CFR member[102] and long-time Bilderberg member,[103] was U.S. Secretary of State in 1961 when he commissioned Lincoln Bloomfield, an M.I.T. professor and fellow CFR member,[104] to study how the UN could be effectively transformed into world government — how they could take the experiment live. In his report, known as *Study Memorandum No.7*, Bloomfield concluded,

> "A world effectively controlled by the United Nations is one in which 'world government' would come about through the establishment of supranational institutions, characterized by mandatory universal membership and some ability to employ physical force. Effective control would thus entail a preponderance of political power in the hands of a supranational organization rather than in individual national units, and would assume the effective operation of a general disarmament agreement. While this supranational organization — the United Nations — would not necessarily be the organization as it now exists, the present UN Charter could theoretically be revised in order to erect such an organization equal to the task envisaged, thereby codifying a radical rearrangement of power in the world."[105]

The United Nations is a facade. Posturing as wise, dignified pioneers of the coming Golden Age of humanity, the UN is actually a modern-day Trojan horse. According to Greek legend, after years of battle during the Trojan War, the ancient Greeks devised an ingenious plan to enter and subsequently destroy the city of Troy. They built a large, wooden horse, which secretly held inside several dozen men. The Greeks left the horse outside the city's walls and sailed away, apparently conceding defeat. Interpreting the horse as a victory trophy, the Trojans pulled it inside the city. That night, the concealed Greeks crept out and opened the city's gates to the Greek army, which had turned back from sea. The Greeks stormed and destroyed the city, effectively ending the long war.

After two long, devastating World Wars, the United Nations emerged as a symbol of peace and universal brotherhood. Its accompanying propaganda blamed *nationalism* as the root cause of these wars. Supposedly national sovereignty was outdated and obsolete, while the UN was progressive and forward thinking. Consequently, the UN and its supporters have been indoctrinating generation after generation that only world government can save humanity. Because the UN *does* perform many peace-building and peacekeeping missions, it has successfully convinced many unsuspecting, peace-loving people that world government is both necessary and beneficial.

From the outside, the UN looks benevolent, triumphant even. However, hidden inside this Trojan horse are treacherous Rockefeller-CFR agents. While humanity sleeps, these agents emerge, prying open the minds of men and subsequently bombarding them with New World Order propaganda. UNESCO (United Nations Educational, Scientific and Cultural Organization) is the primary coordinator and disseminator of such propaganda within the United Nations structure.

The Real UNESCO

Since its inception, UNESCO has committed itself to reengineering society through steady indoctrination, targeting children especially. In 1949, for example, they published *Towards World Understanding*. This book demonized nationalism while glorifying globalization. For example, UNESCO asserted, "One of the chief aims of education today should be to prepare children and young people in schools throughout the world to become intelligent members of a world society."[106] This sounds reasonable, yet according to UNESCO, "intelligent" people consider national sovereignty a threat to, not a pillar of, freedom and security. Furthermore, "intelligent" people consider the family an *obstacle* to intelligence and "world-mindedness." Under the heading, "Nationalism," they wrote, "As long as the child breathes the poisoned air of nationalism, education in world-mindedness can produce only rather precarious results. As we have pointed out, it is frequently the family that infects the child with extreme nationalism."[107]

Since its early years, UNESCO has drawn harsh criticism, especially from Americans aware of its subversive nature. In 1955, for example, Congressman Lawrence H. Smith of Wisconsin described the United Nations and UNESCO as "a permanent international snake pit where Godless Communism is given a daily forum for hate, recrimination, psychological warfare against freedom, and unrelenting moral aggression against peace."[108] The following year, a Senate Judiciary Committee report deemed UNESCO, "by far the worst danger spot, from the standpoint of disloyalty and subversive activity among Americans employed by international organizations."[109]

By the 1980s, UNESCO's hostilities towards the personal freedoms and liberties protected by the U.S. Constitution climaxed with the proposed New World Information Order (NWIO), a plan to license and regulate reporters and journalists worldwide. In June 1981, Senator Dan Quayle introduced legislation opposing NWIO, noting, "No matter how they try to disguise it, what UNESCO is trying to do is to codify and legitimize worldwide censorship."[110] Finally in 1984, Ronald Reagan withdrew the United States from UNESCO. State Department spokesman Alan Romberg explained, "UNESCO has extraneously politicized virtually every subject it deals with, has exhibited hostility toward the basic institutions of a free society, especially a free market and a free press, and has demonstrated unrestrained budgetary expansion."[111]

Eighteen years later, regarding UNESCO, President George Bush declared, "This organization has been reformed and America will participate fully in its mission to advance human rights and tolerance and learning."[112] Despite Bush's assurances, several years earlier, an audit by the Canadian government revealed rampant corruption throughout the organization. The *London Guardian* reported, "[UNESCO] has reformed little in the decade since its name became a byword for inefficiency, nepotism and corruption — practices that caused the United States and Britain to pull out as members, although Britain has since rejoined."[113] They also noted, "Cronyism seems all but endemic, with about 40% of the organisation's appointments and promotions failing to meet UNESCO's own criteria for fair competition."[114] Congressman Ron Paul vehemently opposed Bush's decision. He explained,

- "UNESCO meddles in the education affairs of its member-countries and has sought to construct a UN-based school curriculum for American schools."
- "UNESCO has been fully supportive of the United Nations' Population Fund (UNFPA) in its assistance to China's brutal coercive population control program."
- "UNESCO has designated 47 UN Biosphere Reserves in the United States covering more than 70 million acres, without Congressional consultation."
- "UNESCO effectively bypasses Congressional authority to manage federal lands, by establishing management policies without Congressional consultation or approval."[115]

UNESCO's scandalous history is hardly surprising considering the organization's philosophical roots. In 1946, Julian Huxley became UNESCO's founding Director. Huxley was a longtime member of the British Eugenics Society, holding the Vice-Presidency from 1937-1944 and the Presidency from 1959-1962. Eugenics, detailed in Chapter 9 of this book, is the pseudoscience of race-based superiority made infamous by Adolph Hitler. In 1948, Huxley wrote UNESCO's manifesto, *UNESCO: Its Purpose And Its Philosophy*, which revealed that UNESCO's primary goal is altering culture worldwide in preparation for world government. Huxley explained, the primary task "is to help the emergence of a single world culture, with its own philosophy and background of ideas, and with its own broad purpose. This is opportune, since this is the first time in history that the scaffolding and the mechanisms for world unification have become available."[116]

Huxley left no doubts that world government was the objective. He only regretted that he would not live long enough to see it come to fruition. He explained, "World political unity is, unfortunately, a remote ideal, and in any case does not fall within the field of UNESCO's competence. This does not mean that UNESCO cannot do a great deal towards promoting peace and security. Specifically, in its educational programme it can stress the ultimate need for world political unity and familiarise all peoples with the implications of the transfer of full sovereignty from separate nations to a world organisation. But, more generally, it can do a great deal to lay the foundations on which world political unity can later be built. It can help the peoples of the world to mutual understanding and to a realisation of the common humanity and common tasks which they share, as opposed to the nationalisms which too often tend to isolate and separate them."[117]

To achieve this common world culture, Huxley suggested two primary methods: (1) UNESCO should explore and exploit multiple propaganda and indoctrination techniques, especially scientific techniques, and (2) UNESCO should foster a new philosophy for humanity, one that supports and embraces eugenics. Huxley viewed science as a mechanism for control. Recognizing that science had given man, to some degree, control over natural phenomena, Huxley was eager to use science to achieve greater control over human behavior. He explained,

"The scientific method has firmly established itself as the only reliable means by which we can increase both our knowledge of and our control over objective natural phenomena. It is now being increasingly applied, though with modifications made necessary by the different nature of the raw material, to the study of man and his ways and works, and in the hands of the social sciences is likely to produce an increase in our knowledge of and control over the phenomena of human and social life, almost as remarkable as that which in the hands of the natural sciences it has brought about and is still bringing about in regard to the rest of nature."[118]

The scientific method, Huxley reasoned, could produce enormously effective propaganda. He recommended: "Taking the techniques of persuasion and information and true propaganda that we have learnt to apply nationally in war, and deliberately bending them to the international tasks of peace, if necessary utilising them, as Lenin envisaged, to 'overcome the resistance of millions' to desirable change. Using drama to reveal reality and art as the method by which, in Sir Stephen Tallent's words, 'truth becomes impressive and a living principle of action,' and aiming to produce that concerted effort which, to quote Grierson once more, needs a background of faith and a sense of destiny. This must be a mass philosophy, a mass creed, and it can never be achieved without the use of the media of mass communication. UNESCO, in the press of its detailed work, must never forget this enormous fact."[119] Writing before the Internet and television eras, Huxley recognized art as an indispensible channel for widespread manipulation. He explained, "Whatever the details, it remains true that one of the social functions of art is to make men feel their destiny, and to obtain a full comprehension, emotional as well as intellectual, of their tasks in life and their role in the community. Rightfully used, it is one of the essential agencies for mobilising society for action."[120] Therefore, the government should control and regulate art. According to Huxley, "The physical provision of beauty and art must, in the world of to-day, be largely an affair of government."[121]

Huxley further acknowledged that UNESCO's task was to use media and mass communications to *scientifically* propagandize for "a mass philosophy, a mass creed." This new philosophy would feature an updated morality, one that rationalized and even encouraged the brutality of eugenics. On the socialization of ethics, Huxley explained, "In general, we may say, it is becoming necessary to extend our personal ethical judgments and responsibilities to many collective and apparently impersonal actions — in other words to undertake a considerable socialisation of ethics. It will be one of the major tasks of the Philosophy division of UNESCO to stimulate, in conjunction with the natural and the social scientists, the quest for a restatement of morality that shall be in harmony with modern knowledge and adapted to the fresh functions imposed on ethics by the world of to-day. Still more generally, it will have to stimulate the quest, so urgent in this time of over-rapid transition, for a world philosophy, a unified and unifying background of thought for the modern world."[122]

The "fresh function imposed on ethics" was an allusion to eugenics. Elsewhere in the document, Huxley specifically addressed eugenics and UNESCO's commitment, despite recent setbacks, to supporting and advancing the eugenics movement. He explained, "Thus even though it is quite true that any radical eugenic policy will be for many years politically and psychologically impossible, it will be important for UNESCO to see that the eugenic problem is examined with the greatest care, and that the public mind is informed of the issues at stake so that much that now is unthinkable may at least become thinkable."[123] His mention of political impossibility refers to the reputational damage done to eugenics by Hitler. Huxley and the Anglo-American eugenicists supported Hitler philosophically. They supported his recognition of superior and inferior types and the need to prevent inferiors from reproducing. They merely disagreed with his conspicuous tactics. The Anglo-Americans preferred discreet, subversive tactics.

Since the eugenics ideology clashed with the democratic spirit enshrined in the U.S. Constitution, UNESCO's new philosophy for humanity would have to fix supposedly outdated conceptions of "democratic equality." Huxley explained,

"There remains the second type of inequality. This has quite other implications; for, whereas variety is in itself desirable, the existence of weaklings, fools, and moral deficients cannot but be bad. It is also much harder to reconcile politically with the current democratic doctrine of equality. In face of it, indeed, the principle of equality of opportunity must be amended to read 'equality of opportunity within the limits of aptitude'...To adjust the principle of democratic equality to the fact of biological inequality is a major task for the world, and one which will grow increasingly more urgent as we make progress towards realising equality of opportunity. To promote this adjustment, a great deal of education of the general public will be needed as well as much new research; and in both these tasks UNESCO can and should co-operate."[124]

According to Huxley, democracy leads to mediocrity: "The Age of the Common Man: the Voice of the People: majority rule: the importance of a large population—ideas and slogans such as these form the background of much of our thinking, and tend, unless we are careful, towards the promotion of mediocrity, even if mediocrity in abundance, and at the same time, towards the discouragement of high and unusual quality."[125] Therefore, the new philosophy would eliminate antiquated idealism. According to UNESCO's founding document, democracy and equal opportunity are impractical because not all people are equal. The new philosophy would acknowledge genetically superior people and, for the betterment of the species, give them the best opportunities. As for the inferiors, they would be encouraged to graciously serve their betters (for the improvement of the species of course). The above has merely been a brief introduction to UN chicanery. Part II of this book details the UN's integral role in false environmentalism, especially the depopulation and New World Religion agendas. Part III examines their leadership role in advancing the fraudulent manmade global warming theory.

The World Federalist Movement

While establishing the UN as the prototype for future global government, the Council on Foreign Relations also formulated an alternative model, the United World Federalists (UWF). In 1947, prominent CFR members Norman Cousins, James Warburg,[126] and Cord Meyer, Jr.,[127] established the UWF with the goal of eroding national sovereignty and advancing world government. According to Cousins, "The bugaboo of sovereignty and nationalism [is] by far the heaviest and most deeply anchored of the obstacles facing world citizenship."[128] The UWF advocated world government by way of ever-expanding federations. Under this model, nations would gradually yield more and more sovereignty to regional federations. Then, after many decades, these various regional federations would merge into a world government. By this time, individual nations would be too weak to effectively protect their sovereignty.

The UN and the UWF were merely different paths towards the same destination. Nevertheless, to sell the public on world government, the globalists created the *illusion* of choice. This is a standard sales technique. Effective salespeople always show their customers multiple products and then ask, "Which do you prefer?" Psychologically, customers feel obliged to choose and reluctant to say, "I don't want any." The globalists were saying, "Which model of world government do you prefer, the UN or the UWF?" Consequently, people surmise that world government is inevitable, desirable even. Instead of asking, "Do I actually want world government," they ask, "What *kind* of world government do I want?"

During the late 1940s, the CFR greatly influenced public opinion. According to an August 1946 *Roper* poll, 63 percent of Americans supported world government while only 20 percent opposed.[129] An August 1947 *Gallup* poll had 56 percent supporting, and 30 percent opposing.[130] In two states, Massachusetts and Connecticut, voters overwhelmingly approved world government referenda, 9:1 and 11:1, respectively.[131] In early 1949, nine states and approximately 50 cities including Chicago, Minneapolis, and Miami, celebrated World Government Week.[132] On June 7, 1949, ninety-one Congressmen endorsed House Concurrent Resolution 64 calling for "the United States to support and strengthen the United Nations and to seek its development into a world federation, open to all nations, with defined and limited powers adequate to preserve peace and prevent aggression through the enactment, interpretation, and enforcement of world law."[133] On July 26, nineteen Senators introduced an identical resolution in the Senate.

Congressman Glen H. Taylor persistently supported world government. In 1945, he boldly declared, "I want to raise my voice to say that at last the time has come for a world government. We should surrender enough of our sovereignty to insure world peace."[134] Five years later, Taylor urged President Truman to "take the initiative in requesting a general conference of the United Nations pursuant to Article 109 for the purpose of establishing a true world government through adoption of such a constitution; and if such a general conference is not called within one year after the adoption of this resolution, the President of the United States should then call a world constitutional convention of delegates elected directly by the people for the purpose of adopting a world government constitution."[135]

UWF Leadership and Ideology

Cord Meyer became the first President of UWF in 1947.[136] In an October 1949 article for the *Bulletin of the Atomic Scientists*, Meyer admitted that UWF was actively pressuring the States to apply for "the convening of a national constitutional convention to make such amendments to our Constitution as would be necessary to enable the United States to enter a world government."[137] Furthermore, they were using heavy propaganda "to gain public understanding and acceptance of the need for effective limitation of absolute national sovereignty." Meyer's propensity for subterfuge would later serve him well. In 1951, Allen Dulles recruited Meyer into the CIA, where he worked on Operation Mockingbird, an infamous CIA initiative to manipulate and control the American and foreign press.[138]

Years later, according to legendary CIA agent E. Howard Hunt, Meyer was a key figure in the JFK assassination. Hunt recorded audio and video confessions describing the roles of Meyer, Lyndon Johnson and others, and his own role as "a benchwarmer." These recordings have been publicly available since Hunt's death in 2007. Hunt admitted, "I heard from Frank [Morales] that LBJ had designated Cord Meyer, Jr. to undertake the larger organization while keeping it totally secret. I think that LBJ settled on Meyer as an opportunist like himself, a man who had very little left to him in life ever since JFK had taken Cord's wife as one of his mistresses."[139] Notably, on October 12, 1964, less than a year after Kennedy's assassination, Mary Meyer (Cord's wife) was gunned down in Georgetown in a still-unsolved murder.[140]

James Warburg was another CFR member and co-founder of the UWF.[141] His father, Paul Warburg, was a prominent banker and key architect of the unconstitutional Federal Reserve banking system. In his 1959 book *The West in Crisis*, Warburg wrote, "It has long been obvious that there can be no durable peace in a world of international anarchy; that a world order without world law is an anachronism; and that, since war now means the extinction of civilization, a world which fails to establish the rule of law over the nation-states cannot long continue to exist. We are living in a perilous period of transition from the era of the fully sovereign nation-state to the era of world government."[142]

On February 17, 1950, James Warburg testified before the Senate Committee on Foreign Relations during hearings entitled, "Revision of United Nations Charter." Heated debates ensued regarding global government and the threat of atomic war. Representative Clare E. Hoffman warned against global government, stating, "Stripped of deceptive and confusing language, the present proposals, naked, stand forth as a clear attempt to induce us to scrap our Constitution, surrender our sovereignty, our right to plot our course in world affairs." Warburg's testimony perfectly encapsulated the arrogant, ruling-class mentality. He declared, "The great question of our time is not whether or not one world can be achieved, but whether or not one world can be achieved by peaceful means. We shall have world government whether or not we like it. The only question is whether world government will be achieved by conquest or consent."[143]

Rockefeller Backing

In 1962, New York Governor Nelson Rockefeller, a CFR member[144] and future U.S. Vice-President, published *The Future of Federalism*, a diatribe against national sovereignty. He proclaimed,

- "And so the nation-state, standing alone, threatens, in many ways, to seem as anachronistic as the Greek city-states eventually became in ancient times."[145]
- "No nation today can defend its freedom, or fulfill the needs and aspirations of its own people, from within its own borders or through its own resources alone."[146]
- "The nation-state is becoming less and less competent to perform its international political tasks."[147]

Rockefeller's solution was a "new world order" that would undermine national sovereignty. He explained, "All these, then, are some of the reasons — economic, military, political — pressing us to lead vigorously toward the true building of a new world order. And it urgently requires that the United States take the leadership among all free peoples to make the underlying concepts and aspirations of national sovereignty truly meaningful through the federal approach."[148]

While not an official member of the UWF, known then as the World Association of World Federalists (WAWF), Rockefeller unabashedly supported the central tenets of federalism. In February 1962, prior to the publication of his book, he delivered a series of lectures at Harvard University on "The Future of Federalism." He declared, "The critical political decisions in government are, and must be, primarily shaped and made by elected officials. It is with this particular perspective on our democratic processes that I underline my deep personal conviction that the future of freedom lies in the federal idea. I refer to the federal idea broadly as a concept of government by which a sovereign people — for their greater progress and protection—yield a portion of their sovereignty to a political system that has more than one center of sovereign power, energy and creativity."[149]

During this lecture, Rockefeller also reinforced the false dichotomy between the UN model of world government and the federalism model. He stated, "The UN lacks the strength to master or control the forces that it confronts. I believe the historic answer to the problems the free world confronts can be found in the federal idea. I have long felt that the road toward the unity of free nations lay through regional confederations. But I have come to the conviction that events are driving us rapidly beyond even the limits of regional concepts—to the logic of applying the federal idea wherever possible. What our common danger—and our common aspirations—imperatively require, then, is a common commitment to some basic principles and purposes [leading] ultimately to the gradual devising of political forms of unity."[150] Rockefeller also reproached the UN in his book, writing, "The United Nations, repository of so much hope, has not been able—nor can it be able—to shape a new world order as events now so compellingly command."[151]

Later, while campaigning for the 1968 Republican presidential nomination, Rockefeller reaffirmed his New World Order aspirations. The *Associated Press* reported, "New York Gov. Nelson A. Rockefeller says as president he would work toward international creation of 'a new world order' based on East West cooperation instead of conflict." Rockefeller contended, "We have no need to be mesmerized by our perils, the possibility to build the new world order we all seek is limited only by our imagination and dedication."[152]

World Federalism Today

UWF now calls itself the World Federalist Movement (WFM), a non-governmental organization with Consultative Status with the United Nations. According to their website, "Members of WFM often work closely with UN officials and those of other international institutions to further our mission. We are committed to reforming the UN and ensuring effective democratic global governance."[153] They further explain, "The ultimate goal of world federalists is world federation, but this goal can be reached only after intermediate stages. The challenge that world federalism now must face is to show that it is capable of taking the lead in the process of transition towards world government."[154] In other words, the WFM favors gradualism and obfuscation, standard strategies of the New World Order's architects. John J. Logue, co-founder of UWF's Philadelphia chapter,[155] confirmed,

> "But while early world federalists were urging comprehensive reform of the United Nations, some recent and present leaders of the world federalists have become sincere champions of the gradual, partial approach, a strategy very similar to that of the liberal internationalists. The present leadership of the World Federalist Association, the largest U.S. world federalist group, frequently uses the vague term 'global change' and the obscure term 'global governance' rather that the words 'world government' or 'world federation.'"[156]

The modern era of world federalism has had many prominent supporters including Strobe Talbott and Walter Cronkite. Arch-globalist Talbott has been CFR Director, a Trilateral Commission Trustee, and a key architect of U.S. foreign policy while serving as Deputy Secretary of State from 1994 to 2000.[157] He later became the inaugural Director of Yale University's Center for the Study of Globalization.[158] In 1992, Talbott penned an article for *Time* magazine anticipating the day when, "Nationhood as we know it will be obsolete; all states will recognize a single global authority. A phrase briefly fashionable in the mid-20th century—'citizen of the world'—will have assumed real meaning by the end of the 21st."159 Tracing history from prehistoric times, he attempted to demonstrate both the inevitability of global government. He also rehashed the tired argument that only world government can prevent future world wars: "But it has taken the events in our own wondrous and terrible century to clinch the case for world government...The price of settling international disputes by force was rapidly becoming too high for the victors, not to mention the vanquished." Talbott concluded, "Perhaps national sovereignty wasn't such a great idea after all."

On June 24, 1993, Strobe Talbott received the Norman Cousins Global Governance Award from the World Federalist Association (formerly UWF). At the award ceremony, WFA President John Anderson, a CFR and Trilateral Commission member, read a letter from Bill Clinton congratulating Talbott and praising Cousins.160 In the United States, all appointed and elected government officials take an oath of office stating, "I do solemnly swear that I will support and defend the Constitution of the United States against all enemies, foreign and domestic." How could Talbott, the man who publicly declared that national sovereignty "wasn't such a great idea," possibly ascend to Deputy Secretary of State? How could the President of the United States publicly support world federalism and global governance? Who are the enemies, foreign and domestic?

According to 1970s opinion polls, veteran CBS anchorman Walter Cronkite was "the most trusted man in America,"161 an ironic designation for a man who would later reject American sovereignty and openly call for world government. In his 1997 book *A Reporter's Life*, Cronkite derided the American patriots who blocked the globalists' post-WWI League of Nations. He wrote, "From the end of the *War to End War*, as World War I was delusively described, the nations proved that the lessons of the bloody conflict had been lost upon them. This was no more apparent than in the United States Senate, where the American chauvinists, blindly jealous of meaningless sovereignty, rejected Wilson's dream of a League of Nations, a mild first step toward world government. Whatever chances that body had of preventing another war were diminished to near oblivion by the failure of the United States to join."162

On the subject of nuclear devastation, Cronkite later remarked, "If we are to avoid that catastrophe, a system of world order—preferably a system of world government—is mandatory. The proud nations someday will see the light and, for the common good and their own survival, yield up their precious sovereignty, just as America's thirteen colonies did two centuries ago."163 On October 19, 1999, Cronkite received the WFA's Norman Cousins Global Governance Award. During his acceptance speech, he stated, "We Americans are going to have to yield up some of our sovereignty. That's going to be, to many, a bitter pill. It would take a lot of courage, a lot of faith, a lot of persuasion for them to come along with us on this necessity."164 During the ceremony, Hillary Clinton honored Cronkite via satellite feed, thanking him for his continual leadership.

Cronkite, Talbott and the World Federalist Movement advocate capitulation, like advising children to submit to bullies. They imply, "Always give the bullies your lunch money, otherwise they'll beat you up again." By their twisted logic, to attain world peace, society must appease the aggressors who perpetuate war, terrorism, and economic hardship. The aggressors inflict beating after beating, each time insisting that only world government will prevent the next beating. They demand sovereignty (the people's lunch money) in exchange for peace. Lasting peace, however, cannot arise by way of capitulation. Those who believe otherwise are still naive children, the perpetual victims of bullies.

Chapter 5

The Trilateral Commission

According to David Rockefeller in his 2002 book *Memoirs*, "No organization with which I have played a founding role has attracted as much public scrutiny and attention as the Trilateral Commission…Trilateral, like Bilderberg, is a much more benign organization than the conspiracy theorists have depicted. It is a broadly based effort to bridge national differences and, in this case, invite the Japanese into the international community."[1] According to Trilateral Commission co-founder Zbigniew Brzezinski (during a question and answer session at Colombia University in 2008), "If you want to indulge in conspiracy theories, let me suggest another topic for your investigation and that is the Trilateral Commission. The Trilateral Commission is even larger than Bilderberg, and includes the Japanese." With his distinct, mocking sarcasm, Brzezinski also added, "The purpose of these secret meetings is to devise more effective ways of sucking blood out of poor people, exploiting them, and subverting their identity and independence to foreign control."[2] Despite Brzezinski and Rockefeller's attempts to stigmatize, ridicule, and otherwise discredit those who scrutinize their elitist organizations, the Trilateral Commission is a group of highly influential businessmen and politicians who, according to their own documents, are committed to undermining the U.S. Constitution and advancing the New World Order agenda. Trilateral Commission co-founder Brzezinski, for example, has called for amending the Constitution to facilitate European-style regional government. In his 1970 book *Between Two Ages*, he remarked,

> "The approaching two-hundredth anniversary of the Declaration of Independence could justify the call for a national constitutional convention to re-examine the nation's formal institutional framework. Either 1976 or 1989—the two-hundredth anniversary of the Constitution—could serve as a suitable target date for culminating a national dialogue on the relevance of existing arrangements, the workings of the representative process, and the desirability of imitating the various European regionalization reforms and of streamlining the administrative structure. More important still, either date would provide a suitable occasion for redefining the meaning of modern democracy—a task admittedly challenging but not necessarily more so than when it was undertaken by the founding fathers—and for setting ambitious and concrete goals."[3]

For the Trilateral Commission, however, "modern democracy" is merely a euphemism for classical tyranny. Brzezinski acknowledged that direct Constitutional reform was unattainable. Therefore, he recommended covert subversion: "Realism, however, forces us to recognize that the necessary political innovation will not come from direct constitutional reform, desirable as that would be. The needed change is more likely to develop incrementally and less overtly."[4]

The "Frequently Asked Questions" page of the Trilateral Commission's website asks, "Is the Trilateral Commission trying to establish a world government?" to which they respond, "No. The Trilateral Commission encourages international cooperation on many issues, but does not promote a world government. No Commission report proposes that national governments be dissolved and a world government be created. Individuals or organizations who believe the Trilateral Commission supports or intends to form a world government are misinformed."[5] Despite their patronization, the Commission's founders have been remarkably frank about their intentions for world government. David Rockefeller, as previously discussed, has *admitted* to "working against the best interests of the United States."[6] He has *admitted* to "conspiring with others around the world to build a more integrated global political and economic structure—one world, if you will."[7] Zbigniew Brzezinski has offered numerous similar admissions. In October 1995, for example, he delivered a speech at Mikhail Gorbachev's State of the World Forum, in San Francisco, where he observed, "We cannot leap into world government in one quick step…The precondition for eventual globalization—genuine globalization—is progressive regionalization, because thereby we move toward larger, more stable, more cooperative units."[8]

According to the Commission's website, "David Rockefeller was the principal founder of the Commission. He has served on the Executive Committee from the beginning in mid-1973 and was North American Chairman from mid-1977 through November 1991. Zbigniew Brzezinski played an important role in the formation of the Commission. He was its first Director (1973-76) and its major intellectual dynamo in those years. Dr. Brzezinski rejoined the Commission in 1981 and now serves on the Executive Committee."[9] Rockefeller proposed the Trilateral Commission concept to Brzezinski during the 1972 Bilderberg meeting. Rockefeller wrote, "Zbigniew Brzezinski, then teaching at Columbia University, was a Bilderberg guest that year, and we spoke about my idea on the flight to Belgium for the meeting…after a lengthy discussion we determined to set up the new organization. Zbig agreed to serve as director, and Benjy Franklin, my college roommate and colleague at the Council on Foreign Relations, agreed to help with organizational matters."[10]

Barry Goldwater, a United States Senator in office from 1953 to 1965, and then again from 1969 to 1987, had a unique vantage point from which he observed the Trilateral Commission during its most influential years. Goldwater equated the Commission to a modern day Praetorian Guard. The Praetorian Guard was an elite contingent of the Roman Empire established by emperor Augustus. Goldwater explained, "Over the years the Praetorians accumulated sufficient power to destroy any emperor they opposed. They blocked efforts to reestablish a true republic. They were able to select and elevate their candidate to the position of emperor. The Trilateral Commission is the modern Praetorian Guard."[11] He also observed, "It is intended to be the vehicle for multinational consolidation of the commercial and banking interests by seizing control of the political government of the United States."[12] Goldwater was not shy about naming names: "The Trilateral organization created by David Rockefeller was a surrogate—its members selected by Rockefeller, its purpose defined by Rockefeller, its funding supplied by Rockefeller."[13]

Anthony Sutton, an economics professor at California State University (Los Angeles) and research fellow at Stanford University's Hoover Institution, agreed that the Trilateral Commission is comprised, primarily, of administrators—busy-bees and front men. The members themselves are not power players. Rather, they execute agendas set by the power players above them. In the 1994 edition of his book *Trilaterals Over America*, Sutton explained, "David Rockefeller is essentially the only power center in the entire Trilateral catalog. Politicians, lawyers, bureaucrats, media types, trade unionists come and go in the Trilateral halls of power—they are transient administrators. They retain administrative positions only as long as they are successful in using political power to gain political objectives. Operators do not, by and large, create objectives—this is an important point."[14] Ironically, in his 1970 book *Between Two Ages*, Brzezinski commended Sutton for his excellent scholarship.[15]

The Crisis of Democracy

In 1975, the Trilateral Commission published a lengthy treatise called *The Crisis of Democracy*. Its authors, Michel Crozier, Samuel P. Huntington, and Jôji Watanuki, explored the following questions: "Is political democracy, as it exists today, a viable form of government for the industrialized countries of Europe, North America, and Asia? Can these countries continue to function during the final quarter of the twentieth century with the forms of political democracy which they evolved during the third quarter of that century?"[16] They concluded that democracy is indeed viable, providing that populations submit to and never question their governments. For the Trilateral Commission, democracy is viable only when governments are authoritative and not too focused on domestic issues. Huntington explained, "A government which lacks authority and which is committed to substantial domestic programs will have little ability, short of a cataclysmic crisis, to impose on its people the sacrifices which may be necessary to deal with foreign policy problems and defense."[17] Huntington also argued that people must believe in their governments, unconditionally. "In a democracy," he asserted, "the authority of governmental leaders and institutions presumably depends in part on the extent to which the public has confidence and trust in those institutions and leaders."[18] However, according to the Declaration of Independence, when a government abuses its people, "it is their right, it is their duty" to question and challenge such government and, if necessary, "to throw off such government, and to provide new guards for their future security."[19]

Underlying Huntington's analysis was the false presumption that America's government is necessarily legitimate, and necessarily immune to corruption. All scrutiny is therefore unwarranted. For example, he condemned the "more politically active citizenry" of the 1960s who "took more extreme positions on policy issues" and "in turn, tended to become more distrustful of government."[20] According to Huntington, "The sequence and direction of these shifts in public opinion dramatically illustrates how the vitality of democracy in the 1960s (as manifested in increased political participation) produced problems for the governability of democracy in the 1970s (as manifested in the decreased public confidence in government)."[21] He asked, "If American citizens don't trust their government, why should friendly foreigners? If American citizens challenge the authority of American government, why shouldn't unfriendly governments."[22]

Huntington decisively encapsulated the elitist perspective that democracy must never undermine the authority of the ruling class. "Democracy is more of a threat to itself in the United States," Huntington wrote, "than it is in either Europe or Japan where there still exist residual inheritances of traditional and aristocratic values. The absence of such values in the United States produces a lack of balance in society which, in turn, leads to the swing back and forth between creedal passion and creedal passivity. Political authority is never strong in the United States, and it is peculiarly weak during a creedal passion period of intense commitment to democratic and egalitarian ideals."[23] This is what the Trilateral Commission means by a "crisis" of democracy—an intense commitment to democratic and egalitarian ideals.

Noam Chomsky, an M.I.T. professor and prominent critic of U.S. foreign policy, equates the Trilateral Commission conception of democracy to the feudal system:

"On the one hand, we have the King and Princes (the government). On the other, the commoners. The commoners may petition and the nobility must respond to maintain order. There must, however, be a proper 'balance between power and liberty, authority and democracy, government and society.' 'Excessive swings may produce either too much government or too little authority.' In the 1960s, Huntington maintains, the balance shifted too far to society and against government. 'Democracy will have a longer life if it has a more balanced existence,' that is, if the peasants cease their clamor. Real participation of 'society' in government is nowhere discussed, nor can there be any question of democratic control of the basic economic institutions that determine the character of social life while dominating the state as well, by virtue of their overwhelming power. Once again, human rights do not exist in this domain."[24]

For the Trilateral Commission, proper democracy is characterized by obedience. Not so long ago, Americans actually referred to government employees as "civil servants." Today, they have become "officials" or "authorities." Does the state exist to serve the people? Or do the people exist to serve the state? The *Triumph of Consciousness* suggests that attentiveness, awareness, and involvement largely determine both the measure and quality of freedom attained by individuals and societies. Samuel Huntington and his Trilateral-friends have a vision of democracy that, absent opposition, they will certainly achieve. Therefore, they encourage and promote apathy. "The effective operation of a democratic political system," Huntington asserted, "usually requires some measure of apathy and noninvolvement on the part of some individuals and groups."[25] Obviously these "individuals and groups" are those who oppose the ruling class.

According to Thomas Jefferson, "When the government fears the people, there is liberty. When the people fear the government, there is tyranny."[26] The Trilateral Commission, however, tips the balance of power decidedly towards the government. According to Huntington, "expertise, seniority, experience, and special talents may override the claims of democracy as a way of constituting authority."[27] In other words, leave politics to the politicians. When too many people are politically active, the government's authority weakens. The most dangerous people, according Huntington, are intellectuals who recognize and oppose tyranny. These "adversary intellectuals" constitute "a challenge to democratic government which is, potentially at least, as serious as those posed in the past by the aristocratic cliques, fascist movements, and communist parties."[28] This Huntington-style demonization is disturbingly pervasive today as the Department of Homeland Security has classified Ron Paul supporters, and people who oppose the New World Order, the North American Union and the Federal Reserve as potential terrorists.[29]

Elites greatly fear an educated and informed, politically active population. Thus Huntington advocated political self-restraint to prevent overloading the system: "Marginal social groups, as in the case of the blacks, are now becoming full participants in the political system. Yet the danger of overloading the political system with demands which extend its functions and undermine its authority still remains. Less marginality on the part of some groups thus needs to be replaced by more self-restraint on the part of all groups."[30] In other words, submit to your betters and keep your egalitarian ideals to yourself.

Even the pro-establishment *Washington Post* criticized *The Crisis of Democracy*. In 1977, William Greider reported,

"Huntington's section on American democracy offered a provocative interpretation of recent history: The great dislocations in American politics over the last 15 years did not stem primarily from Vietnam or Watergate, nor from deceptions by Presidents or law-breaking by government agencies. The problem was in the people themselves. 'A democratic distemper,' Huntington called it. 'An excess of democracy' which threatens the authority of government, 'creedal passions' which must be tempered or the United States will become ungovernable...He suggested several ways to restore authority to American government and reduce popular excesses. One is to trim back on higher education. Another is to regulate the news media, something like the way the Interstate Commerce Act attempted to regulate corporations in the 19th century."[31]

Thomas Hughes, president of the Carnegie Endowment for International Peace, observed,

"Instead of criticizing the last two Presidents for grievously misgoverning the country, the burden of the Huntington message is to criticize the country for not submitting to the misgoverning. Recently, we have escaped the excesses of an immoral and unwinnable war and a cynical and criminal President. One might have thought that was the purpose of democracy — to make criminal government unmanageable...But the wars and crimes are brushed over lightly while the 'democratic surge' that ended them is condemned as 'excessive.'"[32]

The Carter Presidency

The Trilateral Commission organized Jimmy Carter's presidential campaign and subsequently dominated his presidency. Twenty-two years later, Barack Obama's presidential campaign bore striking similarities. Carter, like Obama, was a political nobody who emerged from obscurity and rose to sudden prominence. Former Senator Barry Goldwater commented, "Despite [his] unimpressive record, the *New York Times* and *Time* magazine and most of the nation's media projected Jimmy Carter as the invincible, super-popular candidate for the Democratic nomination."[33] The media raised Carter, like Obama, to near messianic heights. Goldwater explained, "As a newcomer on the national scene Jimmy Carter was free of the burden of defending any past errors. His managers carefully cultivated the public image of the unsophisticated farm boy, cloaked in innocence and righteousness, prepared to slay the evil dragons responsible for our discontent. The blurred, imprecise, often contradictory image Jimmy Carter offered to the public in the 1976 presidential campaign was no accident. The basic strategy Carter followed was to tell each particular audience what it wanted to hear."[34]

Carter, like Obama, campaigned during a time when most Americans were frustrated by years of reckless military adventurism and when Americans were growing increasingly distrustful of their government. Carter, like Obama, positioned himself as an outsider. *Time* magazine reported, "At a time when voters' distrust of Washington, runs deep, Carter considers his status a campaign advantage. 'I have been accused of being an outsider,' he says. 'I plead guilty. Unfortunately, the vast majority of Americans are also outsiders. We are not going to get changes by simply shifting around the same groups of insiders, the same tired old rhetoric, the same unkept promises and the same divisive appeals to one party, one faction, one section of the country, one race or religion or one interest group. The insiders have had their chances and they have not delivered. Their time has run out.'"[35]

Carter, like Obama, ran his campaign on empty slogans including, "The people are better than the government," "Trust me, I will never lie to you," "Why not the best?" and "A leader, for a change."[36] And most significantly, Carter, like Obama, was a puppet controlled by Brzezinski and Wall Street. Goldwater explained,

"David Rockefeller and Zbigniew Brzezinski found Jimmy Carter to be their ideal candidate. They helped him win the nomination and the presidency. To accomplish this purpose, they mobilized the money power of the Wall Street bankers, the intellectual influence of the academic community—which is subservient to the wealth of the great tax-free foundations—and the media controllers represented in the membership of the CFR and the Trilateral."[37]

After winning the presidency, Carter, like Obama, broke all his campaign promises and filled his administration with Trilateral Commission and Council on Foreign Relations members. Even David Rockefeller admitted,

"Carter's campaign was subtly anti-Washington and antiestablishment, and he pledged to bring both new faces and new ideas into government. There was a good deal of surprise, then, when he chose fifteen members of Trilateral, many of whom had served in previous administrations, for his team, including Vice President Walter Mondale, Secretary of State Cyrus Vance, Secretary of Defense Harold Brown, Secretary of the Treasury Michael Blumenthal, and Zbigniew Brzezinski as national security advisor…Predictably, I was accused of trying to take control of Carter's foreign policy."[38]

The *Washington Post* observed, "At the very least, Carter's heavy reliance on the Trilateral membership list demonstrates what has long been true—that U.S. foreign policy is shaped by a very exclusive circle of people. This is not going to change under Carter, campaign rhetoric to the contrary notwithstanding."[39]

Overall, 291 members of either the CFR or the Trilateral Commission (or both) held posts during Carter's presidency.[40] Carter, like Obama, also defied his supporters' anti-war sentiments by retaining notorious war hawks (Obama retained Bush's Secretary of Defense, Gates). *New York Magazine* remarked, "When Carter campaigned as 'a leader for a change,' most Americans did not think that by 'change' he meant bringing back the men who gave us Vietnam."[41] They were referring to Cyrus Vance, Carter's pick for Secretary of State. Vance was the former Secretary of the Army and Deputy Secretary of Defense under presidents Kennedy and Johnson. Carter, like Obama, did nothing to challenge and everything to maintain the establishment. Goldwater explained,

"We believed the outsider—the non-Washington politician, the anti-establishment peanut farmer—would challenge the bureaucracy, end unnecessary waste, strengthen national defense. In short, we thought he would turn the rascals out and lead an administration of fresh faces. All the old coaches and all the old players would be sent to the showers. There would be a new team, unfettered by past failures."[42]

The crowning achievement of the Carter administration was the implementation of Continuity of Government, which Obama is now poised to exploit.[43] In 1977 and 1978, Samuel Huntington served as White House Coordinator of Security Planning for the National Security Council. There, he drafted Presidential Review Memorandum 32 (PRM 32), the precursor to Carter's Executive Order 12127, which created the Federal Emergency Management Agency (FEMA) on June 19, 1979. Kathleen Klenetsky and Herbert Quinde of *Executive Intelligence Review* explained,

"FEMA was established in March 1979 by Presidential Review Memorandum 32, with the mandate to maintain 'the continuity of government' (COG) during a national security emergency. PRM 32 bypassed the U.S. Constitution, and awarded power to the unelected officials at the National Security Councils to direct U.S. government operations by emergency decree. By placing FEMA under the NSC's control, Huntington, Brzezinski, et al., turned the NSC into a shadow technocratic dictatorship, waiting for a real or manufactured crisis to seize control of the country."[44]

In his 1979 book, Goldwater poignantly concluded, "Jimmy Carter said he would never lie to us. We believed him."[45]

Zbigniew Brzezinski

In 1976, former *New York Times* correspondent and current Council on Foreign Relations President Emeritus Leslie Gelb described Zbigniew Brzezinski as a man who "portrays himself as the Good Kissinger."[46] While the "Kissinger" part is accurate, the "good" is entirely subjective. Brzezinski has been extremely good for the New World Order, but his enthusiasm for war and his disdain for the U.S. Constitution have made him anything but good for society. A CFR member, frequent Bilderberg attendee, and co-founder of the Trilateral Commission, Brzezinski has dutifully served the New World Order since the early 1970s. From 1977-1981, as Jimmy Carter's National Security Advisor, he effectively dictated U.S. foreign policy. By his own admissions, his exploits included supporting Pol Pot (the genocidal butcher of Cambodia), fostering and enabling Islamic fundamentalism, and provoking the devastating Soviet-Afghan war. Furthermore, throughout his books, Brzezinski has repeatedly expressed scorn and contempt for democracy. At 82 years of age, he remains active and influential, especially through his puppet, Barack Obama.

In his 2006 book *The Audacity of Hope*, Obama acknowledged, "I am new enough on the national political scene that I serve as a blank screen on which people of vastly different political stripes project their own views. As such, I am bound to disappoint some, if not all, of them."[47] Obama, however, has not disappointed Brzezinski, the man who advised him during his campaign and, by all indications, groomed him since the early 1980s. According to the *New York Times*, "Barack Obama does not say much about his years in New York City. The time he spent as an undergraduate at Columbia College and then working in Manhattan in the early 1980s surfaces only fleetingly in his memoir…He barely mentions Columbia, training ground for the elite, where he transferred in his junior year, majoring in political science and international relations and writing his thesis on Soviet nuclear disarmament."[48] About his Columbia years, Obama curtly recalled, "I spent a lot of time in the library. I didn't socialize that much. I was like a monk."[49] Curiously, however, Obama refuses to release his Columbia transcript. The *Associated Press* reported, "The Obama campaign declined to discuss Obama's time at Columbia and his friendships in general. It won't, for example, release his transcript or name his friends."[50]

What is Obama hiding? According to the *New York Times*, "Columbia was a hotbed for discussion of foreign policy…The faculty included Zbigniew Brzezinski, the former national security adviser, and Zalmay Khalilzad, now the American ambassador to the United Nations."[51] Presidential biographer Webster Tarpley observed, "The public is being urged to regard Obama as a politician of phenomenal organizational ability because of his ability to game the absurd rules of the Democratic Party. But what if Obama had been a protected asset of Zbigniew Brzezinski and the Trilateral Commission since about 1981-1983, and a man whose entire career has been fostered and promoted by the Trilateral-Bilderberg Wall Street group? What if Obama's campaign ran on Rockefeller-Soros Trilateral cash, with the backing of the matchless Trilateral network of media whores and agents of influence?"[52]

Brzezinski claims he first met Obama in 2007. He then endorsed Obama for President, compared him to John F. Kennedy, and even advised him on foreign policy during his campaign.[53] Brzezinski explained, "I met him last year, and he made the best impression on me of anyone since John F. Kennedy. He is better equipped in intellect and temperament for the highest office than anyone I can think of in recent memory. He is very different from most American politicians."[54] A closer look a Brzezinski's ideology chillingly foreshadows his plans for Obama.

The Grand Chessboard

In his 1998 book *The Grand Chessboard*, Brzezinski argued that America should continue to geopolitically dominate the world while simultaneously upgrading the United Nations in preparation for global government, or what he called "peaceful global management." He explained,

"The U.S. policy goal must be unapologetically twofold: to perpetuate America's own dominant position for at least a generation and preferably longer still; and to create a geopolitical framework that can absorb the inevitable shocks and strains of social-political change while evolving into the geopolitical core of shared responsibility for peaceful global management. A prolonged phase of gradually expanding cooperation with key Eurasian partners, both stimulated and arbitrated by America, can also help to foster the preconditions for an eventual upgrading of the existing and increasingly antiquated UN structures. A new distribution of responsibilities and privileges can then take into account the changed realities of global power, so drastically different from those of 1945."[55]

Brzezinski derided nationalism and praised the European Union for undermining national sovereignty across Europe: "By pioneering in the integration of nation-states into a shared supranational economic and eventually political union, Europe is also pointing the way toward larger forms of post-national organization, beyond the narrow visions and the destructive passions of the age of nationalism."[56] But Brzezinski's vision was not without significant obstacles — primarily democracy. He argued that America has *too much* democracy to effectively dominate other nations:

"It is also a fact that America is too democratic at home to be autocratic abroad. This limits the use of America's power, especially its capacity for military intimidation. Never before has a populist democracy attained international supremacy. But the pursuit of power is not a goal that commands popular passion, except in conditions of a sudden threat or challenge to the public's sense of domestic well-being."[57]

Brzezinski is clearly indifferent to the will of the people. It doesn't matter that his "pursuit of power" through "military intimidation" has no popular support. According to Brzezinski, "imperial mobilization" is imperative. Therefore, democracy must be subverted, and human sacrifices and economic costs must be incurred. He explained,

"The window of historical opportunity for America's constructive exploitation of its global power could prove to be relatively brief, for both domestic and external reasons. A genuinely populist democracy has never before attained international supremacy. The pursuit of power and especially the economic costs and human sacrifice that the exercise of such power often requires are not generally congenial to democratic instincts. Democratization is inimical to imperial mobilization."[58]

How do Brzezinski and his fellow imperialists plan to overcome these "democratic instincts?" He acknowledged that fear motivates people to support aggression they would otherwise reject. People must recognize some massive external threat, whether real or illusory: "Moreover, as America becomes an increasingly multicultural society, it may find it more difficult to fashion a consensus on foreign policy issues, except in the circumstances of a truly massive and widely perceived direct external threat."[59] Three years later, on a brisk September morning, that truly massive and widely perceived threat braced America, and, to Brzezinski's great pleasure, enabled truly massive restraints on democracy.

Brzezinski's Past Adventurism

From 1975 through 1978, Pol Pot's genocidal Khmer Rouge regime decimated Cambodia. According to Yale University's Cambodian Genocide Program, "Pol Pot combined extremist ideology with ethnic animosity and a diabolical disregard for human life to produce repression, misery, and murder on a massive scale."[60] Overall, the regime slaughtered at least 1.7 million people, or 21% of the country's population. The catalyst for Pol Pot's rise to power was Richard Nixon and Henry Kissinger's devastating, illegal bombing campaign from 1969 to 1973, which claimed 500,000 Cambodian lives.[61] Pol Pot's ensuing reign finally ended on Christmas day, 1978, when Vietnam defeated the Khmer Rouge (Pol Pot escaped to Thailand).

Vietnam's victory greatly embarrassed U.S. war hawks like Brzezinski. Later in 1989, Brzezinski bragged that he supported the exiled Pol Pot. He told *Time* magazine, "I encouraged the Chinese to support Pol Pot. Pol Pot was an abomination. We could never support him. But China could."[62] However, Brzezinski further admitted that the U.S. "winked semipublicly" while the Chinese funneled arms to the Khmer Rouge. Furthermore, despite Brzezinski's claim that "We could never support him," the U.S. *was* supporting Pol Pot to the tune of $85 million between 1980 and 1986.[63]

In 1998, Brzezinski again boasted about his murderous adventurism. In an interview with *Le Nouvel Observateur*, he proudly admitted that he helped initiate the decade-long Soviet-Afghan war. Brzezinski explained,

"According to the official version of history, CIA aid to the mujahideen began during 1980, that is to say, after the Soviet army invaded Afghanistan on December 24, 1979. But the reality, closely guarded until now, is completely otherwise: Indeed, it was July 3, 1979, that President Carter signed the first directive for secret aid to the opponents of the pro-Soviet regime in Kabul. And that very day, I wrote a note to the president in which I explained to him that in my opinion this aid was going to induce a Soviet military intervention."[64]

The interviewer responded, "You don't regret any of this today?" Brzezinski snapped back, "Regret what? That secret operation was an excellent idea. It had the effect of drawing the Russians into the Afghan trap and you want me to regret it? The day that the Soviets officially crossed the border, I wrote to President Carter, essentially: We now have the opportunity of giving to the USSR its Vietnam War." The interviewer countered, "And neither do you regret having supported Islamic fundamentalism, which has given arms and advice to future terrorists?" Brzezinski rationalized, "What is more important in world history? The Taliban or the collapse of the Soviet empire? Some agitated Moslems or the liberation of Central Europe and the end of the Cold War?" Brzezinski's callous remarks demonstrate his total disregard for more than 1 million Afghans and 14,000 Soviets slaughtered in the nine-year conflict.[65]

Between Two Ages

In 1970, Brzezinski published *Between Two Ages: America's Role in the Technetronic Era.* He observed, "The transformation that is now taking place, especially in America, is already creating a society increasingly unlike its industrial predecessor. The post-industrial society is becoming a 'technetronic' society: a society that is shaped culturally, psychologically, socially, and economically by the impact of technology and electronics—particularly in the area of computers and communications."[66] Brzezinski foresaw two possible future scenarios: (a) a nightmarish future whereby psychopathic elites would dominate humanity through sophisticated electronic and biochemical manipulation, or (b) a pleasant future whereby all nations would agree to gradually relinquish their sovereignties and slowly merge into a Marxist New World Order. Brzezinski was essentially saying, 'Nationhood is obsolete. We (the elite) have prepared two New World Order scenarios for you (the public). We can do this the *hard way* or the *easy way*. The choice is yours.'

In the words of Brzezinski, the *hard way* would result in a society "dominated by an elite whose claim to political power would rest on allegedly superior scientific know-how. Unhindered by the restraints of traditional liberal values, this elite would not hesitate to achieve its political ends by using the latest modern techniques for influencing public behavior and keeping society under close surveillance and control. Under such circumstances, the scientific and technological momentum of the country would not be reversed but would actually feed on the situation it exploits."[67] The *easy way*, on the other hand, would "encourage the spread of a more personalized rational humanist world outlook that would gradually replace the institutionalized religious, ideological, and intensely national perspectives that have dominated modern history."[68] But under no circumstances, according to Brzezinski and his cohorts, will national sovereignty remain viable. He explained,

> "The doctrine of sovereignty created the institutional basis for challenging the secular authority of established religion, and this challenge in turn paved the way for the emergence of the abstract conception of the nation-state...Nationalism did not seek to direct the individual toward the infinite, but to activate the impersonal masses for the sake of immediately proximate goals. Paradoxically, these concrete goals were derived from the still intangible and transcendental, though new, object of worship: the nation."[69]

Between Two Ages was extremely manipulative, though not atypical, propaganda. Brzezinski's technique was one of false choices—here are two possible New World Order scenarios, which do you prefer? It's the same technique employed during every presidential election—here are two establishment candidates, which do you prefer? This is called counterfeit democracy, the freedom to choose between preapproved choices. According to Brzezinski, retaining national sovereignty is *not* a preapproved choice. He explained, "Rivalry between nations is inherent in an international system that functions without global consensus—the result of centuries of the conditioning of man's outlook by competitive nations that insisted on their individual superiority, and particular values. Such rivalry is not likely to be terminated by anything short of a fundamental reconstruction in the nature of relations between nations—and hence in the character of national sovereignty itself."[70]

Despite Brzezinski's assertions, sovereignty is not the problem. The problem is the secrecy imbedded in the political system and the lack of political awareness, interest, and participation by the populace. Sovereignty—the diffusion of power—is the greatest protection against tyranny. Brzezinski and the ruling class, however, insist upon diluting sovereignty and further concentrating power. Just who are these ego-inflated elites? Brzezinski candidly admitted, "Today we are again witnessing the emergence of transnational elites, but now they are composed of international businessmen, scholars, professional men, and public officials. The ties of these new elites cut across national boundaries, their perspectives are not confined by national traditions, and their interests are more functional than national. These global communities are gaining in strength and, as was true in the Middle Ages, it is likely that before long the social elites of most of the more advanced countries will be highly internationalist or globalist in spirit and outlook."[71] What plans do they have for humanity?

The Technetronic Society

Brzezinski's false choice between the benign New World Order and the nightmarish version only applies to the beginning, transitional phases of globalization. If allowed to emerge, the New World Order may indeed initially appear benevolent. However, as history has repeatedly demonstrated, when power consolidates, tyranny accelerates. The U.S. federal government, for example, over many generations, has slowly usurped power from the states, resulting in today's steadily increasing tyranny. Brzezinski never addresses this issue. Where are the checks and balances for his envisioned New World Order? Where is the separation of powers? Instead, he implies that "the new global consciousness" will somehow insulate world leaders from tyrannical ambitions.[72] This is simply deceptive marketing, for the New World Order is manipulative to its core. Brzezinski admitted,

"In the technetronic society the trend seems to be toward aggregating the individual support of millions of unorganized citizens, who are easily within the reach of magnetic and attractive personalities, and effectively exploiting the latest communication techniques to manipulate emotions and control reason. Reliance on television—and hence the tendency to replace language with imagery, which is international rather than national, and to include war coverage or scenes of hunger in places as distant as, for example, India—creates a somewhat more cosmopolitan, though highly impressionistic, involvement in global affairs."[73]

The following passages demonstrate the extent of Brzezinski and his fellow globalists' psychopathology:

• "Man is increasingly acquiring the capacity to determine the sex of his children, to affect through drugs the extent of their intelligence, and to modify and control their personalities. Speaking of a future at most only decades away, an experimenter in intelligence control asserted, 'I foresee the time when we shall have the means and therefore, inevitably, the temptation to manipulate the behavior and intellectual functioning of all people through environmental and biochemical manipulation of the brain.'"[74]

- "In addition, it may be possible—and tempting—to exploit for strategic-political purposes the fruits of research on the brain and on human behavior. Gordon J.F. MacDonald, a geophysicist specializing in problems of warfare, has written that accurately timed, artificially excited electronic strokes 'could lead to a pattern of oscillations that produce relatively high power levels over certain regions of the earth. In this way, one could develop a system that would seriously impair the brain performance of very large populations in selected regions over an extended period."[75]

The Transition

As documented throughout this book, the New World Order begins as benign global socialism and progresses towards oppressive global tyranny. Therefore, Brzezinski's affection for Marx is hardly surprising. He sees Marxism as the natural progression from nationalism. He explained, "That is why Marxism represents a further vital and creative stage in the maturing of man's universal vision...Though it may be argued that this intellectually rigorous method was eventually subverted by its strong component of dogmatic belief, Marxism did expand popular self-awareness by awakening the masses to an intense preoccupation with social equality and by providing them with both a historical and a moral justification for insisting upon it...In this sense, Marxism has served as a mechanism of human 'progress,' even if its practice has often fallen short of its ideals...Marxism represents as important and progressive a stage as the appearance of nationalism and of the great religions."[76]

To advance his Marxist New World Order vision, Brzezinski linked nationalism to environmental degradation. He explained, "The concern with ideology is yielding to a preoccupation with ecology. Its beginnings can be seen in the unprecedented public preoccupation with matters such as air and water pollution, famine, overpopulation, radiation, and the control of disease, drugs, and weather, as well as in the increasingly non-nationalistic approaches to the exploration of space or of the ocean bed. There is already widespread consensus that functional planning is desirable and that it is the only way to cope with the various ecological threats."[77] However, despite this alleged "widespread consensus," the transition away from nationalism would require significant American sacrifices and attitude adjustments:

- "Finally, the opportunities and dangers inherent in the scientific-technological age require subtle but important changes in American attitudes and organization."[78]
- "In the economic-technological field some international cooperation has already been achieved, but further progress will require greater American sacrifices. More intensive efforts to shape a new world monetary structure will have to be undertaken, with some consequent risk to the present relatively favorable American position."[79]

In 2010, society is now progressing through the transitional phase towards Brzezinski's New World Order. The following statement, written in 1970, predicted with stunning accuracy the current situation: "Persisting social crisis, the emergence of a charismatic personality, and the exploitation of mass media to obtain public confidence would be the steppingstones in the piecemeal transformation of the United States into a highly controlled society."[80]

Chapter 6

The Bilderberg Group

In his 2002 book *Memoirs*, David Rockefeller sardonically remarked, "If the Council on Foreign Relations raises the hackles of conspiracy theorists, the Bilderberg meetings must induce apocalyptic visions of omnipotent international bankers plotting with unscrupulous government officials to impose cunning schemes on an ignorant and unsuspecting world."[1] Rockefeller's arrogance notwithstanding, Bilderberg is an annual, private meeting of ultra-elite bureaucrats, CEOs and finance oligarchs from around the world, though primarily from Europe and the United States. By their own admissions, they do indeed "impose cunning schemes on an ignorant and unsuspecting world." Bilderberg Chairman Prince Bernhard, for example, revealed in his biography that Bilderberg was responsible for ushering into existence, without public approval, the European Union.

The first Bilderberg meeting took place in Holland in 1954. It included many prominent Americans including Joseph E. Johnson of the Carnegie Endowment for International Peace, Dean Rusk, then the head of the Rockefeller Foundation, as well as David Rockefeller and H.J. Heinz II.[2] Alden Hatch, Bernhard's authorized biographer wrote,

> "At a small hotel near Arnhem in the deeply wooded uplands of eastern Holland on May 29, 30, and 31, 1954, a group of eminent statesmen, financiers, and intellectuals from the principal nations of Europe and the United States met together in, perhaps, the most unusual international conference ever held until then. There was absolutely no publicity. The hotel was ringed by security guards, so that not a single journalist got within a mile of the place. The participants were pledged not to repeat publicly what was said in the discussions. Every person present—Prime Ministers, Foreign Ministers, leaders of political parties, heads of great banks and industrial companies, and representatives of such international organizations as the European Coal and Steel Community, as well as academicians—was magically stripped of his office as he entered the door, and became a simple citizen of his country for the duration of the conference. Thus everybody could and did say what he really thought without fear of international, political, or financial repercussions."[3]

All New World Order globalists share the same goal of eliminating national sovereignty and implementing world government. Bilderberg offers international powerbrokers the opportunity to coordinate and synergize their efforts. Peter Thompson, in Holly Sklar's book *Trilateralism* explained,

> "Bilderberg itself is not an executive agency. However, when Bilderberg participants reach a form of consensus about what is to be done, they have at their disposal powerful transnational and national instruments for bringing about what it is they want to come to pass…Bilderberg is not the only means of Western collective management of the world order; it is part of an increasingly dense system of transnational coordination. The foreign policies of nation-states, particularly economic and monetary policies, have always been a highly elitist matter. Policy options are proposed, reviewed, and executed within the context of a broad bipartisan consensus that is painstakingly managed by very small circles of public and private elites. Democratic interference in foreign policy is avoided, in so far as possible, throughout the Western capitalist democracies. Where necessary, a

consensus is engineered on issues which must get congressional/parliamentary approval, but wherever possible executive agreements between governments are used to avoid the democratic process altogether."[4]

Planning Europe

Prince Bernhard of the Netherlands was Bilderberg's chairman from 1954 until 1976 when he resigned in disgrace facing bribery charges. According to Joseph Retinger, Bilderberg's permanent Secretary until his death in 1960, "the Prince centralizes all Bilderberg activities, appoints all the members of the steering committee and, after consultation with these members, decides on those to be invited to the yearly conferences."[5] Hatch commented, "From Prince Bernhard's point of view the Bilderberg group gives him an opportunity to work in private, without violating the parliamentary taboo against royalty mixing in politics, for the unification of Europe and, indeed, the Atlantic Community as well."[6] Democracy is a nuisance for globalists because most people don't share their enthusiasm for tyranny. Thus the privacy afforded by Bilderberg was essential for Bernhard's grand vision for Europe.

Writing in 1962, Hatch admitted, "[Bernhard] foresees a United States of Europe in which borders are reduced to an absolute minimum, and there is a common currency, a common financial policy, a common foreign policy, and a common policy of trade. The nations will give up so much of their sovereignty as is necessary to implement this."[7] Despite his determined resolve, Bernhard lamented the hassle of having to "re-educate" the people of Europe. Bernhard explained,

"One thing we need for free exchange of goods is complete interchangeability of money, a common currency. I'm flat out for that. And this implies a certain political unity. Here comes our greatest difficulty. For the governments of the free nations are elected by the people, and if they do something the people don't like they are thrown out. It is difficult to re-educate people who have been brought up on nationalism to the idea of relinquishing part of their sovereignty to a supra-national body."[8]

As discussed in Chapter 10, the globalists have always feared third-world economic independence. Therefore they have used, among other tactics, false environmentalism to suppress third-world economic development. In 1961, Bernhard became President of the World Wildlife Fund (WWF) where his "conservation" efforts served this larger agenda. A European union, argued Bernhard, would also serve this agenda by forcing Africa and Asia to participate in European and American style "free trade." He explained,

"If we could all agree beforehand in principle it would result, without doubt, not in Utopia, but in an extremely strong and healthy Europe. This in turn would bring the United States into the economic community. It would encourage a great deal of free trade throughout the world. Now, the more free trade you have the more difficult you make it for the new countries of Africa and Asia to set up an autarchy and live in economic isolation, to adopt trade barriers and quotas which after a hundred years or more we are finding out don't pay. From sheer necessity these people will have to join in free trade."[9]

As discussed in Chapter 7, early planning for the European Union started in America during the late 1940s. Bilderberg, however, championed EU implementation, the first step being the 1957 Treaty of Rome. George McGhee, another early Bilderberg member, candidly admitted, "I believe you could say the Treaty of Rome, which brought the Common Markets into being, was nurtured at these meetings and aided by the main stream of our discussions there. Prince Bernhard is a great catalyst."[10] In March 2009, current Bilderberg Chairman Etienne Davignon admitted that during the 1990s, Bilderberg planned Europe's currency, the euro. About the current economic crisis, Davignon lamented, "People understand confusedly that there is a change [in the air], but no government will satisfy the reactions of the people. They have the greatest reticence and cynicism against anybody who holds responsibility."[11] Perhaps this is because these so-called *responsible* leaders have been planning the future behind closed doors, without public input or approval.

Bernhard's Nazi Past

Shortly after his 1961 nomination for WWF president, Bernhard published *H.R.H. Prince Bernhard of the Netherlands: An Authorized Biography*, a shoddy attempt to downplay and distort his embarrassing past. As a university student, Bernhard became a Nazi SS officer, supposedly out of necessity. His biographer, Alden Hatch, explained, "In order to finish his education quickly Bernhard had to make some compromises with the monstrous political system that was fastening its grip on Germany…At the beginning of his serious studies Bernhard learned that a new sort of test had been decreed for every one graduating from the universities—a written and oral 'political attitude' examination. With his ideals and high temper he knew that was one examination he could not pass."[12] Rather than simply faking his way through the test, Bernhard determined that he would have to *join* the Nazi party. According to Hatch, however, his membership was short-lived: "At the end of their studies Bernhard's whole group, with one exception, left the SS and severed all connection with the party."[13]

Upon graduating in 1935, Bernhard went immediately to Paris. Hatch wrote, "Had he been older and his character more hardened by adversity he might have considered remaining to oppose the regime, hopeless as opposition seemed."[14] Yet for someone supposedly opposed to the regime, Bernhard's decision to work in the Paris office of I.G. Farben was certainly curious. The fourth-largest company in the world at that time, I.G. Farben, was the economic and industrial backbone of the Nazi machine.[15] Nevertheless, Bernhard insisted that his Nazi episode was over. He recalled, "Soon after I began working for I.G. Farben, the German Ambassador sent a man to ask me if I would join the organization of Germans living abroad. It was, of course, a party organization, so I said, 'No.' They gave me no further trouble, but I was never invited to the German Embassy."[16]

Bernhard's younger brother, Prince Aschwin von Lippe-Biesterfeld, was also a member of the Nazi party. Hatch portrayed Aschwin as a young, idealistic German aristocrat who was "swept away by [Hitler's] incendiary oratory." At the impressionable age of 16, according to Hatch, Aschwin "was caught up in the tide of enthusiasm and lent his name to the movement, though he soon repented."[17] With the biography, Hatch and Bernhard effectively swept Bernhard's Nazi connections under the rug, where they remained for several decades.

In 1991, newly declassified Nazi records exposed Hatch's fraudulent, romanticized history. The records revealed that Bernhard, Aschwin, and *nine more* Lippe royals were all Nazis, and that Aschwin was actually 23 when he joined, not the naive 16 year-old from Hatch's fictional biography. These records also revealed that Bernhard did *not* cancel his membership immediately after his studies. He instead waited two more years until the time of his engagement with Princess Juliana of the Netherlands.[18] He then canceled only out of necessity. The *London Telegraph* explained, "Prince Bernhard was questioned in an official inquiry about his political ideas because of fears about Holland's Nazi neighbours. He convinced the Queen and the Dutch government that he was completely opposed to Hitler, despite having been a member of the SS when he was a student. He had joined, he said, because had he not been a member of such an organisation it would have been made harder for him to pass his law exams; moreover, membership brought with it the free use of a garage. He resigned from the party in 1937."[19] Hatch and Bernhard clearly had no aversion to liberally reinventing the prince's past. But was Bernhard truly opposed to Nazi ideology? The *Telegraph* provided the answer in no uncertain terms. Before he could marry Juliana, Bernhard had to renounce, in writing, his Nazi membership. "His letter," the *Telegraph* reported, "now in the National Archives in Washington, DC, ended with the words 'Heil Hitler.'"[20]

Hatch and Bernhard also embellished the I.G. Farben story. They failed to mention, for example, that Bernhard was working for NW7, the office responsible for directing intelligence, espionage and propaganda activities.[21] During the Nuremberg Military Tribunals, in the case of *U.S. v. Krauch*, the prosecution declared, "Farben's foreign agents formed the core of Nazi intrigue throughout the world. Financed and protected by Farben, and ostensibly acting only as business men, Farben officials carried on propaganda, intelligence, and espionage activities indispensable to German preparation for, and waging of, aggressive war. In Germany, Farben's Berlin NW7 office was transformed into an economic intelligence arm of the Wehrmacht. The Nazi party relied upon Farben as one of its main propaganda machines."[22] In 1934, the House of Representatives Special Committee on Un-American Activities, in its "Investigation of Nazi Propaganda Activities," even named Bernhard as "an important Nazi spy."[23] Bernhard worked under Max Ilgner, chief of Farben's Berlin NW7 office, who was convicted for war crimes and sentenced to three years in prison.[24] After serving his sentence, Ilgner went to work for Lockheed Martin, today the world's largest defense contractor. Ilgner and Bernhard would later reconnect in 1959.[25]

By 1962, Bernhard accepted $1.2 million in bribes from Lockheed in return for helping them secure contracts from the Dutch government. However, the story didn't break until 1976 when the Securities Exchange Commission and the Church Committee were seeking documents pertaining to certain Lockheed payments. Lockheed agreed to relinquish the documents, but with all names blacked out. Nevertheless, Lockheed's auditor, Arthur Young and Company, mistakenly gave the Church Committee the documents with the names visible. Ensuing investigations suggested that Bernhard had indeed accepted the bribes.[26]

The Dutch and International press covered the scandal extensively. Bernhard denied wrongdoing yet he accepted the findings of an official inquiry ordered by Prime Minister Joop den Uyl. The inquiry concluded that Bernhard had "shown himself open to dishonourable requests and offers" and "allowed himself to be tempted to take initiatives which were completely unacceptable."[27] Nevertheless, because Bernhard was married to Queen Juliana, Joop den Uyl decided against prosecuting him. Instead he stripped Bernhard of his rank as Inspector General of the Armed Forces and forced him to resign from many business commitments.[28] Bernhard later arranged for a posthumous confession through a series of newspaper interviews. He told the weekly *Groene Amsterdammer* magazine, "I have always earned plenty of money, so I didn't need the million from Lockheed. How could I have been so stupid? I don't know what had gotten into me. I have plenty of money. My money is spread across three countries: America, the Netherlands and Switzerland. In Switzerland alone I have more than six million in the bank."[29]

Recent Bilderberg Antics

Jim Tucker is a veteran reporter who has been tracking Bilderberg for decades. His journalistic integrity has earned him the trust of moles and insiders who consistently leak information and documents from the private Bilderberg meetings. Their information has always been accurate. In 2006, for example, Ottawa, Canada was the scene of the annual meeting. According to Tucker's sources, attendees were discussing the upcoming implosion of the sub-prime housing market and how they would raise interest rates, causing Americans to lose their homes to the banks. On June 19, 2006, two years before this actually started happening, Tucker reported for *American Free Press*,

> "Bilderberg expects interest rates to rise and many Americans to lose their homes in the months ahead...Timothy Geithner, president of the Federal Reserve Bank of New York, predicted rising interest rates and difficulties for families that have obtained adjustable rate mortgages, or 'variable' interest rates. Many are likely to lose their homes as rising home mortgage rates add hundreds of dollars to their monthly payments, he said...When home construction peaks and prices start downward, many will find they owe more on their home than it is worth in the marketplace. They will also find their mortgage—even 'interest-only' payments—are unaffordable. The banks will get the homes back and sell them again. Again, the term 'stupid Americans' was heard among clucks of sympathy or silent indifference. According to one source, no concern was expressed by Allan Hubbard, assistant to President Bush for economic policy."[30]

Daniel Estulin, another veteran Bilderberg reporter with inside sources, reported about Bilderberg 2006,

> "One of the American delegates—I wasn't told who exactly it was—was talking about the concern that the American citizens have had with the housing prices going down, so they're not investing that money. So what they needed to do was they needed to create the illusion that everything is going well. So what they're going to do over the next year or year and a half is to bring the market back up to its 1998-1999 levels, they're going to get all the suckers to invest what little money they have left over, and that's when they're going to make the economy's bottom drop out. They need to destroy the economy because as we're running out of oil, at least that's what they're saying, when people don't have money, they don't travel, which means they don't waste a lot of oil and natural gas."[31]

The seemingly incredible fluctuations in oil prices over the past several years have not been coincidental. At the 2006 meeting, Estulin reported, "Last year they said the oil prices were going to go up to $150, at the time is was $39, it went up to $76, it basically doubled, if it doubles again it's going to be back to were these people say it's going to be."[32] In July 2007, oil prices peaked at a record-high of $147. Asked about his sources, Estulin explained that information always leaks from "members of the Bilderberg club who for years and years and years have been going and participating in all their conversations. They've always been dead on—always."[33]

Chapter 7

Regional Government

On January 24, 1988, the *New York Times* published an interview with George W. Ball, the former Under Secretary of State during the Kennedy and Johnson administrations and a prominent CFR member.[1] Drawing on his insider knowledge, Ball foreshadowed the regional integration of the European Union and the North American Union, both significant steppingstones towards world government. He stated, "On the economic side, we have to take account of the fact that our political structure is totally inadequate for a world where technology has assured that capital flows move around without regard to national boundaries. Sooner or later we're going to have to face restructuring our institutions so that they're not confined merely to the nation states. Start first on a regional, and ultimately you can move to a world, basis."[2]

The European Union

Most Europeans would be surprised, and perhaps angered, to learn that elite U.S. foundations, together with the U.S. intelligence community, began planning the European Union in the 1940s. Citing declassified government documents, Ambrose Evans-Pritchard reported in 2001 for the *London Telegraph*, "The documents confirm suspicions voiced at the time that America was working aggressively behind the scenes to push Britain into a European state. One memorandum, dated July 26, 1950, gives instructions for a campaign to promote a fully-fledged European parliament. It is signed by Gen William J. Donovan, head of the American wartime Office of Strategic Services, precursor of the CIA."[3] Donovan was the chairman of the American Committee for a United Europe (ACUE), created in 1948 to plan European integration. During these years, European puppets such as Joseph Retinger fronted the integration agenda while U.S. puppet-master pulled the strings. Evans-Pritchard noted, "The U.S. role was handled as a covert operation. ACUE's funding came from the Ford and Rockefeller foundations as well as business groups with close ties to the U.S. government. The head of the Ford Foundation, ex-OSS officer Paul Hoffman, doubled as head of ACUE in the late Fifties. The State Department also played a role. A memo from the European section, dated June 11, 1965, advises the vice-president of the European Economic Community, Robert Marjolin, to pursue monetary union by stealth. It recommends suppressing debate until the point at which 'adoption of such proposals would become virtually inescapable.'"[4]

Just like the impending North American Union, EU implementation began decades ago with various treaties and economic agreements. The 1951 Paris Treaty established the European Coal and Steel Community (ECSC). The 1957 Treaty of Rome established the European Economic Community (EEC). The 1965 Merger Treaty established the predecessors of several current EU institutions including the EU Council and the EU Commission. The 1987 Single European Act (SEA) revised the Paris and Rome treaties. And the 1992 Maastricht Treaty created the modern European Union.[5] Gradually, the EU emerged from simple economic agreements into a dominant political institution. Today, the European Union is attempting to further consolidate power and subsequently control every aspect of life, just like past totalitarian regimes.

Vladimir Bukovsky grew up in Soviet Russia where, due to his political dissidence, from age 22 until 34 he endured Soviet prisons, labor camps and psychiatric institutions. Finally, in 1976, the USSR expelled him to the West. After the USSR fell, the Russian government called upon Bukovsky in 1992 to provide expert testimony during a trial examining criminality within the Soviet Communist Party. To prepare for his testimony, the government granted him access to archives of classified Soviet documents. In 2006, Bukovsky visited the European Parliament in Brussels on the invitation of the Hungarian political party Fidesz to commemorate the 50th anniversary of the 1956 Hungarian Uprising. During this visit, Paul Beliën, founder of the *Brussels Journal* and occasional contributor to such publications as *The Wall Street Journal* and the *Washington Times*, interviewed Bukovsky about the parallels between the USSR and the European Union.

Bukovsky revealed rare glimpses of a shadow government in action. He described a 1989 meeting between then-President Gorbachev and representatives of the Trilateral Commission, including David Rockefeller and Henry Kissinger:

"In January of 1989, a delegation of the Trilateral Commission came to see Gorbachev. It included [former Japanese Prime Minister] Nakasone, [former French President] Giscard d'Estaing, Rockefeller and Kissinger. They had a very nice conversation where they tried to explain to Gorbachev that Soviet Russia had to integrate into the financial institutions of the world, such as GATT, the IMF and the World Bank. In the middle of it, Giscard d'Estaing suddenly takes the floor and says: 'Mr. President, I can't tell you exactly when it will happen—probably within 15 years—but Europe is going to be a federal state and you have to prepare yourself for that. You have to work out with us, and the European leaders, how you would react to that, how would you allow the other East-European countries to interact with it or become a part of it. You have to be prepared.' Now, 1989, January, by that time even the Maastricht Treaty was not drafted yet. How the hell did Giscard d'Estaing know what was going to happen in 15 years? And surprise, surprise, how did he become the author of the European Constitution? A very good question, It does smell of conspiracy, doesn't it?"[6]

Bukovsky compared the European Parliament to the Supreme Soviet, stating, "It looks like the Supreme Soviet. When you look at the [European] Commission, it looks like the Politburo, exactly, except it has 25 members and they usually had 13 or 15 members in the Politburo, but they are exactly the same, unaccountable to anyone, electing each other, not directly elected by anyone at all. So it's the same thing…when you look into all this bizarre activity of the European Union with its 80,000 pages of regulations it looks like Gosplan. We used to have an organization which was planning everything in the economy, to the last nut and bolt for five years in advance. Exactly the same thing is happening here. When you look at the type of corruption, it is exactly the Soviet type of corruption, going from top to bottom rather than going from bottom to top. And you go through all the structures and features of this emerging European monster, and it more and more resembles the Soviet Union."[7] Beliën questioned, "But some people would say that all these countries that joined the European Union did so voluntarily." Bukovsky answered assertively, "No they didn't. Look at Denmark, which resisted the Maastricht. Look at Ireland. Look at many other countries. They are under enormous pressure, almost blackmail. Switzerland was forced to vote five times in a referendum, five times! All five times they have rejected it. But who knows what will happen the sixth time, the seventh time. It is always the same thing. It's a trick for idiots. We vote in referendums until people vote what we want."[8]

Even Jean-Claude Juncker, the Prime Minister of Luxembourg, has acknowledged this trick for idiots. Days before the 2005 French referendum on the EU Constitution, while Juncker concurrently held the rotating EU Presidency, he arrogantly declared, "If it's a Yes, we will say 'on we go,' and if it's a No we will say 'we continue.'"[9] Unfortunately, populations fall for this trick time after time. In 1992, the people of Denmark voted against the Maastricht Treaty. The next year, forced to vote again, they voted yes. In 2001, the people of Ireland voted against the Nice Treaty. The next year, forced to vote again, they voted yes.

In 2005, the people of France and the Netherlands both voted against the EU Constitution, a grandiose document that would, for the first time, give the European Union a legal personality whereby its laws would trump those of national parliaments. The EU Constitution would end national veto rights in 60 vital areas of policy. It would also establish a powerful EU Presidency, replacing the six-month rotating position. According to the *BBC*, "The new President will therefore be a permanent figure with much greater influence and symbolism."[10] Rather than allowing the people of Europe to vote for the new President, the European Council would simply *appoint* the leader of their choice. The constitution would also establish an unelected EU Foreign Minister. Regarding defense, the proposed Constitution stated, "The Union shall have competence to define and implement a common foreign and security policy, including the progressive framing of a common defense policy."[11] The constitution would also make EU treaties self-amending. Originally, amendments required the convening of an Intergovernmental Conference (IGC), unanimous voting by each member state, and ratification by each member state according to its own Constitutional traditions. Under the EU constitution, however, the EU could incrementally amend treaties by itself, thereby bypassing national parliaments and all public scrutiny. Finally, the Constitution would remove, or greatly inhibit, the ability of any member nation to withdraw from the EU. In its totality, the European Constitution would concentrate power and control while further eroding national sovereignty.

The Lisbon Treaty

With the French and Dutch rejections, according to EU rules, the EU Constitution was dead. However, the EU simply repackaged the Constitution as the Lisbon Treaty. There were no significant differences between the two documents. However, by calling the Constitution a "treaty," most governments could bypass otherwise obligatory referendums and simply ratify the Lisbon Treaty without public approval. Former Italian Prime Minister Giuliano Amato acknowledged this sleight of hand during a speech at the London School of Economics. "The good thing about not calling it a Constitution," Amato stated, "is that no one can ask for a referendum on it."[12] In a historic and riveting speech before the EU Parliament in Strasbourg on February 20, 2008, Nigel Farage, a member of the European Parliament and leader of the U.K. Independence Party lambasted EU elites for their incessant chicanery. Farage proclaimed,

"What the European Parliament is engaged in here today, with this debate and subsequent vote, is nothing less than a massive exercise in deceit. A pack of lies is being told, and they're being told because *you do not want the people of Europe to have the referendums that you promised them*. It is the imposition of the will of the political class upon the citizens. We all know the truth: the Lisbon Treaty *is* the same as the EU Constitution—*exactly* the same number of new competences, *exactly* the same number of surrenders of veto. It is virtually identical in every regard—and it *is* constitutional, because it gives this European Union a full legal personality, but worse still, it gives it the ability to amend itself in the future without having to refer to more intergovernmental conferences. It gives the EU the ability to legislate over literally *every single aspect* of our lives. I am told though I shouldn't worry, because the flag and the anthem have been dropped. Well pull the other one! There is a great big flag at the front there. It's nonsense. It's all part of the lie. The truth is that you are too chicken to have a referendum. You do not want to hear the voice of the people, and now you resort to totalitarian means to get this Treaty through. We heard Danny Cohn-Bendit say that those who oppose this Treaty are mentally ill. We heard Mr. Schulz say that those of us who objected to the sheer dishonesty of this project were behaving like Nazis back in the 1930s. Well, I think the ordinary decent citizens of Europe will work out for themselves who the extremists are. And I must say that I wish the people of Ireland—I wish those that believe in democracy every success in a couple of months' time. And I hope they send out a large, resounding *no*. And I hope the voice of the peoples of Europe will be heard, not that of just you, the political class. And you may be complacent today, but you are *increasingly being held in contempt*."[13]

Dozens of European leaders candidly acknowledged that the Lisbon Treaty was simply an incognito EU Constitution. German Chancellor Angela Merkel, for example, in a speech before the European Parliament, admitted, "The substance of the Constitution is preserved. That is a fact."[14] Spanish Prime Minister José Zapatero admitted, "We have not let a single substantial point of the constitution go."[15] Danish Prime Minister Anders Fogh Rasmussen stated, "The good thing is that all the symbolic elements are gone, and that which really matters—the core—is left."[16] Irish Minister of Foreign Affairs Dermot Ahern acknowledged, "The substance of what was agreed in 2004 has been retained. Really, what is gone is the term constitution."[17]

Since the substance remained, French and Dutch voters presumably remained opposed. Nevertheless, the Dutch Council of State, the highest governmental advisory body, ruled that a new referendum was not *legally* required because the Lisbon Treaty lacked sufficient "constitutional" elements. The Dutch government had brazenly defied the people.[18] In France, opinion polls suggested that most voters wanted a referendum,[19] however in November 2007, French President Nicolas Sarkozy declared, "A referendum now would bring Europe into danger. There will be no Treaty if we had a referendum in France, which would again be followed by a referendum in the UK."[20] During a closed-door meeting between senior MEPs, Sarkozy showcased his unabashed contempt for democracy by confessing, "It would happen in all member states if they have a referendum. There is a cleavage between people and governments."[21]

About Sarkozy, Nigel Farage objected, "Not only does he stop his own people from having a say but he is trying to block Britain from having the referendum which our government promised."[22] Farage was referring to Former Prime Minister Tony Blair's promise that, "The new Constitutional Treaty ensures the new Europe can work effectively. We will put it to the British people in a referendum."[23] Blair broke this promise.

Current Prime Minister Gordon Brown has also disdainfully blocked a referendum. In May 2008, the *London Times* reported that millionaire businessman Stuart Wheeler won a High Court challenge to force Brown to hold a referendum on the Lisbon Treaty.[24] Nevertheless, Wheeler lost the judgment and on July 16, the U.K. ratified the Lisbon Treaty.

Only the 3 million people of Ireland, because of explicit provisions under Irish law, would be allowed to decide for themselves the direction of their own nation. As for the EU's remaining 26 member states, comprising 487 million people, their so-called leaders had decided for them. Despite heavy pressure and heavy EU propaganda, Irish voters rejected the Lisbon Treaty with a 53.4% majority on June 12, 2008.[25] Predictably, EU elites reacted by demanding yet *another* vote. The *London Telegraph* reported, "Mr. Giscard d'Estaing told the *Irish Times* that Ireland's referendum rejection would not kill the Treaty, despite a legal requirement of unanimity from all the EU's 27 member states."[26]

Giscard d'Estaing, the former French President, was the author of the original European Constitution. "Mr. Giscard d'Estaing also admitted," wrote the *London Telegraph*, "that, unlike his original Constitutional Treaty, the Lisbon EU Treaty had been carefully crafted to confuse the public." He proudly confessed, "What was done in the Treaty, and deliberately, was to mix everything up. If you look for the passages on institutions, they're in different places, on different pages. Someone who wanted to understand how the thing worked could with the Constitutional Treaty, but not with this one."[27] Belgian Foreign Minister Karel de Gucht also confirmed this deception. He stated, "The aim of the Constitutional Treaty was to be more readable; the aim of this treaty is to be unreadable. The Constitution aimed to be clear, whereas this Treaty had to be unclear. It is a success."[28]

Danish MEP Jens-Peter Bonde, during a January 2008 seminar, explained how EU elites crafted the Lisbon Treaty. According to Bonde, the Prime Ministers who endorsed and pushed the Lisbon Treaty never read it themselves because, "It can't be read." He explained, "This is not a treaty. This is 300 pages of amendments to 3,000 other pages of treaties, and you can only read it if you take one amendment by one and then look it up in the existing treaties and insert it."[29] He continued, "They have decided in the Council that it's not *allowed* for any institution in the European Union to print a consolidated version which can be read before it has been approved in all 27 member states. This is a *decision*."[30] Elected MEPs like Bonde demanded that the public receive a reader-friendly version, but higher powers refused. Bonde explained, "We agreed unanimously in the Constitutional Affairs Committee that we wanted a reader-friendly edition, a consolidated version which could be read—*unanimously*. We will not have it. Because higher powers decided we cannot have it. This is an instruction from some Prime Ministers that do not want the text to be read. The order is *sign*, read afterword."[31]

Both d'Estaing and Sarkozy insisted that Ireland would vote again. Immediately following the Irish rejection, when the treaty should have officially been dead, Sarkozy called upon member nations to continue their ratification procedures. He arrogantly declared, "The others must continue ratification so that the Irish incident does not become a crisis."[32] Despite such rampant disdain for democracy among the EU's leadership, Czech President Václav Klaus rightly asserted, "The EU cannot ignore its own rules. The Lisbon Treaty has been roundly and democratically rejected by Ireland, and it therefore cannot come into force. Any attempt to ignore this fact and make recourse to pressure and political manipulation to move the treaty forward would have disastrous consequences."[33] William Hague, England's Shadow Foreign Secretary, also observed, "The only point in other countries continuing to ratify the treaty is to put pressure on the Irish, to bully the Irish—a kind of preparatory move to saying to the Irish 'You're going to have to vote again on this.' Let's treat this as a wake up call and stop this centralising agenda and abandon this treaty."[34]

In October 2009, EU elites again performed their reliable "trick for idiots," forcing a second referendum upon the Irish people. Nothing had changed in the document. "Not one comma has changed," observed Declan Ganley, founder of the political party Libertas.[35] Nevertheless, against heavy propaganda, which insisted that the Lisbon Treaty would save Ireland from financial ruin, the people voted yes. On November 3, 2009, Czech President Václav Klaus reluctantly signed the treaty, thus completing the ratification process. The EU quickly appointed Belgium Prime Minister Herman Van Rompuy as EU President and British noble Baroness Ashton as Foreign Minister. At his first press conference, Van Rompuy hailed 2009 as "the first year of global governance." He went on, praising the UN's forthcoming Copenhagen Climate Conference as "another step towards the global management of our planet."[36]

Undemocratic Structure of the European Union

The European Union is comprised of the Commission, the Council of Ministers, the Parliament, and the Court of Justice. The Commission is one of the EU's executive bodies and performs both legislative and administrative functions.[37] Michele Cini, a Senior Lecturer in Politics at the University of Bristol has remarked, "There is no doubt that the vast majority of EU laws and EU rules originate from within the Commission."[38] Commission members are appointed, not elected representatives: "In response to the question 'is the European Commission an unelected legislator?' the answer must be 'yes', the Commission is unelected…It is interesting because to identify the Commission as an unelected legislator is to imply that law is somehow being made in an unaccountable manner and that our expectations about how an executive (one which includes both a political leadership and an administrative dimension) ought to behave have in some way been challenged."[39]

Another concentration of unelected legislators is the Working Groups, members of which advise the Commission. Professor Roger Scully, the Director of Research at Aberystwyth University and a prolific author on EU political representation, explained, "One important contribution by van Schendelen is simply to make clear the scale of the phenomenon, encompassing some 50,000 or so participants in the roughly 1,000 expert committees which act in an advisory capacity to the Commission, another 450 comitology committees, and a further 300 or so working groups which do much of the real work in developing the legislation formally passed by the Council of Ministers."[40]

The Council of Ministers is comprised of 27 national ministers, one per state. The Council is a legislative body drawing support from hundreds of Working Groups. The Council and the Commission have far greater power than the Parliament. Danish MEP Jens-Peter Bonde explained, "85% of all laws in Brussels are decided by civil servants in 300 secret Working Groups in the Council, prepared by 3,000 other secret Working Groups in the Commission. As a Member of Parliament, I have no access to the documents, I have no access to what they are doing."[41]

The European Parliament is the only EU institution whose membership is elected by the people. The other institutions, the Commission, the Council, and the Court of Justice, all have appointed members. The elected members of Parliament have no power to initiate legislation. Only the unelected Commission members can do so.[42] Bonde explained, "The European Commission have a monopoly to propose a law, no one else. You have to be non-elected to be able to propose a law in the European Union…You can propose amendments. And then if the Commission says it has been a good amendment, then it has a chance. If the Commission of unelected says it was a bad amendment, for passing it has no chance."[43]

The European Union is an insidious steppingstone towards centralized, unaccountable, command-and-control global government. Just days after the ratification of the Lisbon Treaty, World Trade Organization Director General Pascal Lamy praised the evolving union as "a new paradigm of global governance," "the laboratory of international governance," and "the most ambitious experiment to date in supranational governance."[44] He further acknowledged and praised, "The fact that Community law takes precedence over national law. The creation of a supranational body such as the European Commission that has been given the monopoly of initiating legislation. A European Court of Justice whose decisions are binding on national judges."[45] Although his address was prepared for an audience of approving globalists, those who value national sovereignty and self-determination should heed Lamy's words. For the European Union is merely the blueprint for further regional consolidation projects across the globe.

The North American Union

The United States and its ingeniously devised Constitution stands as the biggest obstacle against tyrannical globalization. The globalists, therefore, are intent on systematically dismantling the Constitution while gradually transferring sovereignty to international institutions. In 2009, German Chancellor Angela Merkel plainly admitted, "We Europeans are used to this. We have voluntarily given up many of our powers to Brussels and to the European Union." Merkel lamented, however, "But our American partners find it much more difficult to hand over powers to the International Monetary Fund or to any other international organisation."[46]

On May 16, 2002, Mexican President Vicente Fox admitted that ruling-class elites intend to merge the United States, Canada and Mexico into a European Union-style super-state. Fox confirmed, "Eventually, our long-range objective is to establish with the United States, but also with Canada, our other regional partner, an ensemble of connections and institutions similar to those created by the European Union…The new framework we wish to construct is inspired in the example of the European Union."[47] Like the European Union, the North American Union (NAU) began with trade agreements, especially the North American Free Trade Agreement (NAFTA).

NAFTA, Precursor to the NAU

In September 1993, *Insight on the News* (a sister publication of the *Washington Times*) accurately described NAFTA as a viable threat to national sovereignty. Veteran columnist Samuel Francis wrote, "Negotiated in secret, slapped on a legislative 'fast track' that won't even permit its amendment in Congress, pushed by millions of Mexican dollars to

Washington public relations firms and lobbyists, NAFTA essentially points a shotgun at this country and commands marriage with a global regime in which our own identity and interests as a nation would vanish before the honeymoon was over."[48] Francis concluded, "NAFTA really is 'the first vital step' toward the political ratification of a global politico-economic regime, but what's vital to the new world order is lethal to America and any other nation that wants to remain independent. Without NAFTA or something like it, the 'global community' remains a trend and a wish, not a legally and politically enforceable fact."[49]

Francis' observations were neither radical nor controversial. Even arch-globalist Henry Kissinger openly admitted that NAFTA would facilitate the New World Order agenda. In July 1993, he wrote an article for the *Los Angeles Times* aptly called, "With NAFTA, U.S. finally creates a new world order." Kissinger wrote the article to garner support ahead of a crucial congressional vote. He stated, "What Congress will have before it is not a conventional trade agreement but the architecture of a new international system."[50] NAFTA's architects envisioned the agreement as a steppingstone towards a North American Union and, eventually, "It will represent the most creative step toward a new world order taken by any group of countries since the end of the Cold War, and the first step toward an even larger vision of a free-trade zone for the entire Western Hemisphere."[51]

Throughout 1992 and 1993, a time of intense NAFTA debate, William A. Orme served as an establishment mouthpiece, writing numerous pro-NAFTA articles for major publications. The Council on Foreign Relations, for example, featured his article "NAFTA: Myths versus Facts," in *Foreign Affairs* magazine.[52] During this time, Orme also published the book *Continental Shift* (later republished as *Understanding NAFTA*). On November 14, 1993, just days before the Congressional vote, Orme published a curious article in the *Washington Post*. At the time, NAFTA critics were fastidiously studying the larger implications of the agreement and warning about hidden agendas. Surprisingly, however, Orme openly admitted, "the critics were essentially right."[53] He explained, "When NAFTA was first proposed, critics in all three countries claimed that its hidden agenda was the development of a European-style common market...NAFTA's defenders said no. They argued that the agreement is designed to dismantle tariff barriers, not build a new regulatory bureaucracy." He continued, "Yet the critics were essentially right. NAFTA lays the foundation for a continental common market, as many of its architects privately acknowledge." Finally, he revealed NAFTA's true purpose: "Part of this foundation, inevitably, is bureaucratic: The agreement creates a variety of continental institutions — ranging from trade dispute panels to labor and environmental commissions — that are, in aggregate, an embryonic NAFTA government."[54]

Sixteen years later, the success or failure of NAFTA depends entirely upon one's perspective. From the perspective of its globalist architects, it has successfully laid the groundwork for further concentration of wealth and power, and further integration into a European Union-style North American Union. For the people of the United States, Canada and Mexico, however, the story is much different. In 2006, the Economic Policy Institute (EPI) published a detailed, sixty-page report called, "Revisiting NAFTA: Still Not Working For North America's Workers." They demonstrated that despite the promises of increased jobs and economic growth, "In each nation, workers' share of the gains from rising productivity fell and the proportion of income and wealth going to those at the very top of the economic pyramid grew."[55] They observed, "supporters claimed the gains would be so large as to more than compensate for the erosion of the average workers' bargaining power and the weakening of citizens' rights to use government to protect themselves against the insecurities of unregulated markets."[56] Nevertheless, in the U.S. alone, trade deficits have skyrocketed and more than one million jobs that otherwise would have been created have been lost.[57] The bottom line is, "NAFTA was supposed to make things a great deal better for workers, not—even a little—worse."[58] But of course, that was the plan all along.

Catherine Austin Fitts, the Assistant Secretary for Housing-Federal Housing Commissioner from 1989-1990, witnessed first-hand this betrayal of the American people during closed-door NAFTA meetings. She has since blown the whistle, stating on national radio, "I was part of a group that was saying with GATT and NATFA we're going to ship all the jobs abroad and we need to do something or else the American people are going to be left high and dry and the middle class will be wiped out…and literally a decision was made that rather than address the real problems and help the American people, we would bubble the economy and pull massive amounts of capital and reinvest abroad." Interviewer Alex Jones replied, "So this was a conscious decision and your were there on the inside witnessing this?" "Absolutely," Fitts replied.[59]

During his presidential campaign, Barack Obama promised to renegotiate NAFTA, yet three months into his presidency, he broke this promise. Reuters reported, "It is not necessary to renegotiate the North American Free Trade Agreement to honor President Barack Obama's campaign promise to add stronger labor and environmental provisions, the top U.S. trade official said on Monday."[60] Canadian President Stephen Harper was "delighted" by this decision because, "Once we reopen it, it would be very hard to get that cat back in the bag."[61]

From SPP to NAU

On March 23, 2005, President Bush, Paul Martin (Prime Minister of Canada), and Vicente Fox (President of Mexico) signed the Security and Prosperity Partnership of North America in Waco, Texas, a treasonous agreement designed to eventually usurp national sovereignty and force the countries into a North American Union.[62] The SPP website claims, "The SPP outlines a comprehensive agenda for cooperation among our three countries while respecting the sovereignty and unique cultural heritage of each nation."[63] The SPP, however, is merely a steppingstone between NAFTA and the NAU. On June 21, 2006, Representative Marcy Kaptur, addressed this progression during a speech before the House of Representatives. She stated, "Today, I want to talk a little bit about super NAFTA and what the Bush administration is planning to lock NAFTA in even tighter in this country and across the continent. There is something called the Agreement on Security and Prosperity [SPP] that is being negotiated by the Bush administration very quietly. No hearings are being held in this Congress. Most Americans have never even heard the term, but it really is the successor to NAFTA."[64] Kaptur further observed that the SPP "is being negotiated by the very same elites that negotiated NAFTA," and, "They don't care about you, they don't care about this country, they don't care about where you come from and, my friends, they don't care about democracy."[65]

Judicial Watch, the educational foundation promoting, "transparency, accountability and integrity in government, politics and the law,"[66] filed Freedom of Information Act

(FOIA) requests to obtain documents from SPP meetings held in Banff, Canada, from September 12-14, 2006. The meetings were between hundreds of bureaucrats and private-sector representatives from Chevron, General Motors, Merck, Wal-Mart, Lockheed Martin, Ford Motor Company, New York Life, UPS, FedEx, General Electric and many others.[67] Prior to the FOIA request, the SPP was keeping private the names of all SPP working group participants. Geri Word, head of the SPP office within the U.S. Department of Commerce, admitted to *World Net Daily*, "We did not want to get the contact people of the working groups distracted by calls from the public."[68] The FOIA documents revealed an incredibly intricate North American integration agenda spanning transportation, law enforcement, agriculture, economic integration, manufacturing, construction, education, immigration, and military functions. The "Detailed Agenda" from September 13, for example, included discussions concerning the planned exploitation of climate change fears to impose North American integration: "There is significant interest however in climate change and this would be an issue ripe for a North American integrated approach."[69] They also discussed implementing carbon taxes: "Long term solution cannot be a Cap & Trade system, rather a carbon tax or a technology solution."[70] Regarding security, they determined, "Today there is not necessarily a shared view of the risks but the dialog needs to continue on what the shared vision is for security of North America."[71] Most disturbingly, to overcome opposition, they proposed a strategy described as "evolution by stealth."[72]

By 2007, North American integration was progressing rapidly. Texas Congressman Ted Poe stated, "NAFTA superhighway plans exist to move goods from Mexico through the United States to Canada. It appears to be another one of the open-border philosophies that chips away at American sovereignty, all in the name of so-called trade."[73] In January 2007, Jeffrey Shane, Undersecretary of the Department of Transportation, lied during Congressional testimony regarding the NAFTA superhighway. He claimed he was "not familiar with any plan at all." Later, he dismissed the project as "an urban legend." Being from Texas, where the highway was already under construction, Poe objected, "Mr. Shane was either blissfully ignorant or he may have been less than candid with the committee."[74] Less than five months later, Kaptur revealed that NASCO (North American SuperCorridor Coalition), the organization tasked with implementing the NAFTA superhighway, "received $2.5 million in earmarks from the U.S. Department of Transportation to plan this NAFTA superhighway as a 10-lane, limited-access road, plus passenger and freight rail lines running alongside pipelines originally laid for oil and natural gas."[75]

Official Denials

During the 2007 SPP meetings in Montebello, Quebec, *Fox News* successfully confronted President Bush regarding the NAU. Their reporter asked, "Can you say today that this is not a prelude to a North American Union, similar to a European Union?" Bush dodged the question and responded with ridicule: "I'm amused by the difference between what actually takes place in the meetings and what some are trying to say takes place. It's quite comical actually when you realize the difference between reality and what some people are talking on TV about."[76]

The following year, presidential candidate Barack Obama also ridiculed NAU speculation. On the campaign trail in Lancaster, Pennsylvania, on March 31, 2008, an attendee asked Obama if he was a member of the CFR and whether or not he supported the NAU. Obama awkwardly mumbled, "Well first of all, uh, I'm not uh, the Council on Foreign Relations, uh I don't know if I'm an official member. I've spoken there before, uh, it basically is just a forum where a bunch of people talk about foreign policy." He then lied, claiming the CFR has no official membership: "Uh, and, so there's nothing, uh there's no official membership, I don't have a card." Then Obama shamefully played innocent regarding the NAU: "In terms of this North America, uh, what do you call it? Union. This has been something I know Ron Paul's talked about and people have been talking about. I have to say, with all due respect that, I see no evidence of this actually taking place. I think this has been something that has been ginned up in certain blogs on the Internet. It was based partly on the plan that there's this highway being built in Texas that will facilitate more transportation travel between Mexico and the intercontinental United States on up to Canada. And so people have perceived that this potentially means that somehow there's going to be this Union like the European Union. There's no evidence that that's taking place…I don't think there's some conspiracy to create this one, you know, continental government between Canada and Mexico. I see no evidence of that. Alright?"[77]

Opposition to the NAU

In April 2008, the American Policy Center published a study on Americans' attitudes concerning North American integration. 97 percent of respondents answered 'No' to the question, "Should the Bush administration be allowed to move forward with its plans to create a 'North American Community' without congressional approval?" 92 percent answered 'No' to the question, "Would you support efforts to replace the U.S. dollar with a common North American currency some call the Amero?"[78] While Congress has largely ignored these vital issues, the NAU is not without Congressional critics. Representative Virgil Goode, for example, publicly stated, "It will lead to a union between Canada, the United States, and Mexico. And it will greatly harm the sovereignty of the United States. It is part of the open borders philosophy to do away with borders. And I vigorously oppose it."[79] On January 22, 2007, Goode introduced House Concurrent Resolution 40, "Expressing the sense of Congress that the United States should not engage in the construction of a North American Free Trade Agreement (NAFTA) Superhighway System or enter into a North American Union with Mexico and Canada."[80] The resolution also stated that "regulatory and border security changes implemented and proposed by the SPP violate and threaten United States sovereignty." 52 Representatives signed on as co-sponsors.

In Canada, in April 2007, MP Peter Julian began organizing public hearings to expose the secretive SPP meetings. Julian stated, "We must have a full and accountable public debate and expose the entire agenda around the SPP…These issues all have a tremendous impact on Canadians and once it becomes a public debate, Canadians will be deeply concerned. That's why a lot of this has been taking place behind closed doors. It shows a basic mistrust of democracy and democratic debate."[81]

Congressman Ron Paul has also addressed the North American Union and the general New World Order ideology. During the November 28, 2007 *CNN/YouTube* Republican Presidential debate, he remarked,

"The CFR exists, the Trilateral Commission exists, and it's a 'conspiracy of ideas.' This is an ideological battle. Some people believe in globalism. Others of us believe in national sovereignty. And there is a move on toward a North American Union, just like early on, there was a move on for a European Union, and it eventually ended up. So we had NAFTA and moving toward a NAFTA highway. These are real things. It's not that somebody made these up. It's not a conspiracy. They don't talk about it, and they might not admit it, but there's been money spent on it."

The camera then panned to John McCain as he arrogantly rolled his eyes. Paul continued,

"There was legislation passed in the Texas legislature, unanimously, to put a halt on it. They're planning on millions of acres taken by eminent domain for an international highway from Mexico to Canada, which is going to make the immigration problem that much worse. So it's not so much a secretive conspiracy, it's a contest between ideologies, whether we believe in our institutions here, our national sovereignty, our Constitution, or are we going to further move into the direction of international government, more UN. You know, this country goes to war under UN resolutions. I don't like big government in Washington, so I don't like this trend toward international government. Knowledge is out there. If you look for it, you'll realize that our national sovereignty is under threat."[82]

For his remarks, Paul received a sustained, triumphant applause. And according to CNN's Election Center 2008 Debate Scorecard, Ron Paul trounced the competition.[83] A dire threat to the ruling class, they tried desperately to shun him. Yet despite paltry media attention, Paul won every debate, out-fundraised the entire field during the 4th quarter of 2007,[84] led in campaign contributions from military personnel,[85] and set the single-day fundraising record of raising $6 million on December 16, 2007.[86] Furthermore, he sparked a massive grassroots movement to audit the Federal Reserve, reform the banking system, end the wars, and restore the U.S. Constitution.

James P. Hoffa, President of the International Brotherhood of Teamsters, also disbelieved the Bush administrations assurances regarding the SPP. Speaking with *New York Times* best-selling author and Harvard Ph.D. Jerome Corsi, Hoffa stated, "I think the Bush administration has a master plan to erase all borders and to have a super-government in North America…Bush won't speak openly about the Security and Prosperity Partnership because he doesn't want anybody to know anything about it. If the American people knew openly about the plan to build superhighway toll roads with foreign money or the plan to create a North American common market and regional government, we would rally to stop it. The Bush administration hasn't even openly consulted Congress about SPP."[87] Hoffa concluded, "We've got to stop this now. Nothing is inevitable. The problem is that there is an agenda of these globalists and these free traders to ship every good job out of America."

In the mainstream media, only Lou Dobbs of CNN has worked to expose the SPP and NAU agendas. On January 18, 2007, he described ongoing efforts by "the Bush administration and corporate elites…to create a North America Union…all the while trying to deny it." Dobbs reported, "The Bush administration is pushing and pushing hard a partnership between the United States and Mexico and Canada, with a goal of what it calls integration by 2010. This partnership among three nations is being discussed at the highest levels of the three governments at the urging of the largest multinational corporations. But it is barreling ahead with absolutely no congressional oversight, no voter approval, out of sight completely of the American people, and as far as we can determine, without any constitutional authority whatsoever."[88] Regarding the SPP, fellow CNN correspondent Christine Romans stated, "The president promised no matter who the players are in Washington, Mexico City and Ottawa, this bureaucracy is meant to endure."[89] Regarding SPP secrecy, Dobbs commented, "And some mindless and uninformed people are suggesting that this is not even occurring. I saw a number of articles suggesting this North American Union just isn't happening. It's a fiction, as some wild conspiracy theory, yet it is absolutely documented. This administration continues to deny what is happening right in front of us. Although it is happening with stealth and with secrecy, it's happening."[90] Romans added, "They will not come on camera to talk about it. But they will refer us to a website. But they won't come on camera to talk about it."

Further CFR Treason

Just days after the March 2005 signing of the SPP agreement, Rafael Fernandez de Castro and Rossana Fuentes Berain, editors of the CFR's *Foreign Affairs en Español*, penned an op-ed for the *New York Times* openly calling for a European Union-style North American Union: "In time, the idea would be to create a permanent North American commission with cabinet-level representatives from each country; it would be charged not only with firming up nuts-and-bolts agreements on trade and security, but also with working toward an eventual goal of a true North American union."[91] The CFR did more than just support the North American Union—they helped plan it. In 2004, for example, the CFR launched "an independent task force on the future of North America to examine regional integration since the implementation of the North American Free Trade Agreement ten years ago."[92] CFR President Richard Haass remarked, "Ten years after NAFTA, it is obvious that the security and economic futures of Canada, Mexico, and the United States are intimately bound. But there is precious little thinking available as to where the three countries need to be in another ten years and how to get there. I am excited about the potential of this task force to help fill this void."[93] The task force was chaired by three high-level, former government officials: former Canadian Deputy Prime Minister and Minister of Finance John Manley, former Finance Minister of Mexico Pedro Aspe, and former Governor of Massachusetts and Assistant Attorney General William Weld.

In May 2005, the CFR published the task force's final report, *Building a North American Community*. A virtual blueprint for regional government, the report called for dissolving the borders between the three countries in favor of a continental security perimeter: "The Task Force proposes the creation by 2010 of a North American community to enhance security, prosperity, and opportunity...Its boundaries will be defined by a common external tariff and an outer security perimeter within which the movement of people, products, and capital will be legal, orderly, and safe."[94] The Vice-Chair of this report was Robert Pastor, a longtime Trilateral Commission operative who served as U.S. National Security Advisor on Latin America and the Caribbean during the TC-dominated Carter administration. Before that, from 1975-1977, Pastor was Executive Director of the Linowitz Commission, an organization comprised almost exclusively of Trilateral Commission members, credited with persuading President Carter to relinquish control of the Panama Canal.[95]

In 2002, Pastor appeared before a plenary session of the Trilateral Commission where he presented his paper, "A Modest Proposal To the Trilateral Commission." His so-called "modest proposal" was nothing short of a complete restructuring of the entire continent. His proposals included, for example, the introduction of "North American passports," the creation of a "Permanent Court on Trade and Investment," and the establishment of a "North American Commission." He explained, "There is much more that a North American Commission could propose—a continental plan for infrastructure and transportation, a plan for harmonizing regulatory policies, a customs union, a common currency."[96] He also advocated manipulating, and eventually dissolving, national identity: "Our three governments should sponsor Centers for North American Studies in each of our countries to help the people of all three understand the problems and the potential of North America and begin to think of themselves as North Americans."[97] His final advice to the Trilateral Commission contained glaring overtones of classical fascist propaganda: "Despite this convergence and a popular desire to experiment, the three governments have devoted so much effort to defining differences that the people have not had a chance to expand what they have in common. That is the challenge of the Trilateral Commission—to sketch an alternative future for the entire continent that the people will embrace and the politicians will feel obligated to accept."[98]

Testifying on June 9, 2005 before the U.S. Senate Foreign Relations Committee, Pastor declared, "The U.S. Congress should also merge the US-Mexican and US-Canadian inter-parliamentary groups into a single 'North American Parliamentary Group.'"[99] He also advocated the building of "North American highways" such as the aforementioned NAFTA superhighway: "The North American Council should develop an integrated continental plan for transportation and infrastructure that includes new North American highways and high-speed rail corridors."[100] Another aspect of Pastor and the CFR's regional integration plans was a mandatory National ID card and North American border pass. Pastor explained:

- "Instead of stopping North Americans on the borders, we ought to provide them with a secure, biometric Border Pass that would ease transit across the border like an E-Z pass permits our cars to speed through toll booths."
- "The passage of the Real ID Act shows that there is a growing and grudging recognition that some form of National Identification Card may be needed. Congress really ought to address this issue head-on. We should not use the driver's license, the social security card, the Medicare card, or our credit cards for anything other than the purpose for which they are intended. These cards are not intended to judge immigration status or citizenship. We will not only fail if we use them for that purpose; we will also undermine their real purpose."
- "What we need is a new approach to jointly police the perimeter, a North American border pass to facilitate travel, and a Customs Union to allow inspectors to concentrate on terrorists rather than tariffs on goods."[101]

The CFR's *Building a North American Community* recommended a North American Border Pass "with biometric identifiers."[102] Further recommendations from this report included military integration, the exploitation of fraudulent carbon trading schemes, and the use of taxpayer funds for indoctrinating students:

- "In the area of military cooperation, collaboration can proceed more slowly, especially between U.S. and Mexican militaries. However, the ultimate goal needs to be the timely sharing of accurate information and intelligence and higher levels of cooperation."[103]

- "Canada has signed the Kyoto Protocol on global climate change, which requires significant reductions in emissions of greenhouse gases, but that agreement does not cover Mexico, and Washington has opted out. A North American energy and emissions regime could offer a regional alternative to Kyoto that includes all three countries. Such a regime should include a tradable voucher system for emissions trading within the region analogous to the Clean Development Mechanism."[104]

- "We recommend that the three governments open a competition and provide grants to universities in each of the three countries to promote courses, education, and research on North America and assist elementary and secondary schools in teaching about North America."[105]

Pastor, in his book *Toward a North American Community*, also argued for the creation of a Central Bank of North America and a new North American monetary unit—the amero. "In the long run," he declared, "the amero is in the best interests of all three countries."[106] He also asked, "But are governments ready to give up their sovereignty and develop common approaches?" He answered, "A majority of the public in all three countries is willing to support North American political union if they could be convinced that it would improve their lives and not threaten their cultural identity."[107] However, Pastor lamented, "Every step toward integration seems to encounter an almost insurmountable obstacle labeled 'sovereignty.' And yet in the experience of the European Union, such debates on sovereignty have impeded negotiations only for limited periods. Eventually, the member-states find ways to deal pragmatically with a problem, laying aside the ideological debate—or rather redefining it—so as to permit a new transnational approach."[108]

In 2007, the CFR's *Foreign Affairs* published an article by Benn Steil called, "The End of National Currency." Steil argued, "The right course is not to return to a mythical past of monetary sovereignty, with governments controlling local interest and exchange rates in blissful ignorance of the rest of the world. Governments must let go of the fatal notion that nationhood requires them to make and control the money used in their territory...In order to globalize safely, countries should abandon monetary nationalism and abolish unwanted currencies, the source of much of today's instability."[109]

While the NAU is not yet complete, the globalists have already taken steps to merge the North American Union with the European Union. On April 30, 2007, the BBC reported on agreements to create a single US-EU common market: "The United States and the European Union have signed up to a new transatlantic economic partnership...The two sides agreed to set up an 'economic council' to push ahead with regulatory convergence in nearly 40 areas, including intellectual property, financial services, business takeovers and the motor industry."[110] These agreements also included provisions for carbon taxes and other carbon emission reduction schemes. The alleged threat of global warming, covered extensively in upcoming chapters, will be among the primary justifications for further regional, and eventually, global governance.

Chapter 8

The Bush Administration and Beyond

The Project for a New American Century (PNAC) was a think-tank, established in 1997, comprised of self-described "neoconservatives" including Jeb Bush (brother of George W. Bush), Dick Cheney, Donald Rumsfeld, Paul Wolfowitz, Elliot Abrams, Richard Perle, John Bolton, Scooter Libby and other Bush administration insiders. Their neoconservative ideology stemmed from the philosophy of Leo Strauss, an iconic Zionist and former University of Chicago professor of political philosophy.[1] Organized through PNAC, the Bush administration was a bees' nest, swarming with Straussian ideologues. The *New York Times* observed,

> "To intellectual-conspiracy theorists, the Bush administration's foreign policy is entirely a Straussian creation. Paul D. Wolfowitz, the deputy secretary of defense, has been identified as a disciple of Strauss; William Kristol, founding editor of *The Weekly Standard*, a must-read in the White House, considers himself a Straussian; Gary Schmitt, executive director of the Project for the New American Century, an influential foreign policy group started by Mr. Kristol, is firmly in the Strauss camp."[2]

Strauss fused the ideas of Plato, Nietzsche, and Machiavelli. His elitist doctrine was one of *might makes right* combined with *the ends justify the means*. Abram Shulsky, a Strauss disciple who played a vital role in directing Iraq prewar intelligence, published an article in 1999 called "Leo Strauss and the World of Intelligence." Shulsky remarked, "Strauss is of course famous for his doctrine (or, rather, his discovery) of 'esoteric' writing, i.e., the idea that, at least before the Enlightenment, most serious writers wrote so as to hide at least some of their thought from some of their readers." He further observed, "Strauss's view certainly alerts one to the possibility that political life may be closely linked to deception. Indeed, it suggests that deception is the norm in political life, and the hope, to say nothing of the expectation, of establishing a politics that can dispense with it is the exception."[3] Shulsky was certainly not the exception.

Historian Shadia Drury, a Fellow of the Royal Society of Canada, has studied Strauss intently and written several books decoding his cryptic philosophy. On his book *Natural Right and History*, Drury commented, "Although the Declaration of Independence appears on the cover, the thesis of the book subverts the whole idea of equality as well as the idea of natural rights. Instead, Strauss defends the natural right of the superior few to rule over the inferior many."[4] Drury asserted, "Strauss thinks that a political order can be stable only if it is united by an external threat; and following Machiavelli, he maintains that if no external threat exists, then one has to be manufactured."[5] This point is pivotal. For in September 2000, PNAC published a now-infamous report *welcoming* a massive terror attack against America to justify preplanned invasions of Afghanistan and Iraq. The report, entitled "Rebuilding America's Defenses," was an audacious outline for perpetual American hegemony. It was the antecedent to the Bush doctrine of preemptive war, with or without proof of purpose. Bush introduced this doctrine in June 2002, while addressing graduating cadets at the U.S. Military Academy. He declared, "We cannot defend America and our friends by hoping for the best. We cannot put our faith in the word of tyrants, who solemnly sign non-proliferation treaties, and then systemically break them. If we wait for threats to fully materialize, we will have waited too long."[6] But Bush was merely parroting an agenda planned years in advance and detailed in PNAC's September 2000 report. According to this report, America's "core mission" included four vital objectives:

1. "Defend the American homeland"
2. "Fight and decisively win multiple, simultaneous major theater wars"
3. "Perform the 'constabulary' duties associated with shaping the security environment in critical regions"
4. "Transform U.S. forces to exploit the 'revolution in military affairs'"[7]

PNAC argued that middle-eastern nations, especially Iraq and Iran, have no right to sovereignty. Therefore, America must dominate the entire region. Should these countries refuse this arrangement, American military forces would ensure their compliance: "We cannot allow North Korea, Iran, Iraq or similar states to undermine American leadership...Adversaries like Iran, Iraq and North Korea are rushing to develop ballistic missiles and nuclear weapons as a deterrent to American intervention in regions they seek to dominate."[8] Imagine the audacity of these nations, seeking to "dominate" their *own* land.

The report also made clear that America's impending Iraq invasion, attributed to the "threat" of Saddam Hussein, would simply be a means to an end, that end being permanent occupation: "While the unresolved conflict with Iraq provides the immediate justification, the need for a substantial American force presence in the Gulf transcends the issue of the regime of Saddam Hussein."[9] Furthermore, following the swift ousting of Hussein, a long, violent, military occupation, or what PNAC called "constabulary missions," would begin: "Thus, facing up to the realities of multiple constabulary missions will require a permanent allocation of U.S. armed forces...Further, these constabulary missions are far more complex and [more] likely to generate violence than traditional 'peacekeeping' missions."[10]

PNAC had just one obstacle—the American people. Would Americans really go along with such savagery, the wholesale slaughter of millions of Iraqi and Afghani lives for cheap oil and imperial bravado? PNAC's options were two:

1. They could boldly execute their mission without public support by adopting Madeline Albright's infamous "the price is worth it" justification. On May 12, 1996, Secretary of State Albright showcased her repugnant moral depravity on CBS's *60 Minutes*. Speaking about the aftermath of the first Iraq war, Lesley Stahl asked Albright, "We have heard that a half million children have died. I mean that's more children than died in Hiroshima. And, you know, is the price worth it?" Albright responded coldly, "I think this is a very hard choice, but we think the price is worth it."[11]
2. According to Straussian doctrine, they could stage a terror attack to justify the mission. Indeed, the PNAC document stated, "The process of transformation, even if it brings revolutionary change, is likely to be a long one, absent some catastrophic and catalyzing event—like a new Pearl Harbor."[12]

Were the events of September 11, 2001 simply a magnificent coincidence? Gustave Gilbert was the prison psychologist for Nazi officers during the Nuremberg Trials. In 1947, he published *Nuremberg Diary*, documenting numerous conversations and interviews. Speaking with arch-Nazi Hermann Goering, Gilbert asked if the German people wanted war. "Why, of course, the *people* don't want war," Goering admitted. He continued,

"Why would some poor slob on a farm want to risk his life in a war when the best that he can get out of it is his farm in one piece? Naturally, the common people don't want war, neither in Russia nor in England nor in America, nor for that matter in Germany. That is understood. But, after all, it is the leaders of the country who determine the policy and it is always a simple matter to drag the people along, whether it is a democracy or a fascist dictatorship or a Parliament or a Communist dictatorship."[13]

Gilbert interjected, "There is one difference. In a democracy the people have some say in the matter through their elected representatives and in the United States only Congress can declare war." Goering replied, "Oh, that is all well and good, but, voice or no voice, the people can always be brought to the bidding of the leaders. That is easy. All you have to do is tell them they are being attacked and denounce the pacifists for lack of patriotism and exposing the country to danger. It works the same way in any country."[14]

The core of the PNAC cabal had been festering in Washington for decades. Ray McGovern, a 27-year CIA veteran who prepared Presidential Daily Briefs for both the Reagan and H.W. Bush White Houses, worked around these individuals. McGovern explained that during the 1980s, in certain Washington circles, the PNAC boys were known as "the crazies." In an article published by the *Asia Times*, he wrote,

"Heady in the afterglow of victory in the Gulf War of 1991, the 'crazies' stirred up considerable controversy when they articulated their radical views. Their vision, for instance, became the centerpiece of the draft 'Defense Planning Guidance' that Paul Wolfowitz, de facto dean of the neo-conservatives, prepared in 1992 for then-defense secretary Dick Cheney. It dismissed deterrence as an outdated relic of the Cold War and argued that the U.S. must maintain military strength beyond conceivable challenge — and use it in preemptive ways in dealing with those who might acquire 'weapons of mass destruction.' Sound familiar?"[15]

McGovern continued, "In 2001, the new Bush brought the neo-cons back and put them in top policymaking positions…They are much more dangerous now. Unlike in the 1980s, they are the ones crafting the adventurous policies our sons and daughters are being called on to implement."

Thus long before the 2000 election, Bush's handlers were already planning the invasion of Iraq. In the months and weeks leading up to the March 2003 invasion, the world endured ceaseless posturing from Bush administration officials regarding the imminent threat posed by Saddam Hussein. During an August 26, 2002 speech, for example, Dick Cheney confidently asserted, "Simply stated, there is no doubt that Saddam Hussein now has weapons of mass destruction. There is no doubt that he is amassing them to use against our friends, against our allies and against us."[16] On February 5, 2003, Secretary of State Colin Powell addressed the United Nations Security Council in New York. His speech included the following statements:

- "Every statement I make today is backed up by sources, solid sources. These are not assertions. What we're giving you are facts and conclusions based on solid intelligence."
- "Our conservative estimate is that Iraq today has a stockpile of between 100 and 500 tons of chemical weapons agent."
- "Saddam Hussein is determined to get his hands on a nuclear bomb."

- "When we confront a regime that harbors ambitions for regional domination, hides weapons of mass destruction and provides haven and active support for terrorists, we are not confronting the past, we are confronting the present. And unless we act, we are confronting an even more frightening future."[17]

And just three days before the invasion, during a televised address to the nation, Bush confidently asserted, "Intelligence gathered by this and other governments leaves no doubt that the Iraq regime continues to possess and conceal some of the most lethal weapons ever devised. This regime has already used weapons of mass destruction against Iraq's neighbors and against Iraq's people. The regime has a history of reckless aggression in the Middle East. It has a deep hatred of America and our friends and it has aided, trained and harbored terrorists, including operatives of Al Qaeda."[18]

Such firm convictions and unfettered resolve can only arise from one of two sources: truth or lies. Such rhetoric does not allow for any middle ground. Yet the administration would later claim that intelligence breakdowns generated faulty data. If this were true, it would be the biggest "intelligence failure" ever. Or was that September 11, 2001? The absurdity of experiencing *the two biggest* intelligence failures in history just 18 months apart should have been reason enough to convene an independent investigation, sufficiently funded and with full subpoena powers. But since that never happened, the American public can only evaluate the available evidence—from an analytical, logical and historical perspective—and use common sense to understand the past and its undeniable influence upon the present and future. Are these men and women, as they claim, innocent of war crimes? They claim to have acted based on faulty intelligence. So from where did that intelligence arise?

Office of Special Plans

In the days immediately following the September 11 attacks, Wolfowitz and fellow neoconservative Douglas Feith, the Under Secretary of Defense for Policy, established for the Pentagon an agency called the Office of Special Plans (OSP). Its purported objective was to determine what threat Saddam Hussein might pose to America and to develop preliminary Iraq invasion plans. They selected the aforementioned Abram Shulsky, the self-confessed advocate of political deception, as OSP Director. The fallacy of the OSP was its redundancy. It was completely unnecessary. The two existing intelligence agencies, the CIA and the DIA (the Pentagon's Defense Intelligence Agency) were already adequate and sufficient.

So why create another intelligence agency? Quite simply, the existing structure was problematic for the PNAC cabal because they could not fully control these agencies. Writing for the *New Yorker*, veteran journalist Seymour Hersh explained, "Their operation, which was conceived by Paul Wolfowitz, the Deputy Secretary of Defense, has brought about a crucial change of direction in the American intelligence community. These advisers and analysts, who began their work in the days after September 11, 2001, have produced a skein of intelligence reviews that have helped to shape public opinion and American policy toward Iraq."[19] Citing an anonymous Pentagon source who had worked with OSP, Hersh explained, "Special Plans was created in order to find evidence of what Wolfowitz and his boss, Defense Secretary Donald Rumsfeld, believed to be true—that Saddam Hussein had close ties to Al Qaeda, and that Iraq had an enormous arsenal of chemical, biological, and possibly even nuclear weapons that threatened the region and, potentially, the United

States." Former CIA Director of Intelligence Programs Vincent Cannistraro confirmed Hersh's report. Referring to Shulsky and his PNAC cohorts, Cannistraro stated,

> "His politics were typical for his group—the Straussian view. They reinforce each other because they're the only friends they have, and they all work together. This has been going on since the nineteen-eighties, but they've never been able to coalesce as they have now. September 11th gave them the opportunity, and now they're in heaven. They believe the intelligence is there. They want to believe it. It *has* to be there."[20]

Another former CIA agent who worked with OSP told Hersh, "One of the reasons I left was my sense that they were using the intelligence from the CIA and other agencies only when it fit their agenda. They didn't like the intelligence they were getting, and so they brought in people to write the stuff. They were so crazed and so far out and so difficult to reason with—to the point of being bizarre. Dogmatic, as if they were on a mission from God. If it didn't fit their theory, they don't want to accept it." Furthermore, another former CIA analyst, Larry Johnson, told Scotland's *Sunday Herald* that OSP was "dangerous for U.S. national security and a threat to world peace." He added, "[It] lied and manipulated intelligence to further its agenda of removing Saddam. It's a group of ideologues with predetermined notions of truth and reality. They take bits of intelligence to support their agenda and ignore anything contrary. They should be eliminated"[21]

Caught Red Handed

While the above anecdotal accounts cast great suspicion upon the official prewar narrative, two pieces of physical evidence, unfortunately discovered *after* the invasion, completely destroyed the Bush administration's credibility. One is known as the Downing Street memo and the other, the White House memo. On May 1, 2005, the *London Times* published a leaked memo featuring the minutes of a July 23, 2002 Downing Street meeting between Tony Blair and several top advisors. Richard Dearlove, then head of Britain's MI6 intelligence agency, had just been to Washington to meet with Bush administration officials. At the Downing Street meeting, Dearlove summarized the Bush administration's position on Iraq. The memo read, "There was a perceptible shift in attitude. Military action was now seen as inevitable. Bush wanted to remove Saddam, through military action, justified by the conjunction of terrorism and WMD. But the intelligence and facts were being fixed around the policy. The NSC had no patience with the UN route, and no enthusiasm for publishing material on the Iraqi regime's record. There was little discussion in Washington of the aftermath after military action."[22] Fixing intelligence? Fixing the facts? When facts are "fixed," they cease to be facts. Was the Bush administration manufacturing evidence to justify its predetermined war of aggression? The Blair government did not deny the memo's authenticity; it was legitimate. Blair, however, claimed it was just a big misunderstanding. Not convincing anyone, he claimed that while the memo said *fixing*, it actually meant *not fixing*. "No, the facts were not being fixed in any shape or form at all," Blair asserted.[23]

Bush's defense, even more pathetic, also completely avoided the obvious question—why did the memo explicitly say, "The intelligence and the facts were being fixed around the policy?" Bush reasoned, "Look, both of us didn't want to use our military. Nobody wants to commit military into combat. It's the last option." Bush downplayed the memo's significance, contending that the leak was politically motivated because "they dropped it out in the middle of his race."[24] Indeed, Blair was facing an election just four days before the story broke. Unquestionably, it *was* politically motivated. With such incriminating evidence, this was the only prudent and responsible course of action. The intent could be nothing short of destroying Tony Blair politically, and Bush as well, and hopefully arresting and trying them both for conspiracy to commit war crimes. When one discovers that the Prime Minister of his country has colluded with the President of the United States to commit genocide, the immediate objectives must be impeach, indict and arraign.

While the lapdog media apathetically licked its paws, the widespread animosity generated by the memo, especially across the Internet and alternative media, did force some congressional action. On July 16, 2005, more than thirty members of Congress convened a public hearing to investigate the memo. Representative John Conyers stated, "The Downing Street minutes outline this administration's deliberate intention to manipulate intelligence in order to justify a predetermined policy of war against Iraq."[25] Representative Maxine Waters stated, "We all believe — many of us believe that there's been a manipulation of intelligence information, that they had to make the intelligence fit the conclusion that they were indeed going to invade Iraq."[26] Ray McGovern stated,

"Now in a 27 year career in intelligence, one encounters many examples of attempts to trim the truth, or as the British minutes put it, 'Fix the intelligence and facts around the policy.' It's in the woodwork. It's part of the political scene. But I had never known fixing to include the Vice-President arrogating the right to turn a key piece of intelligence on its head. Nor had I in all those years ever known a sitting Vice-President to make multiple visits to CIA headquarters to make sure the fix was in. And this is just one example. There is no word to describe the reaction of us intelligence officers, active and retired, to the reality that our intelligence community managers were eager to participate in this deceit and to this deliberate subversion of the oath we all take to protect and defend the constitution of the United States."[27]

McGovern had previously written an op-ed declaring, "The intelligence was not simply mistaken; it was manufactured, with the president of the United States awarding foreman George Tenet the Medal of Freedom for his role in helping supervise the deceit. The British documents make clear that this was not a mere case of 'leaning forward' in analyzing the intelligence, but rather mass deception — an order of magnitude more serious. No other conclusion is now possible."[28]

The Downing Street memo stands on its own as significantly incriminating evidence. Coupled with the numerous accounts from intelligence community insiders, the PNAC crowd and the Bush Cabinet are certainly culpable for war crimes. But in February 2006, major international news outlets reported on yet another memo, equally incriminating, proving premeditated intelligence fixing. Britain's *Channel 4* news acquired the minutes to a January 31, 2003 meeting at the White House between Bush and Blair. The memo made clear that Bush had already decided on going to war. David Manning, Blair's chief foreign policy adviser at the time, wrote in the memo, "The start date for the military campaign was now penciled in for 10 March. This was when the bombing would begin."[29]

Although Bush was resolute, he was also highly distressed. The *New York Times* reported, "The memo also shows that the president and the prime minister acknowledged that no unconventional weapons had been found inside Iraq. Faced with the possibility of not finding any before the planned invasion, Mr. Bush talked about several ways to provoke a confrontation."[30] As one possibility, Bush suggested, "The U.S. was thinking of flying U2 reconnaissance aircraft with fighter cover over Iraq, painted in UN colors. If Saddam fired on them, he would be in breach."[31] Bush also stated, "The U.S. might be able to bring out a defector who could give a public presentation about Saddam's W.M.D." Manning asserted, "Our diplomatic strategy had to be arranged around the military planning."[32] The *Times* confirmed the authenticity of the memo with two senior British officials who declined to comment further, citing Britain's Official Secrets Act.

The Case for War Crimes

Another important factor in the case for war crimes is the perpetrators' change in rhetoric *after* the invasion began. In May 2006, McGovern confronted Donald Rumsfeld during a speech in Atlanta. McGovern stated, "I would like to ask you to be upfront with the American people—why did you lie to get us into a war that was not necessary and that has caused these kinds of casualties? Why!" Rumsfeld callously replied, "I'm not in the intelligence business. They gave the world their honest opinion. It appears that there were not weapons of mass destruction there." McGovern rebutted, "You said you knew where they were." Rumsfeld replied, "I did not. I said I knew where suspect sights were." McGovern stormed back, "You said you knew where they were—near Tikrit, near Baghdad, and north, east, south and west of there. Those are *your* words." Rumsfeld stumbled, but managed to divert the conversation.[33]

Two years before this confrontation, Bush, the undisputed master of imperial hubris, used the nonexistent WMDs as material for a shameful stand-up comedy routine. The *New York Times* reported, "Back in March 2004 President Bush had a great time displaying what he felt was a hilarious set of photos showing him searching the Oval Office for the weapons of mass destruction that hadn't been found in Iraq. It was a spoof he performed at the annual dinner of the Radio and Television Correspondents' Association."[34] While he displayed pictures of himself looking behind curtains and under furniture, Bush commented, "Those weapons of mass destruction have got to be somewhere. Nope, no weapons over there—maybe under here?"

In November 2003, Pentagon hawk and PNAC member Richard Perle confessed that, while the invasion certainly was *illegal*, it was nevertheless the right thing to do. "I think in this case international law stood in the way of doing the right thing," Perle told a London audience.[35] And on July 23, 2003, Wolfowitz, one of the chief-architects of the entire PNAC agenda, appeared on NBC's *Meet the Press* to confess that prewar intelligence was "murky." The *New York Times* reported, "Wolfowitz, defending the Bush administration's justification of the Iraq war, said today that intelligence on terrorism is by its nature 'murky,' and that the United States may have little choice in the future but to 'act on the basis of murky intelligence' if terror attacks are to be prevented."[36]

The Erosion of Civil Liberties

Just three days before Wolfowitz's appearance on *Meet the Press*, former CIA director John Deutch appeared before the House Intelligence Committee to speak about prewar intelligence. "It is an intelligence failure in my judgment of massive proportions," Deutch asserted. "It means our leaders and the American public based its support for the most serious foreign policy judgments—the decision to go to war—on an incorrect intelligence judgment."[37] Host Tim Russert asked Wolfowitz to comment on Deutch's statements.

Wolfowitz replied, "I think people should be a little careful about throwing around words like intelligence failure. It's easy to go around and play this blame game. I mean, let's stop and realize that in a country like Iraq, and let me repeat, where children are tortured to make their parents talk, secrets are kept in a way we can't even imagine."[38]

Despite Wolfowitz's emotional appeal, the Bush administration itself would soon assume the authority to torture children. The administration *explicitly* declared that interrogators, having received presidential approval, are permitted to crush the testicles of children to acquire information about their parents. On December 1, 2005, John Yoo, the former legal advisor to the President, debated Doug Cassel, the director of Notre Dame Law School's Center for Civil and Human Rights. Cassel asked Yoo, "If the president deems that he's got to torture somebody, including by crushing the testicles of the person's child, there is no law that can stop him?"[39] Yoo confirmed there was no law, adding, "I think it depends on why the president thinks he needs to do that."[40]

Yoo was Bush's most influential legal advisor and the author of portions of the Patriot Act,[41] and numerous unconstitutional memos authorizing torture,[42] warrantless domestic spying,[43] and the denial of protections guaranteed under the Geneva Conventions. According to the administration, these atrocities could be justified simply by labeling someone, American citizens included, as an "enemy combatant."[44] Geoffrey Stone, a law professor at the University of Chicago, writing about Yoo for the *Washington Post*, stated, "His conception of our Constitution—and that of the Bush administration—must be resoundingly repudiated by Congress, the courts and the American people."[45]

The Bush administration's contempt for the Constitution is almost unfathomable. And Obama, having failed to repeal Bush's revisions, exhibits the same contempt for America's political bedrock. Nine days after the terror attacks of 9/11, Bush appeared before an extraordinary joint session of Congress to declare, "Tonight, we are a country awakened to danger and called to defend freedom."[46] The terrorists, he contended, attacked America because "they hate our freedoms." In a perverse way, he was not lying. If a rogue, criminal network within the government planned and executed the attacks to justify multiple "theatre wars," the steady dismantling of the country, and the march towards global government then yes, absolutely, "they hate our freedoms." But Bush was implying that some mysterious Islamic cavemen were the haters. Bush asserted, "They hate what they see right here in this chamber: a democratically elected government. Their leaders are self-appointed. They hate our freedoms: our freedom of religion, our freedom of speech, our freedom to vote and assemble and disagree with each other."[47]

At the same time, however, Bush and his team were preparing an unprecedented *eradication* of freedoms guaranteed under the Constitution of the United States, an Orwellian removal of freedoms under the pretense of *protecting* freedoms. The so-called Patriot Act, signed into law October 26, 2001, would become the first flagrant abuse. Pulitzer Prize winning *New York Times* columnist William Safire, a former presidential speechwriter and recipient of the Presidential Medal of Freedom, declared that Bush "has just assumed what amounts to dictatorial power to jail or execute aliens. Intimidated by terrorists and inflamed by a passion for rough justice, we are letting George W. Bush get away with the replacement of the American rule of law with military kangaroo courts."[48] Safire dismissed the standard rhetoric that extreme circumstances call for extreme measures:

> "Those are the arguments of the phony-tough. At a time when even liberals are debating the ethics of torture of suspects—weighing the distaste for barbarism against

the need to save innocent lives — it's time for conservative iconoclasts and card-carrying hard-liners to stand up for American values."

In the immediate aftermath of the attacks, Safire was an isolated voice, barely audible against the deafening backdrop of timidity and submission. Hermann Goering's words were never more true: "The people can always be brought to the bidding of the leaders — all you have to do is tell them they are being attacked." Of course he also advised, "Denounce the pacifists for lack of patriotism and exposing the country to danger."[49]

Simply put, the 4th quarter of 2001 was history repeating itself, blind to the lessons of the past. The Patriot Act passed both houses of Congress by a combined 455-67 vote. Predictably, the nay voters were indeed denounced for lack of patriotism. Ron Paul observed, "The insult is to call this a 'patriot bill' and suggest I'm not patriotic because I insisted upon finding out what is in it and voting no. I thought it was undermining the Constitution, so I didn't vote for it — and therefore I'm somehow not a patriot. That's insulting."[50] As Paul explained, he insisted upon finding out what was actually *in* the bill. The executive branch and the House and Senate leadership, however, were firmly against allowing members of Congress to read the bill. The urgency was supposedly so great that time did not permit for such inconsequential meandering. Paul explained, "The bill wasn't printed before the vote — at least I couldn't get it. They played all kinds of games, kept the House in session all night, and it was a very complicated bill. Maybe a handful of staffers actually read it, but the bill definitely was not available to members before the vote."

What does the Patriot Act say? According to the American Civil Liberties Union, the Patriot act assaults the 1st, 4th and 5th amendments of the Constitution. Perhaps the most abhorrent part was section 213, the so-called "sneak and peak" provision of the act, which directly violated the 4th Amendment. With this provision, the government assumed the power to enter private homes and business without prior warning and without obtaining search warrants. Nadine Strossen, President of the ACLU and professor of law at New York University, told *Insight*,

> "The sneak-and-peek provision is just one that will be challenged in the courts. We're not only talking about the sanctity of the home, but this includes searches of offices and other places. It is a violation of the Fourth Amendment and poses tremendous problems with due process. By not notifying someone about a search, they don't have the opportunity to raise a constitutional challenge to the search."[51]

Anthony Romero, Executive Director of the ACLU, explained, "The sneak-and-peek searches allow them to come into your home, search your personal effects, download information from your computer and not tell you until after the fact. The law states that notice must be provided within 'a reasonable time frame,' but it provides no guidance as to what is reasonable."[52]

Section 215 violated both the 1st and 4th Amendments. It granted the FBI sweeping powers to obtain citizens' personal and business records. They could peruse and collect library records, information on Web-surfing habits, medical records, all business, financial and employment records, and even genetic information.[53] Additionally, the government could place gag orders on those parties who were forced to surrender the records. In other words, if the government wanted to know what books you had been reading, they could acquire your library records, and concurrently forbid your librarian from notifying you about the acquisition. Caroline Fredrickson, Director of ACLU's Washington Legislative

Office, explained, "Innocent people will likely never learn that their sensitive financial, medical, Internet or library records have been gathered. Even if a person were charged with a crime, he or she might never learn the basis for the search in the first place."[54]

Sections 202 and 216 enabled law enforcement agencies to monitor, without obtaining warrants, all Internet, email and phone activities of any citizen suspected of a crime.[55] Bush apologists argued that suspects *should* be monitored, even without a warrant. However, the crafty legalese of Section 802 ensured that *everyone* could be a suspected "terrorist." This provision defined terrorism as any "activity that involves acts dangerous to human life that violate the laws of the United States or any state and appear to be intended: (i) to intimidate or coerce a civilian population; (ii) to influence the policy of a government by intimidation or coercion; or (iii) to affect the conduct of a government by mass destruction, assassination or kidnapping." The subjectivity of words like *appear* and *intimidate* are particularly troubling.

Preparing for Martial Law

Two more bills, each signed into law October 17, 2006, effectively put the United States under martial law. The Military Commissions Act of 2006 (MCA06) completely undermined the Constitution by revoking the writ of habeas corpus, an 800-year tradition stretching back to the Magna Carta, established in 1215 in England. The Latin meaning of habeas corpus is "you shall have the body." Legally, habeas corpus guarantees due process; anyone accused of a crime can defend himself in court. Judge Andrew Napolitano, a life-tenured Superior Court Judge of New Jersey and Fox News analyst, explained,

> "The guarantee of habeas corpus is the part of the Constitution that prevents the government from swooping in, under the cover of night, abducting us out of our beds and confining us in a cell for the rest of our lives, or even for a day, without ever knowing what we did or having the opportunity to compel the government to prove our guilt. It is the foundation of all other rights, because without it, we would have no right to challenge unlawful infringements upon our privacy, free speech, religion, and due process."[56]

The Bush administration ended habeas corpus by declaring that the President can label *anyone*, citizens included, as an "enemy combatant," effectively stripping all other rights protected under the Constitution and international law. Napolitano explained,

> "Enemy combatants are exempt from the Geneva Convention protections against torture and inhumane treatment. MCA06 gives the military commissions exclusive jurisdiction over unlawful enemy combatants. A Combatant Status Review Tribunal 'or another tribunal established under the authority of the President or Secretary of Defense' decides whether a prisoner is an unlawful enemy combatant. The bill defines unlawful combatants as *anyone* who has 'engaged in hostilities, or purposefully and materially supported hostilities against the United States.' So, anyone a tribunal convened by the president decides has 'materially supported hostilities' against the United States may be detained indefinitely without charge at Gitmo and has no right to challenge the detention in the U.S. courts. And if the 'support' was mental in nature only, you, my dear reader, can lose your freedom because of your thoughts: no charges, no judge, no lawyer, no trial, no witness, no evidence. Just permanent punishment because of *what the president claims are your illegal thoughts*."[57]

The *New York Times* editorialized, "The law does not apply to American citizens, but it does apply to other legal United States residents."[58] Constitutional law scholar and George Washington University professor Jonathan Turley, however, confirmed that MCA06 *does indeed* apply to American citizens. On MSNBC's Countdown, Keith Olbermann asked Turley, "Does that not basically mean that if Mr. Bush or Mr. Rumsfeld say so, anybody in this country, citizen or not, innocent or not, can end up being an unlawful enemy combatant?" Turley replied, "It does. And it's a huge sea change for our democracy. The framers created a system where we did not have to rely on the good graces or good mood of the president. In fact, Madison said that he created a system essentially to be run by devils, where they could not do harm, because we didn't rely on their good motivations. Now we must. And people have no idea how significant this is. What, really, a time of shame this is for the American system. What the Congress did and what the president signed today essentially revokes over 200 years of American principles and values."[59]

In fact, years before MCA06 even became official law, it was already policy under the Bush administration. For example, José Padilla, an American citizen, was arrested in 2002 for allegedly plotting terror attacks. In 2005, PBS reported, "President Bush labeled Padilla an 'enemy combatant,' allowing the U.S. military to detain him indefinitely with limited access to attorneys. Padilla then sat in a Navy brig in South Carolina for more than three years without ever being formally charged with a crime, until yesterday, when Attorney General Alberto Gonzales announced the U.S. Government was indicting Padilla for supporting terror campaigns in Afghanistan and elsewhere between 1993 and 2001."[60] Padilla is now serving a 17-year sentence, essentially for thought crime. Paul Craig Roberts, former Editor of the *Wall Street Journal* and Assistant Secretary of the Treasury in the Reagan Administration, commented,

> "José Padilla's conviction on terrorism charges on August 16 was a victory, not for justice, but for the U.S. Justice Department's theory that a U.S. citizen can be convicted, not because he committed a terrorist act but for allegedly harboring aspirations to commit such an act. By agreeing with the Justice Department's theory, the incompetent Padilla Jury delivered a deadly blow to the rule of law and opened Pandora's Box…The U.S. Constitution and Anglo-American legal tradition prevent indictments, much less convictions, based on a prosecutor's theory that a person wanted to commit a crime in the past or might want to in the future. Padilla has harmed no one. There is no evidence that he made an agreement with any party to harm anyone whether for money or ideology or any reason."[61]

The John Warner Defense Authorization Act (DAA) killed yet another longstanding tradition, that of deploying the military only for foreign operations. The Insurrection Act of 1807 severely restricted the president's ability to deploy the military *within* the United States. The Posse Comitatus Act of 1878 tightened even further these restrictions. But the DAA struck down both of these acts, enabling the President to declare a public emergency and station troops anywhere in America to "restore pubic order." What constitutes a public emergency was extremely flexible including, "a natural disaster, epidemic, or other serious public health emergency, terrorist attack or incident, or other condition."[62] What is an "other condition?" The DAA also enabled the president to mobilize state National Guard units without the consent of state governors in order to "suppress public disorder." Senator Patrick Leahy protested that martial law provisions were slipped into the bill without allowing adequate time to study their implications. "Other congressional committees with jurisdiction over these matters had no chance to comment, let alone hold hearings on, these proposals," he explained.[63] Leahy opposed the bill, noting, "we certainly do not need to make it easier for Presidents to declare martial law."[64]

A National Security Presidential Directive (NSPD) is an executive order issued directly by the President on the advice of the National Security Council. On May 4, 2007, Bush signed NSPD-51 to ensure "continuity of government" in the case of a "catastrophic emergency." Commenting on the directive, former Congressman Dan Hamburg explained, "Should the president determine that such an emergency has occurred, he and he alone is empowered to do whatever he deems necessary to ensure 'continuity of government.' This could include everything from canceling elections to suspending the Constitution to launching a nuclear attack. Congress has yet to hold a single hearing on NSPD-51."[65] Congress was not even allowed to see the document. The White House released the cover page only. Even members of the Congressional Homeland Security Committee were denied. One such member, Peter DeFazio, addressed the Congress regarding the administration's reckless behavior. He stated,

"Most Americans would agree that it would be prudent to have a plan to provide for the continuity of government and the rule of law in case of a devastating terrorist attack or natural disaster, a plan to provide for the cooperation, the coordination and continued functioning of all three branches of the government. The Bush administration tells us they have such a plan. They have introduced a little sketchy public version that is clearly inadequate and doesn't really tell us what they have in mind, but they said, don't worry; there's a detailed classified version. But now they've denied the entire Homeland Security committee of the United States House of Representatives access to their so-called detailed plan to provide for continuity of government. They say, trust us. Trust us? The people who brought us Katrina to be competent in face of a disaster? Trust us? The people who brought us warrantless wiretapping and other excesses eroding our civil liberties? Trust us?"[66]

Why did Germans *allow* the Nazis to become so oppressive? Why did they submit to such atrocious tyranny? They submitted for the same reasons sheep submit to shepherds who move them from field to field. Everything progresses so gradually and that the sheep fail to recognize what's happening. Laws are written into the books, collecting dust for years, until people forget they ever existed. Ironically, dictators are usually not criminals— according to the laws they write for themselves. Hitler preemptively rewrote German laws years before he ever enacted them. Americans should question why Bush enacted such laws, and more importantly, why Obama hasn't retracted them? Power passes slowly between generations of ruling-class would-be tyrants. They are working through a long-term, intergenerational business plan, what Freemasonry refers to as "the great work."[67]

Training the Sheep

In May 2006, the *Alex Jones Show* obtained internal Federal Emergency Management Agency (FEMA) documents detailing the existence of so-called "Clergy Response Teams." FEMA had been secretly recruiting and training thousands of preachers to pacify their congregations in preparation for martial law. In August 2007, Shreveport, Louisiana local news station KSLA-12 confirmed the story and further revealed that it was already operational during the aftermath of hurricane Katrina. The KSLA-12 reporter stated, "If martial law were enacted here at home, like depicted in the movie *The Siege*, easing public fears and quelling dissent would be critical and that's exactly what the Clergy Response Team, as it's called, helped accomplish in New Orleans."[68] The reporter further remarked, "For the clergy, one of the biggest tools that they will have in helping calm the public down or obey the law is the Bible itself, specifically Romans, Romans 13." Romans 13 allows for creative interpretations, which justify the supreme authority of the state. Durell Tuberville, a pastor enrolled in FEMA's manipulative program, reasoned, "The government's established by the lord, and that's what we believe in the Christian faith, that's what's stated in the scripture."[69] Not surprisingly, Hitler also invoked Romans 13 against German Christians.[70]

Another aspect of the martial law takeover broke in March 2008 when *The Progressive* published a story about InfraGard, a public-private partnership between the FBI and 23,000 business executives. InfraGard, on its website, describes itself as "an information sharing and analysis effort serving the interests and combining the knowledge base of a wide range of members. At its most basic level, InfraGard is a partnership between the FBI and the private sector."[71] Its stated goal is "to promote ongoing dialogue and timely communication between members and the FBI. InfraGard members gain access to information that enables them to protect their assets and in turn give information to government that facilitates its responsibilities to prevent and address terrorism and other crimes." A whistleblower from within the organization, however, revealed that InfraGard is much more than it publicly discloses. In the event of martial law in America, according to *The Progressive*, the FBI has given InfraGard members "shoot to kill" permission to protect their assets.[72] To join, prospective members must be sponsored by "an existing InfraGard member, chapter, or partner organization." InfraGard members also receive sensitive intelligence information, unavailable to the public. Phyllis Schneck, Chairman of the Board of Directors of the InfraGard National Members Alliance, explained, "We get very easy access to secure information that only goes to InfraGard members. People are happy to be in the know."

Martial law plans further escalated in September 2008 when the *Army Times* announced that starting in October, an active-duty army brigade (of 4,700 troops)[73] would be stationed within the United States to "help with civil unrest and crowd control or to deal with potentially horrific scenarios such as massive poisoning and chaos in response to a chemical, biological, radiological, nuclear or high-yield explosive, or CBRNE, attack."[74] But "unruly individuals" need not worry; the army doesn't want to kill you, only subdue you with nonlethal weapons. The article explained, "The 1st BCT's soldiers also will learn how to use 'the first ever nonlethal package that the Army has fielded,' 1st BCT commander Col. Roger Cloutier said, referring to crowd and traffic control equipment and nonlethal weapons designed to subdue unruly or dangerous individuals without killing them."[75] What does the package include? "The package includes equipment to stand up a hasty road block; spike strips for slowing, stopping or controlling traffic; shields and batons; and, beanbag bullets." Does the army know something is coming? Are they preparing to engage the American people? While the initial brigade has a 12-month tour, "expectations are that another, as yet unnamed, active-duty brigade will take over and that the mission will be a permanent one."[76]

Shortly after the *Army Times* article, the *Washington Post* reported on the army's intention to station 20,000 troops inside the country by 2011. The ACLU and the Cato Institute sharply criticized these so called "rapid-reaction forces," calling them "just the first example of a series of expansions in presidential and military authority," and the "creeping militarization" of homeland security, respectively.[77] By August 2009, the military sought even more power by approaching Congress with a bill "authorizing the Secretary of Defense to order any unit or member of the Army Reserve, Air Force Reserve, Navy Reserve, and the Marine Corps Reserve, to active duty for a major disaster or emergency."[78] According to Paul Stockton, Assistant Secretary of Defense for Homeland Defense and America's Security Affairs, the proposed legislation would enable the deployment of "more than 379,000 military personnel in thousands of communities across the United States."

One month later, during the G20 Summit in downtown Pittsburgh, the military openly clashed with the American people. As reported by the Department of Defense, "More than 2,500 Pennsylvania National Guardsmen have been called up to support the international G-20 Summit."[79] During the summit, the military supported police in their savage attacks against peaceful protesters. According to the *Associated Press*, "Police used all the nonlethal tools at their disposal to thwart protesters at the Group of 20 summit this week, firing bean bags, hurling canisters of smoke and pepper spray, using flash-bang grenades and batons and deploying a high-tech sound-blasting device meant to push back crowds."[80] The sound weapon unleashed upon the public was the Long Range Acoustical Device (LRAD). According to the American Speech-Language-Hearing Association, the LRAD can exceed 150 decibels, the equivalent of "peak rock music, firearms, or a jet engine."[81] Any sound greater than 80 decibels can damage hearing. According to the Vic Walczak, legal director of the American Civil Liberties Union, "the deployment of police seems to be more geared toward suppressing lawful demonstrations than actually preventing crime." He added, "In a week when we need freedom of speech more than ever, free speech died in Pittsburgh this week."[82]

In October 2008, the country slid dangerously close to martial law when the banking establishment and the Secretary of Treasury pressured Congress into passing the Troubled Asset Relief Program (TARP), the $700 billion banker bailout. The House of Representatives rejected the first version of the bill (H.R. 3997) on September 29. Senate leaders, however, simply attached the bailout to an existing House bill (H.R. 1424), thus forcing the House to vote again. The bailout was overwhelmingly unpopular with the public, and rightly so. The banks had created the crisis, so why not let them fail? Americans disbelieved the corporate spin that irresponsible *borrowers* had caused this crisis. The real reasons were, among others, the repeal of Glass-Steagall, fraudulent inducement, derivatives, and credit-default swaps. Even if the average American hadn't figured out the scam, common sense alone told them that either (a) these bankers were grossly incompetent, or (b) they were criminals. Certainly neither case warranted granting them billions or trillions of bailout dollars. *USA Today* estimated that public opposition was at least 10-1 against the bailout bill.[83] Representative Brad Sherman told CNBC, "My calls are coming in about 300 to 2 against this."[84] And Representative Adam Smith reported that phone calls, emails and letters coming into his office were running 1,000 to 1 against.[85] Facing an upcoming election, conventional wisdom held that Congress would respect the will of the people. So why did they so clearly capitulate to Wall Street? One reason: Secretary of the Treasury Henry Paulson stuck a gun to the Constitution and said pass the bill or I'll shoot.

On October 2, Brad Sherman addressed the Congress, stating, "The only way they can pass this bill is by creating and sustaining a panic atmosphere. That atmosphere is not justified. Many of us were told in private conversations that if we voted against this bill on Monday that the sky would fall, the market would drop two or three thousand points the first day, another couple of thousand the second day, and a few members were even told that there would be martial law in America if we voted no."[86] While Sherman didn't reveal who these financial terrorist were, several weeks later Representative James Inhofe appeared on Tulsa Radio 1170 KFAQ. Host Pat Campbell asked him, "Somebody in DC was feeding you guys quite a story prior to the bailout, a story that if we didn't do this we were going to see something on the scale of the depression, there were people talking about martial law being instituted, civil unrest. Who was feeding you guys this stuff?" "That was Henry Paulson," Inhofe replied, explaining that the threats came during a September 19 conference call.[87] The point bears repeating. Paulson, the jackbooted thug, with the full weight of Wall Street behind him, *threatened* the United States Congress—either hand over the money or we'll unleash the army on you.

Despite the unpopularity of the bailout, Senator Obama was determined to push the bill through Congress. On September 30, Forbes reported, "U.S. presidential candidate Barack Obama on Tuesday urged lawmakers to return to the negotiating table."[88] And on October 5, the *International Herald Tribune* reported, "Obama not only supported the bill but made phone calls to members of the Congressional Black Caucus, many of whom voted against the bill Monday. By Friday, many of them had switched sides."[89]

In concurrence with all the overt martial law preparations, FEMA has been quietly planning and preparing detention centers, or camps, to house dissidents during declared national emergencies. Despite repeated denials by the mainstream media about such plans, on January 22, 2009, Representative Alcee Hastings (a former judge who was impeached and removed from the bench) introduced House Resolution 645, the National Emergency Centers Establishment Act. If passed, the bill would authorize FEMA to build camps "to provide temporary housing, medical and humanitarian assistance to individuals and families dislocated due to an emergency or major disaster."[90] Ron Paul responded,

> "Apparently, the fusion centers, militarized police, surveillance cameras and a domestic military command are not enough. Even though we know that detention facilities are already in place, they now want to legalize the construction of FEMA camps on military installations using the ever-popular excuse that the facilities are for the purposes of a national emergency. With the phony debt-based economy getting worse and worse by the day, the possibility of civil unrest is becoming a greater threat to the establishment. One need only look at Iceland, Greece and other nations for what might happen in the United States next."[91]

What might happen in the United States next depends mostly on the people. The United States Constitution is the New World Order's primary impediment. If the American people educate themselves and take action to defend the Constitution, the New World Order *will* crumble. If not, the United States will relinquish its sovereignty to regional government, and eventually to tyrannical global government.

PART II: FALSE ENVIRONMENTALISM

"All religions are founded on the fear of the many and the cleverness of the few."[1] – *Marie-Henri Beyle (Stendhal)*

The New World Order derives power and authority from ancient methods of social control, primarily banking, religion, and war. Part I of this book has addressed banking and war, so what about religion? Thomas Jefferson observed, "Religion is a matter which lies solely between Man and his God." The First Amendment of the Constitution, he explained, is "a wall of separation between Church and State."[2] The United States Constitution has insulated Americans, to some degree, from control-freak social engineers. As a social control mechanism, religion has generally been more benign during the past century than during previous eras. Nevertheless, when governments began separating Church from State, social engineers began separating God from Church. They were determined to create a new "God," outside the Church, through which they could manipulate society. By controlling the environmental movement, they are now attempting to impose a New World Religion based on Earth worship and New Age ideology. The United Nations, Al Gore, Mikhail Gorbachev and many other self-ordained environmental leaders openly admit they are preparing the world for dramatic moral and spiritual shifts. This preparation, they believe, will facilitate widespread acceptance of Mother Earth as the new God. Like all other religious systems, they are infusing the environmental movement with faith and dogma of which the most important beliefs are (a) the planet is overpopulated, and (b) humans are causing global warming. Both these beliefs are false.

Understanding the environmental movement requires first understanding its actors and directors. Who started this movement? Who funds it? Who are its leaders and what are their core beliefs? The New World Order globalists discussed in Part I are the same people behind the environmental movement. The environment is to them what Christianity was to the Romans. Environmentalism is the ideological fabric through which they weave their global government. For their plans to succeed, two things must happen: (a) a large portion of humanity must willingly accept global government, and (b) they must drastically reduce world population so as to better manage those who resist. The globalists realized long ago that by manipulating the environmental movement, they could fulfill both these criteria. Most environmental activists, and most employees and supporters of environmental groups are neither aware of nor complicit with this covert agenda. The proverbial men behind the curtain, however, know exactly what they are doing. As discussed in Part I, defeating the New World Order does not mean overthrowing the government. It means *restoring* the government and *restoring* the Constitution. Regarding the environmental movement, the same principle applies. False leaders must be expelled, false issues like depopulation and global warming must be dismissed, and genuine environmentalism, focused on legitimate issues, must be embraced.

This study begins by examining the eugenics movement and its connections to the environmental movement. It then turns to the depopulation agenda and how and why the New World Order has made overpopulation an environmental issue. The next focal points are the elite foundations that fund and control the environmental groups, and the pivotal role of the United Nations. The next topic is the New Age movement, its connections to the environmental movement, and its larger New World Religion implications. Upon this historical foundation, readers can effectively examine and understand the current hype concerning manmade global warming, and how and why the ultra-elite Club of Rome, over several decades, has been preparing society to accept this massive fraud. The New World Order is not invincible. It depends on widespread ignorance and apathy. The documentation and proof of what the globalists have done and what they are planning is readily available, hidden in plain view. Humanity now faces many uncomfortable and inconvenient truths. The time is ripe, for those who care, to stand up and do something. All actions and all decisions have consequences. Even doing nothing still is doing something.

Chapter 9

Eugenics: The Roots of False Environmentalism

Merging Charles Darwin's concepts of *survival of the fittest* and *superior and inferior types*, eugenics is an ideology whereby men profess to be Gods, thereby claiming the moral authority to kill, maim and sterilize. According to Sir Francis Galton, the cousin of Charles Darwin,

> "We greatly want a brief word to express the science of improving stock, which is by no means confined to questions of judicious mating, but which, especially in the case of man, takes cognisance of all influences that tend in however remote a degree to give to the more suitable races or strains of blood a better chance of prevailing speedily over the less suitable than they otherwise would have had. The word eugenics would sufficiently express the idea."[1]

As for the plight of these so-called inferiors, Galton asserted, "That the members of an inferior class should dislike being elbowed out of the way is another matter; but it may be somewhat brutally argued that whenever two individuals struggle for a single place, one must yield, and that there will be no more unhappiness on the whole, if the inferior yield to the superior than conversely, whereas the world will be permanently enriched by the success of the superior."[2]

Hitler's genocidal attempt to eliminate 'inferiors' while preserving the Aryan race was history's most infamous example of eugenics. However, eugenics was not Hitler's creation. He adopted this philosophy from its British and American pioneers who had already been practicing eugenics for decades. For example, regarding American sterilization laws, Hitler remarked, "Now that we know the laws of heredity, it is possible to a large extent to prevent unhealthy and severely handicapped beings from coming into the world. I have studied with great interest the laws of several American states concerning prevention of reproduction by people whose progeny would, in all probability, be of no value or be injurious to the racial stock."[3]

Edwin Black, in his book *War Against the Weak*, explained, "The intellectual outlines of the eugenics Hitler adopted in 1924 were strictly American. He merely compounded all the virulence of long-established American race science with his fanatic anti-Jewish rage. Hitler's extremist eugenic science, which in many ways seemed like the logical extension of America's own entrenched programs and advocacy, eventually helped shape the institutions and even the machinery of the Third Reich's genocide."[4] Furthermore, the Nazi death machine depended heavily on funding from American financiers, especially the Rockefeller Foundation.

From 1922 to 1926, the Rockefeller foundation granted $410,000 to German eugenicists, an astounding sum during those hyper-inflationary Weimar Republic years. A large portion went to Berlin's Kaiser Wilhelm Institute for Psychiatry whose lead psychiatrist was Ernst Rüdin, the President of the World Eugenics Federation in 1932 and the architect of the 1933 Nazi Sterilization Law. In 1929, the Rockefeller Foundation granted another $317,000 to the Kaiser Wilhelm Institute for Brain Research and in 1931, another $89,000 for Rüdin's Institute for Psychiatry. This money was used for eugenics experiments targeting Jews, gypsies, the mentally handicapped, and others.[5]

Throughout the 1930s, the Rockefeller Foundation continued funding Hitler's eugenics. Black remarked, "In June of 1939, when the Rockefeller Foundation tried to convince protestors that it was not financing Nazi science, [Rockefeller Foundation President Raymond] Fosdick was forced to remind his colleagues that such denials were 'of course hardly correct.' Rockefeller money was still flowing through the Emergency Fund for German Science, now the German Research Society."[6]

Eugenics is an ideology combining power, subordination and conquest. Understanding eugenics is crucial to understanding the false environmental movement and the larger New World Order agenda. Eugenics is the common thread connecting wars, depopulation initiatives, genetically modified foods, and manmade global warming. The organizations that created, supported, and funded the eugenics movement, the Rockefeller and Ford Foundations for example, are the same organizations that created, supported and funded the environmental movement. What was their rationale? Why did these particular ideas capture their interest?

Quite simply, psychopaths crave power. Global government, they reasoned, would help secure enduring power. Implementing global government, however, would necessitate killing or otherwise neutralizing all sources and all *potential* sources of opposition. Eugenics would facilitate this objective. Their sadistic pleasure notwithstanding, eugenics is primarily utilitarian for these control-freaks. Of course for most people, eugenics is an abomination. So how would they sell eugenics to the public? False environmentalism has become the marketing department for eugenics. Green slogans and perceived Earth stewardship make eugenics palatable, unnoticeable even. Whereas eugenics is the bitter pill, false environmentalism is the accompanying spoonful of sugar. The eugenics movement and the environmental movement are fully intertwined. Together, they facilitate the New World Order agenda.

Eugenics Comes to America

Inspired by the work of his cousin Charles Darwin, Francis Galton began formulating his eugenics theories in England during the 1860s and 1870s. By the turn of the century, Galton had inspired many prominent Americans including Charles Davenport and Henry Fairfield Osborn. Davenport, seeking to establish a permanent eugenics research facility, solicited the Carnegie Institution for funding. In 1903, Davenport wrote to John Billings, Carnegie's Chairman, with ideas for improving America's racial hygiene:

"We have in this country the grave problem of the negro, a race whose mental development is, on the average, far below the average of the Caucasian. Is there a prospect that we may through the education of the individual produce an improved race so that we may hope at last that the negro mind shall be as teachable, as elastic, as original, and as fruitful as the Caucasian's? Or must future generations, indefinitely, start from the same low plane and yield the same meager results? We do not know; we have no data. Prevailing 'opinion' says we must face the latter alternative. If this were so, it would be best to export the black race at once."[7]

Davenport added, "I propose to give the rest of my life unreservedly to this work." Billings and the Carnegie board were impressed. On January 19, 1904, the Carnegie Institution established, under Davenport's direction, the Station for Experimental Evolution at Cold Spring Harbor on Long Island, New York.

Interest in eugenics quickly spread among America's elite. And by 1907, Indiana passed the world's first forced sterilization legislation, paving the way for similar laws in thirty more states and at least twelve more countries.[8] Twenty years later, the U.S. Supreme Court would officially endorse eugenics with a precedent-setting decision in the 1927 case of *Buck v. Bell*. The Court decided that forced sterilization was Constitutional, "for the protection and health of the state." Justice Oliver Wendell Holmes ruled, "It is better for all the world, if instead of waiting to execute degenerate offspring for crime, or to let them starve for their imbecility, society can prevent those who are manifestly unfit from continuing their kind. The principle that sustains compulsory vaccination is broad enough to cover cutting the Fallopian tubes. Three generations of imbeciles are enough."[9]

America's wealthiest families began enthusiastically donating for eugenics research. Black explained, "The eugenics visions offered by Davenport and [Harry] Laughlin pleased the movement's wealthy sponsors. On January 19, 1911, Andrew Carnegie doubled the Carnegie Institution's endowment with an additional ten million dollars for all its diverse programs, including eugenics. Mrs. Harriman increased her enthusiastic grants. John D. Rockefeller's fortune also contributed to the funding…Eugenics was nothing less than an alliance between biological racism and mighty American power, position and wealth against the most vulnerable, the most marginal and the least empowered in the nation. The eugenic crusaders had successfully mobilized America's strong against America's weak. More eugenic solutions were in store."[10]

International Congresses of Eugenics

The First International Congress of Eugenics convened in London in 1912. Leonard Darwin, the son of Charles, was the event's President. The roster included Alexander Graham Bell, Winston Churchill, Charles Davenport, Charles Eliot (Emeritus President of Harvard University), and Gifford Pinchot (founding Director of the United States Forest Service).[11]

The Second International Congress convened in New York in 1921. The *New York Times* reported, "The theory held by some eminent anthropologists that all races have an equal capacity for development and that all race questions, even the negro question, is to be solved in the long run by race mixture, was vigorously combated."[12] Professor Henry Fairfield Osborn, President of the Congress, openly touted his racist views: "Dr. Osborn argued that the State's right to prevent disease implied a similar right to prevent the multiplication of feeble-minded, idiocy and moral and intellectual diseases." In his own word, Osborn asserted,

> "In the United States we are slowly awaking to the consciousness that education and environment do not fundamentally alter racial values. We are engaged in a serious struggle to maintain our historic republican institutions through barring the entrance of those who are unfit to share the duties and responsibilities of our well-founded Government. The true spirit of American democracy, that all men are born with equal rights and duties, has been confused with the political sophistry that all men are born with equal character and ability to govern themselves and others, and with the educational sophistry that education and environment will offset the handicap on ancestry."[13]

On the relationship between science and government, he stated, "As science has enlightened government in the prevention and spread of disease, it must also enlighten government in the prevention of the spread and multiplication of worthless members of society." Osborn also addressed the perennial anxiety of elites regarding losing power due to negligent breeding: "In New England a century has witnessed the passage of a many-child family to a one-child family. The purest New England stock is not holding its own. The next stage is the no-child marriage and the extinction of the stock which laid the foundations of the Republican institutions of this country."[14]

The Third International Congress of Eugenics convened again in New York in 1932. Both Darwin and Osborn were Honorary Presidents. Neither attended but both wrote letters for the event. Darwin wrote,

"My firm conviction is that if widespread eugenic reforms are not adopted during the next hundred years or so, our Western civilization is inevitably destined to such a slow and gradual decay as that which has been experienced in the past by every great ancient civilization. The size and the importance of the United States throws on you a special responsibility in your endeavors to safeguard the future of our race. Those who are attending your congress will be aiding in this endeavor, and though you will gain no thanks from your generation, posterity will, I believe, learn to realize the great debt it owes to all the workers in this field."[15]

Another paper presented to the Congress was Sir Bernard Mallet's "The Reduction of the Fecundity of the Socially Inadequate." Mallet wrote, "It must, unfortunately, be recognized that little progress has yet been made in the solution of the problem of the reduction in the fertility of the undesirable elements in our population."[16] Osborn, writing about eugenics in general, asserted, "It aids and encourages the survival and multiplication of the fittest; indirectly it would check and discourage the multiplication of the un-fittest. As to the latter, in the United States alone it is widely recognized that there are millions of people who are acting as dragnets or sheet anchors on the progress of the ship of State."[17]

On the poor and unemployed, Osborn wrote, "In nature these less-fitted individuals would gradually disappear, but in civilization we are keeping them in the community in the hopes that in brighter days they may all find employment. This is only another instance of humane civilization going directly against the order of nature and encouraging the survival of the un-fittest...Only by some wise and selective means of limiting the number of births can the world find a solution for its disturbed economics."[18] Osborn also again addressed overpopulation. The *New York Times* reported, "Discussing the reality of overpopulation, Dr. Osborn said that even the lives sacrificed in the World War are entirely negligible compared with the natural increase of mankind when no longer checked by disease, by infant mortality and by internecine wars."[19]

Margaret Sanger

A virulent racist and eugenics enthusiast, Margaret Sanger founded the birth control movement and the Planned Parenthood Federation. Today, Planned Parenthood recognizes her as "one of the movement's great heroes," noting, "Sanger's early efforts remain the hallmark of Planned Parenthood's mission."[20] Since its 1920s origins, the birth control movement has simply been an extension of the eugenics movement. In 1921, with funding from the Rockefeller Foundation, Sanger established the American Birth Control League, the precursor to Planned Parenthood.[21] In 1923, Sanger invited many prominent eugenicists to the Birth Control Conference in Chicago. To Harry Laughlin, she wrote, "I believe that this conference is going to do much to unite the Eugenic Movement and the Birth Control movement, for after all they should be and are the right and left hand of one body."[22]

In her writings, Sanger also frequently praised eugenics. In October 1921, for example, she wrote an article for her journal *Birth Control Review* called, "The Eugenic Value of Birth Control Propaganda." Sanger wrote, "As an advocate of Birth Control, I wish to take advantage of the present opportunity to point out that the unbalance between the birth rate of the 'unfit' and the 'fit,' admittedly the greatest present menace to civilization, can never be rectified by the inauguration of a cradle competition between these two classes...Birth Control propaganda is thus the entering wedge for the Eugenic educator."[23] In her 1922 book *The Pivot Of Civilization*, Sanger continued her eugenics rants while also showcasing her abhorrently elitist views on personal liberty:

> "The actual dangers can only be fully realized when we have acquired definite information concerning the financial and cultural cost of these classes to the community, when we become fully cognizant of the burden of the imbecile upon the whole human race; when we see the funds that should be available for human development, for scientific, artistic, and philosophic research, being diverted annually, by hundreds of millions of dollars, to the care and segregation of men, women, and children who never should have been born. The advocate of Birth Control realizes as well as all intelligent thinkers the dangers of interfering with personal liberty...But modern society, which has respected the personal liberty of the individual only in regard to the unrestricted and irresponsible bringing into the world of filth and poverty an overcrowding procession of infants foredoomed to death or hereditable disease, is now confronted with the problem of protecting itself and its future generations against the inevitable consequences of this long-practiced policy of laissez-faire."[24]

H.G. Wells, the Fabian Society eugenicist and New World Order darling, wrote the introduction to this book. He proclaimed, "Birth Control, Mrs. Sanger claims, and claims rightly, to be a question of fundamental importance at the present time...The New Civilization is saying to the Old now: 'We cannot go on making power for you to spend upon international conflict. You must stop waving flags and bandying insults. You must organize the Peace of the World; you must subdue yourselves to the Federation of all mankind. And we cannot go on giving you health, freedom, enlargement, limitless wealth, if all our gifts to you are to be swamped by an indiscriminate torrent of progeny. We want fewer and better children who can be reared up to their full possibilities in unencumbered homes, and we cannot make the social life and the world-peace we are determined to make, with the ill-bred, ill-trained swarms of inferior citizens that you inflict upon us."[25]

In 1939, Sanger wrote a letter to Clarence Gamble (the heir to the Proctor and Gamble fortune and an enthusiastic birth control supporter) regarding the so-called "Negro Project." This project was an attempt by the Birth Control Federation of America to encourage contraceptive use among Southern blacks. The project sought "colored Ministers, preferably with social-service backgrounds, and with engaging personalities" to travel around propagandizing for birth control. Sanger chose this strategy because "the most successful educational approach to the Negro is through a religious appeal."[26] She also candidly revealed her penchant for genocide: "We do not want word to go out that we want to exterminate the Negro population and the minister is the man who can straighten out that idea if it ever occurs to any of their more rebellious members."[27]

In 1952, the Rockefeller Foundation financed the founding of Sanger's International Planned Parenthood Federation (IPPF). In his book *Seeds of Destruction*, William Engdahl explained, "Following this initial cash infusion, her IPPF was soon backed by a corporate board which included DuPont, U.S. Sugar, David Rockefeller's Chase Manhattan Bank, Newmont Mining Co., International Nickel, RCA, Gulf Oil and other prominent corporate members. The cream of America's corporate and banking elite were lining up behind Rockefeller's vision of population control on a global scale."[28]

Julian Huxley

The founder of the World Wildlife Fund and the founding Director of UNESCO, Julian Huxley was a hardcore eugenicist. Consequently, he infused these organizations with eugenics ideology. Huxley grew up very close to the eugenics movement. His grandfather, Thomas Henry Huxley, was known as "Darwin's Bulldog" for his steadfast advocacy of Charles Darwin's theory of evolution. Thomas Huxley was also a mentor and teacher of H.G. Wells. Julian, in his 1941 book *Man Stands Alone*, proposed restricting health care from "inferiors" so as to quicker kill them off: "The lowest strata, allegedly less well-endowed genetically, are reproducing relatively too fast. Therefore birth-control methods must be taught them; they must not have too easy access to relief or hospital treatment lest the removal of the last check on natural selection should make it too easy for children to be produced or to survive; long unemployment should be a ground for sterilization."[29]

Huxley also favored genocide: "Thus no really rapid eugenic progress would come of encouraging the reproduction of one class or race against another: striking and rapid eugenic results can be achieved only by a virtual elimination of the few lowest and truly degenerate types and a high multiplication-rate of the few highest and truly gifted types."[30] On February 17, 1936, in an address known as *The Galton Lecture*, Huxley made this ominous statement: "It may be that, as a scientist myself, I overrate the importance of the scientific side. At any rate, it is my conviction that eugenics cannot gain power as an ideal and a motive until it has improved its position as a body of knowledge and a potential instrument of control."[31]

Rockefellers in Puerto Rico

Beginning in the 1950s, John D. Rockefeller III, through his newly formed Population Council initiated massive sterilization campaigns throughout Puerto Rico. The U.S. Department of Health, Education and Welfare, where brother Nelson Rockefeller was Under-Secretary, also participated. Despite claims that sterilization would protect women's health and stabilize incomes, the program nevertheless devastated Puerto Rico's population. In 1965, a Department of Health study found that fully 35% of Puerto Rican women of childbearing age had been sterilized. Many women were forcibly sterilized in hospitals. Doctors were ordered to sterilize any woman who had already given birth twice.[32] Such Rockefeller malfeasance, however, was nothing new in Puerto Rico.

In 1931, the Rockefeller Institute for Medical Research, later renamed Rockefeller University, funded cancer research in San Juan, Puerto Rico by Doctor Cornelius Rhoads. In terms of depravity, his research rivaled that of Nazi "Angel of Death" Joseph Mengele. Rhoads, for example, purposely infected patients with cancer viruses to study the effects, killing eight people in the process. Rhoads was proud of his murderous deeds. He wrote in a letter, "Porto Ricans are beyond doubt the dirtiest, laziest, most degenerate and thievish race of men ever inhabiting this sphere. What the island needs is not public health work but a tidal wave or something to totally exterminate the population. It might then be livable. I have done my best to further the process of extermination by killing off 8."[33]

Unfortunately for Rhoads, Pedro Albizu y Campos, President of the Nationalist Party of Puerto Rico, acquired the letter and passed it on to local newspapers. Upon its publication, the entire island became enraged. *Time* magazine even took notice and reprinted the letter for American readers. However, *Time* shamefully defended Rhoads who insisted he actually meant the *opposite* of everything he wrote. "Of course nothing in the document was ever in-tended to mean other than opposite of what was stated," Rhoads declared.[34] *Time* characterized the Puerto Rican upheaval as "an agitation typical of the prejudice with which the Foundation is obliged to contend in many backward countries." Consequently, Rhoads and the Rockefeller Institute were exonerated and left to pursue further eugenics.

Fallout From Hitler

In 1946, Frederick Osborn, the nephew of Henry Fairfield Osborn, became President of the American Eugenics Society (AES). At the time, the eugenics movement was still reeling from the embarrassment brought about by Hitler. While they of course philosophically agreed with Hitler, they lamented his over-aggressive approach, resulting in massive publicity damage. Thus in 1946, in an article called, "Eugenics and Modern Life: Retrospect and Prospect," Osborn admitted, "We do not want to repeat in some new form the mistake of the earlier eugenicists who declared for race and social class, and thereby set back the cause of eugenics for a generation."[35]

The following year, Osborn and the AES Board of Directors determined, "The time was not right for aggressive eugenics propaganda." Thus during the ensuing years, while continuing to receive funding from Rockefeller's Population Council, AES worked steadily towards improving the public image of eugenics.[36] By 1965, Osborn was gorging himself on self-congratulation. To a colleague at Duke University, he proudly wrote, "We have struggled for years to rid the word eugenics of all racial and social connotations and have finally been successful with most scientists, if not with the public." He also wrote, "The term medical eugenics has taken the place of the term negative eugenics."[37]

Osborn and other leading eugenicists, however, were only interested in changing the *appearance* of eugenics. At their cores they were all still racists and genocidal psychopaths. From 1947 to 1956, for example, Osborn also secretly held the presidency of the racist Pioneer Fund. According to the *New York Times*, the Pioneer Fund "supported highly controversial research by a dozen scientists who believe that blacks are genetically less intelligent than whites."[38] The Pioneer Fund would later grant over $1 million to Stanford University Nobel laureate William Shockley, a man who believed that whites are inherently more intelligent than blacks and who advocated forcibly sterilizing any person with an IQ score lower than 100.[39] The Rockefeller's were no different. Engdahl explained, "Rockefeller's Population Council gave grants to leading universities including Princeton's Office of Population headed by Rockefeller eugenicist, Frank Notestein, a long-time friend of

Osborn, who in 1959 became President of the Rockefeller Population Council in order to promote a science called demography. Its task was to project horrifying statistics of a world overrun with darker-skinned peoples, to prepare the ground for acceptance of international birth control programs. The Ford Foundation soon joined in funding the various Population Council studies, lending them an aura of academic respectability, and above all, money."[40]

The Next Million Years

Charles Galton Darwin, the grandson of Charles Darwin, was a physicist and former Director of London's National Physical Laboratory. He was also an arrogant spokesman for eugenics and social engineering. In 1952, Darwin wrote *The Next Million Years*, an abhorrent blueprint for scientifically designed tyranny. Like Bertrand Russell, Darwin also envisioned a species split whereby genetically superior elites would rule inferior slaves. The elites, he reasoned, would genetically engineer their slaves to serve their planned utopia. He proposed three options for future world management:

"If the history of the future is not regarded as the automatic unfolding of a sequence of uncontrollable events—and few, of us would accept this inevitability—then anyone who has decided what measures are desirable for the permanent betterment of his fellows will naturally have to consider what is the best method of carrying his policy through. There are three levels at which he might work. The first and weakest is by direct conscious political action; his policy is likely to die with him and so to be ineffective. The second is by the creation of a creed, since this has the prospect of lasting for quite a number of generations, so that there is some prospect of really changing the world a little with it. The third would be by directly changing man's nature, working through the laws of biological heredity, and if this could be done for long enough it would be really effective."[41]

Darwin was clearly excited by the prospects of genetically manipulating and engineering humans:

"Imagine that through new discoveries in biology, say by suitably controlled doses of X-rays, it becomes possible to modify the genes in any desired direction, so that heritable changes can be produced in the qualities of some members of the human race...The first success might be in some physical attribute, for example, by making a breed with longer and stronger legs so that it could jump a good deal higher than anyone can at present. But passing to more important matters, there might be created a breed which could think more abstractly, say a breed of mathematicians, or one that could think more judiciously, say a breed of higher civil servants. These would be of great value, but they would not be the master breed, and the question arises of a more precise prescription for what the qualities of the master breed are to be."[42]

Darwin thought of human beings as wild animals capable of domestication so as to better serve the masters. The biggest obstacle, he reasoned, was man's inherent sexual drive. Yet by freeing humans from their sexual instincts, they could be transformed into perfect slaves:

"Why cannot man set up a community like an ants' nest? This would be the ideal of the anarchist, and hitherto it has held no promise at all of success, but with the help of recent and probable future biological discoveries, some sort of imitation by man of the ants' nest cannot be quite excluded from consideration. Thus the control of the numbers of the two sexes may become possible, and with the knowledge of the various sexual hormones it might also become possible to free the majority of mankind from the urgency of sexual impulse, so that they could live contented celibate lives, instead of the unsatisfied celibate lives that are the compulsory lot of such a large fraction of the present population of the world. If these discoveries should be made—and this is really by no means impossible—man would be able to carry out the sex revolution which is the typical characteristic of the insect civilizations. The detail would of course have to be quite different, for instead of one queen there would have to be large numbers of fertile women to renew the population, whereas there might be one king, literally the father of his country."[43]

Darwin toyed with the idea of scientifically designing a master breed to rule the slaves. In the end, however, he determined that the masters must remain wild:

"The reason for the impossibility of making a prescription for the master breed is that it is not a breed at all; to call it so is to change the sense of the word. Breeds are specialized for particular purposes, but the essence of masters is that they must not be specialized. They have to be able to deal with totally unforeseen conditions, and this is a quality of wild, not of tame, life. No prescription for the master breed is possible."[44]

Darwin also wrote about world government. He expected resistance from an informed minority, the "goats" as he called them, but was considerably less worried about the masses of "sheep." He explained,

"If then there is ever to be a world government, it will have to function as governments do now, in the sense that it will have to coerce a minority—and indeed it may often be a majority—into doing things they do not want to do."[45]

Finally, Darwin wrote about overpopulation and various population reduction strategies:

"It would have to be [a] world-wide [policy], because if any nation were recalcitrant, its population would increase relatively to the rest, so that sooner or later it would dominate the others...It is clear from all this that the world policy would need to be supported by international sanctions, and the only ultimate sanction must be war. Present methods of warfare would not be nearly murderous enough to reduce populations seriously, and even so they would take a nearly equal toll of victims from the unoffending nations. So after the war the question would arise of how to reduce the excess population of the offending nation. It is not possible to be humane in this, but the most humane method would seem to be infanticide together with the sterilization of a fraction of the adult population. Such sterilization could now be done without the brutal methods practised in the past, but it would certainly be vehemently resisted."[46]

As previously discussed, eugenics is the underlying ideology of the New World Order's architects. Hitler was extreme, but he pales in comparison to today's top eugenicists. Hitler killed millions. Today's top eugenicists intend to kill billions. Euphemistically known as "population control," the eugenics movement is alive and well.

Chapter 10

Applied Eugenics: The Depopulation Agenda

Thanks to heavy propaganda and calculated indoctrination, most people blindly accept the notion that Earth is overpopulated, or soon will be. This idea appears both subliminally and overtly in advertising, movies, television and many other outlets. This chapter includes numerous statements from environmentalists, economists, scientists, philanthropists and politicians all of whom insist that overpopulation is among humanity's most urgent and serious problems. Such alarmism, however, is nothing new. For centuries, elites (and their dupes, stooges, and agents) have been making such claims. Incessantly, these ideologues repeat their overpopulation mantra, effectively muffling all dissenting opinions. Nevertheless, scores of prominent scientists and economists contend that Earth can support many billions *more* people. Furthermore, many argue that a larger population would be highly *beneficial*. The world, after all, is much bigger than commonly perceived. University of San Francisco biologist Francis P. Felice has observed, "All the people in the world could be put into the state of Texas, forming one giant city with a population density less than that of many existing cities, and leaving the rest of the world empty."[1] Carrying capacity, however, is not only a matter of geography. Many other factors, especially technology and ingenuity are involved. In their book *The Third World: Premises of U.S. Policy*, Peter Bauer and Basil Yamey, both world-acclaimed economists and former professors at the London School of Economics, concluded that the overpopulation theory,

- "Relies on misleading statistics."
- "Envisages children exclusively as burdens."
- "Misunderstands the determinants of economic progress."
- "Misinterprets the causalities in changes in fertility and changes in income."[2]

Colin Clark, the former Director of the Institute of Agricultural Economics at Oxford University and one of the world's most prominent economic demographers, published a book in 1972 called *Population Growth: The Advantages*. Clark rebutted the pseudoscience of Paul Ehrlich's *The Population Bomb,* a masterpiece of alarmism and fear-mongering. Clark was so revered that John Maynard Keynes, the father of Keynesian economics, commented, "Indeed, Clark is, I think, a bit of a genius: almost the only economic statistician I have ever met who seems to me quite first-class."[3] Clark determined that Earth could support 35.1 billion people eating the very rich, energy intensive American diet, or up to 105.3 billion people eating the Japanese standard diet.[4] Furthermore, his calculations allowed for the preservation of nearly half the earth's arable land for conservation and recreation areas, and for the preservation of wildlife.[5]

Roger Revelle, Al Gore's mentor and the founder and former Director of the Center for Population Studies at Harvard University, determined that Earth could support 40 billion people eating 2,500 calories per day. This would require less than one-forth of Earth's arable land.[6] Furthermore, he based his assumptions on low average crop yields. Higher yields would obviously support many more than 40 billion. Africa alone, according to Revelle, is capable of producing enough food to feed 10 billion people.[7] Furthermore, American economist Julian Simon, a former professor at the University of Maryland and

Senior Fellow at the Cato Institute, has written extensively about the economic *benefits* of population growth, supported by technological progress.[8]

Because world population over the past two centuries has increased from 1 billion to 6.5 billion, many people naively fear population will continue to double and double and double until Earth eventually exceeds its carrying capacity. "But that," observed Max Hudson, founder of the Hudson Institute, "is like fearing that your baby will grow to 1,000 pounds because its weight doubles three times in its first seven years."[9] Almost invariably, after experiencing sustained industrial development, population rates *decline*. In 2005, a United Nations population study concluded, "Since 1990-1995, fertility decline has been the rule among most developed countries."[10] According to the UN, from now until 2050,

- World population will stabilize at 9.1 billion.
- The population of the developed world, currently 1.2 billion, will remain virtually unchanged.
- Fifty-one countries including Germany, Italy and Japan will experience declining populations.
- In the least developed countries, the current fertility rate of 5.0 will drop to 2.57.
- In the rest of the developing world, the current rate of 2.58 will decline to 1.92.
- In developed countries, the current rate of 1.56 will increase slightly to 1.84.[11]

Therefore, in all but the world's 50 poorest countries, fertility rates in 2050 will be lower than the zero-growth rate of 2.1 children per women. According to Hania Zlotnik, director of the UN Population Division, "It doesn't seem that there is a crisis coming."[12]

Perhaps the most serious population issue is the *declining* populations of Europe and other western nations. In 2004, a Human Rights International (HRI) study, based on UN data, concluded, "For the first time since the Black Death ravaged the European population in 1347-1351, an entire continent is experiencing a population collapse due to natural causes."[13] HRI also observed, "The weighted TFR [total fertility rate] of Europe's 47 nations has plunged from an already-low 2.6 in 1965 to 1.3 in 2004, far below replacement."[14] Europe is losing 1.6 million people per year and by 2050, the population will decline to 565 million, a 25 percent reduction. The average age of the European population by 2050 will be 52 years old, whereas the average age in Africa will be 31 years old.[15]

As documented throughout this book, historically, elites suppress third-world economic development because, in the words of John D. Rockefeller, "Competition is a Sin."[16] Instead of empowering and encouraging third-world nations, elites prefer to coerce and corrupt their leaders, and then pillage their natural resources. Such meddling has stifled genuine economic development, thus triggering some genuine overpopulation problems. Posing as saviors for the very problems they engineered, these elites typically offer "population assistance," through USAID for example, including birth control, abortion and sterilization. These are merely false solutions to contrived problems.

As documented in upcoming chapters, many prominent environmental think tanks, the Club of Rome for example, have mystified the public with "limits to growth" computer modeling, suggesting that global carrying capacity is fixed and thus population cannot exceed natural limitations. However, many informed, aware people have recognized this deception. For example, David Hopper, former Regional Vice President of the World Bank, observed,

"The world's food problem does not arise from any physical limitation on potential output or any danger of unduly stressing the environment. The limitations on abundance are to be found in the social and political structures of nations and in the economic relations among them. The unexploited global food resource is there, between Cancer and Capricorn. The successful husbandry of that resource depends on the will and actions of men."[17]

Roger Maduro and Ralf Schauerhammer, in their book *Holes in the Ozone Scare*, explained,

"There is no correlation between 'natural resources' and human population potential, for the simple reason that resources are not really 'natural.' The resources for human existence are defined by human science and technology, and the development of science and technology defines whole new arrays of 'resources' for the societies that avail themselves of such progress…This means two things. First there are no 'limits to growth.' There are only limits within the confines of a given array of technology…Second, were modern agricultural and industrial capabilities, even as they exist in industrialized nations today, diffused throughout the Third World, we would discover that not only do we have ample resources…but we also have too few people to operate advanced agro-industrial facilities at optimum capacity. If we took account of in-sight technological advances, we would discover that under-population is the main problem we face."[18]

Thomas Malthus

Thomas Malthus was a professor of Political Economy at East India Company College in England during the early 19th century. He later became a Fellow of the Royal Society. Malthus is best known for his population studies and his theory that population will inevitably outstrip available food and natural resources. Malthus is also an enduring icon of ruling-class elites who respect his callous disregard for the meager and weak. In his book, *Political Economy*, Barry Stewart Clark summarized the Malthusian doctrine:

"In contrast to Enlightenment thinkers, Malthus believed that human misery was caused by nature rather than badly organized institutions. In his book *An Essay on the Principle of Population (1798)*, Malthus claimed that population grows at a faster rate than do food supplies because of the limited availability of fertile land. Population growth could be restrained either by 'positive checks' such as famines, plagues, and wars, or by 'preventive checks' such as delayed marriages and 'moral restraint.' Malthus had little hope that humans would be capable of exercising the restraint required to control population growth, so positive checks would be the effective controls on overpopulation. Moreover, Malthus believed that government should not attempt to interfere with the operations of these positive checks. He proposed that raw sewage be permitted to flow in the streets, that insect-infested swamps remain undrained, and that cures for disease be suppressed, thereby allowing nature to carry out the grisly task of limiting population growth."[19]

Malthus, in his own words, stated,

"Instead of recommending cleanliness to the poor, we should encourage contrary habits. In our towns we should make the streets narrower, crowd more people into the houses, and court the return of the plague. In the country, we should build our villages near stagnant pools, and particularly encourage settlement in all marshy and unwholesome situations. But above all we should reprobate specific remedies for ravaging diseases; and restrain those benevolent, but much mistaken men, who have thought they are doing a service to mankind by protecting schemes for the total extirpation of particular disorders."[20]

Charles Darwin acknowledged that Malthus' influenced him. Likewise, Malthus also profoundly influenced scores of eugenicists and social Darwinists. Clark explained,

"Malthus's ideas would resurface as 'social Darwinism.' Relying on the evolutionary theory of Charles Darwin, Herbert Spencer (1820-1903) in England and William Graham Sumner (1840-1910) in America claimed that the human species evolves according to the principle of 'survival of the fittest.' Any attempts by government to aid the poor would cause deterioration of the human gene pool by allowing unfit members of the human species to survive and reproduce. This biological argument for hierarchy and inequality became a powerful ideological force in Europe and the United States."[21]

For the past 200 years, at least, elites have insisted that Earth is overpopulated. In 1800, global population was about one billion people.[22] Although the planet now carries nearly seven times the 1800 population, in 1798 Thomas Malthus confidently declared, "The period when the number of men surpasses their means of easy subsistence has long since arrived."[23] As previously discussed, carrying capacity depends mostly on agricultural and energy technology, and geopolitical harmony. The Earth could potentially carry many times the current population. Malthus and his successors, however, have determined by fiat that one billion was far too many, and the current seven billion is dangerous and precarious. Of course they never mention that the industrial revolution and the current state of technology would never have been possible *without* this increased population. Malthusian economics and politics amount to inciting panic about population and directing that panic towards oppressive, anti-human eugenic and depopulation policies.

The Rebirth of Malthus

William Vogt, an ecologist and ornithologist, was the Secretary of the Conservation Foundation, Chief of Conservation of the Pan American Union, and the National Director of the Planned Parenthood Federation of America from 1951-1961.[24] In his 1948 book *Road to Survival*, Vogt welcomed the prospects of widespread death by starvation in China: "There is little hope that the world will escape the horror of extensive famines in China within the next few years. But from the world point of view, these may be not only desirable but indispensable. A Chinese population that continued to increase at a geometric rate could be a global calamity. The [peace] mission of General Marshall in this unhappy land was called a failure. Had it succeeded, it might well have been a disaster."[25] Vogt also declared, "There are too many people in the world for its limited resources to provide a high standard of living…The handwriting on the wall on five continents now tells us that the Day of Judgment is at hand."[26] As reported by *Time* magazine, "Vogt suggests that the U.S. should help no country with food or anything else unless it first agrees to limit its birth rate. One method he favors: a bonus to males who allow themselves to be sterilized."[27]

Time also recognized and reported on the resurgence of Malthusian ideology then beginning to infest the environmental movement:

"After more than a century of intermittent haunting, the ghost of a gloomy British clergyman, Thomas Robert Malthus, was on the rampage last week. Cresting a wave of postwar pessimism, it flashed through the air on the radio, rode through the mails in magazines. Publishers opened their arms and presses to 'Neo-Malthusian' manuscripts prophesying worldwide overpopulation and hunger. Two 'scarce books'—*Our Plundered Planet*, by Fairfield Osborn, and *Road to Survival* (a Book-of-the-Month selection), by William Vogt—were glowingly reviewed and selling like hot cakes. Their influence has already reached around the world...The Neo-Malthusians want to warn man of danger; but their alarm is so loud that it may have the effect of deafening the world to its opportunities. To the real agricultural scientists, close to the soil and its sciences, such pessimism sounds silly or worse. Every main article of the Neo-Malthusian creed, they say, is either false or distorted or unprovable."[28]

Time then intelligently debunked the Malthusian fear mongering:

"A country is overpopulated when its people cannot get enough food. This is seldom because they have too little land. Usually it is because their social organization and farming methods are ineffective. India is a hungry country, but it is not permanently overpopulated. It has much potentially good land whose present yields are pathetically low. India averages only ten bushels of wheat an acre while Denmark gets 50. India's rice yield is only 750 lbs. an acre, one-quarter as good as Japan's. A little fertilizer and some simple improvements in agricultural technique would make a huge difference to India's food supply."[29]

Vogt's book became the best-ever selling conservation book until Rachel Carson's *Silent Spring*, published in 1962 (see Chapter 16).[30]

In 1953, the *New York Times* editorialized, "In the last ten years we have witnessed a revival of the Malthusian doctrine that the world's population is increasing more rapidly than its supply of food, minerals and other commodities considered necessary for the maintenance of a high standard of living. We owe this revival largely to Fairfield Osborn (*Our Plundered Planet*) and to William Vogt (*Road to Survival*), who have been followed by economists, public health officials and governments with predictions of misery."[31] In 1951, Vogt delivered a speech in which he declared overpopulation a "deadlier and more insidious killer than the atomic bomb."[32]

Lester Brown and Al Gore are among the many fake environmentalists who today *still* regard Thomas Malthus as an esteemed, visionary thinker. In his 1992 book *Earth in the Balance*, Gore portrayed Malthus as an insightful economist who simply miscalculated the potential of agricultural technology. "But with the scientific revolution in the seventeenth and eighteenth centuries," Gore noted, "the human population began surging, and for the first time it seemed possible that the population might soon outstrip the ability of the environment to yield enough food. This fear was articulated at the beginning of the nineteenth century by the English political economist Thomas Malthus; that he was famously wrong has been due to a series of remarkable innovations in the science of agricultural production. Malthus was right in predicting that the population would grow geometrically, but he didn't foresee our ability to make geometric improvements in agricultural technology."[33] Apparently Gore doesn't mind Malthus' disdain for the poor and his recommendations for eliminating them.

Lester Brown founded the Worldwatch Institute in 1974 with financial assistance from the Rockefeller Brothers Fund.[34] He earned the Rockefellers' trust after promoting Malthusian depopulation in his 1972 book *Man and His Environment: Food*. Brown wrote,

"Thomas Malthus was probably the first to detect worldwide population pressure and to identify world population growth as a problem. When he published his essay on *The Principle of Population* in 1798, he defined the population problem primarily in terms of food supplies and the threat of famine. For almost 200 years men have perceived the population-food problem in these terms...The relevant question is no longer, 'Can we produce enough food?' but 'What are the environmental consequences of attempting to do so?'"[35]

In 1997, Brown participated in an event called, "Reinventing Malthus for the 21st Century: A Bicentennial Event on Malthus' Original Population Essay."[36] Held at the National Press Club in Washington, the event's sponsor was Negative Population Growth (NPG), an organization describing itself as "a national membership organization founded in 1972 to educate the American public and political leaders about the detrimental effects of overpopulation on our environment, resources and quality of life."[37] They claim, "The scientific consensus is that 150-200 million is the ideal population size for the U.S." They also state, "NPG believes that a world population size of two to three billion would be optimal."[38] Their motto is, "Any cause is a lost cause without a reduction in population."[39] In his address, Brown stated, "Although Malthus was controversial, he has not been dismissed as a person, as a thinker."[40] Dismissed by whom? Certainly not by psychopathic elites—in such circles, genocide is just as popular today as it was 200 years ago.

Government and Foundations: Partnering for Genocide

In her thoroughly researched book *The War Against Population*, Jacqueline Kasun determined that "Since 1965 the United States contributed more to foreign population-control programs than all other countries combined and pressured other countries and international agencies to back the programs. In addition to billions of dollars in explicit AID 'population assistance' appropriations to various countries and international organizations such as the United Nations Fund for Population Activities, the United States has made donations to the World Bank and to United Nations organizations—including the World Health Organization, the Food and Agriculture Organization, UNESCO, UNICEF, and the International Labor Organization—that have been used for population control, with a degree of enthusiasm and dedication equal to that of the AID bureaucracy."[41]

In 1931, John D. Rockefeller III joined the board of the Rockefeller Foundation where Raymond Fosdick and Frederick Osborn mentored him and fostered his interest in population control.[42] Two decades later, in 1952, JDR III founded the Population Council with $1.4 million of his own funds along with generous infusions from the Rockefeller and Ford Foundations.[43] According to the Population Council website, Rockefeller was "Impressed by the complexities of population issues, convinced of their fundamental significance for human well-being, and undeterred by the sensitivity then associated with birth control."[44] Frederick Osborn also helped found the Council, served as the first Vice-President, and later became President in 1957.[45] Over the next 25 years, the Population Council would become the world leader in population control, spending a staggering $173 million during this period.[46]

Starting in 1959, establishment journalists began prodding John F. Kennedy to publicly endorse population control programs abroad. Kennedy, however, was strictly opposed to such intervention. In 1959, James Reston, writing for the *New York Times*, lamented, "Kennedy opposes advocacy by U.S. of birth control."[47] Shortly thereafter, John Fischer, a CFR member, trustee of the Brookings Institution and then Editor-in-Chief of *Harper's* magazine, pressured Kennedy to endorse coercive birth control.[48] Later, in June 1963, full-page advertisements appeared in both the *New York Times* and *Washington Post* with the headline, "Population Explosion Nullifies Foreign Aid: An Appeal to the President of the United States." Signatories included members of the Osborn, Rockefeller, Vanderbilt, Aldrich, Scripps, Chase and other elite families.[49]

Nevertheless, Kennedy was firmly against coercive population control. He stated, "I believe it is a judgment which the countries and the people involved must make as to whether they wish to limit their population. Since it involves so personal a decision, I think it would be unwise for the United States to intervene."[50] Donald Gibson, in his book *Environmentalism: Ideology and Power*, concluded,

> "Kennedy, like Franklin Roosevelt, had a primarily human use and resource development orientation to conservation. Kennedy was opposed to the population control movement, making only tiny concessions to it, like agreeing to share information on birth control with other countries. The whole thrust of the Kennedy Presidency was to expand production, support scientific progress, and develop resources for human use. In the six years following Kennedy's death, the environmental movement, with a strong preservationist and Malthusian tone, would burst upon the national scene, putting its imprint on both major political parties and on most of the political spectrum."[51]

Kennedy didn't cooperate with the New World Order. For his opposition to the depopulation agenda and for many other reasons, he was eliminated.

Under Lyndon B. Johnson, depopulation enthusiasts had a friend in the White House. In his 1965 State of the Union address, Johnson spoke about "the explosion in world population and the growing scarcity in world resources."[52] With his 1966 address, he alluded to a new U.S. foreign aid policy with population control as its primary objective. Soon thereafter, he committed federal funding for this purpose. "Let us act on the fact," stated Johnson, "that less than five dollars invested in population control is worth a hundred dollars invested in economic growth."[53]

Nixon, Kissinger and NSSM 200

In 1969, President Nixon delivered his Special Message to the Congress, calling for the creation of the Commission on Population Growth and the American Future. He stated, "It is our belief that the United Nations, its specialized agencies, and other international bodies should take the leadership in responding to world population growth. The United States will cooperate fully with their programs...I have asked the Secretary of State and the Administrator of the Agency for International Development to give population and family planning high priority for attention, personnel, research, and funding among our several aid programs."[54] Nixon concluded, "One of the most serious challenges to human destiny in the last third of this century will be the growth of the population. Whether man's response to that challenge will be a cause for pride or for despair in the year 2000 will depend very much on what we do today."[55]

Congress endorsed the Special Message and subsequently named John D. Rockefeller III to chair the Commission on Population Growth, a two-year project culminating in the publication of *Population and the American Future*. This report included over 70 recommendations including the passage of a Population Education Act to establish population education programs in schools. It also called for widespread sex education in schools, the dissemination of contraceptives by the government, government-sponsored abortions, and vastly expanded research pertaining to population control.[56]

After enthusiastically commissioning this study, Nixon sharply condemned the final report just six months before the 1972 presidential election.[57] Commission member James Scheuer commented, "Our exuberance was short-lived. Then-president Richard Nixon promptly ignored our final report. The reasons were obvious—the fear of attacks from the far right and from the Roman Catholic Church because of our positions on family planning and abortion. With the benefit of hindsight, it is now clear that this obstruction was but the first of many similar actions to come from high places."[58]

Consequently, the report remained dormant until the 1974 UN Population Conference in Bucharest, Romania. The U.S. delegates had proposed a plan, prepared by JDR III and the Rockefeller Foundation, calling for drastic global population reduction.[59] Again the plan was met with considerable resistance, forcing these depopulation enthusiasts to consider alternatives methods. William Engdahl, in his book *Seeds of Destruction*, explained, "A fierce resistance from the Catholic Church, from every Communist country except Romania, as well as from Latin American and Asian nations, convinced leading U.S. policy circles that covert means were needed to implement their project."[60]

The continual unpopularity of depopulation initiatives prompted Nixon, on the advice of John D. Rockefeller III, to commission Henry Kissinger to draft a secret policy paper. The now infamous National Security Study Memorandum 200 (NSSM 200), dated December 10, 1974, was subtitled, "Implications of Worldwide Population Growth For U.S. Security and Overseas Interests."[61] NSSM 200 remained secret for 15 years until a lawsuit by organizations associated with the Catholic Church finally prompted its declassification.[62] This chilling report called for massive reductions of world population growth rates. The report also directly linked U.S. national security to third-world population reduction. NSSM 200 further acknowledged that third-world population increases might enable these nations to gain control over their own natural resources. Yet since the U.S. and other first-world nations depended upon third-world natural resources, the populations of these countries would have to be reduced. Kissinger wrote,

- "The U.S. economy will require large and increasing amounts of minerals from abroad, especially from less developed countries. That fact gives the U.S. enhanced interest in the political, economic, and social stability of the supplying countries. Wherever a lessening of population pressures through reduced birth rates can increase the prospects for such stability, population policy becomes relevant to resource supplies and to the economic interests of the United States."[63]

- "The world is increasingly dependent on mineral supplies from developing countries, and if rapid population frustrates their prospects for economic development and social progress, the resulting instability may undermine the conditions for expanded output and sustained flows of such resources."[64]

- "The location of known reserves of higher-grade ores of most minerals favors increasing dependence of all industrialized regions on imports from less developed countries. The real problems of mineral supplies lie, not in basic physical sufficiency, but in the politico-economic issues of access, terms for exploration and exploitation, and division of the benefits among producers, consumers, and host country governments…Although population pressure is obviously not the only factor involved, these types of frustrations are much less likely under conditions of slow or zero population growth."[65]

- "The political consequences of current population factors in the LDCs [Least Developed Countries]—rapid growth, internal migration, high percentages of young people, slow improvement in living standards, urban concentrations, and pressures for foreign migration—are damaging to the internal stability and international relations of countries in whose advancement the U.S. is interested, thus creating political or even national security problems for the U.S. In a broader sense, there is a major risk of severe damage to world economic, political, and ecological systems and, as these systems begin to fail, to our humanitarian values."[66]

Kissinger favored victimizing financially vulnerable men and women. His proposed programs included paying for abortions and vasectomies:

- "Pay women in the LDCs to have abortions as a method of family planning or to pay persons to perform abortions or to solicit persons to undergo abortions."[67]

- "Similarly, there have been some controversial, but remarkably successful, experiments in India in which financial incentives, along with other motivational devices, were used to get large numbers of men to accept vasectomies."[68]

NSSM 200 called for drastic fertility rate reductions resulting in 3 billion fewer people by 2050 as compared to UN projections. Kissinger proposed two action plans, one based on voluntary reductions and aggressive leadership, the other based on mandatory reductions and coercive measures. For the first plan, Kissinger wrote,

- "World policy and programs in the population field should incorporate two major objectives: (a) actions to accommodate continued population growth up to 6 billions by the mid-21st century without massive starvation or total frustration of developmental hopes; and (b) actions to keep the ultimate level as close as possible to 8 billions rather than permitting it to reach 10 billions, 13 billions, or more."[69]

- "While specific goals in this area are difficult to state, our aim should be for the world to achieve a replacement level of fertility, (a two-child family on the average), by about the year 2000. This will require the present 2 percent growth rate to decline to 1.7 percent within a decade and to 1.1 percent by 2000. Compared to the UN medium projection, this goal would result in 500 million fewer people in 2000 and about 3 billion fewer in 2050. Attainment of this goal will require greatly intensified population programs."[70]

For his mandatory plan, Kissinger proposed using food as a weapon to coerce foreign nations into forcibly reducing their populations. He wrote,

- "There is an alternate view which holds that a growing number of experts believe that the population situation is already more serious and less amenable to solution through voluntary measures than is generally accepted. It holds that, to prevent even more widespread food shortage and other demographic catastrophes than are generally anticipated, even stronger measures are required and some fundamental, very difficult moral issues need to be addressed."[71]

- "The conclusion of this view is that mandatory programs may be needed and that we should be considering these possibilities now. This school of thought believes the following types of questions need to be addressed: Should the U.S. make an all out commitment to major limitation of world population with all the financial and international as well as domestic political costs that would entail? Should the U.S. set even higher agricultural production goals which would enable it to provide additional major food resources to other countries? Should they be nationally or internationally controlled? On what basis should such food resources then be provided? Would food be considered an instrument of national power? Will we be forced to make choices as to whom we can reasonably assist, and if so, should population efforts be a criterion for such assistance? Is the U.S. prepared to accept food rationing to help people who can't/won't control their population growth? Should the U.S. seek to change its own food consumption patterns toward more efficient uses of protein? Are mandatory population control measures appropriate for the U.S. and/or for others? Should the U.S. initiate a major research effort to address the growing problems of fresh water supply, ecological damage, and adverse climate?"[72]

Kissinger anticipated opposition to such policies, especially from within his targeted countries. He was particularly concerned with the young people in these countries, those who might be capable of sparking revolution against imperial oppression. Kissinger wrote,

"The young people, who are in much higher proportions in many LDCs, are likely to be more volatile, unstable, prone to extremes, alienation and violence than an older population. These young people can more readily be persuaded to attack the legal institutions of the government or real property of the 'establishment,' 'imperialists,' multinational corporations, or other—often foreign—influences blamed for their troubles."[73]

To mitigate the inevitable accusations of imperialism, Kissinger recommended phony propaganda insisting that third-world population reduction was mutually beneficial:

"The U.S. can help to minimize charges of an imperialist motivation behind its support of population activities by repeatedly asserting that such support derives from a concern with: (a) the right of the individual couple to determine freely and responsibly their number and spacing of children and to have information, education, and the means to do so; and (b) the fundamental social and economic development of poor countries in which rapid population growth is both a contributing cause and a consequence of widespread poverty. Furthermore, the U.S. should also take steps to convey the message that the control of world population growth is in the mutual interest of the developed and developing countries alike."[74]

In other words, deny the reality of the situation simply by claiming that forced sterilization and other coercive depopulation measures are beneficial. According to Kissinger, these countries should be *thanking* him for maiming their men and women and killing their unborn.

After the Watergate scandal forced Nixon to resign in disgrace in 1975, the incoming Ford administration acted promptly to make NSSM 200 official U.S. government policy. On November 26, 1975 National Security Adviser Brent Scowcroft, on behalf of President Ford, signed National Security Decision Memorandum 314 (NSDM 314), approving virtually all of NSSM 200's recommendations.[75] In NSSM 200, Kissinger targeted 13 countries for immediate depopulation:

"In order to assist the development of major countries and to maximize progress toward population stability, primary emphasis would be placed on the largest and fastest growing developing countries where the imbalance between growing numbers and development potential most seriously risks instability, unrest, and international tensions. These countries are: India, Bangladesh, Pakistan, Nigeria, Mexico, Indonesia, Brazil, The Philippines, Thailand, Egypt, Turkey, Ethiopia, and Colombia."[76]

These countries, among the most mineral-rich in the world, have suffered greatly thanks to NSSM 200. In the late 1980s in Brazil, for example, a coalition of 165 legislatures representing every political party in the Brazilian government ordered an official investigation into forced sterilization. The Brazilian Ministry of Health determined that 44 percent of Brazilian women between 14 and 55 years of age had been sterilized. The older women were sterilized in the mid-1970s, just when NSSM 200 became official U.S. policy.[77] The U.S. Agency for International Development (USAID) oversaw the sterilization program while various surrogate groups including the International Planned Parenthood Federation, the U.S. Pathfinder Fund and Family Health International carried out the sterilizations.[78]

In 1977, Reimert Ravenholt, the vicious director of USAID from 1965 until 1979, declared that USAID's goal was to sterilize fully one-quarter of the world's fertile women, or about 100,000,000 women.[79] Ravenholt's approach to population control was "by any means necessary." If foreign governments refused to participate, he would simply hire non-governmental organizations (NGOs) to implement USAID policies.[80] Ravenholt also argued *against* supplying Africa with food, water, and antibiotics assistance. He claimed that these items are actually harmful because they enable too many people to live. Ravenholt wrote, "How could these be harmful? Quite simply, they are enormously harmful to African societies when the deaths prevented thereby are not balanced by prevention of a roughly equal number of births."[81] He also surmised that children rescued by such intervention would likely grow up to be criminals: "It is the population excesses resulting from well-intentioned but population-unbalancing interventionist activities which are largely driving today's killing fields in Africa. Many infants and children rescued from preventable disease deaths by interventionist programs during the 1970s and 1980s have become machete-wielding killers."[82]

According to Human Life International, "The U.S. government has never renounced NSSM-200, but has only amended certain portions of its policy. NSSM-200, therefore, remains the foundational document on population control issued by the United States government."[83] HLI further observed, "From the very beginning, the 'population explosion' concept was an ideologically motivated false alarm. The resulting push for population control in LDCs has borne absolutely no positive fruit in its decades of implementation. In fact, population control ideologies and programs make it even more difficult to respond to the impending grave crisis looming in the form of a disastrous worldwide 'population implosion.'"[84] They continued,

"NSSM-200 does not emphasize the rights or welfare of individuals or of nations, just the 'right' of the United States to have unfettered access to the natural resources of developing nations. The United States and the other nations of the developed world, as well as ideologically motivated population control NGOs, should be supporting and guiding authentic economic development that allows the people of each nation to use their resources for their own benefit, thereby leading to an enhancement of human rights worldwide and healthier economies for all."[85]

Global 2000

In his May 23, 1977 Environmental Message to the Congress, President Jimmy Carter directed both the Council on Environmental Quality and the Department of State to work with various federal agencies to study the "probable changes in the world's population, natural resources, and environment through the end of the century." He wanted this endeavor to be "the foundation of our longer-term planning."[86] The study culminated in July 1980 with publication of *The Global 2000 Report to the President*. The report began,

"If present trends continue, the world in 2000 will be more crowded, more polluted, less stable ecologically, and more vulnerable to disruption than the world we live in now. Serious stresses involving population, resources, and environment are clearly visible ahead. Despite greater material output, the world's people will be poorer in many ways than they are today. For hundreds of millions of the desperately poor, the outlook for food and other necessities of life will be no better. For many it will be worse. Barring revolutionary advances in technology, life for most people on Earth will be more precarious in 2000 than it is now—unless the nations of the world act decisively to alter current trends."[87]

The report cast dire warnings concerning its projected 6.35 billion world-population by 2000, and 10 billion by 2030. Such numbers, said the report, would excessively strain the world's food production capacity and fuel resources. Additionally, *Global 2000* claimed,

- "Regional water shortages will become more severe."
- "Significant losses of world forests will continue over the next 20 years."
- "Serious deterioration of agricultural soils will occur worldwide."
- "Atmospheric concentrations of carbon dioxide and ozone-depleting chemicals are expected to increase at rates that could alter the world's climate and upper atmosphere significantly by 2050."
- "Extinction of plant and animal species will increase dramatically."[88]

Consequently, "Unless this circle of interlinked problems is broken soon, population growth in such areas will unfortunately be slowed for reasons other than declining birth rates. Hunger and disease will claim more babies and young children, and more of those surviving will be mentally and physically handicapped by childhood malnutrition."[89] As solutions to these supposedly dire circumstances, *Global 2000* recommended various eugenic depopulation schemes including "family planning" and "reduced fertility." They also recommended "bold and imaginative" measures such as,

- "The need for family planning is slowly becoming better understood."[90]
- "Here it must be emphasized that, unlike most of the Global 2000 Study projections, the

population projections assume extensive policy changes and developments to reduce fertility rates. Without the assumed policy changes, the projected rate of population growth would be still more rapid. Unfortunately population growth may be slowed for reasons other than declining birth rates."[91]

- "The time for action to prevent this outcome is running out. Unless nations collectively and individually take bold and imaginative steps toward improved social and economic conditions, reduced fertility, better management of resources, and protection of the environment, the world must expect a troubled entry into the twenty-first century."[92]

Henry Kissinger: Portrait of a War Criminal

It was perhaps the greatest irony of the 20th century. In 1973, Henry Kissinger won the Nobel *Peace* Prize. More than three decades later, his wanted status for crimes against humanity makes travel abroad difficult for the elder statesman. In 2002, for example, before a scheduled trip to London, Spanish judge Baltasar Garzon issued a court order requesting that British authorities detain Kissinger for questioning regarding his involvement, among other things, in the 1973 coup d'état in Chile (see Chapter 25).[93] French judge Sophie-Helene Chateau also sought to detain Kissinger for questioning.[94] In 2009, the *New Statesman* reported, "Henry Kissinger, who was largely responsible for bombing 600,000 peasants to death in Cambodia in 1969-73, is wanted for questioning in France, Chile and Argentina."[95]

Kissinger's enthusiasm for eugenics, depopulation and genocide is not surprising considering his long-running association with the Rockefeller family. While still a Harvard graduate student in the early 1950s, the Rockefeller Foundation sponsored several of Kissinger's initiatives including the foreign policy magazine *Confluence*, and the Harvard International Seminar.[96] Then in 1955, Nelson Rockefeller invited Kissinger to serve as Study Director for the Council of Foreign Relations' Nuclear Weapons and Foreign Policy study. From 1956-1958 Kissinger worked for the Rockefeller Brothers Fund as Director of the Special Studies Project.[97]

The evidence against Kissinger could fill volumes. To illustrate his psychopathic New World Order mentality, however, requires just a cursory glance into his depravity. For example, in their 1976 book *The Final Days*, Pulitzer Prize winning reporters Bob Woodward and Carl Bernstein reported on Kissinger's belief that military men are *subhuman slaves*. Specifically, Kissinger called them "dumb, stupid animals to be used" as pawns for foreign policy.[98] Woodward and Bernstein also reported on Kissinger's curious admissions regarding his own integrity. Kissinger told his aides, "You systems-analysis people have too much integrity. This is not an honorable business conducted by honorable men in an honorable way. Don't assume I'm that way and you shouldn't be."[99] Kissinger's abrasive personality guided his decisions both as National Security Advisor (1968-1975) and Secretary of State (1973-1977). His involvement with the 1975 slaughter in East Timor is telling.

On December 6, 1975, Kissinger and President Gerald Ford visited Indonesia to meet with their ally General Suharto, the ruthless dictator and head of the country's military junta. The small nation of East Timor, a Portuguese colony since the 16th century, had recently declared independence. On December 7, having received the blessings of Kissinger and Ford, Suharto's forces invaded East Timor, slaughtering 200,000 people, and subsequently occupying and controlling the territory for 26 oppressive years. In his 1994 book *Diplomacy*, Kissinger documented his diplomatic experiences throughout his career, yet wrote nothing about East Timor. Consequently, while promoting his book on July 11, 1995, at the Park Central Hotel in New York, he drew some unsettling questions.

A Timorese woman named Constancio Pinto asked him why he was in Indonesia the day before the invasion. Kissinger claimed he was heading to China with Ford, but since Mao Tse-tung happened to be sick, they instead went to Indonesia. He claimed they were not discussing East Timor and only briefly heard about the planned invasion just before boarding their plane. "Timor was never discussed with us," Kissinger claimed, "when we were in Indonesia. At the airport as we were leaving, the Indonesians told us that they were going to occupy the Portuguese colony of Timor. To us that did not seem like a very significant event, because the Indians had occupied the Portuguese colony of Goa ten years earlier, and to us it looked like another process of decolonization."[100] Kissinger thought he had escaped embarrassment, but the next questioner, Alan Nairn, exposed his lies and deception:

Nairn: Mr. Kissinger, my name is Allan Nairn. I'm a journalist from the United States. I'm one of the Americans who survived the massacre in East Timor on November 12, 1991, a massacre during which Indonesian troops armed with American M-16s gunned down at least 271 Timorese civilians in front of the Santa Cruz Catholic cemetery as they were gathered in the act of peaceful mourning and protest. Now you just said that in your meeting with Suharto on the afternoon of December 6, 1975, you did not discuss Timor, you did not discuss it until you came to the airport. Well, I have here the official State Department transcript of your and President Ford's conversation with General Suharto, the dictator of Indonesia. It was obtained through the Freedom of Information Act. It has been edited under the Freedom of Information Act so the whole text isn't there. It's clear from the portion of the text that is here, that in fact you did discuss the impending invasion of Timor with Suharto, a fact which was confirmed to me by President Ford himself in an interview I had with him. President Ford told me that in fact you discussed the impending invasion of Timor with Suharto and that you gave the US...

Kissinger: Who? I or he?

Nairn: That you and President Ford together gave U.S. approval for the invasion of East Timor. There is another internal State Department memo, which is printed in an extensive excerpt here, which I'll give to anyone in your audience that's interested. This is a memo of a December 18, 1975, meeting held at the State Department. This was held right after your return from that trip and you were berating your staff for having put on paper a finding by the State Department legal advisor Mr. Leigh that the Indonesian invasion of Timor was illegal, that it not only violated international law, it violated a treaty with the U.S. because U.S. weapons were used and it's clear from this transcript which I invite anyone in the audience to peruse that you were angry at them first because you feared this memo would leak, and second because you were supporting the Indonesian invasion of East Timor, and you did not want it known that you were doing this contrary to the advice of your own people in the State Department. If one looks at the public actions, sixteen hours after you left that meeting with Suharto the Indonesian troops began parachuting over Dili, the capital of East Timor. They came ashore and began the massacres that culminated in [the slaughter of] a third of the Timorese population. You announced an immediate doubling of U.S. military aid to

Indonesia at the time, and in the meantime at the United Nations, the instruction given to Ambassador Daniel Patrick Moynihan, as he wrote in his memoirs, was to, as he put it, see to it that the UN be highly ineffective in any actions it might undertake on East Timor…

[Shouts from the audience]

Kissinger: Look, I think we all got the point now…

Nairn: My question, Mr. Kissinger, my question, Dr. Kissinger, is twofold. First will you give a waiver under the Privacy Act to support full declassification of this memo so we can see exactly what you and President Ford said to Suharto? Secondly, would you support the convening of an international war crimes tribunal under UN supervision on the subject of East Timor and would you agree to abide by its verdict in regard to your own conduct?

Kissinger: I mean, uh, really, this sort of comment is one of the reasons why the conduct of foreign policy is becoming nearly impossible under these conditions. Here is a fellow who's got one obsession, he's got one problem, he collects a bunch of documents, you don't know what is in these documents…

Nairn: I invite your audience to read them.[101]

To this day, Kissinger continues to deny wrongdoing. Yet Phillip Liechty, a former CIA operations officer in Indonesia, has admitted, "Suharto was given the green light to do what he did. There was discussion in the Embassy and in traffic with the State Department about the problems that would be created for us if the public and Congress became aware of the level and type of military assistance that was going to Indonesia at that time."[102]

Paul Ehrlich's Creative Depopulation

Paul Ehrlich, professor of Biology at Stanford University, is a hero of fake environmentalism, an ultra-extremist and alarmist, and a cold-blooded eugenicist. Ehrlich is a Fellow of the American Association for the Advancement of Science, the American Academy of Arts and Sciences, and the American Philosophical Society, and a member of the National Academy of Sciences. Additionally, he has won numerous environmental awards including the John Muir Award of the Sierra Club, the Gold Medal Award of the World Wildlife Fund International, the Crafoord Prize of the Royal Swedish Academy of Sciences, the 1993 Volvo Environmental Prize, the 1994 United Nations' Sasakawa Environment Prize, the 1995 Heinz Award for the Environment, the 1998 Tyler Prize for Environmental Achievement, the 1999 Blue Planet Prize, the 2001 Eminent Ecologist Award of the Ecological Society of America and the 2001 Distinguished Scientist Award of the American Institute of Biological Sciences.[103]

Despite his numerous awards, Ehrlich's research has proven to be remarkably inept — or intentionally deceptive. In 1970 for example, while speaking to a symposium in England, he declared, "If I were a gambler, I would take even money that England will not exist in the year 2000, and give 10 to 1 that the life of the average Briton would be of distinctly lower quality than it is today."[104] As a propagandist, however, Ehrlich has been much more effective. In 1968, the Sierra Club published his doomsday book *The Population Bomb*. The book, which sold more than 1 million copies in less than two years, incited and encouraged widespread fear about overpopulation.[105]

Ehrlich, like many environmentalists, declared that humans are a cancer upon the Earth. He explained, "The battle to feed all of humanity is over. In the 1970s and 1980s, hundreds of millions of people will starve to death in spite of any crash programs embarked upon now…We can no longer afford merely to treat the symptoms of the cancer of population growth; the cancer itself must be cut out."[106] He advocated massive depopulation, ideally down to 500 million worldwide: "But with a human population of, say, one-half billion people, some minor changes in technology and some major changes in the rate of use and equity of distribution of the world's resources, there would clearly be no environmental crisis."[107] Trying not to appear too extreme, Ehrlich reasoned that 1 billion would also suffice. "But at a minimum," he wrote, "it seems safe to say that a population of one billion people could be sustained in reasonable comfort for perhaps 1000 years if resources were husbanded carefully."[108]

Just how would Ehrlich and his cohorts implement such ambitious plans? Ideally, people would eagerly volunteer to be sterilized. Perhaps a clever ribbon campaign as per the Red Cross would encourage participation. But should volunteerism fail, they would opt for *forced* sterilization. Ehrlich explained, "We must have population control at home, hopefully through changes in our value system, but by compulsion if voluntary methods fail."[109] Ehrlich also offered many creative solutions. For example, he suggested putting sterilization chemicals in the water. "Many of my colleagues," he wrote, "feel that some sort of compulsory birth regulation would be necessary to achieve such control. One plan often mentioned involves the addition of temporary sterilants to water supplies or staple food. Doses of the antidote would be carefully rationed by the government to produce the desired population size."[110]

According to Ehrlich, government-issued pregnancy licenses might also be effective. He explained, "A governmental 'first marriage grant' could be awarded each couple in which the age of both partners was 25 or more. 'Responsibility prizes' could be given to each couple for each five years of childless marriage, or to each man who accepted irreversible sterilization (vasectomy) before having more than two children. Or special lotteries might be held—tickets going only to the childless."[111] And of course, he also supported the "weapon" of abortion. "Abortion," he asserted, "is a highly effective weapon in the armory of population control. It is condemned by many family planning groups, which are notorious for pussyfooting about methodology, despite their beginning 60 years ago as revolutionary social pioneers."[112]

And of course, doesn't every man secretly yearn to have his vasa deferentia snipped? Ehrlich asked, "What about vasectomies? A few years ago, there was talk in India of compulsory sterilization for all males who were fathers of three or more children…we should have encouraged the Indian government to go ahead with the plan. We should have volunteered logistical support in the form of helicopters, vehicles, and surgical instruments. We should have sent doctors to aid in the program by setting up centers for training para-medical personnel to do vasectomies. Coercion? Perhaps, but coercion in a good cause."[113]

Regarding India, Ehrlich got his wish. In the early 1970s, the Indian government set up vasectomy camps in Kerala and Gujarat. Angela Franks, in her book *Margaret Sanger's Eugenic Legacy*, explained, "hundreds of thousands of men received $6, a bag of supplies, and a new sari for the wife in exchange for sterilization."[114] USAID, under Kissinger's NSSM 200, supplied the funding. Franks commented, "In 1975, the situation worsened when Prime Minister Indira Gandhi declared Emergency Rule, and forced sterilizations became routine.

Among the pressures put on local officials was the suspension of their salaries unless sterilization quotas were met. In the last six months of 1976, 6.5 million people were sterilized, including men (often elderly) forcibly given vasectomies. Among the Americans who voiced their support for the program were World Bank president Robert McNamara and Paul Ehrlich."[115]

In his 1971 *Revised and Expanded* edition of *The Population Bomb*, Ehrlich commended the various organizations and individuals actively participating in depopulation programs: "The UN has greatly increased its family planning activities, operating through several agencies including WHO, UNICEF, and UNESCO. Secretary General U Thant has been urged by a study group to establish a special 'world population institute' promptly to take practical action against population growth. Robert McNamara, president of the World Bank, has put population projects high on the bank's list of priorities. The Organization for Economic Cooperation and Development (OECD) is also getting into the field."[116] He also recognized the upstanding, philanthropic foundations. Ehrlich noted,

- "In March 1970, a two-year Commission on Population Growth and the American Future was established under the chairmanship of John D. Rockefeller."[117]
- "Aside from government contributions, private foundations such as Ford and Rockefeller are becoming more involved in programs, both for research and overseas family planning projects."[118]

Ehrlich was also enthusiastic about new depopulation groups. He wrote, "A new organization exists — Zero Population Growth — whose mission is to educate the public and politicians to the necessity for stopping population growth as soon as possible, to lobby for legislation, and to work for politicians who support the same goals."[119] Ehrlich failed to mention that he founded ZPG, with the help of prominent Sierra Club members.[120] The organization still operates today, now under a somewhat more auspicious name, the Population Connection. Since the 1970s, Ehrlich's corrupting influence has permeated the environmental movement.

False Environmentalists Seek Real Depopulation

While foundations and psychopathic politicians have commandeered the depopulation movement, scores of fake environmentalists have played supporting roles. The United Nations, for example, with their 1996 *Global Biodiversity Assessment*, suggested that North American lifestyles would necessitate world population being cut to one billion people. They wrote, "An 'agricultural world' in which most human beings are peasants, should be able to support 5 to 7 billion people, probably more if the large agricultural population were supported by an industry-promoting agricultural activity. In contrast, a reasonable estimate for an industrialized world society at the present North American material standard of living would be one billion. At the more frugal European standard of living, 2-3 billion would be possible."[121]

In 1991, the *UNESCO Courier* published an interview with famous environmentalist Jacques-Yves Cousteau. He advocated eliminating 350,000 people per day to save the planet. "Overpopulation," wrote Cousteau, "is our planet's number one problem. Of the 5.7 billion people on Earth, less than 2 billion live in decent conditions. This figure will soon double. Perhaps we shall manage to feed the expected 10 or 12 billion. But that's just about all we shall be able to do…It's terrible to have to say this. World population must be stabilized and to do that we must eliminate 350,000 people per day. This is so horrible to contemplate that we shouldn't even say it. But the general situation in which we are involved is lamentable."[122]

In 1969, Alan Guttmacher, President of Planned Parenthood and a top official of the International Planned Parenthood Federation (IPPF), bluntly demanded, "Each country will have to decide on its own form of coercion, determining when and how it should be employed…The means presently available are compulsory sterilization and compulsory abortion."[123] Guttmacher was also Vice-President of the American Eugenics Society. According to Paul Marx, the founder of Human Life International, "The United Nations Fund for Population Activities and the International Planned Parenthood Federation have the blood of millions of innocent babies worldwide on their hands."[124]

Prince Philip, the longtime President of the World Wildlife Fund and current President Emeritus, has been fear mongering about overpopulation for decades. In his 1988 book *Down to Earth*, Philip recalled, "I don't claim to have had any special interest in natural history, but as a boy I was made aware of the annual fluctuations in the number of game animals and the need to adjust the 'cull' to the size of the surplus population."[125] Of course he wasn't speaking about animals. He made clear his stance on human "herd management" in 1981 when asked by *People* magazine, "What do you consider the leading threat to the environment?" Philip responded,

> "Human population growth is probably the single most serious long-term threat to survival. We're in for a major disaster if it isn't curbed—not just for the natural world, but for the human world. The more people there are, the more resources they'll consume, the more pollution they'll create, the more fighting they'll do. We have no option. If it isn't controlled voluntarily, it will be controlled involuntarily by an increase in disease, starvation and war."[126]

And of course, who could forget Philips's ode to the deadly virus? He brazenly remarked, "In the event that I am reincarnated, I would like to return as a deadly virus, in order to contribute something to solve overpopulation."[127]

Philip is now an old man. Nevertheless, he maintains his voracious enthusiasm for depopulation. In 2008, for example, he explained to the *London Times* the supposed connection between overpopulation and rising food costs. "Everyone thinks it's to do with not enough food," Philip observed, "but it's really that demand is too great—too many people. Basically, it's a little embarrassing for everybody. No one quite knows how to handle it. Nobody wants their family life to be interfered with by the government."[128] In Europe, however, birth rates have been falling for decades and the average birth rate is now far below the 2.1 replacement level.[129] So what's the problem, Philip? Everyone is weary of government intrusion, but somebody ought to reassure the Prince that he'll be okay. The British government will undoubtedly leave him and his four children and his eight grandchildren alone.

Paul Watson, a co-founder of Greenpeace, former Sierra Club board member, and one of *Time* magazine's 20th century environmental heroes is another depopulation extremist. Besides having described mankind as "the AIDS of the Earth," the *Boston Globe* quoted him as saying, "We need to radically and intelligently reduce human populations to fewer than one billion."[130] Watson also announced his New World Order sentiments by stating, "We need to eliminate nationalism and tribalism and become Earthlings." He concluded with the assertion that, "Curing a body of cancer requires radical and invasive therapy, and therefore, curing the biosphere of the human virus will also require a radical and invasive approach."[131]

Ted Turner, former media mogul and founder of the United Nations Foundation, would love to see global population cut by 99 percent. "Personally," Turned asserted, "I think the population should be closer to when we had indigenous populations, back before the advent of farming. Fifteen thousand years ago, there was somewhere between 40 and 100 million people."[132] In May 2009, Turner attended a secret meeting of billionaire depopulation enthusiasts in New York. The *London Times* reported, "Described as the Good Club by one insider, it included David Rockefeller Jr., the patriarch of America's wealthiest dynasty, Warren Buffet and George Soros, the financiers, Michael Bloomberg, the mayor of New York, and the media moguls Ted Turner and Oprah Winfrey."[133] Bill Gates, the former Microsoft CEO, planned and organized The Good Club, which held its inaugural meeting at the Manhattan home of Sir Paul Nurse, the president of Rockefeller University.

On the secretive nature of the meeting, Stacy Palmer, the Editor of the *Chronicle of Philanthropy*, commented, "Normally these people are happy to talk good causes, but this is different—maybe because they don't want to be seen as a global cabal."[134] An unnamed attendee of the meeting told the *Times*, "They wanted to speak rich to rich without worrying anything they said would end up in the newspapers, painting them as an alternative world government."[135] The "cabal" discussed a host of issues, yet overpopulation was their focus. The *Times* reported, "Taking their cue from Gates they agreed that overpopulation was a priority...A consensus emerged that they would back a strategy in which population growth would be tackled as a potentially disastrous environmental, social and industrial threat."[136]

Incidentally, since stepping down as Microsoft CEO in 2008, Gates spends most of his time pushing vaccines upon African and other impoverished countries. In January 2010, his Bill and Melinda Gates Foundation allocated $10 billion to vaccine research and distribution. Gates declared, "We must make this the decade of vaccines."[137] Have vaccines ever included sterilizing additives? Are vaccines linked to cancer, Alzheimer's, autism and other diseases? The reader may investigate further and draw his own conclusions.

As discussed above, when populations industrialize and attain higher living standards, population growth rates start to decline. Yet depopulation enthusiasts never consider, as an option for stabilizing third world populations, actually *helping* these countries to industrialize. They claim that such solutions would detrimentally impact the environment due to increased carbon emissions. Therefore, their so-called philanthropy involves Malthusian eugenics strategies, which always target poor and supposedly racially inferior populations.

Chapter 11
The Environmental Movement: Of, By and For Elites

In 1990, two prominent sociologists, Mayer Zald and John McCarthy, published an incisive book called *Social Movements in an Organizational Society*. Zald was formerly the Chairman of the sociology department at Vanderbilt University. McCarthy currently heads the sociology department at Pennsylvania State University. In their book, they argued that social movements are generally planned, organized, funded, and directed by elites. They are not idealistic grassroots uprisings. Quoting Ralph Turner, the former President of the American Sociological Association, they explained, "We are willing to assume 'that there is always enough discontent in any society to supply the grass-roots support for a movement if the movement is effectively organized and has at its disposal the power and resources of some established elite group.'"[1] Zald and McCarthy, however, went even further, stating, "Grievances and discontent may be defined, created, and manipulated by issue entrepreneurs and organizations." Grassroots, collective experiences and grievances certainly fuel social movements, but nevertheless, elites captain the ship. Zald and McCarthy concluded, "Recent empirical work, however, has led us to doubt the assumption of a close link between preexisting discontent and generalized beliefs in the rise of social movement phenomena."[2]

The modern environmental movement is precisely what Zald and McCarthy had described. Starting in the 1950s, building on the momentum of the eugenics movement and the Malthusian depopulation agenda, various oligarchs and plutocrats began seizing control of the environmental movement and reinventing environmentalism according to their lofty plans for global tyranny. Their new-school false environmentalism came to rationalize and justify both depopulation and global governance while simultaneously injecting plenty of fear for motivational purposes. The environmental movement was the perfect vehicle for achieving their New World Order. They have since duped millions of people across the world. Most people have never learned this important history. And without such knowledge, most will continue believing what the self-declared environmental leaders tell them to believe.

Elite Foundations Orchestrate the Environmental Movement

Pervasive stereotypes notwithstanding, idealistic hippies, impassioned tree-huggers and ardent birdwatchers don't run the environmental movement. The environmental movement runs them. In their book *Trashing the Economy: How Runaway Environmentalism is Wrecking America*, an extensive study of sixty environmental groups based on insider documents and three years of financial research, Ron Arnold and Alan Gottlieb concluded,

"The mainstream environmental leadership is lily-white, highly educated, overwhelmingly professional, and well versed in the ways of power. The environmental leadership is not only found heading up well known organizations such as the Sierra Club, Audubon Society, or National Wildlife Federation, those very same people are also found in government, on the boards of directors of top corporations, in high society, and wherever else they damn well please to be. These people are accustomed to wielding political power, many coming from Eastern families long listed in the Social Register…Many of them live in Washington and saunter back and forth between leadership positions in environmental public interest groups and government agencies…So we are looking at the higher if not elite ranks of American society when we look at the organized environmental movement's members: better educated people of ample money and liberal privilege who benefit from—and are totally oblivious to—all the productive resource enterprises they try to destroy."[3]

Founded in 1956, the Foundation Center is a national nonprofit organization "recognized as the nation's leading authority on organized philanthropy."[4] They maintain "the most comprehensive database on U.S. grantmakers and their grants."[5] Based on Foundation Center data, Donald Gibson, in his book *Environmentalism: Ideology and Power*, detailed the surge in foundation funding of the environmental movement starting in the early 1960s. Gibson, a University of Pittsburgh professor and former U.S. Air Force intelligence analyst, explained,

"In 1963, a total of 957,000 dollars was given to various conservation groups by eight foundations. For the two years 1970 and 1971, 34.2 million dollars in grants are listed under the heading of 'Recreation and Conservation' and 14.9 million for 'Environmental Studies.' In 1984 the Grants Index reports 89.1 million dollars given for 'Environmental Law, Protection and Environmental Education,' 25.9 million for 'Abortion, Birth Control and Family Planning' and 27.5 million for 'Energy.' The vast majority, if not all, of the grants promoted activities or ideas which are antagonistic to industrial production, economic growth and population growth. What was favored was population control, conservation, labor intensive production and less material consumption."[6]

The Ford Foundation and the Rockefeller Brothers Fund are among the leading donors to environmental and population control groups. From 1974 to 1978, the Ford Foundation contributed over $14 million, and the Rockefeller Brothers Fund over $7 million, to such organizations, including the Conservation Foundation, the Environmental Defense Fund, the Natural Resources Defense Council, and the Sierra Club.[7] During this time the Rockefeller Brothers Fund board included six Rockefeller family members and the notorious Henry Kissinger. On the Ford Foundation board were numerous elites including McGeorge Bundy (CFR), Andrew F. Brimmer (CFR, Trilateral Commission, and board member of Bank of America, International Harvester, United Airlines, and DuPont), Hedley Donovan (CFR, *Time* magazine Editor-in-Chief), J. Irwin Miller (CFR, AT&T board member, Union Bank Chairman of the Board), and Robert S. McNamara (CFR, President of World Bank, Brookings Institution board member, former Secretary of Defense).[8]

In 1968, the Ford Foundation granted $3.9 million to seven universities to spur training in ecology. The *New York Times* reported, "The science of ecology is getting a multimillion-dollar grant from the Ford Foundation on the assumption that it can help rescue mankind from the dire consequences of one of its own follies. That folly is the exploitation of the environment without regard for the other creatures and the plants in it."[9] Gordon Harris, a spokesman for the Fund, explicitly linked the donation to the supposed overpopulation problem. "The precipitous increase," he stated, "in human population has begun all over the world to put unprecedented demands on the natural resources to feed and clothe the multiplying generations."[10]

Indoctrinating students and young scientists has been a consistent tactic for these foundations. In 1990, for example, the Rockefeller Foundation announced a $50 million global environment program. The *New York Times* reported, "As an initial step, the five-year program will assist hundreds of young scientists and policy makers in developing countries to create a worldwide network of trained environmental leaders, who will meet regularly at workshops, sharing information and discussing strategy. Through the international network, the foundation wants to encourage efforts to build environmental protection into governments' long-range economic planning. Other major elements would promote the drafting of international treaties to deal with forest, land, and water preservation, and hazardous waste disposal."[11] In other words, the Rockefeller Foundation was seeking to indoctrinate and manipulate young, impressionable, would-be environmentalists.

Wealthy elites have also bankrolled and directed the conservation movement, an important branch of the environmental movement. Starting in 1915, John D. Rockefeller, Jr. granted $25 million for conservation.[12] JDR's son, Laurence, later worked closely with Henry Fairfield Osborn, Jr. on many conservation projects during the 1940s and 1950s. Osborn, notably, was the nephew of notorious globalist J.P. Morgan.[13] In 1948, Rockefeller and Osborn founded the Conservation Foundation. Like most elite environmentalists, they viewed national sovereignty as an impediment to environmental wellbeing. Rockefeller's authorized biographer explained,

> "[Osborn] saw that it was impossible to consider the environment in a nationalistic and isolated way, telling his readers (especially in a second important book, *The Limits of Earth*, published in 1953) that sustainable natural resources could be achieved only in an international context. To achieve his goals, Osborn built institutions—the Bronx Zoo, the Conservation Foundation, which he initiated with help from Rockefeller and others in 1948, and Resources for the Future, an environmental and economic think tank in which LSR also was involved—which would outlast him. He viewed society much as Carnegie and John D. Rockefeller had done: an organism best reached through organized institutional change and best helped through centralized, efficient sources of information and philanthropy."[14]

Environmental Groups as New World Order Pawns

Most prominent environmental groups are simply pawns of elite foundations. The seemingly radical groups are no exception. Maduro and Schauerhammer explained, "Almost every one of today's land-trust, environmental, animal-rights, and population-control groups was created with grants from one of the elite foundations, like the Ford Foundation and the Rockefeller Foundation. These 'seed grants' enable the radical groups to become established and start their own fundraising operations. These grants are also a seal-of-approval for the other foundations."[15] This section examines two prominent environmental groups, Greenpeace and the Worldwatch Institute, so as to better understand whom they are working for and whose interests they represent. After this, the focus turns to Conservation Biology and Deep Ecology, two branches of the environmental movement. Additionally, the next chapter provides extensive analysis of the leading Conservation Biology group, the World Wildlife Fund.

Greenpeace

In a 1991 article called, "The Not So Peaceful World of Greenpeace," *Forbes* magazine quoted Greenpeace cofounder David McTaggart on the secret of the organization's success. McTaggart explained, "It doesn't matter what is true, it only matters what people believe is true...You are what the media defines you to be. [Greenpeace] became a myth, and a myth-generating machine."[16] With more than 5 million members and offices in over twenty countries, Greenpeace is the world's largest environmental group. Their primary objective is fighting climate change because, they claim, "Climate change is the biggest threat now facing our planet."[17] While they deserve credit for opposing genetically modified foods, Greenpeace allocates the majority of its funds to climate change. "The climate imperative," they explain, "underpins all of our international campaigns. We will continue to increase substantially the proportion of our resources spent on tackling climate change."[18]

Greenpeace claims that its financiers do not influence its objectives: "Greenpeace does not solicit or accept funding from governments, corporations or political parties. Greenpeace neither seeks nor accepts donations that could compromise its independence, aims, objectives or integrity."[19] Instead, they proudly accept money from foundations: "Greenpeace relies on the voluntary donations of individual supporters, and on grant support from foundations." Greenpeace exploits the public's ignorance regarding foundations. They know most people don't understand that foundations are creations of elites, designed to serve elite interests. Despite their claims of independence, Greenpeace dutifully serves the New World Order. Not surprisingly, the bulk of their funding comes from the Rockefeller Brothers Fund, the Turner Foundation, and other elite foundations.[20]

Greenpeace cofounder Robert Hunter, reflecting on the environmental movement in his 1979 book *Warriors of the Rainbow*, stated, "Machiavellianism and mysticism alike played their parts in the shaping of the consciousness this movement expressed. It embodied at times a religious fervor, at other times, a ruthlessness that bordered on savagery."[21] He went on to admit, "Corruption and greatness both played their part and both took their tolls." Greenpeace has certainly been no stranger to corruption or to the Machiavellian "ends justify the means" philosophy. For example, in the early 1980s, Greenpeace produced a documentary called *Goodbye Joey*, purportedly an exposé on animal cruelty in Australia. For the film's pivotal, heart-wrenching scenes, however, they *hired* kangaroo shooters to *torture* kangaroos. In 1983, an Australian court confirmed that the scenes were staged and fined Greenpeace for animal torture.[22]

In another documentary, *Bitter Harvest*, a Greenpeace actor dragged a baby seal away from its protesting mother before skinning it alive. In a Norwegian court, journalist Magnus Gudmundsson proved that Greenpeace staged the scene and had deliberately falsified information throughout the documentary. Consequently membership in Greenpeace Norway collapsed from 15,000 members down to just 35. The Chairman of the Norwegian chapter, Bjørn Økern, resigned in disgust. Økern acknowledged that Greenpeace never used its funding for "environmental protection." Furthermore, he said Greenpeace was "an eco-fascist group."[23] Today, Greenpeace characterizes itself as a "pressure group," according to its recently replaced Executive Director, Gerd Leipold. During an August 2009 interview on the BBC, Leipold defended his organization's alarmism, stating, "We, as a pressure group, have to emotionalize issues, and we're not ashamed of emotionalizing issues."[24]

The Worldwatch Institute

With Rockefeller Brothers Fund cash, Lester Brown founded the Worldwatch Institute in 1974.[25] Today, Worldwatch continues to receive generous funding from the Rockefeller Brothers Fund as well as many other elite foundations including the Ford Foundation, the Andrew W. Mellon Foundation and the Turner Foundation.[26] Incidentally, they also receive funding from the Foundation for Deep Ecology, a foundation started by Douglas Tompkins, founder of both Esprit and The North Face clothing companies.[27] Regarding Worldwatch, Arnold and Gottlieb observed, "The fundamental premise of Worldwatch Institute's output is that development is using up all the resources and our economies are collapsing as a result. They never seem to notice that new technologies make resources out of what was once useless."[28]

Lester Brown is a hardcore globalist, firmly opposed to national sovereignty. For example, in his 1972 book *World Without Borders*, Brown called for the establishment of a "world environmental agency." He claimed, "Arresting the deterioration of the environment does not seem possible within the existing framework of independent nation-states."[29] Following the 1992 Earth Summit, Worldwatch Institute published *After the Earth Summit: The Future of Environmental Governance*, by Hilary French. She wrote,

- "National sovereignty — the power of a country to control events within its territory — has lost much of its meaning in today's world, where borders are routinely breached by pollution."[30]

- "Nations are in effect ceding portions of their sovereignty to the international community, and beginning to create a new system of international environmental governance as a means of solving otherwise-unmanageable problems."[31]

- "Paradoxically, one way to make environmental agreements more effective is in some cases to make them less enforceable — and therefore more palatable to negotiators who may initially feel threatened by any loss of sovereignty. So-called 'soft law' — declarations, resolutions, and action plans that nations do not need to formally ratify and are not legally binding — can help to create an international consensus, mobilize aid, and lay the groundwork for the negotiation of binding treaties later."[32]

Besides advocating global government, Worldwatch also supports depopulation and the manmade global warming myth. Their mission statement explains, "Worldwatch focuses on the 21st century challenges of climate change, resource degradation, population growth, and poverty by developing and disseminating solid data and innovative strategies for achieving a sustainable society."[33]

Conservation Biology

Conservation biology is the study of biodiversity and its environmental implications. Seeking to legitimize and institutionalize this discipline, Michael Soulé founded the Society for Conservation Biology (SCB) in 1985. SCB describes itself as "an international professional organization dedicated to promoting the scientific study of the phenomena that affect the maintenance, loss, and restoration of biological diversity."[34] Soulé, however, has admitted that conservation biology is pseudoscientific, with no grounding in traditional science. In his book *Conservation Biology*, he wrote, "In many situations conservation biology is a crisis discipline. In crisis disciples, in contrast to 'normal' science, it is sometimes imperative to make an important tactical decision before one is confident in the sufficiency of the data...Warfare is the epitome of a crisis disciple. On a battlefield, if you observe a group of armed men stealthily approaching your lines, you are justified in taking precautions, which may include firing on the men."[35]

Conservation biology groups insist that humans are ravaging the world's ecosystems. For example, in their 2008 *Living Planet Report*, the World Wildlife Fund (WWF) stated, "Humanity's demand on the planet's living resources, its Ecological Footprint, now exceeds the planet's regenerative capacity by about 30 percent. This global overshoot is growing and, as a consequence, ecosystems are being run down and waste is accumulating in the air, land and water."[36] According to the WWF, the problem is so severe that, "If we continue with business as usual, by the early 2030s we will need two planets to keep up with humanity's demands for goods and services."[37] Such rhetoric, typical of conservation biology groups, serves to demonize humanity, victimize nature and pave the way for false solutions and draconian legislation. The Wildlands Project is one striking example.

In 1991, Michael Soulé and David Foreman founded the Wildlands Project to "provide a Conservation Strategy that is Continental, Regional and Local in Scale."[38] Now called the Wildlands Network (WN), the organization seeks to enlarge and connect protected wilderness areas because, "The science of conservation biology tells us that existing protected areas are too small and too isolated from one another to allow for the natural migrations of animals, plants and ecological processes."[39]

The Wildlands Network believes that nature is more important than humanity. They see themselves as nature's defense force against selfish and wicked humans. For example, Reed Noss, a current scientific advisor and former WN board member explained, "Most conservation biologists agree that compatible human uses of the landscape must be considered...however, the native ecosystem and the collective needs of non-human species must take precedence over the needs and desires of humans."[40] John Davis, former editor of *Wild Earth*, the now defunct official journal of the Wildlands Project further asserted,

> "*Wild Earth* exists in part to remind conservationists that in the long run all lands and waters should be left to the whims of Nature, not to the selfish desires of one species which chose for itself the misnomer Homo sapiens. Does the foregoing mean that *Wild Earth* and the Wildlands Project advocate the end of industrial civilization? Most assuredly. Everything civilized must go. Humanizing of landscapes must stop now and be reversed."[41]

At the 1993 SCB annual meeting, Soulé and Foreman introduced the ultra-radical Wildlands Project. *Science* magazine observed that their "controversial plan to protect North American Biodiversity calls for nothing less than resettling the entire continent."[42] Soulé and Foreman wanted one-quarter of America to become wilderness areas, strictly off-limits to human activity. They also called for "severely restricted" human activity across another one-quarter of the country. They called for "a network of wilderness reserves, human buffer zones, and wildlife corridors stretching across huge tracts of land — hundreds of millions of acres; as much as half the continent."[43] *Science* equated the plan to the creation of "an archipelago of human-inhabited islands surrounded by natural areas."[44]

In his 1991 book *Confessions of an Eco-Warrior*, Foreman further revealed the mentality of the conservation biology leadership:

"We should demand that roads be closed and clearcuts rehabilitated, that dams be torn down, that wolves, grizzlies, cougars, river otters, bison, elk, pronghorn, bighorn sheep, caribou and other extirpated species be reintroduced to their native habitats. We must envision and propose the restoration of biological wildernesses of several million acres in all of America's ecosystems, with corridors between them for the transmission of genetic variability. Wilderness is the arena for evolution, and there must be enough of it for natural forces to have free rein."[45]

In 1996, the United Nations *endorsed* the Wildlands Project. Their *Global Biodiversity Assessment*, under the heading 'Protection and Management of Fragments,' stated,

"This means that representative areas of all major ecosystems in a region need to be reserved, that blocks should be as large as possible, that buffer zones should be established around core areas, and that corridors should connect these areas. This basic design is central to the recently-proposed Wildlands Project in the US."[46]

In the same report, the UN also proposed draconian legislation, purportedly to protect biodiversity:

"We should accept biodiversity as a legal subject, and supply it with adequate rights. This could clarify the principle that biodiversity is not available for uncontrolled human use. However, this non-availability should not turn into an unrealistic conservation. Contrary to current custom, it would therefore become necessary to justify any interference with biodiversity, and to provide proof that human interests justify the damage caused to biodiversity."[47]

Despite its radical anti-humanism, the Wildlands Network is making steady progress towards its objectives. Reed Noss, former SCB President and Wildlands Network's science chief, bragged, "Many international, national, regional, and local conservation groups—both mainstream and grassroots—have adopted the general goals and reserve design model of the Wildlands Project. What once was radical is now almost status quo. This profound influence on the conservation movement is the Wildlands Project's most enduring achievement."[48] The infamous Paul Ehrlich also bragged, "Although the Wildlands Project's call for restoring keystone species and connectivity was met, at first, with amusement, these goals have now been embraced broadly as the only realistic strategy for ending the extinction crisis."[49]

Deep Ecology

The Deep Ecology movement is the most extreme, anti-human branch of the environmental movement. Michael Coffman, in his book *Saviors of the Earth? The Politics and Religion of the Environmental Movement*, explained,

"Biocentric beliefs have divided environmental activism into two camps, New Age (Shallow Ecology), and Deep Ecology. The New Age, or Shallow Ecology approach holds that although man is of no greater 'value' than other aspects of nature, humans have attained a higher level of conscious evolution. Once a person finally sees this 'light of truth' through full 'self-realization' he is supposedly able to become part of nature and work harmoniously with her. Through proper use of technology and mystic communion with nature, man becomes a benevolent caretaker of nature."[50]

Most false environmentalists are more closely aligned with the New Age camp, which enables them to justify their depopulation and anti-industrial growth agendas. Deep Ecologists, though fewer in number, are ideologically far more dangerous. They see humanity as a cancer infecting the Earth. Therefore, they advocate drastic population reduction. Some, like David Graber, the chief scientist for the Pacific West region of the National Park Service, even call for human extinction. In 1989, Graber wrote a *Los Angeles Times* review for Bill McKibben's book *The End of Nature*. Graber stated,

"Human happiness, and certainly human fecundity, is not as important as a wild and healthy planet. I know social scientists who remind me that people are part of nature, but it isn't true. Somewhere along the line—at about a billion years ago, maybe half that—we quit the contract and became a cancer. We have become a plague upon ourselves and upon the earth. It is cosmically unlikely that the developed world will choose to end its orgy of fossil-energy consumption, and the Third World its suicidal consumption of landscape. Until such time as Homo sapiens should decide to rejoin nature, some of us can only hope for the right virus to come along."[51]

Eric Pianka, a biology professor from the University of Texas, is currently the de facto leader of the Deep Ecology movement. In 2006, while accepting the Distinguished Texas Scientist award from the Texas Academy of Science, Pianka sparked controversy for his anti-human sentiments including his assertion that "the world will be much better off when there's only 10 percent of us left."[52] In a phone interview before his speech, Pianka asserted, "[Disease] will control the scourge of humanity. We're looking forward to a huge collapse."[53] Local Texas newspaper the *Seguin Gazette-Enterprise* reported, "The professor weighed the killing power of various diseases such as bird flu and HIV, insisting neither yield the needed results."[54] Pianka further asserted, "HIV is too slow. It's no good."[55]

According to Pianka, Ebola is the perfect remedy to cure the Earth's human problem. The *Seguin Gazette-Enterprise* reported, "Pianka began reciting the merits of an Ebola pandemic to accomplish the requisite 'collapse.'"[56] He stated, "Good terrorists would be taking [Ebola] so that they had microbes they could let loose on the Earth that would kill 90 percent of people."[57] Pianka also wrote a paper, published on the University of Texas website, called *The Vanishing Book of Life on Earth*. He declared,

"Now it is only a matter of time until Ebola Zaire evolves and mutates a little, it will eventually become airborne, and then we might finally see it spread. And if it does, when it does finally sweep across the world—we're going to have a lot of dead people. Every one of you that is lucky enough to survive gets to bury nine. Think about that. However, I doubt Ebola is going to be the one that gets us, I think it will be something else. Did you ever wonder why things like SARS and now the Avian Flu are continually cropping up? They're arising because we were dumb enough to make a perfect epidemiological substrate for an epidemic. We bred our brains out, and now we're going to pay for it. The microbes are going to take over. They're going to control us again as they have in the past. Think about that."[58]

In the same paper, Pianka also warned about human CO_2 emissions and wrote about his pet bull, which he affectionately calls Lucifer.

Besides mad-scientists like Pianka, numerous organizations also support the Deep Ecology movement. Earth First! is the undisputed leader. Through various conduits, especially the Fund for Wild Nature and the Trees Foundation, Earth First! receives funding from the Turner Foundation, the Tides Foundation, the Natural Resources Defense Council and many others.[59] In 1980, before founding the Wildlands Project, David Foreman founded Earth First! Their motto is "No Compromise in the Defense of Mother Earth!"[60] They declare, "Our actions are tied to Deep Ecology, the spiritual and visceral recognition of the intrinsic, sacred value of every living thing."[61] On the roots of the organization, Arnold and Gottlieb reported,

> "Defectors from the environmental movement have told us that Earth First! founder Dave Foreman was approached by the Sierra Club and his employer, the Wilderness Society, in 1979 with an offer to fund a new extremist point group for the movement. It would serve the function of making their own demands look more reasonable…Defectors say that Foreman made the deal by himself in a comfortable Wilderness Society office, and accepted the offer on the condition that funding would be steady and adequate, and that his participation was a limited 10-year deal, and only then began looking for a model that would attract an appropriate cult following."[62]

This report was substantiated by David Brower, founder of the Sierra Club, who later admitted, "The Sierra Club made the Nature Conservancy look reasonable. I founded Friends of the Earth to make the Sierra Club look reasonable. Then I founded Earth Island Institute to make Friends of the Earth look reasonable. Earth First! now makes us look reasonable. We're still waiting for someone else to come along and make Earth First! look reasonable."[63] This will likely never happen. When it comes to depravity, Earth First! has no rivals.

In 1987, Australian magazine *Simply Living* interviewed David Foreman. At the time, he was miffed because people were rejecting his genocidal depopulation theories. Foreman explained, "When I tell people how the worst thing we could do in Ethiopia is to give aid—the best thing would be to just let nature seek its own balance, to let the people there just starve—they think this is monstrous."[64] Foreman has since stated, "We advocate biodiversity for biodiversity's sake. It may take our extinction to set things straight."[65]

Also in 1987, Earth First! member Christopher Manes wrote an article for the organization's journal praising AIDS as an effective depopulation mechanism. Manes suggested,

> "If radical environmentalists were to invent a disease to bring human population back to ecological sanity, it would probably be something like AIDS…Barring a cure, the possible benefits of this to the environment are staggering. If, like the Black Death in Europe, AIDS affected one-third of the world's population, it would cause an immediate respite for endangered wildlife on every continent. More significantly, just as the Plague contributed to the demise of feudalism, AIDS has the potential to end industrialism, which is the main force behind the environmental crisis."[66]

Tom Stoddard, another Earth First! member, wrote an article entitled, "Oh, What a Wonderful Famine!" Stoddard suggested, "We may someday look back at the 1985 famine and say it was a historical turning point."[67] The article was a monstrous diatribe against population growth. "Indeed, this could be a wonderful famine," Stoddard wrote, "if we convince all environmental organizations that most of their concerns are directly related to human population and that including the population problem in their programs and publications is fundamental to advancing their cause." He concluded his article by suggesting, "We should change our slogan from Save The Children to Stop the Children."[68]

In 1989, Foreman and four other Earth First! members were arrested for terrorism after destroying thirty-four power lines leading to uranium mines in Wenden, Arizona.[69] Undercover FBI agent Michael Fain, having infiltrated their group, made over 800 hours of audio recordings. After Wenden, Earth First! was planning a massive, coordinated attack targeting power lines at three separate nuclear facilities in California, Arizona, and Colorado. Foreman pled guilty but got off with a misdemeanor and probation. The other four were convicted of felonies and imprisoned.[70] Foreman then quit Earth First! to start the aforementioned Wildlands Project. Then in 1995, he received his payoff from the Sierra Club—a cushy position on their board of directors. The Sierra Club, incidentally, receives funding from the Rockefeller Brothers Fund, and the Turner Foundation, and the Tides Foundation, among others.[71]

The CFR as Auxiliary Environmentalists

As documented in Part I, the Council on Foreign Relations is a New World Order steering committee intent on destroying national sovereignty and implementing world government. In accordance with the New World Order's heavy investment in false environmentalism, the CFR postures as an auxiliary environmental group, providing analysis and publicity for global warming, overpopulation and other contrived environmental issues. Chapter 24 examines the CFR's dominant influence over mainstream, corporate media and thus their role in fostering sustained media hype concerning global warming. The CFR has also published scores of articles and white papers supporting the manmade global warming theory. In 2007, for example, they published *Climate Change and National Security*, a document in which CFR President Richard Haass argued that climate change poses not only economic threats, but national security threats as well. Haass declared, "Domestically, the effects of climate change could overwhelm disaster-response capabilities. Internationally, climate change may cause humanitarian disasters, contribute to political violence, and undermine weak governments."[72]

The CFR's website houses over 3,300 articles and publications concerning climate change. Also their journal *Foreign Affairs* has always supported New World Order environmentalism. For example, in his infamous 1974 article "The Hard Road to World Order," Richard Gardner endorsed the globalists' plan to siphon off national sovereignty by establishing an authoritative global environmental protection agency. He explained,

"The next few years should see a continued strengthening of the new global and regional agencies charged with protecting the world's environment. In addition to comprehensive monitoring of the earth's air, water and soil and of the effects of pollutants on human health, we can look forward to new procedures to implement the principle of state responsibility for national actions that have transnational environmental consequences, probably including some kind of 'international environmental impact statement' procedure by which at least some nations agree to have certain kinds of environmental decisions reviewed by independent scientific authorities. At the same time, international agencies will be given broader powers to promulgate and revise standards limiting air and ocean pollution."[73]

Gardner then saddled environmental degradation onto overpopulation, noting, "We are entering a wholly new phase of international concern and international action on the population problem, dramatized by the holding this year of the first World Population Conference to take place at the political level. By the end of this decade, a majority of nations are likely to have explicit population policies, many of them designed to achieve zero population growth by a specific target date."[74]

George Kennan, the former U.S. Ambassador to the Soviet Union and Yugoslavia, wrote a 1970 article for *Foreign Affairs* called "To Prevent a World Wasteland." He asserted, "Not even the most casual reader of the public prints of recent months and years could be unaware of the growing chorus of warnings from qualified scientists as to what industrial man is now doing—by overpopulation, by plundering of the earth's resources, and by a precipitate mechanization of many of life's processes—to the intactness of the natural environment on which his survival depends."[75] Kennan acknowledged that restoring the environment might involve unpopular policies. "If the present process of deterioration is to be halted," he wrote, "things are going to have to be done which will encounter formidable resistance from individual governments and powerful interests within individual countries. Only an entity that has great prestige, great authority and active support from centers of influence within the world's most powerful industrial and maritime nations will be able to make headway against such recalcitrance."[76]

Once again, the incestuous ménage à trois of overpopulation, environmentalism, and globalization had come to the forefront. Kennan's proposed "entity of great prestige" was the United Nations. According to Kennan, this entity would benevolently represent all of humanity. "What is lacking in the present pattern of approaches," he wrote, "would seem to be precisely an organizational personality—part conscience, part voice—which has at heart the interests of no nation, no group of nations, no armed force, no political movement and no commercial concern, but simply those of mankind generally, together—and this is important—with man's animal and vegetable companions, who have no other advocate."[77]

But once this entity acquires power, admitted Kennan, sacrifices would inevitably follow. He explained, "The process of compromise of national interests will of course have to take place at some point in every struggle against environmental deterioration at the international level. But it should not occur in the initial determination of what is and is not desirable from the conservational standpoint."[78] When nations eventually lose their sovereignty, Kennan acknowledged, the environmental enforcement entity would become increasingly tyrannical. "But one could hope," he wrote, "that eventually, as powers were accumulated and authority delegated under multilateral treaty arrangements, the Agency could gradually take over many of the functions of enforcement for such international arrangements as might require enforcement in the international media, and in this way expand its function and designation from that of an advisory agency to that of the single commanding International Environmental Authority which the international community is bound, at some point, to require."[79]

By consistently propping up Al Gore, Mikhail Gorbachev and other false environmental spokespersons, the CFR also functions as an auxiliary. In a 1990 speech, Gorbachev declared, "The ecological crisis we are experiencing today—from ozone depletion to deforestation and disastrous air pollution—is tragic but convincing proof that the world we all live in is interrelated and interdependent. This means that we need an appropriate international policy in the field of ecology. Only if we formulate such a policy shall we be able to avert catastrophe. True, the elaboration of such a policy poses unconventional and difficult problems that will affect the sovereignty of states."[80] He further advocated "making ecology a part of education and instruction from an early age, molding a new contemporary attitude by which we recover a sense of being a part of nature."

Flora Lewis, a prominent establishment journalist and Council on Foreign Relations member, later wrote a gushing article for the *New York Times* entitled, "Gorbachev Turns Green." Lewis enthusiastically reported, "Soviet diplomacy is preparing a dramatic leap in the concept of 'new world order' that will leave President Bush back in the primeval sludge if he doesn't move."[81] She continued, "Moscow suggests a convention for all states to sign. It would have an aspect of world government, because it would provide for the World Court to judge states. The Court, which has no enforcement power now, could recommend sanctions. This is a breathtaking idea, beyond the current dreams of ecology militants. It is meant to show that the Soviets really take seriously their 'integration' in the world economy. And it is fitting that the environment be the topic for what amounts to global policing."[82]

Chapter 12

The WWF: A Case Study in False Environmentalism

The World Wide Fund for Nature (WWF), formerly known as the World Wildlife Fund, is the preeminent ultra-elite, false environmental organization. While posturing as stewards of the Earth, the WWF exists primarily to support elite objectives including eugenics, depopulation, suppression of third world economies, and the implementation of global government. On their website they state, "From our start in 1961, WWF has worked toward the protection of endangered species. Our mission is to use the best conservation science available and work with people to find solutions to save the marvelous array of life on our planet."[1] WWF founder Julian Huxley, however, made clear that "protecting endangered species" really means controlling human populations. Furthermore, according to Huxley, animals are more important than humans. He explained,

> "The recognition of the idea of an optimum population-size (of course relative to technological and social conditions) is an indispensable first step towards that planned control of populations which is necessary if man's blind reproductive urges are not to wreck his ideals and his plans for material and spiritual betterment. The recognition of the fact that the wildlife of the world is irreplaceable, but that it is being rapidly destroyed, is necessary if we are to realise in time that areas must be set aside where, in the ultimate interests of mankind as a whole, the spread of man must take second place to the conservation of other species."[2]

WWF's founders and leaders arouse instant suspicion. Of course there is Huxley, the racist eugenicist and primary founder. Then there is Prince Bernhard, the former Nazi, founder of Bilderberg and founding President of WWF from 1962-1976. Then there is Prince Philip, the genocidal depopulation enthusiast and third WWF President from 1981-1996. Then there is Ruud Lubbers, the Club of Rome member and former Dutch Prime Minister who was found guilty by an official UN investigation of a "pattern of sexual harassment."[3] And then, finally, there is John Loudon, the former head of Dutch Royal Shell. Under Loudon's watch, starting in 1958, Shell devastated the Niger Delta and swindled local populations for decades. Nigerian activist Ken Saro-Wiwa observed, "What Shell and Chevron have done to Ogoni people, land steams, creeks and atmosphere amounts to genocide."[4] In 1995, Nigeria's military regime hung Saro-Wiwa for protesting Shell's environmental practices. Facing trial in 2009 for decades of human rights abuses, Shell claimed innocent yet nevertheless paid a $15.5 million settlement, which they termed a "humanitarian gesture."[5] The WWF today is quite proud of this legacy. On their website, they boast,

- "One of the most important figures in WWF's early history was the renowned British biologist, Sir Julian Huxley."[6]
- "Known as the 'Flying Prince of Conservation,' HRH Prince Bernhard of the Netherlands is the Founding President of WWF, a position he held from 1962 to 1976."[7]
- "Better known as 'the Grand Old Man of Shell,' John H. Loudon, a Dutchman, headed Royal Dutch Shell from 1951 to 1965…He was President of WWF from 1976 to 1981."[8]

- "HRH The Duke of Edinburgh served as International President of WWF for 16 years until his retirement at the end of 1996…His Royal Highness' contribution to the organization has been inestimable."[9]
- "[Ruud Lubbers] became President of WWF International on 1 January 2000, but only served for one year as he was appointed United Nations High Commissioner for Refugees from 2001-2005."[10]

The WWF's history stretches back to a 1940s club known as Tots and Quots, the place where co-founders Huxley and Max Nicholson first met. According to the *Oxford Dictionary of National Biography*, "Tots and Quots was a London-based dining club composed of scientific and other intellectuals who sought to influence ways in which science and its practitioners could better address economic, social, and political concerns in inter-war and wartime Britain"[11] Solly Zuckerman, an Oxford zoology professor and later the chief science advisor to the British Ministry of Defense, founded the club in 1931. He invited various Marxists, communists and socialists to join: "All held close British Communist Party links; other members embraced varied progressive socialist views, with centralized scientific planning a common cause."[12] Journalist Kevin Dowling, a writer for the *London Guardian*, *London Times*, and *Daily Telegraph*, also noted, "But to suggest that this developing scientific elite was dominated by Communists is to miss the point. In fact the group included several well-known fascists as well. When he founded Tots and Quots, Zuckerman was already an intimate member of the social circle revolving around Alex and John Spearman and John Strachey, co-founder of Oswald Mosley's New Party, which was later renamed the British Union of Fascists."[13] After three years of initial activity, the club was dormant from 1934 until 1939.

The *Oxford Dictionary* states, "In the interval before Zuckerman resurrected the Tots and Quots in November 1939, its returning communist contingent had accelerated their political activism through writing, lecturing, and organizational initiatives—all in response to heightened international and domestic tensions. While such activities, and their Stalinist underpinnings, dissuaded former members like Hogben and (probably) Gaitskell from returning, they helped give the club's next phase a new clarity of purpose…Zuckerman's expanding elite society connections and the club's growing status in intellectual circles were reflected in the high social standing and political connections of invited guests during this phase."[14]

Guests representing the scientific elite included Royal Society executives Alfred Egerton, A.V. Hill, Sir Henry Tizard, Churchill's science adviser F.A. Lindemann, and Victor Rothschild. Guests representing the political elite included Home Secretary Herbert Morrison, Minister of Works and Buildings Lord Reith, Minister of Food Lord Woolton, the Labour Party's leading Zionist Richard Crossman, Soviet ambassador Ivan Maisky, and American ambassador John Winant. Guests from the intellectual elite included the racist eugenicist H.G. Wells, the founder of Penguin books Allen Lane, and influential Labour Party member and closet occultist Tom Driberg.[15] Incidentally, documentary evidence shows that Driberg, a close friend of Aleister Crowley, swore his allegiance to the "Great Work" and the "Beast 666."[16]

During this second phase of Tots and Quots, Zuckerman also recruited Julian Huxley, Max Nicholson and other scientific elites as members. Zuckerman admitted in his own autobiography, "The point was again made that the first thing we had to do was to devise sensible propaganda directed not only at the scientific community, but also at the general public."[17] Dowling explained, "What united this extraordinary group was a common commitment to the 'scientific management' of nature and of human society. Their goal was a corporatist utopia: just what flag flew over it was of little concern, so long as they had a hand in building it."[18] This was the incubatory period of the WWF, a time during which Huxley and Nicholson were brainstorming future projects.

In 1945, Huxley became the founding Director-General of UNESCO. Three years later, under UN auspices, he established the International Union for Conservation of Nature (IUCN). Dowling explained, "Nicholson immediately arranged for the Foreign Office to draft a new constitution for the IUCN, which would ensure it a unique autonomy. It was to be in, but not of, the new United Nations system, immune from peer review by the Association of Scientific Unions, and to be governed by a self-perpetuating Board of insiders."[19] According to Nicholson,

> "The Union was an uncommon if not unique animal, having as members governments, official agencies and many non-governmental bodies such as societies. Its first impulse was to meet, with unrealistic frequency, at a widely spread series of cities, and to pass with carefree abandon a copious set of urgent resolutions calling for action on varied controversial issues by bodies who were in no way obliged to give heed to them."[20]

In his book *The New Environmental Age*, Nicholson lamented, "As I had feared, the launch of the Union in 1948 was ill-prepared and premature."[21] Nevertheless, it paved the way for the creation of the WWF. Huxley later reflected, "My proudest achievement was at UNESCO, where I succeeded, against considerable opposition, in setting up IUCN, which has achieved so much in its short life, and other bodies with similar aims—the World Wildlife Fund."[22]

In 1960, the IUCN commissioned Huxley, then 73 years old, to assess the state of the environment in Africa.[23] Traveling through ten African countries, he visited twenty-five wildlife preserves and scientific institutions. Along with several other prominent elites, Huxley was planning the formation of the World Wildlife Fund. According to Dowling,

> "[Huxley] came up with a thunderously pessimistic report, in which a great deal of invective was aimed at the tribespeople of Africa, who he claimed were denuding the continent of its natural resources. In the report, he emphasized the need to preserve the land in its pristine condition, but was dismissive of the need to protect particular species such as the African elephant and the rhino. He suggested that all Africans were potential poachers, and that if Africa was to be saved, it would be saved by Europeans."[24]

The *Observer* published regular installments of Huxley's findings, a shrewd ploy to garner public support for the idea that Africa must be saved from the Africans. The WWF website notes, "The articles hit home, alerting readers to the fact that nature conservation was a serious issue. Huxley received a number of letters from concerned members of the public. Among these was a letter from businessman Victor Stolan, who pointed out the urgent need for an international organization to raise funds for conservation."[25]

Max Nicholson, at the time, was a trustee for the *Observer* and Director General of the British Nature Conservancy. He also had previously helped with the founding of the MI6-sponsored Information Research Department of the Foreign Office. Nicholson recruited two other conservationists with intelligence ties, Peter Scott and Guy Mountford, for the

WWF project. Dowling noted, "And then there was Scott, the first swashbuckling birdwatcher the world had seen, who founded The Wildfowl Trust and staffed it with cryptographers and other former members of Britain's wartime intelligence service. More infamously, he was a founder of the Primitive People's Fund, together with right-wing evolutionists John Aspinall and Sir James Goldsmith. And yes, Mountford too was an agent of MI6—in France in the thirties, and in various theatres, from Africa to Europe to the East during WWII."[26]

With the British royal family endorsing the WWF, and with powerful corporations and wealthy elites providing financial backing, the organization cleverly preempted public suspicions regarding its true purpose by launching an emotional media blitz. The *Daily Mirror*, for example, printed a six-page "shock issue" with plenty of pictures of the supposedly doomed black rhino.[27] Prior to this media blitz, the pioneering television program *Look* had been molding public sentiments by predicting the extinction of the black rhino within 30 years. The co-host of *Look* was media celebrity Maxwell Knight.[28] In 1926, long before his media career, Knight was a council member of the British Fascists Limited.[29] He also associated with pro-Nazi groups including the British Union, the Imperial Fascist League, and the Link. Knight later became head of MI5's B5 (b), a counter-intelligence unit operating independently from the rest of MI5.[30] The Security Executive, the body overseeing all of Britain's intelligence agencies, later rescued Knight's image by arranging for him to host nature programs for children on BBC radio. Adopting the persona of "Uncle Max," he quickly became a celebrated icon.[31]

Huxley, Scott, Nicholson and Mountford officially established the World Wildlife Fund on September 11, 1961, "the day after the IUCN's Special Africa Project conference drew to a close in Arusha, capital of then Tanganyika. The conference had used as its keynote document Huxley's damning screed. The World Wildlife Fund was to be a collection of money-gathering agencies, which in association with 'competent organisations'—such as the IUCN—would develop strategies for the spending of funds collected."[32]

After becoming WWF president in 1961, ex-Nazi Prince Bernhard recruited several more Nazis into the organization including Bernard Timm, the personal assistant of IG Farben chief Carl Bosch, and Herman Abs. Dowling commented, "During the war, Abs represented Nazi interests on the board of the Bank of International Settlements in Switzerland, which was happy to receive gold plundered from Jews. He also sat on the supervisory board of IG Auschwitz, which used slave labour to make the artificial rubber that kept the Wehr-macht on wheels."[33] Abs later served on the board of trustees for WWF-International. Also in 1961, the WWF launched several satellite groups including WWF National Organizations in several western nations, and the African Wildlife Foundation (AWF).

At least half of AWF's founding trustees had links to U.S. intelligence services. Of the ten people chosen for AWF's first board, five had been senior officers in the wartime Office of Strategic Services (OSS), the predecessor of the CIA. Kermit Roosevelt, ringleader of the CIA-sponsored 1953 Iran coup, was one of them.[34] Dowling reported, "The other spooks amongst the AWF's trustees included Arthur 'Nick' Arundel, fresh from his service in a psychological warfare team in Vietnam, and Philip Crowe, U.S. Ambassador to South Africa (he had spied on the British in wartime India). Hal Coolidge, the only zoologist among them, was an old OSS China hand, while AWF founder Russell Train had been one of Coolidge's junior officers. Another trustee, James Sweeney, was a member of the Subversive Activities Control Board, closely linked to both the CIA and the FBI. A seventh was Maurice Stans, later named (but not convicted) as a Watergate conspirator."[35]

Prince Philip became the first President of WWF-England. At least half of the founding trustees of this chapter had links to British intelligence.[36] Dowling observed, "In addition to Mountford, the WWF's first trustees included Kenneth Keith of MI6, and veteran spy Rex Benson, first cousin to the head of MI6, Stewart Menzies…It soon became apparent that the British and American spooks had decided to co-ordinate their activities on the conservation front. All the CIA men amongst the founding trustees of the African Wildlife Foundation, had, before the year was out, become directors of WWF's U.S. chapter. And when Hurcomb came to retire as Vice-President of the IUCN, the American former OSS man, Coolidge, was elected its President."[37]

Sufficiently staffed with spooks and elites, the WWF, together with the United Nations, quickly and efficiently mobilized to dominate the African continent. In his book *Reclaiming Paradise*, John McCormick explained, "In December 1962 the UN adopted a resolution supporting the argument that natural resources were vital to economic development and that economic development in less developed countries could jeopardize natural resources if it took place 'without due attention to their conservation and restoration.'"[38] At its 1963 General Assembly in Nairobi, the IUCN discussed the balance between population and environment and the effects of population pressures on the land. IUCN concluded that "if the purely political considerations which have dominated the African scene during the last few years can now be subordinated to the more precise consideration of primary human and animal needs—which are usually complementary—plans can be prepared which, if realistically implemented, will lead to a reasonable and improving standard of living for the African peoples."[39] Speaking at the Assembly, U.S. Secretary of the Interior Stewart Udall stated, "The burgeoning of populations jeopardizes the conservation idea, and furthers the obsession of over-exploitation. This is simply a 'plunder now, pay later' policy. The expansion of populations everywhere makes all the more necessary a wise policy of conservation. Conservation is not a luxury; it is a necessity."[40]

Under the guise of conservation, the WWF and the IUCN worked to restrain economic growth and development in Africa. For example, in 1965, the WWF pressured the Ugandan government to abandon plans for constructing a hydroelectric power station in the Murchison Falls National Park.[41] The only "development" WWF encouraged was the development of national parks and conservation projects. From 1963 to 1967, the IUCN steadily prepared and revised drafts for an African Convention. In September 1968, 33 African nations adopted the final draft, the African Convention for Conservation of Nature and Natural Resources, purportedly founded for the "conservation, utilization and development of soil, water, flora, and faunal resources in accordance with scientific principles and with due regard to the best interests of the people."[42] To date, westerners have imposed 741 protected areas in 50 African countries, spanning 2 million square kilometers—an area larger than Kenya, Ethiopia and Uganda combined.[43]

Environmental Elites Wield Economic Handcuffs

WWF's Africa strategy typified the ingrained attitudes of the British establishment. In 1918 for example, Philip Henry Kerr, also known as Lord Lothian, wrote, "I need not of course expound our view as to the necessity for some civilised control over politically backward peoples...the inhabitants of Africa and parts of Asia have proved unable to govern themselves...they were quite unable to withstand the demoralising influences to which they were subjected in some civilised countries, so that the intervention of a European power is necessary in order to protect them from those influences."[44] Lothian was a British aristocratic, an elite politician and diplomat, and a member of Lord Milner's round table groups.[45] When Lothian met with Hitler on January 29, 1935, Hitler recommended, "The countries which have relatively common interests are Germany, England, France, Italy, America and Scandinavia; [they] should arrive at some agreement whereby they would prevent their nationals from assisting in the industrialising of countries such as China, and India. It is suicidal to promote the establishment in the agricultural countries of Asia of manufacturing industries."[46] Julian Huxley also had considerable contempt for Africans. In 1924 he wrote for *The Spectator*,

> "You have only to go to a nigger camp-meeting to see the African mind in operation — the shrieks, the dancing and yelling and sweating, the surrender to the most violent emotion, the ecstatic blending of the soul of the Congo with the practice of the Salvation Army. So far, no very satisfactory psychological measure has been found for racial differences: that will come, but meanwhile the differences are patent."[47]

Huxley continued,

> "Then there is the undoubted fact that by putting some of the white man's mind into the mulatto you not only make him more capable and more ambitious (there are no well-authenticated cases of pure blacks rising to any eminence), but you increase his discount and create an obvious injustice if you continue to treat him like any full blooded African. The American negro is making trouble because of the American white blood that is in him."[48]

While perhaps more diplomatic than Huxley, Nicholson was no less extreme. In his pompously titled 1970 book *The Environmental Revolution: A Guide for the New Masters of the World*, Nicholson showcased his Machiavellian ideology. "If the ultimate goal," he wrote, "is to be full harmony between man and nature, all those influences and vested interests which lead mankind into postures and situations inconsistent with that goal must in turn be converted or otherwise brought into line, or in extreme cases neutralised. This process cannot be left to chance. Being above all concerned with the ends, the conservation movement must be untiring in devising and applying the means to achieve them as fully and as swiftly as possible."[49]

Other WWF leaders have also expressed their opposition to third-world self-determination. Thomas Lovejoy, for example, has been a board member of WWF-USA, an advisor to the president of the World Bank, senior advisor to Ted Turner's United Nations Foundation, head of the Heinz Center, and President of WWF-USA from 1973 to 1987. In an early 1980s interview, he expounded on WWF's most serious challenges. "The biggest problem," Lovejoy remarked, "is the damn national sectors of these developing countries. These countries think that they have the right to develop their resources as they see fit.

They want to become powers, sovereign states."[50] He continued, "We thought that we could control things better by reasoning with these leaders, these nationalist fools. We have overestimated our ability to control the people and are going to have to adjust. It will be a painful adjustment, indeed. No, the real problem is this stupid nationalism and the plans for development it leads to."[51]

With unabashed enthusiasm, Lovejoy also outlined the WWF's economic handcuffs program. He explained, "The Brazilians—and I know that from 17 years of experience— think that if they develop the Amazon, they can become a superpower. They have swollen heads about this. So you have to be careful. You buy them off with less. Let them develop bauxite and some other things, but restructure the plans to scale back energy development for environmental reasons. They can't get money now. So we have some friendly banks tell them that they can get money for what we are suggesting. Then we have some of our friends in the development ministry say that this is a good idea."[52]

Arne Schiøtz, former WWF International Director of Conservation from 1980-1983, also supports the WWF Malthusian ideology. He stated, "Malthus has been vindicated; reality is finally catching up with Malthus. We are running out of space. The Third World is overpopulated, it's an economic mess, and there is no way they could get out of it with this fast-growing population. Our philosophy is: Back to the village. We have reached the end of the era of projects whose environmental consequences we do not know. Things in the Third World must not be run any more along these large, blind, grandiose projects, but return to the village, and appropriate technologies. Without appropriate technologies, they could not overcome the economic gap. But it is more and more a question of population. It is a question of space; do we have space on Earth for villages, for so many villages?"[53] He also acknowledged, "WWF/IUCN is part of a worldwide intervention: The idea is that of a supranational intervention into the policies of nation-states."[54]

The WWF masquerades as a benevolent conservation organization. But when the disguise comes off, they are simply a gaggle of racists, eugenicists and depopulation enthusiasts with significant ties to intelligence agencies. As discussed in previous chapters, third-world depopulation became official U.S. policy during the mid-1970s for reasons including first-world dependency on third-world natural resources. In centuries past, third-world oppression and economic suppression was obvious. Today, while the policy has not changed, the execution has become more refined. Today's imperialists cloak their aggression and domination behind seemingly benign conservation and environmental activism.

Chapter 13

The UN: Epicenter of False Environmentalism

In 1972, the United Nations organized the Conference on the Human Environment in Stockholm, Sweden, the first major international environmental conference. Following the conference, they launched the United Nations Environment Programme (UNEP), "To provide leadership and encourage partnership in caring for the environment by inspiring, informing, and enabling nations and peoples to improve their quality of life without compromising that of future generations."[1] Since then, the UN has assumed various leadership and coordinating roles within the environmental movement. They have been the primary nexus between environmental groups and governments. With their second major environmental conference in 1992, the UN symbolically ushered in the current phase of the environmental movement, characterized by unsubstantiated alarmism and scientific aloofness.

The 1992 Conference on Environment and Development (UNCED), commonly known as The Earth Summit, was a massive 12-day conference held in Rio de Janeiro, attended by representatives of 172 nations, including 108 heads of state, and some 35,000 participants.[2] The event sparked a worldwide media frenzy of absolutism. The experts had spoken; Earth was in dire straits and the UN's "sustainable development" initiative was the only solution. Maurice Strong, the UNCED Secretary General, told reporters, "As a bureaucrat, I don't always use the hottest words to describe things, but now we have to push like hell to make sure implementation takes place."[3]

He was referring to Agenda 21, the centerpiece document of the conference, "an 800-page blueprint for environmental action through the 21st century in a number of areas, including pollution, poverty, population and waste management."[4] Strong and his UN cohorts staged the conference, with all its hype and drama, to appeal for worldwide implementation of this radically draconian document. *The Nation* reported, "Called Agenda 21, it sets the goal of bringing world population growth to a stop in the next century at a number that the finite resources of the planet can sustain in health and dignity. To that end, it maps out and prices some 2,500 tasks that need doing to restore and conserve the environment and to extend the chance of a human existence to every child born."[5]

In 1994, environmental attorney Daniel Sitarz published *Agenda 21: The Earth Summit Strategy to Save Our Planet*, an abridged version of the full-length document, which received endorsement and praise from Maurice Strong.[6] The following analysis is based on Sitarz's book. Agenda 21 promoted the standard UN claim that humans are provoking environmental catastrophe and that the "scientific consensus" agrees. Since Rio, they have repeated this claim incessantly, especially regarding manmade global warming. Agenda 21 claimed, "There is strong evidence from the world's scientific community that humanity is very, very close to crossing certain ecological thresholds for the support of life on Earth…the global scientific consensus is that if the current levels of environmental deterioration continue, the delicate life-sustaining qualities of this planet will collapse. It is a stark and frightening potential."[7] The following statements demonstrate the scope and magnitude of Agenda 21:

- "As its sweeping programs are implemented world-wide, it will eventually impact every human activity on our planet."[8]
- "It is a blueprint for action in all areas relating to the sustainable development of our planet into the 21st century. It calls for specific changes in the activities of all people."[9]
- "Effective execution of Agenda 21 will require a profound reorientation of all human society, unlike anything the world has ever experienced—a major shift in the priorities of both governments and individuals and an unprecedented redeployment of human and financial resources. This shift will demand that a concern for the environmental consequences of every human action be integrated into individual and collective decision-making at every level."[10]
- "The successful implementation of the far-ranging actions proposed by Agenda 21 will require active participation by people throughout the world, at the local, national and global levels. There are measures that are directed at all levels of society—from international bodies such as the United Nations and the World Bank to local groups and individuals. There are specific actions which are intended to be undertaken by multinational corporations and entrepreneurs, by financial institutions and individual investors, by high tech companies and indigenous people, by workers and labor unions, by farmers and consumers, by students and schools, by governments and legislators, by scientists, by women, by children—in short, by every person on Earth."[11]

In accordance with New World Order ideology, Agenda 21 argued that overpopulation is the primary cause of environmental degradation and, therefore, any sensible solution must necessarily involve culling the excess, useless eaters:

- "Population pressures are placing increasing stress on the ecological systems of the planet. All countries must improve their ability to assess the environmental impact of their population growth rates and develop and implement appropriate policies to stabilize populations."[12]
- "There is an immediate need to develop strategies aimed at controlling world population growth. There is an urgent demand to increase awareness among decision-makers of the critical role that population plays in environmental protection and development issues."[13]
- "Existing plans for sustainable development have generally recognized that population is a vital factor which influences consumption patterns, production, lifestyles and long-term sustainability. Far more attention, however, must be given to the issue of population in general policy formulation and the design of global development plans."[14]

Agenda 21 also foreshadowed the planned global government: "The transition to a global civilization in balance with nature will be an exceedingly difficult task, but Agenda 21 is the collective global alert that there is no alternative."[15] The stench of globalization heavily permeated the 1992 Earth Summit. For example, former French Prime Minister Michel Rocard boldly proclaimed,

"Let's not deceive ourselves. It is necessary that the community of nations exert pressure, even using coercion, against countries that have installations that threaten the environment. International instruments must be transformed into instruments of coercion, of sanctions, of boycott, even—perhaps in 15 years' time of outright confiscation of any dangerous installation. What we seek, to be frank, is the legitimacy of controlling the application of the international decisions. We need a real world authority, to which should be delegated the follow up of the international decisions, like the treaties signed [in Rio, June 1992]…this authority must have the capacity to have its decisions obeyed. Therefore, we need means of control and sanctions. I well know the nervousness of some countries when they consider their sovereignty to be threatened. But we are not dealing with national problems. These problems are international. Pollution knows no borders and the sea level cannot change in one place without changing in another…Obviously, this supranational authority must be a world authority."[16]

Lester Brown was another arch-globalist who called for world government during the Rio conference. Brown is the founder and President of Earth Policy Institute whose website brags, "The *Washington Post* called Lester Brown 'one of the world's most influential thinkers.' The *Telegraph of Calcutta* refers to him as 'the guru of the environmental movement.'"[17] He's also a longtime Council on Foreign Relations member and founder of the Worldwatch Institute, another elitist environmental think-tank. At the 1992 Earth Summit, Brown endorsed world government during an interview with *Terraviva*, a special daily newspaper from the conference. He stated,

"One hears from time to time from conservative columnists and others that we, as the United States, don't want to sign these treaties that would sacrifice our national sovereignty. But what they seem to overlook is that we've already lost a great deal of our sovereignty. We can no longer protect the stratospheric ozone layer over the United States. We can't stabilize the U.S. climate without the cooperation of the countries throughout the world. If even one major developing country continues to use CFCs, it will eventually deplete the ozone layer. We can't protect the biological diversity of the planet by ourselves. We have lost sovereignty; we've lost control."[18]

He determined that "ecological sustainability will become the new organising principle, the foundation of the 'new world order,' if you will."[19] Brown concluded with an emblematic statement typifying the mentality of false environmentalists. He declared, "It might take a few more scares to get this country energised."[20]

Brown was one of the driving forces behind the UNCED. The year before the conference, his Worldwatch Institute released its annual report, *State of the World 1991*. In the opening chapter, entitled "The New World Order," Brown asserted, "The battle to save the planet will replace the battle over ideology as the organizing theme."[21] He confidently asserted that with "the end of the ideological conflict that dominated a generation of international affairs, a new world order, shaped by a new agenda, will emerge." The Rio Earth Summit was a cesspool of globalists and fake environmentalists. Nevertheless, some dignified reporters did attend and cover the event.

Dixy Lee Ray, an accomplished scientist and an acclaimed statesman and humanitarian, was a professor at the University of Washington and Stanford, the first woman to chair the U.S. Atomic Energy Commission, the first female Governor of the State of Washington, and the 1978 winner of both the National Freedom Foundation Award and the United Nations Peace Prize. This last award is ironic because after witnessing continual UN duplicity, Ray became one of the UN's harshest critics. She attended the 1992 Earth

Summit as an observer with press credentials. Although most attendees were mesmerized by the UN's ambitious yet determined resolve to save humanity from impending ecological doom, Ray was acutely aware of their ulterior agenda. She shared her impressions about the conference and its larger implications during a 1992 interview with *Religion and Liberty* magazine:

> *R&L:* With the worldwide decline of socialism, many individuals think that the environmental movement may be the next great threat to freedom. Do you agree?

> *Ray:* Yes, I do, and I'll tell you why. It became evident to me when I attended the worldwide Earth Summit in Rio de Janeiro last June. The International Socialist Party, which is intent upon continuing to press countries into socialism, is now headed up by people within the United Nations. They are the ones in the UN environmental program, and they were the ones sponsoring the so-called Earth Summit that was attended by 178 nations.

> *R&L:* Did you have a specific purpose in attending the Earth Summit?

> *Ray:* I was sent there by the Free Congress Committee, headed by Paul Weyrich. Fred Smith and I were sent down as observers, with reporters' credentials, so we could witness the events. One of the main organizers of the program, Prime Minister Gro Harlem Brundtland of Norway was the assistant executive for the conference. She is also the vice-president of the World Socialist Party. When she was questioned by Brazilian reporters after her talk and asked if what they were proposing didn't have a peculiar resemblance to the agenda of the World Socialist Party she said, "Well, of course." That was reported in Brazil but not picked up by the American press.

> *R&L:* Did you see a big influence by the radical environmentalists there?

> *Ray:* Oh yes. No question about that, the radicals are in charge. One of the proposals that did indeed pass as part of Agenda 21 proposes that there be world government under the UN, that essentially all nations give up their sovereignty, and that the nations will be, as they said quite openly, frightened or coerced into doing that by threats of environmental damage.[22]

Ray went even further in her 1993 book *Environmental Overkill*. She reported, "The objective, clearly enunciated by the leaders of UNCED, is to bring about a change in the present system of independent nations. The future is to be world government, with central planning by the UN. Fear of environmental crises, whether real or not, is expected to lead to compliance. If force is needed, it will be provided by a UN green-helmet police force, already authorized by the Security Council."[23] In 1992, the *New York Times* ran an article called, "The New World Army," confirming the UN's shotgun environmental diplomacy. They reported, "For years the United Nations has been notable mostly for its vocal cords. That's changed. Nowadays the UN's muscle—its blue-helmeted soldiers—seems to be everywhere…The Security Council recently expanded the concept of threats to peace to include economic, social and ecological instability."[24] Furthermore, in his 1988 speech to the United Nations, Mikhail Gorbachev also called for a militarized environmental enforcement agency. Maduro and Schauerhammer explained,

> "In his speech to the United Nations, Gorbachev called for a world environmental order, run by an ecological security council operating within the United Nations and deploying its own military, with the power to enforce and oversee the implementation

of environmental policies. This international green police would operate independently of sovereign national governments with the power to intervene militarily against nations accused of polluting the environment."[25]

Prelude to Rio

In 1987, The United Nations World Commission on Environment and Development published *Our Common Dreams*, a report that resuscitated the Club of Rome's *Limits to Growth* doctrine. Published in 1972, *Limits to Growth* argued that increasing population and economic growth would inevitably outstrip natural resources and trigger widespread catastrophe.[26] The UN argued the same, although they used the environmentally friendly buzzword "sustainable development" to describe this theory. According to the UN, "Sustainable development is development that meets the needs of the present without compromising the ability of future generations to meet their own needs."[27] Most UN environmental documents follow an obvious pattern. First they make seemingly incisive observations and benevolent recommendations. Then they cleverly incite fear about overpopulation. *Our Common Dreams* made clear that "sustainable development" means population reduction:

- "Rapidly growing populations can increase the pressure on resources and slow any rise in living standards; thus sustainable development can only be pursued if population size and growth are in harmony with the changing productive potential of the ecosystem."[28]
- "Policies to bring down fertility rates could make a difference of billions to the global population next century."[29]
- "Environmental education should be included in and should run throughout the other disciplines of the formal education curriculum at all levels—to foster a sense of responsibility for the state of the environment and to teach students how to monitor, protect, and improve it."[30]
- "The issue is not just numbers of people, but how those numbers relate to available resources…Urgent steps are needed to limit extreme rates of population growth."[31]

Our Common Dreams was the prelude to the onslaught of Rio and Agenda 21. It also foreshadowed many other depopulation documents, both from the UN and other fake environmental organizations.

Maurice Strong: New World Order Henchman

Strong was both the Secretary General of the 1992 Earth Summit and the driving force behind the 1972 Stockholm conference. After Stockholm, he became Executive Director of the United Nations Environment Programme. At the time, he was also a trustee of the Rockefeller Foundation.[32] Strong has also been an advisor to the World Wildlife Foundation, and a senior advisor to both UN Secretary General Kofi Annan and World Bank President James Wolfensohn. Strong started in the oil business, becoming CEO of the Power Corporation of Canada in the 1960s. In the 1970s he headed Petro-Canada where he "enabled Shell to take over the only remaining all-Canadian oil company by throwing a controlling block of shares in its direction."[33]

Writing for the *National Review*, Ronald Bailey observed, "It is instructive to read Strong's 1972 Stockholm speech and compare it with the issues of Earth Summit 1992. Strong warned urgently about global warming, the devastation of forests, the loss of biodiversity, polluted oceans, [and] the population time bomb. Then as now, he invited to the conference the brand-new environmental NGOs: he gave them money to come; they were invited to raise hell at home. After Stockholm, environment issues became part of the administrative framework in Canada, the U.S., Britain, and Europe."[34] Following the Earth Summit, Strong published *From Stockholm to Rio: A Journey Down a Generation*, an essay candidly calling for the dissolution of national sovereignty and the imposition of global government. This essay foreshadowed his next venture, the establishment of the United Nations' Commission on Global Governance. Strong wrote,

"Strengthening the role the United Nations can play on behalf of its members will require serious examination of the need to extend into the international arena the rule of law and the principle of taxation to finance agreed actions which provide the basis for governance at the national level. But this will not come about easily. Resistance to such changes is deeply entrenched. The concept of national sovereignty has been an immutable, indeed sacred, principle of international relations. It is a principle which will yield only slowly and reluctantly to the new imperatives of global environmental cooperation. It is simply not feasible for sovereignty to be exercised unilaterally by individual nation states, however powerful. The global community must be assured of environmental security."[35]

Starting in 1991, Strong began promoting the concept of "global governance," an obvious euphemism for global government. Riding the momentum of the Earth Summit, he helped establish the UN's Commission on Global Governance (CGG).[36] In their infamous 1995 report, *Our Global Neighborhood*, CGG asserted, "It is our firm conclusion that the United Nations must continue to play a central role in global governance."[37] Regarding national sovereignty, they observed,

"Sovereignty has been the cornerstone of the inter-state system. In an increasingly interdependent world, however, the notions of territoriality, independence, and non-intervention have lost some of their meaning. In certain areas, sovereignty must be exercised collectively, particularly in relation to the global commons...It is time also to think about self-determination in the emerging context of a global neighborhood rather than the traditional context of a world of separate states."[38]

CGG also called for a global taxation scheme: "It is time for a consensus on global taxation for servicing the needs of the global neighborhood."[39] They urged the adoption of global carbon taxes "as a first step towards a system that taxes resource use rather than employment and savings."[40]

Further evidence of Strong's New World Order allegiances comes from a 1990 interview with Daniel Wood of Canada's *West* magazine during which Strong described a fictional apocalyptic scenario involving the collapse of industrialized civilization. He said the scenario came from a novel he was contemplating writing. Strong stated,

"Each year the World Economic Forum convenes in Davos, Switzerland. Over a thousand CEOs, prime ministers, finance ministers, and leading academics gather in February to attend meetings and set the economic agendas for the year ahead. What if a small group of these world leaders were to conclude that the principle risk to the earth comes from the actions of the rich countries? And if the world is to survive, those rich countries would have to sign an agreement reducing their impact on the environment. Will they do it? Will the rich countries agree to reduce their impact on the environment? Will they agree to save the earth? The group's conclusion is 'no.' The rich countries won't do it. They won't change. So, in order to save the planet, the group decides: isn't the only hope for the planet that the industrialized civilizations collapse? Isn't it our responsibility to bring that about? This group of world leaders forms a secret society to bring about a world collapse. It's February. They're all at Davos. These aren't terrorists—they're world leaders. They have positioned themselves in the world's commodity and stock markets. They've engineered, using their access to stock exchanges, and computers, and gold supplies, a panic. Then they prevent the markets from closing. They jam the gears. They have mercenaries who hold the rest of the world leaders at Davos as hostage. The markets can't close. The rich countries...?"[41]

Wood then wrote, "and Strong makes a slight motion with his fingers as if he were flicking a cigarette butt out of the window. I sat there spellbound. This is not any storyteller talking. This is Maurice Strong. He knows these world leaders. He is, in fact, co-chairman of the Council of the World Economic Forum. He sits at the fulcrum of power. He is in a position to do it."[42] Strong later confirmed that the apocalyptic scenario was not strictly fiction. In 1997, *National Review* magazine quoted him as saying, "If we don't change, our species will not survive...Frankly, we may get to the point where the only way of saving the world will be for industrial civilization to collapse."[43]

According to Maurice Strong, "If you put yourself in a larger unit, of course, you get some advantages and you give up some of your freedom. And that's what's happening in Europe, that the states of Europe have decided that overall they're better off to create a structure in which they give up some of their national rights and exercise them collectively through the Union."[44] This statement reveals the fallacy of global government/governance, proponents of which steadily seduce their countrymen with the contrived notion that *increased* peace and prosperity can arise from *decreased* rights and freedoms. The New World Order does not *guarantee* peace and prosperity. It does, however, *guarantee* the discontinuation of cherished rights and freedoms won by the courage and blood of generations past. Under the New World Order there are no Constitutional rights; there are only privileges. Yet on a tyrant's whim, privileges can be restrained, revoked, or altered. Constitutional rights, however, are untouchable.

The New World Order's global government is entirely faith-based and deceptively designed to exploit people's naivety and ignorance. That's how the triumph of consciousness works. But such chicanery is expected from psychopathic elites. They themselves are stumbling through the sub-conscious corridors of human consciousness-potential. So long as humanity passively accepts tyranny, it will surely persist. So long as humanity favors fear-fueled consciousness over love-fueled consciousness, freedom will steadily atrophy. Thus, in the words of William Jennings Bryan, "Destiny is not a matter of chance, it is a matter of choice. It is not a thing to be waited for, it is a thing to be achieved."[45]

Chapter 14

The New World Religion

An integral component of the New World Order agenda is the planned New World Religion. For thousands of years, from ancient Egypt, to Greece, to Rome, elites have been manipulating and controlling people through religion. Realizing that human beings are inherently and magnetically attracted to "the beyond," these opportunists long ago created religious institutions, which claimed proprietary rights to God. God, however, does not depend on religion. Conversely, religion depends on God. Religion institutionalizes that which otherwise would be a personal matter. Nevertheless, absent manipulation, religion *can* facilitate spiritual growth. Yet like all faith-based institutions, religion is extremely susceptible to both infiltration and exploitation.

New World Order elites consider themselves fishermen—fishers of men. Fish are perpetual victims, never perceiving the fishermen's trap. Always biting the worms, they never suspect the hooks. The human situation is remarkably similar. Upon the hook of servitude, the fishers of men dangle the worms of salvation. They patiently wait, knowing most people are not conscious enough (not aware enough) to recognize the trap. Most people never question reality as presented by society's faith-based institutions. They simply accept these presentations as truth. Most people only see the worms, never the hooks. Consequently, like fish, for thousands of years humanity has been 'getting hooked' on religion.

The New World Religion combines New Age philosophy with environmental dogma, modern Pagan Earth-worship with New World Order ideology. Its self-righteous creators and spokesmen have been writing and speaking about the New World Religion for decades. The Earth Charter is but one of many notable examples. According to its official website, "The Earth Charter is a widely recognized, global consensus statement on ethics and values for a sustainable future."[1] The United Nations' 1986 Brundtland Commission called for "a universal declaration" to encapsulate the environmental movement. In 1994, Maurice Strong and Mikhail Gorbachev filled this void by co-founding the Earth Charter. Today, they co-chair the Earth Charter Commission. Earth Charter International (ECI), an offshoot group, receives funding from the Rockefeller Brothers Fund and UNESCO, two organizations with considerable ties to eugenics.[2] Also Steven Rockefeller is a member of the Earth Charter Commission and a co-chair of ECI.

Why does society need an Earth Charter? According to Strong and Gorbachev, the New World Religion requires a *new* Ten Commandments. The Earth Charter, they believe, will fulfill this role. During a 1998 interview, Strong explained, "The real goal of the Earth Charter is that it will in fact become like the Ten Commandments."[3] Likewise, Gorbachev admitted, "My hope is that this charter will be a kind of Ten Commandments, a 'Sermon on the Mount,' that provides a guide for human behavior toward the environment in the next century."[4]

Neo-Spirituality and Neo-Morality

To sell their planned New World Religion, fake environmentalists and New Age swindlers like Al Gore, Steven Rockefeller and Mikhail Gorbachev have been increasingly portraying contrived environmental issues (global warming and overpopulation) as moral and spiritual crises. Al Gore, for example, upon receiving the 2007 Nobel Peace Prize, declared, "We face a true planetary emergency. The climate crisis is not a political issue; it is a moral and spiritual challenge to all of humanity. It is also our greatest opportunity to lift global consciousness to a higher level."[5] For dramatic effect, Gore has even compared society's reluctance to embrace his environmental fear mongering to the widespread passivity of Hitler-era Germany. In his book *Earth in the Balance*, Gore declared,

> "As the prospect of war increased in Europe, many refused to recognize what was about to happen, even when Jews were rounded up and sent to the concentration camps...Now warnings of a different sort signal an environmental holocaust without precedent. But where is the moral alertness that might make us more sensitive to the new pattern of environmental change? Once again, world leaders waffle, hoping the danger will dissipate. Yet today the evidence of an ecological Kristallnacht is as clear as the sound of glass shattering in Berlin."[6]

In his 2009 book, *Our Choice: A Plan to Solve the Climate Crisis*, Gore turned increasingly towards spiritual associations. *Newsweek* portrayed him as an "eco-prophet," because he recognizes "the spiritual dimension of climate change, the idea that God gave man stewardship over the earth, and that preserving it for future generations is a sacred obligation."[7]

Steven Rockefeller, one of the globalists' New World Religion point men, is the son of former U.S. Vice-President Nelson Rockefeller and a longtime trustee of the Rockefeller Brothers Fund.[8] In his 1992 book *Spirit and Nature: Why the Environment is a Religious Issue*, Rockefeller wrote,

- "The global environmental crisis, which threatens not only the future of human civilization but all life on earth, is fundamentally a moral and religious problem."[9]
- "The integration of the moral and religious life with a new ecological worldview, leading to major social transformations, is a fundamental need of our time."[10]
- "If a major transformation is to occur in human values and behavior, it will have to involve a concern and faith that is religious in nature. A religious concern is one that is a matter of fundamental controlling interest to a person. A person is religiously concerned about those values which he or she regards as essential to fulfillment in the deepest sense."[11]
- "In a biocentric approach, the rights of nature are defended first and foremost on the grounds of the intrinsic value of animals, plants, rivers, mountains, and ecosystems rather that simply on the basis of their utilitarian value or benefit to humans."[12]

Mikhail Gorbachev has positioned himself as High Priest of the New World Religion. His book *The Search For A New Beginning: Developing A New Civilization* served as a pulpit upon which he scolded humanity. He declared, "The roots of the current crisis of civilization lie within humanity itself. Our intellectual and moral development is lagging behind the rapidly changing conditions of our existence, and we are finding it difficult to adjust psychologically to the pace of change."[13] Our problem, he claims, is our attitude towards nature: "It is my view that the individual's attitude toward nature must become one of the principal criteria for ensuring the maintenance of morality."[14] His most ominous statement, however, foreshadowed the death and destruction traditionally wrought by elites while passing between two ages. He explained, "Between the old order and the new one lies a period of transition that we must go through."[15] Despite Gorbachev's proclamations, humanity is not obligated to accept the globalists' New World Religion. The New Age is neither fate nor destiny. It is simply the next phase of the New World Order's business plan. Humanity can reject this plan, but to do so requires first understanding the New Age movement and its connections to the New World Religion.

The New Age Movement

In his book *The Emerging Network*, Michael York described the New Age as "a blend of pagan religions, Eastern philosophies, and occult-psychic phenomena. The Euro-American metaphysical tradition and the counterculture of the 1960s together constitute the occult underground or what Campbell refers to as the 'cultic milieu.'"[16] Pierre Teilhard de Chardin was a Jesuit priest whose unorthodox ideas largely influenced the New Age movement. York observed that Teilhard's "evolutionary philosophy and visionary prophecy have become bedrock ideas of the movement."[17] In his book *Earth in the Balance*, Al Gore quoted Teilhard as saying, "The fate of mankind, as well as of religion, depends upon the emergence of a new faith in the future." Gore remarked, "Armed with such a faith, we might find it possible to re-sanctify the earth."[18]

During a 1992 interview, *Religion and Liberty* magazine asked former Washington Governor Dixy Lee Ray, "Much of the current environmental movement is couched in terms of pagan religions, worshiping the Earth, goddess Gaia, equating the value of trees and people, animal rights, etc. Can you account for how this is accepted in the public forum, when traditional Judeo-Christian religious ethics are basically outlawed from policy-making decisions? Do you think the general public is just unaware of the tendency to make environmentalism a religion?" Ray replied,

"I understand what you're asking, and I have to tell you, no, I can't account for it. It is not classified as a pagan religion. The so-called New Age activities and this are not called religions and therefore don't come under the prohibition of mingling church and state that we have in this country. It's almost as if nature worship were accepted without its being considered a religion."[19]

Lindisfarne

The Lindisfarne Association is one of many New Age organizations firmly aligned with the New World Order agenda. Former M.I.T. professor and New Age enthusiast William Irwin Thompson established Lindisfarne in 1972. In 2008, he reflected,

"The Lindisfarne Association that I founded in New York in 1972 was very much a seventies kind of cultural movement in that it sought to avoid the sixties-style mass movements of drugs and revolutionary violence to seek out a more spiritual and intellectual third way to effect the transformation of culture. As a student of cultural history, I had a sense that our Western civilization was at the edge of its kairos and that a newly emerging planetary culture called out for a new formation with which to realize itself, so I sought out others in what became the Lindisfarne Fellowship who seemed to embody this third way."[20]

In the early 1970s when Lawrence Rockefeller heard about Thompson's ideas, he helped secure funding for the fledgling project.[21] Since its inception, the Rockefeller Brothers Fund and the Rockefeller Foundation have been its primary financial backers.[22] Also arch-globalist and fake environmentalist Maurice Strong was the former Director of Finance for Lindisfarne.[23] Their current Roster of Fellows includes New Agers William Irwin Thompson, James Lovelock, David Spangler, and Wendell Berry.[24]

In the summer of 1975, Thompson organized a conference at Lindisfarne concerning the future of mankind. *New York Times* columnist Ted Morgan attended the event and astutely observed, "Not the least irony of Thompson's alternatives to the established order, was its partial funding by the Rockefellers."[25] The conference had many interesting attendees including Jonas Salk and Gregory Bateson. Salk had previously (in the 1950s) developed a polio vaccine contaminated with SV40 live cancer virus. Hundreds of millions of people throughout the world, including 98 million Americans, received this vaccine.[26] Bateson, a co-founder of Lindisfarne, had previously participated in the CIA's notorious MK-ULTRA mind control programs. Douglas Valentine, in his acclaimed book *The Strength of the Wolf: The Secret History of America's War of Drugs*, described Bateson as "a former OSS officer with radical ideas about political and psychological warfare."[27] On Bateson's involvement with MK-ULTRA, Valentine explained,

> "Bateson was there to explain how LSD and narcotics could be used to reconstruct American society. In a memo he sent to William Donovan nine days after the U.S. dropped an atomic bomb on Hiroshima, Bateson predicted an era in which propaganda, subversion, and 'social and ethnic manipulation' would be more critical to national security than guided missiles."[28]

In his *New York Times* article, Morgan wrote,

> "Bateson would emerge as the high priest…The gospel according to Gregory, as it came to be called at the conference, was that such catastrophic errors as pollution, overpopulation, the possibility of melting the Antarctic icecap or destroying the ozone layer, and the threat of nuclear war were all due to a discrepancy between the way the human mind thinks and the way the world really works."[29]

Lindisfarne promotes the standard, apocalyptic New Age vision. Morgan explained,

> "If a single theme could be abstracted from the talks, it was that industrial society is finished…kaput. The System, the conference was told, can no longer deal with global problems such as energy and food supply. It was time to get ready for postindustrial society."[30]

Later in 1975, American Association for the Advancement of Science (AAAS) President Margaret Mead, a New Age anthropologist and wife of Lindisfarne co-founder Gregory Bateson, chaired a conference entitled, *The Atmosphere: Endangered or Endangering*. The official conference proceedings stated, "The session was concluded with the thought that we as a species are trying to maintain ourselves at the expense of other species; there seems to be a conflict between preserving nature and feeding the rapidly increasing population. Is our major objective really to feed the population, or do we realize we cannot continue to feed the world at any price? Where do we strike a balance between preserving nature and feeding the world?"[31] Mead called upon scientists to issue "artificial but effective warnings, warnings which will parallel the instincts of animals who flee before the hurricane."[32] Conference attendees included young scientists like John Holdren (Obama's current White House Science Czar), George Woodwell, and Stephen Schneider, one of today's most prominent global warming alarmists (who during the 1970s was warning about global cooling).

Theosophy: Satanic Roots of the New Age

The New Age movement is largely an outgrowth of Theosophy, the 1870s metaphysical doctrine of Helena Petrovna Blavatsky. When Blavatsky died in 1891, Annie Besant, the Fabian socialist, became the most prominent leader of Blavatsky's Theosophical Society. In 1902, philosopher Rudolph Steiner became the founding President of the German Theosophical Society. Steiner and Besant, however, had numerous ideological conflicts. In 1911, Besant founded the Order of the Star of the East (OSE), a Theosophical sect focused on preparing for the coming "World Teacher." Soon thereafter, they claimed that Jiddu Krishnamurti, a young Indian boy who later became an internationally renowned philosopher, was the reincarnation of Jesus Christ.[33] Steiner vehemently protested this claim and subsequently banned all OSE members from the German Theosophical branch. In 1912 he dissolved the German Theosophical Society and started a separate movement. For many years, Krishnamurti accepted his anointed role. But by 1929, he too finally perceived the absurdity of Theosophy. In a speech before 3,000 people, including Besant, he officially disbanded the Order of the Star. Krishnamurti explained,

"You may remember the story of how the devil and a friend of his were walking down the street, when they saw ahead of them a man stoop down and pick up something from the ground, look at it, and put it away in his pocket. This friend said to the devil, 'What did that man pick up?' 'He picked up a piece of the Truth,' said the devil. 'That is a very bad business for you, then' said his friend. 'Oh, not at all,' the devil replied, 'I am going to help him organize it.'

I maintain that Truth is a pathless land, and you cannot approach it by any path whatsoever, by any religion, by any sect. That is my point of view, and I adhere to that absolutely and unconditionally. Truth, being limitless, unconditioned, unapproachable by any path whatsoever, cannot be organized; nor should any organization be formed to lead or coerce people along a particular path. If you first understand that, then you will see how impossible it is to organize a belief.

A belief is purely an individual matter, and you cannot and must not organize it. If you do, it becomes dead, crystallized; it becomes a creed, a sect, a religion, to be imposed on others. This is what everyone throughout the world is attempting to do: Truth is narrowed down and made a plaything for those who are weak, for those who are only momentarily discontented. Truth cannot be brought down; rather the individual must make the effort to ascend to it."[34]

Krishnamurti's Devil reference was appropriate for reasons probably not obvious to his audience. Blavatsky, and especially Alice Bailey, another prominent leader of the Theosophical Society, frequently praised Satan and Lucifer throughout their writings. Bailey joined the Society in 1917. Two years later, she claimed that deceased Tibetan Master Djwhal Khul had "contacted" her. From 1919 until her death in 1949, she wrote 24 books, supposedly by channeling ancient wisdom directly from Khul. In 1922, Alice Bailey and her husband Foster Bailey (National Secretary of the Theosophical Society) founded the Lucifer Publishing Company to publish her books. Now known as the Lucis Trust, the organization continues to actively promote the New Age movement, mostly through Bailey's Luciferian doctrine. Most alarmingly, they are closely allied with the UN.[35]

Bailey, Blavatsky and Satan

Blavatsky viewed Theosophy as the modern school of ancient occult knowledge. Michael Coffman, in his book *Saviors of the Earth? The Politics and Religion of the Environmental Movement*, observed, "Throughout their books both Bailey and Blavatsky trace the roots of Theosophy to the Rosicrucians, Knights of the Templar, the Jewish Kabalists, Greek and Egyptian mystics, Buddhists, Babylonian and Chaldean mystics, Aryans, back to the superhumans of Atlantis." [36] *The Secret Doctrine* was Blavatsky's magnum opus. Originally published in 1888, it quickly became the preeminent book of Theosophy. Blavatsky claimed she learned her esoteric knowledge from a group of isolated Tibetan masters. Edward James, in his book *Notable American Women*, explained,

"The Theosophical approach to the unknown was through contact with the Mahatmas (Masters), a highly evolved but not supernatural order of mankind. The Mahatmas, isolated in their Egyptian (later Tibetan) retreats, had chosen Madame Blavatsky as their channel of communication with the outer world, and through her transmitted a steady stream of letters to true believers."[37]

The Secret Doctrine was decidedly anti-religion and anti-science. Blavatsky wrote,

"Faulty, materialistic, and biased as the scientific theories may be, they are a thousand times nearer the truth than the vagaries of theology. The latter are in their death agony for every one but the most uncompromising bigot and fanatic. Hence we have no choice but either to blindly accept the deductions of Science, or to cut adrift from it, and withstand it fearlessly to its face, stating what the Secret Doctrine teaches us, being fully prepared to bear the consequences."[38]

Blavatsky claimed that Christianity misrepresents Satan, that Christians worship the wrong God, and that Satan is actually the most divine Spirit. She wrote extensively about Satan in *The Secret Doctrine*. Here are few revealing passages:

- "To make the point clear once for all: that which the clergy of every dogmatic religion—preeminently the Christian—points out as Satan, the enemy of God, is in reality, the highest divine Spirit—occult Wisdom on Earth—in its naturally antagonistic character to every worldly, evanescent illusion, dogmatic or ecclesiastical religions included."[39]

- "However it may be, no Satan could be more persistent in slandering his enemy, or more spiteful in his hatred, than the Christian theologians are in cursing him as the father of every evil."[40]

- "Thus 'Satan,' once he ceases to be viewed in the superstitious, dogmatic, unphilosophical spirit of the Churches, grows into the grandiose image of one who made of *terrestrial* a *divine* man; who gave him, throughout the long cycle of Mahakalpa the law of the Spirit of Life, and made him free from the Sin of Ignorance, hence of death."[41]

Although she promoted "Universal Brotherhood," Blavatsky's reverence for the Aryan race foreshadowed that of Hitler:

"The intellectual difference between the Aryan and other civilized nations and such savages as the South Sea Islanders, is inexplicable on any other grounds. No amount of culture, nor generations of training amid civilization, could raise such human specimens as the Bushmen, the Veddhas of Ceylon, and some African tribes, to the same intellectual level as the Aryans, the Semites, and the Turanians so called. The 'sacred spark' is missing in them and it is they who are the only inferior races on the globe, now happily — owing to the wise adjustment of nature which ever works in that direction — fast dying out. Verily mankind is 'of one blood,' but not of the same essence. We are the hothouse, artificially quickened plants in Nature, having in us a spark, which in them is latent."[42]

Alice Bailey

Through her "spirit-guide" Djwhal Khul, Alice Bailey wrote about the "New World Order," "The New Age," and "The Plan." She clearly understood and supported the elites' planned global system. Using typical New Age rhetoric, she presented the New World Order as benevolent and universally appealing: "The new world order will recognize that the produce of the world, the natural resources of the planet and its riches, belong to no one nation but should be shared by all. There will be no nations under the category 'haves' and others under the opposite category."[43] She also foresaw and promoted regional governance as per the European Union and the impending North American Union: "Such blocs would be cultural and not militaristic, economic and not greedy, and they could provide a normal and progressive movement away from the separative nationalism of the past and towards the distant creation of the One World, and the One Humanity."[44]

According to Bailey, Theosophy has "three grades of workers."[45] The first, by "their own free choice realised the immediate and coming needs of humanity and have pledged themselves to serve."[46] These are the people who elites refer to as "useful idiots." They believe in the idealism of the New World Order, the oneness of humanity for example, yet they are completely ignorant regarding the true agenda. The second grade includes people like Bailey and Blavatsky who "act as intermediaries in the working out of the Plan in the world and they hold themselves in readiness to go anywhere."[47] The third "work primarily upon the inner side. Their activities are confined largely to the mental plane and to the scientific use of thought. Thus they guide their workers and helpers and influence their working disciples and the world disciples."[48]

Additionally, Bailey wrote about a nondescript "New Group of World Servers." She wrote, "They are overseeing or ushering in the New Age. They are present at the birth-pangs of the new civilisation, and the coming into manifestation of a new race, a new culture and a new world outlook."[49] The Lucis Trust website further explains, "The new group of world servers is not an exterior organisation but could be thought of as a subjective organism, a group of people united as one through the synthesising nature of their vision and work. They may not be consciously aware of this affiliation, they may not know or meet each other in person, but members of this worldwide group intuitively recognise and resonate with others who share this vision of human unity and are working to implement it in their chosen field."[50]

According to Theosophy, the Aryan race, through group consciousness, has developed super-human powers, while lower races have shunned group consciousness. Bailey categorized Jews as the latter. She explained, "They constitute, in a strange manner, a unique and distinctly separated world centre of energy. The reason for this is that they represent the energy and the life of the previous solar system. You have often been told how, at the close of this solar system, a certain percentage of the human family will fail to make the grade and will then be held in pralaya, or in solution, until the time for the manifestation of the next and third solar system comes around...The Jews are the descendants of that earlier group which was held in pralaya between the first and the second solar systems."[51] Bailey continued, "The Jew, down through the ages, has insisted upon being separated from all other races...It is this fact which has militated against him down the years and made it possible for the forces of separativeness and of hate, to use the Jewish race to stir up world difficulty and thus bring to a crisis the basic human problem of separation."[52] Bailey even had the audacity to accuse Jews of selfishness for supposedly ignoring non-Jewish Nazi victims: "The Jew fought only for himself, and largely ignored the sufferings of his fellow men in the concentration camps."[53]

Bailey also encouraged human sacrifice, a central tenet of the globalists' psychopathology. Wars and slaughter, after all, inevitably provide an impetus for substantial societal change. Therefore, according to psychopathic elites, death and destruction are both necessary and beautiful. In her 1954 book *Education in the New Age*, Bailey applauded the recent World War and hinted that another was perhaps desirable:

"The World War was in the nature of a major surgical operation made in an effort to save the patient's life. A violent streptococcus germ and infection had menaced the life of humanity (speaking in symbols) and an operation was made in order to prolong opportunity and save life...Another surgical operation may be necessary, not in order to destroy and end the present civilisation, but in order to dissipate the infection and get rid of the fever...But at the same time, let us never forget that it is the Life, its purpose and its directed intentional destiny that is of importance...When a form proves inadequate, or too diseased, or too crippled for the expression of that purpose, it is—from the point of view of the Hierarchy—no disaster when that form has to go. Death is not a disaster to be feared; the work of the Destroyer is not really cruel or undesirable. I say this to you who am myself upon the Ray of Love and know its meaning."[54]

Nazi Environmentalism, Steeped in Theosophy

The occultists and philosophers who influenced Hitler borrowed heavily from Blavatsky, especially her thoughts on the Aryan race. Lanz von Liebenfels was a self-described "racial researcher, philosopher of religion and sexual mystic."[55] In 1907, he established the Order of the New Templars to promote his occult doctrine and racial theories. In 1904, von Liebenfels wrote, "The Aryan hero is on this planet the most complete incarnation of God and of the Spirit."[56] Von Liebenfels claimed he personally consulted with Hitler on at least one occasion. In 1932, he wrote to one of his New Templar brothers, "Do you know that Hitler is one of our pupils? You will still live to see that he, and thereby we, also will triumph and kindle a movement that will make the world tremble."[57] Blavatsky's writings heavily influenced von Liebenfels. He noted, "She was almost a generation ahead of her time and of anthropology. Today for the first time, work on the latest material has brought to light results which show a completely amazing identity with those of the spiritual Theosophist."[58]

Another occultist who influenced Hitler was Guido von List. In 1903, his supporters founded the Guido von List Society in Vienna. It soon became the hub of Viennese occult activity, counting Franz Harmann, a well-known Theosophist, as a member.[59] Blavatsky's conception of Aryan superiority seems to have heavily influenced von List. Furthermore, his plans for an Aryan nation strongly resembled those of Hitler. In their essay, "Hitler's Racial Ideology: Content and Occult Sources," Jackson Spielvogel and David Redles noted, "List made proposals for an Ario-Germanic state. It would be based on the recognition of the superiority of Aryan peoples and the need for lower races to serve the higher race. Only Ario-Germans could hold leadership positions in the state, schools, professions, industry and banks, newspapers, theater and the arts. Racial laws would maintain the purity of the Ario-Germanic race by prohibiting racial intermarriage and by reserving citizenship for Ario-Germans."[60] Although Hitler lived in Vienna from 1907-1913, historians have not established a definite personal relationship between Hitler and von List. Nevertheless, concerning his Vienna years, Hitler recalled in *Mein Kampf*, "In this period there took shape within me a world picture and a philosophy which became the granite foundation of all my acts."[61]

Alfred Baeumler was one of Nazi Germany's most influential philosophers. In his essay *Nietzsche and National Socialism*, Baeumler asserted,

"Consciousness is only a tool, a detail in the totality of life. In opposition to the philosophy of the conscious, Nietzsche asserts the aristocracy of nature. But for thousands of years a life-weary morality has opposed the aristocracy of the strong and healthy. Like National Socialism, Nietzsche sees in the state, in society, the 'great mandatory of life,' responsible for each life's failure to life itself. 'The species requires the extinction of the misfits, weaklings, and degenerates: but Christianity as a conserving force appeals especially to them.' Here we encounter the basic contradiction: whether one proceeds from a natural life context or from an equality of individual souls before God. Ultimately the ideal of democratic equality rests upon the latter assumption. The former contains the foundations of a new policy. It takes unexcelled boldness to base a state upon the race. A new order of things is the natural consequence. It is this order which Nietzsche undertook to establish in opposition to the existing one."[62]

Like their American counterparts, the Nazi eugenicists also justified and legitimized their anti-human agenda through an Earth-based nature religion. Janet Biehl and Peter Staudenmaier, in their insightful book *Ecofascism: Lessons from the German Experience*, describe this religion as "a volatile admixture of primeval teutonic nature mysticism, pseudo-scientific ecology, irrationalist anti-humanism, and a mythology of racial salvation through a return to the land."[63] Biehl explained, "Nazi 'ecologists' even made organic farming, vegetarianism, nature worship, and related themes into key elements not only in their ideology but in their governmental policies. Moreover, Nazi 'ecological' ideology was used to justify the destruction of European Jewry."[64]

Don't Trust The Lucis Trust

Why does the United Nations promote Theosophy and the teachings of Alice Bailey? Is the UN the central hub of fake environmentalism? On the Roster of the United Nations Economic and Social Council, is the infamous Lucis Trust.[65] The UN cannot claim ignorance regarding Lucis Trust's connection to Alice Bailey as the Lucis Trust website plainly reads, "The worldwide activities of the Lucis Trust, founded by Alice and Foster Bailey, are dedicated to establishing right human relations."[66] World Goodwill is an offshoot of the Lucis Trust. They describe themselves as "an organised movement founded in 1932 to help establish right human relations and solve humanity's problems through the constructive power of goodwill."[67]

The UN also works with and supports World Goodwill: "World Goodwill is an accredited non-governmental organisation with the Department of Public Information of the United Nations."[68] Furthermore, World Goodwill's objective is "To support the work of the United Nations and its Specialised Agencies as the best hope for a united and peaceful world."[69] World Goodwill also has declared, "Humanity is not following a haphazard or uncharted course—there is a Plan."[70]

Prior to 1925, as previously mentioned, the Lucis Trust was named the Lucifer Publishing Company.[71] Their website explains, "The Baileys' reasons for choosing the original name are not known to us, but we can only surmise that they, like the great teacher H.P. Blavatsky, for whom they had enormous respect, sought to elicit a deeper understanding of the sacrifice made by Lucifer. Alice and Foster Bailey were serious students and teachers of Theosophy, a spiritual tradition which views Lucifer as one of the solar Angels, those advanced Beings who Theosophy says descended (thus 'the fall') from Venus to our planet eons ago to bring the principle of mind to what was then animal-man. In the theosophical perspective, the descent of these solar Angels was not a fall into sin or disgrace but rather an act of great sacrifice, as is suggested in the name 'Lucifer' which means light-bearer."[72] In 1922, the Baileys established *The Beacon*, a magazine "of esoteric philosophy, presenting the principles of the Ageless Wisdom as a contemporary way of life."[73] In a 1989 edition, Sarah McKechnie attempted to rationalize the worship of Lucifer:

> "The mystery of the descent or 'fall' to Earth of the rebellious angels—the solar angels or agnishvattas—is said to be the mystery hinted at in the Scriptures, and 'the secret of the ages.' Thus it is not surprising that there is so much confusion and misunderstanding concerning the 'fallen angels' of which Lucifer is the best known representative...The role of the solar angels and their sacrifice on behalf of humanity is discussed at length in *The Secret Doctrine* by H.P. Blavatsky. In fact, in 1887 the magazine of the Theosophical Society took 'Lucifer' as its name in an effort to bring clarity to what it regarded as an unfairly maligned sacrificing angel."[74]

Therefore the modern Lucis Trust, and by extension the UN, openly embrace the worship of Lucifer. And they are not ashamed by the racist, anti-human pronouncements of Blavatsky and Bailey discussed above.

By actively pushing Theosophical ideology, the Lucis Trust, World Goodwill and the UN are openly ushering in the New World Religion. World Goodwill, for example, created a mantra known as the Great Invocation. They explain, "The Great Invocation is a world prayer, translated into almost 70 languages and dialects. It is an instrument of power to aid the Plan of God [to] find full expression on Earth. To use it is an act of service to humanity and the Christ. It expresses certain central truths which all people innately and normally accept."[75] It concludes with, "Let Light and Love and Power restore the Plan on Earth."

Why the reference to the Christ? Theosophy, after all, rejects Christianity and worships Lucifer. However, they know their New World Religion will face formidable resistance. Bailey even acknowledged, "This inherent fanaticism (found ever in reactionary groups) will fight against the appearance of the coming world religion and the spread of esotericism."[76] Therefore, they are appealing to Christians by cloaking their Luciferian New World Religion in Christian terms. Bailey used the same deceptive tactic in her book *The Reappearance of the Christ*. She wrote,

> "Today, another recognition is becoming possible. It is the recognition everywhere of the imminent return of Christ (if such a phrase can be true of someone Who has never left us!) and of the new spiritual opportunities which this event will make possible…The Son of God is on His way and He cometh not alone. His advance guard is already here and the Plan which they must follow is already made and clear. Let recognition be the aim."[77]

She even admitted her deception in another book, *The Externalization of the Hierarchy*. Bailey confessed, "It is easier to swing the masses into step and give them the newer light of truth if that light is poured on to familiar ground."[78] In another book, *The Rays and the Initiations*, Bailey even reversed the Bible's warning concerning "the mark of the beast." Just as the Bible says Satan will do, Bailey encouraged people to take the mark of the beast, or what she referred to as "the mark of the Saviour." Bailey explained, "It will embody the mark of indication (the signature as medieval occultists used to call it) of a new type of salvation or salvage."[79] Today, nothing has changed. World Goodwill declares,

> "This is a time of preparation not only for a new civilisation and culture in a new world order, but also for the coming of a new spiritual dispensation. Humanity is not following an uncharted course. There is a divine Plan in the Cosmos of which we are a part. At the end of an age, human resources and established institutions seem inadequate to meet world needs and problems. At such a time, the advent of a Teacher, a spiritual leader or Avatar, is anticipated and invoked by the masses of humanity in all parts of the world. Today the reappearance of the World Teacher, the Christ, is expected by millions…The coming world Teacher will be mainly concerned, not with the result of past error and inadequacy, but with the requirements of a new world order and with the reorganisation of the social structure."[80]

The Aquarian Age Community (AAC) is yet another New Age non-governmental organization (NGO) accredited by the United Nations.[81] The AAC hosts monthly meditation nights in the Meditation Room of UN headquarters in New York. The AAC also pushes the Theosophical doctrine of Blavatsky and Bailey: "The Aquarian Age Community is dedicated in loving service to humanity, the planet and the 'Great Thinker.' It is inspired by the teachings of Master Koot Hoomi, Master Morya, and Master Djwhal Khul as these are set forth in the books of Helena Blavatsky, Alice Bailey and the Agni Yoga Society."[82] The AAC *also* attaches Theosophy to Christianity in order to misdirect Christians towards the New World Religion. They admit, "The Aquarian Age Community cooperates and collaborates with the worldwide community that is actively preparing the way for the reappearance of the World Teacher—the Christed (Anointed) One, the true Aquarian."[83]

The United Nations and the New World Religion

In November 2009, *New York Post* columnist Andrea Peyser wrote, "When did global warming turn into a forced religion?"[84] She was referring to the global warming brainwashing imposed upon her daughter during school lessons. She further observed, "Our children are on the front lines of the warming hysteria, a place where 'experts' from Al Gore to the president leave no room for dissent or even the slightest skepticism." Was Peyser's experience an isolated incident, or part of a larger agenda to indoctrinate the public, especially children, to accept a new world religion? Also in November 2009, Fox News reported on United Nations Environmental Program (UNEP) documents which proposed, "The environment should compete with religion as the only compelling, value-based narrative available to humanity."[85] According to the documents, UNEP "should pioneer a new style of work. This requires going beyond a narrow interpretation of UNEP's stakeholders as comprising its member states—or even the world's governments—and recruiting a far wider community of support, in civil society, the academic world and the private sector."[86] And furthermore, "civil society, including children and youth, and the private sector will be reached through tailor-made outreach products and campaigns."[87]

The United Nations is the central hub from which the new world religion movement emanates. Besides working with various theosophical offshoots like World Goodwill, Lucis Trust, and the AAC, the UN has consistently promoted politically motivated New Age ideology. Donald Keys, for example, is the founder and president of Planetary Citizens, and a former speechwriter for former UN Secretary General U Thant. In his paper, "Spirituality at the United Nations," Keys wrote,

> "We have an informal network at the UN, a humanity underground. It consists of those who are committed, aware, and striving to bring the New World to birth. It consists of people in high places and in low—of the patient secretary who has been 30 years with the UN, but lives with the vision and the spirit; of the professionals, and undersecretaries and heads of departments who are acting out the imperatives that their own inner vision gives them. Some few are conscious of the sources of their inspiration; most are not. They are the Karma Yogis of our time—those whose path of spirituality is to achieve through doing—to grow through serving. They are found not only in the secretariat but also in the delegations to the UN, among the diplomats and their staffs, and also among folks like us, representatives of non-governmental organizations around the UN."[88]

Robert Muller

Of the so-called Karma Yogis described by Keys, Robert Muller has been the most prolific. Now retired from the UN, Muller had a distinguished 40-year career. His highest position was Assistant Secretary General, a post he held under three consecutive Secretary Generals: U Thant, Kurt Waldheim, and Javier Perez de Cuellar. At the UN, Muller's New Age orientation earned him the moniker, "The Prophet of Hope."[89] Currently, he is Chancellor Emeritus of Costa Rica's University for Peace, an institution established under UN auspices in 1980 by Muller and Costa Rican President Rodrigo Carazo.[90] The university has served to indoctrinate future New Age and New World Order servants. Throughout his career, Muller has tirelessly marketed the New World Order to unsuspecting New Age enthusiasts and onlookers. He recently declared, "I have a plan for Earth Government in 2020 which outlines the measures of a better management of our planetary home. I am now writing a plan to achieve Paradise Earth by the year 2050."[91] In his book *New Genesis: Shaping a Global Spirituality*, Muller used the term "the planetary age" to describe the New World Order:

> "While continuing to learn more about our planet and its proper management, we must now pass from the national to the planetary age and from the rational to the moral and spiritual age. We must reinsert ourselves into the total visions of the great religions and prophets."[92]

He further declared,

> "The world's major religions must speed up dramatically their ecumenical movement and recognize the unity of their objectives in the diversity of their cults. Religions must actively cooperate to bring to unprecedented heights a better understanding of the mysteries of life and of our place in the universe. 'My religion, right or wrong,' and 'My nation, right or wrong' must be abandoned in the planetary age."[93]

Like Alice Bailey, Muller also blends Christian terms and concepts with New Age and New World Order rhetoric. A particularly bold example was his rewrite of the biblical story of creation where Muller declared, "And God saw that all nations of the earth, black and white, poor and rich, from North and South, from East and West, and of all creeds were sending their emissaries to a tall glass house on the shores of the River of the Rising Sun, on the Island of Manhattan to study together, to think together and to care together for the world and all its people. And God said: That is good. And it was the first day of the New Age of the earth."[94] The "tall glass house" was of course United Nations headquarters. According to Muller, only the UN and its planned global government can save humanity. On the "seventh day," according to Muller, "God saw humans restore God and the human person as the alpha and omega, reducing institutions, beliefs, politics, governments, and all man-made entities to mere servants of God."[95]

Muller sometimes refers to the New World Order as the "cosmic government." He stated, "I have come to believe firmly today that our future peace, justice, fulfillment, happiness and harmony on this planet will not depend on world government but on divine or cosmic government, meaning that we must seek and apply the 'natural,' 'evolutionary,' 'divine,' 'universal' or 'cosmic' laws which must rule our journey in the cosmos. Most of these laws can be found in the great religions and prophecies, and they are being rediscovered slowly but surely in the world organizations."[96]

According to Muller, the United Nations is a spiritual organization destined to save the world. He has declared, for example,

- "If Christ came back to earth his first visit would be to the United Nations to see if his dream of human oneness and brotherhood had come true."[97]
- "I would also like to see published someday a Bible which would show how the United Nations is a modern biblical institution, bent on implementing world-wide the wise precepts and divine commandments of the Bible."[98]
- "In the middle of my life I discovered that the only true, objective education I had received was from the United Nations where the earth, humanity, our place in time

and the worth of the human being were the overriding concerns. So at the request of educators I wrote the World Core Curriculum, the product of the United Nations, the meta-organism of human and planetary evolution."[99]

Muller's World Core Curriculum (WCC) is a framework for New Age and New World Order indoctrination. In an article on the United Nations' Online website, Catherine Clark explains that the World Core Curriculum "prepares students to be cooperative planetary citizens. This curriculum has as its basic platform the value of the individual and his/her unique place in the One Humanity. Thus students are empowered to be aware of themselves as 'cosmic units,' and one with the whole of humankind. Such a basis for learning is very much in alignment with the growing interdependency which characterizes our present world."[100] For his World Core Curriculum, UNESCO awarded Muller the 1989 Peace Education Prize.[101]

In 1980, Gloria Crook founded The Robert Muller School to "implement Robert Muller's World Core Curriculum and make the results available to educators around the world."[102] The school is fully accredited and "Certified as a United Nations Associated School."[103] The United Nations bookshop in New York promotes and sells many WCC publications. Under the heading, "Foundations of the World Core Curriculum and the Robert Muller School," the WCC website openly admits, "The underlying philosophy upon which The Robert Muller School is based will be found in the Teachings set forth in the books of Alice A. Bailey by the Tibetan Teacher, Djwhal Khul (published by Lucis Publishing Company)."[104]

The UN, Baha'is and Further New World Religion Integration

In 2000, *Insight on the News*, a sister publication of the *Washington Times*, published an article called, "UN Faithful Eye Global Religion." *Insight* reported, "In the name of world peace, the United Nations appears to have embraced a sort of religious universalism that views all religions as equals and is seeking to ban proselytizing."[105] The article focused on the Millennium Peace Summit of Religious and Spiritual Leaders, a UN conference of 1,000 delegates held in New York City in August 2000. Bawa Jain, Secretary-General of the Peace Summit, told delegates, "Religions need to accept the validity of all beliefs to attain world peace." The UN strategically scheduled the Peace Summit just before the UN Millennium Assembly. Jain gloated, "The timing was perfect as it allowed religious leaders to update their political counterparts on how to usher in the peace of the new world order through religious universalism."[106]

Many delegates, however, adamantly opposed what they perceived as obviously subversive tactics. *Insight* reported, "Less than a week after the summit the Vatican released a 36-page declaration rejecting what it said are growing attempts to depict all religions as equally true. A spokesman for the National Association of Evangelicals says they were astonished that a UN endorsed summit would take a stand against proselytizing when the UN charter proposes to guarantee the human right to choose one's own religion."[107]

Three years later, the Vatican's Archbishop Javier Lozano Barragán exposed the UN's plan to impose "a new spirituality that supplants all religions, because the latter have been unable to preserve the ecosystem."[108] After studying UN documents including UNESCO's *A Common Framework for Ethics of the 21st Century*, Barragán determined that the

United Nation is pushing "a new secular religion, a religion without God, or if you prefer, a new God that is the earth itself with the name Gaia." In an article for *L'Osservatore Romano*, Barragán described this effort as the "New Paradigm." He explained, "The series of values that sustain the New Paradigm are values subordinated to this divinity that becomes the supreme ecological value, which they call sustainable development."[109]

As previously discussed, the UN has made numerous attempts to marginalize and subvert existing religions, especially Christianity. By contrast, the UN has incessantly promoted the New Age. The Baha'i religion is another New Age darling of the UN. Founded in 1844 in Iran by Baha'u'llah, the Baha'i religion preaches New Age and New World Order ideology. According to their official website,

"The central theme of Baha'u'llah's message is that humanity is one single race and that the day has come for humanity's unification into one global society. While reaffirming the core ethical principles common to all religions, Baha'u'llah also revealed new laws and teachings to lay the foundations of a global civilization."[110]

They also openly admit they are working with the United Nations to establish a global government:

"The ultimate goal is to unite humanity by eliminating prejudice, promulgating the equality of the sexes, adopting a universal standard of human rights, ensuring education for all, recognizing the harmony between religion and science, and establishing a world federated government. Baha'is work with other organizations, such as the United Nations, to achieve these goals."[111]

In 1948, the United Nations accredited the Baha'i International Community (BIC) as an international nongovernmental organization. In 1970, the UN Economic and Social Council granted them consultative status thus "allowing for a greater degree of interaction with the Council and its subsidiary bodies in efforts to promote social and economic development worldwide."[112] In 1976, the United Nations Children's Fund (UNICEF) also granted them consultative status. BIC also maintains close working relationships with the World Health Organization and the United Nations Environment Program.[113] Furthermore,

"Baha'i representatives provide leadership in a number of UN related bodies, including the NGO/Department of Public Information Executive Committee, the Committee of Religious NGOs, the Values Caucus, the Millennium NGO Network for UN reform, the Commission on the Status of Women and the Commission on Sustainable Development."[114]

In 1992, *Baha'is Magazine* published an article called "Toward the New World Order." They stated, "The new world order, like the Baha'i Faith itself, covers the full range of human activities, from the social and political realm to the everyday relationships in our cultural, spiritual, economic and community lives. It is both an internal and an external re-ordering. This grand vision is what Baha'is both work for and see as imminent."[115] They also asserted,

"This world commonwealth must, as far as we can visualize it, consist of a world legislature, whose members will, as the trustees of the whole of mankind, ultimately control the entire resources of all the component nations, and will enact such laws as shall be required to regulate the life, satisfy the needs and adjust the relationships of all races and peoples. A world executive, backed by an international Force, will carry out the decisions arrived at, and apply the laws enacted by, this world legislature, and will safeguard the organic unity of the whole commonwealth. A world tribunal will adjudicate and deliver its compulsory and final verdict in all and any disputes that may arise between the various elements constituting this universal system."[116]

By their own admission, the Baha'is, together with the UN, envision, "A world federal system, ruling the whole earth and exercising unchallengeable authority over its unimaginably vast resources."[117]

With such close ties to the United Nations, it is hardly surprising that the Baha'is are also avid global warming fear-mongers. In 2007, during the UN Commission on Sustainable Development, the Baha'i International Community organized a panel discussion on the "Ethical Dimension of Climate Change." The *Baha'i World News Service* reported, "As the scientific consensus on global warming grows, it's time to look more closely at how to share the economic, social, and humanitarian burdens that climate change will likely bring."[118] The Baha'is have also participated in the process of drafting the Earth Charter: "Throughout the decade-long initiative, the Baha'i International Community has been an active international partner in the drafting process, giving input, hosting and participating in meetings to solicit comments, and serving on various Earth Charter committees."[119]

Out With The Old, In With The New

The globalists believe that in order to bring about the new, they must simultaneously destroy the old. Therefore, they are attempting to destroy the existing religions to pave the way for their New World Religion. With their false War on Terror, for example, they are trying to discredit and marginalize Islam. Against Christianity, they prefer the Theosophical approach of infusing New World Order and New Age ideology into Christianity so as to seduce Christians away from Christianity and towards the New Age. Additionally, they are slandering Christianity and portraying the religion as antagonistic to nature.

The Assault on Christianity

UCLA professor Lynn White's essay, "The Historical Roots of Our Ecological Crisis," published by *Science* in 1967, was the first prominent publication attacking Christianity on ecological grounds. White claimed that Christianity promotes cultural attitudes hostile to the environment. He also openly called for a new eco-religion. White remarked, "The victory of Christianity over paganism was the greatest psychic revolution in the history of our culture. It has become fashionable today to say that, for better or worse, we live in 'the post-Christian age.'"[120] Implying that the Christian era has already closed, White asked, "What did Christianity tell people about their relations with the environment?" He answered, "Especially in its Western form, Christianity is the most anthropocentric religion the world has ever seen. Christianity, in absolute contrast to ancient paganism and Asia's religions...not only established a dualism of man and nature but also insisted that it is God's will that man exploit nature for his proper ends." White further asserted,

"Christianity made it possible to exploit nature in a mood of indifference to the feelings of natural objects...Christianity bears a huge burden of guilt. More science and more technology are not going to get us out of the present ecological crisis until we find a new religion."

The environmental leadership enthusiastically embraced White's essay. Alston Chase, a philosopher formerly involved with the environmental movement, explained, "The environmental movement now had an epistle for spiritual reform."[121] Michael Coffman, in his book *Saviors of the Earth? The Politics and Religion of the Environmental Movement*, commented, "White's paper gave credibility to what conservationists/environmentalists believed and is undoubtedly the most often quoted article in environmental literature."[122] Summarizing the effect of White's essay, Chase stated, "White had set environmentalists a three-fold challenge: to find a religion replacing Judeo-Christianity which would resolve the question of our place in nature; to find a science, replacing the one that had produced our destructive technology, which would show us how nature could be known; and to construct a social agenda replacing the one based on unlimited growth, which would change our culture before it had destroyed the earth."[123]

The following year, Paul Ehrlich echoed White. In his book *The Population Bomb*, Ehrlich declared,

"Somehow we've got to change from a growth-oriented, exploitative system to one focused on stability and conservation. Our entire system of orienting to nature must undergo a revolution. And that revolution is going to be extremely difficult to pull off, since the attitudes of Western culture toward nature are deeply rooted in Judeo-Christian tradition...Both our present science and our present technology are so tinctured with orthodox Christian arrogance toward nature that no solution for our ecologic crisis can be expected from them alone. Since the roots of our trouble are so largely religious, the remedy must also be essentially religious, whether we call it that or not."[124]

Several years later, prominent British elitist Arnold J. Toynbee continued the assault begun by White. Toynbee was the man who Quigley called the "the most important" member of the Royal Institute for International Affairs, the British branch of the CFR.[125] In a 1973 *New York Times* article, Toynbee isolated one sentence from the Bible as ammunition for his argument. Quoting from Genesis, he wrote, "And God blessed them, and God said unto them, be fruitful, and multiply, and replenish the earth, and subdue it: and have dominion over the fowl of the air, and over every living thing that moveth upon the earth."[126] Although most Christians interpret this passage as God entrusting humans to be stewards of the Earth, Toynbee saw things differently. He explained, "In 1663, this read like a blessing on the wealth of Abraham in children and livestock; in 1973, it reads like a license for the population explosion and like both a license and an incentive for mechanization and pollution."[127] He continued, "Monotheism, as enunciated in the Book of Genesis, has removed the age-old restraint that was once placed on man's greed by his awe. Man's greedy impulse to exploit nature used to be held in check by his pious worship of nature. This primitive inhibition has been removed by the rise and spread of monotheism."[128]

Steven Rockefeller, the UN, and Prince Phillip have also all publicly condemned Christianity. Rockefeller, for example, in his book *Spirit and Nature*, also attacked the Bible. He stated, "Certain aspects of the biblical tradition can be seen as generating anthropocentric, dualistic, hierarchical, and patriarchal ideas and attitudes that are problematic from an ecological as well as a democratic perspective."[129] In 1996, the United Nations officially condemned Christianity and Islam in their *Global Biodiversity Assessment*:

> "Societies dominated by Islam, and especially by Christianity, have gone farthest in setting humans apart from nature and in embracing a value system that has converted the world into a warehouse of commodities for human enjoyment. In the process, not only has nature lost its sacred qualities, but most animal species that have a positive symbolic value in other human cultures have acquired very negative connotations in the European culture. Conversion to Christianity has meant an abandonment of an affinity with the natural world for many forest dwellers, peasants, [and] fishers all over the world."[130]

Also in 1996, Prince Philip chimed in, stating, "The Western Christians seem to have terrible anxieties about associating the 'sacred' with the natural environment. I think there's a sort of folk memory that their predecessors the heathens were worshippers of brooks and trees and stones and streams—so there's a fear that if we show any interest in nature then it implies we're becoming heathens."[131]

Yet when it comes to bashing Christians, nobody does it with more bravado than Ted Turner. At times, he almost parodies himself. If he wasn't so powerful and influential, his comments could be ignored. However, Turner has donated over $1 billion to depopulation and false environmental agendas. He started the United Nations Fund and the Turner Foundation, both heavy donators to depopulation groups and global warming alarmists. Turner's anti-Christian sentiments are part of a larger agenda, and are therefore worth exposing. In 1990, Turner publicly declared that Christianity is "a religion for losers."[132] In 1999, Turner stated, "Christianity is an eco-unfriendly religion."[133]

In 2000, Turner was the honorary chairman of the aforementioned UN Millennium Peace Summit. In his address, he lashed out against Christianity, calling it an "intolerant" religion.[134] On Ash Wednesday 2001, during a meeting with CNN employees, Turner noticed some of them wearing ashes on their foreheads in observance of the holy day. He impudently asked them, "What are you, a bunch of Jesus freaks?"[135] Turner has also mockingly rewritten the 10 Commandments. His version, called the 10 Voluntary Initiatives, calls for population control: "I promise to have no more than two children." It also calls for submission to the UN: "I support the United Nations and its efforts to collectively improve the conditions of the planet." Ironically, Turner habitually violates the second of his Voluntary Initiatives: "I promise to treat all persons everywhere with dignity, respect, and friendliness."[136]

Wolves In Sheep's Clothing

In the introduction of David Spangler's book *Revelation: The Birth of A New Age*, William Irwin Thompson asserted, "The new spirituality does not reject the earlier patterns of the great universal religions. Priest and church will not disappear; they will be absorbed into the existence of the New Age."[137] Thanks to many concerted efforts, this process is now fully underway. For example, in 2008, Ted Turner formed a $200 million partnership with two Christian groups, the United Methodist Church and the Evangelical Lutheran Church, to combat malaria in Africa. Why would he do this? Was Turner seeking to rectify his longstanding contempt for Christians? Actually, by his own admission, he was only using them to advance a separate agenda. He told the *Associated Press*, "The religious community is huge and has a very good reputation for being able to mobilize resources. Why not use them and be thankful?"[138] Use them for what? Turner is an admitted depopulation enthusiast. And as documented in Chapter 16, DDT is the safest and most effective malaria preventative. Turner's campaign, however, utilized far less effective methods, thus ensuring *more* unnecessary deaths. Seen through the eyes of Turner, an *ineffective* malaria campaign is an *effective* depopulation campaign.

In the late 1980s, Prince Philip and the World Wildlife Fund executed plans to coerce religious leaders, and by extension the religious community, into joining the false environmental movement. In 1986, Philip organized a conference in Assisi, Italy, including representatives from the five major religions—Buddhism, Christianity, Hinduism, Islam, and Judaism. As explained by WWF religious advisor Martin Palmer in his book *Faith in Conservation*, Philip believed that "for the conservation movement to have any real chance of success, it needed to find allies who could help spread the message and engage people in the struggle to save the earth."[139] At the conference Philip declared, "I believe that today, in this famous shrine of the saint of ecology, a new and powerful alliance has been forged between the forces of religion and the forces of conservation."[140]

In 1990, WWF-US sponsored the North American Conference on Religion and Ecology. Regarding his address to participants, Russell Train, the WWF-US President from 1978 to 1985 and Chairman from 1985 to 1994, later recalled, "I expressed puzzlement over what had seemed to me to be organized religion's almost total obliviousness to environmental issues."[141] These issues, according to Train, "went to the heart of the human condition, the quality of human life, even humanity's ultimate survival. Here we had problems that could be said to threaten the very integrity of Creation."[142] The biggest problem, according to Train and the WWF, was overpopulation. He continued,

"Almost every significant threat to the environment, I pointed out, has been contributed to and compounded by human numbers. And whatever other adverse effects on the natural environment might result from the growth in sheer human numbers, such growth is necessarily accompanied by a reduction in space for other species, a reduction of opportunity for other forms of life."[143]

Train then attempted to provoke feelings of guilt and responsibility. He explained,

"I did not suggest that the Christian church abandon its concern for humanity; instead, I suggested it give at least equal time to the rest of God's creation...I urged that the church assume major responsibility for teaching that we humans, individually and collectively, are part of the living community of the earth that nurtures and sustains us...Taking a deep breath, I declared that 'these are precepts that could provide the substance for an Eleventh Commandment: Thou shalt cherish and care for the Earth and all within it.'"[144]

This was clever manipulation. Train and his fake environmental cohorts were implicitly telling Christians, "This is your fault," while simultaneously offering the solution: "Welcome to the New Age."

In 1995 Prince Philip, together with the WWF and the World Bank, founded the Alliance of Religions and Conservation (ARC). According to their website, "ARC's strategy is twofold: to help faiths realise their potential to be proactive on environmental issues and to help secular groups recognise this and become active partners."[145] Philip admitted,

"It occurred to me that the people who could most easily communicate with them were their religious leaders. They are in touch with their local population more than anyone else. And if we could get the local leaders to appreciate their responsibility for the environment then they would be able to explain that responsibility to the people of their faith. It didn't seem a particularly bright idea at the time—it was pretty obvious. If your religion tells you (as it does in Christianity anyway) that the Creation of the world was an act of God, then it follows naturally that if you belong to the church of God then you ought to look after His Creation."[146]

World Bank President James Wolfensohn stated quite candidly the reason why they targeted religious groups:

"The reason is simple. The 11 faiths that now make up ARC represent two-thirds of the world's population. They own around 7 percent of the habitable surface of the planet, they have a role in 54 percent of all schools, and their institutional share of the investment market is in the range of 6-8 percent."[147]

Since 1995, ARC has become extremely influential and now collaborates with the United Nations. According to Martin Palmer, "UN officials say they need people who can speak about climate change straight from, and to, the heart."[148] In 2008, *National Public Radio* reported,

"Palmer and his colleagues at the Alliance travel the globe, cajoling religious leaders to learn about climate change and take the message to their followers. And those leaders bring more to the table than faith—Palmer says they control as much as seven percent of the world's forests, forests that help curb global warming. The various official churches are, taken as a single institution, the world's third largest investor."[149]

Environmental groups have had great success in infiltrating religious groups. The Sierra Club, for example, proudly asserted in their 2008 *Faith in Action* report, "Environmental concerns continue to rise in prominence on the agenda of the faithful, with no sign of receding. As the implications of global warming and its disproportionate impact on the world's poor become increasingly clear, prophetic voices are being raised in religious communities around the globe. In the United States, 67 percent of Americans say they care about the environment because it is 'God's creation.'"[150]

In 2009, Al Gore also began targeting religious groups with propaganda. He bragged to *Newsweek*, "I've done a Christian training program; I have a Muslim training program and a Jewish training program coming up, also a Hindu program coming up. I trained 200 Christian ministers and lay leaders here in Nashville in a version of the slide show that is filled with scriptural references. It's probably my favorite version, but I don't use it very often because it can come off as proselytizing."[151]

The Earth Day Network (EDN) brags that on Earth Day 2007, "EDN was successful in creating 12,000 sermons and religious events through outreach to leaders from the Jewish, Muslim, and Christian faiths." And on Earth Day 2008, "We activated 500,000 parishioners in areas of the country that have not responded to the climate crises in support of climate legislation that invests in renewable energy and the creation of green jobs and helps low-income Americans transition to the new green economy."[152] Furthermore, according to EDN,

"Faith leaders from across the U.S. and Canada are preaching on global climate change as a moral issue. If you are a religious leader, we encourage you to sign Earth Day Network's Global Warming in the Pulpit Pledge, and mark your commitment to preach on global warming. Lay leaders can join this movement by inviting your faith leader to sign the Pledge."[153]

Where did this group come from? EDN is chaired by CFR member Gerald Torres,[154] and funded by the Rockefeller Brothers Fund, the Rockefeller Family Fund, and the Turner Foundation.[155]

For false environmentalists attempting to manipulate and exploit the established religions, the most coveted prize is the Vatican. Unfortunately, in recent years, Pope Benedict has repeatedly supported false environmental issues. In 2007, for example, the *London Guardian* reported, "The Vatican yesterday added its voice to a rising chorus of warnings from churches around the world that climate change and abuse of the environment is against God's will, and that the one billion-strong Catholic church must become far greener."[156] And in 2005, the Pope called for a "new world order based on just ethical and economic relationships." He explained,

"A united humanity will be able to confront the many troubling problems of the present time: from the menace of terrorism to the humiliating poverty in which millions of human beings live, from the proliferation of weapons to the pandemics and the environmental destruction which threatens the future of our planet."[157]

Despite these statements, before ascending to the papacy, Benedict (Ratzinger) strongly opposed the New World Order. In 2000, Cardinal Joseph Ratzinger denounced the UN's New World Order vision, specifically condemning their population reduction agenda. Their philosophy, he explained, "no longer hopes that men, used to wealth and well-being, will be disposed to make the necessary sacrifices to attain a general welfare, but rather proposes strategies to reduce the number of guests at the table of humanity, so that the presumed happiness they have attained will not be affected."[158] Although he now encourages Christians to submit to the New World Order, he previously called for mass resistance. "At this stage of the development of the new image of the new world," Ratzinger proclaimed, "Christians—and not just them but in any case even more than others—have the duty to protest."[159]

In recent years, the New World Religion has become increasingly apparent. What was once surreptitious is now overt. In November 2009, in a landmark ruling, a British judge determined that belief in manmade climate change has the same legal status as religion. In March 2009, a British firm had fired an employee for his "philosophical belief about climate change and the environment," including, for example, his refusal to travel by air for company business.[160] The company argued that climate change is a scientific rather than philosophical issue because "philosophy deals with matters that are not capable of scientific proof." Justice Michael Burton, however, ruled that, "belief in man-made climate change…is capable, if genuinely held, of being a philosophical belief for the purpose of the 2003 Religion and Belief Regulations."[161]

Chapter 15
The Club of Rome

Founded in 1968, the Club of Rome is an extremely influential, ultra-elite think tank. Comprised of globalists with an overt distain for democracy and national sovereignty, they nevertheless describe themselves as "a not-for-profit organisation, independent of any political, ideological or religious interests."[1] The Club of Rome, however, is *deeply* aligned with numerous political, ideological, and religious interests. Most conspicuously, they are aligned with the UN's global warming agenda. Club of Rome members include global warming crusaders Al Gore and Mikhail Gorbachev (and many others listed below). Also UNESCO counts the Club as an "NGO maintaining official relations with UNESCO."[2] According to their website, the Club's essential mission is "to act as a global catalyst for change through the identification and analysis of the crucial problems facing humanity and the communication of such problems to the most important public and private decision makers as well as to the general public."[3] Since 1968 they have been doing just that — except the problems they identify are bogus and the recommendations they offer are repugnant.

The Club of Rome's first major book, published in 1972, was *The Limits to Growth*. The premise was that population growth, if left unchecked, would outstrip available resources, leading to mass starvation, political chaos and general catastrophe by mid-21st century. Therefore, they recommended sharply restraining both population and economic growth to avoid such crises. The authors stated, "If the present growth trends in world population, industrialization, pollution, food production, and resource depletion continue unchanged, the limits to growth on this planet will be reached sometime within the next one hundred years. The most probable result will be a rather sudden and uncontrollable decline in both population and industrial capacity."[4] The authors also asserted, "Given the finite and diminishing stock of nonrenewable resources and the finite space of our globe, the principle must be generally accepted that growing numbers of people will eventually imply a lower standard of living and a more complex problematique."[5] The authors also doled out plenty of fear and extremism, both consistent hallmarks of Club of Rome reports:

> "We can talk seriously about where to start only when the message of *The Limits to Growth*, and its sense of extreme urgency, are accepted by a large body of scientific, political, and popular opinion in many countries. The transition in any case is likely to be painful, and it will make extreme demands on human ingenuity and determination. As we have mentioned, only the conviction that there is no other avenue to survival can liberate the moral, intellectual, and creative forces required to initiate this unprecedented human undertaking."[6]

While not elaborating on methodology, they also callously advocated genocide: "There are only two ways to restore the resulting imbalance. Either the birth rate must be brought down to equal the new, lower death rate, or the death rate must rise again."[7] This typical elitist sentiment completely disregards technological and socio-political advances, which enable *increased* human populations. As previously discussed, these elites believe they have both the right and the responsibility to manage and occasionally cull human populations.

So how did the Club of Rome arrive at their sensational conclusions? They based the *Limits to Growth* projections on computer models originally designed by M.I.T. professor Jay Forrester. Dennis Meadows, a 29-year old computer expert and Forrester protégé, used Forrester's models to make predictions regarding population growth, industrial output, natural resources, and pollution. Meadows claimed he tested thousands of hypotheses, yet amazingly, he concluded, "All growth projections end in collapse."[8] He arrived at this conclusion by *rigging* the models—on orders from the Club of Rome—to yield the desired catastrophic predictions. As documented below, the Club later publicly confessed to this fraud. Yet for four years, they aggressively pushed their no-growth doctrine. In 1972, *Time* magazine reported,

> "The Meadows team offers a possible cure for man's dilemma—an all-out effort to end exponential growth, starting by 1975. Population should be stabilized by equalizing the birth and death rates. To halt industrial growth, investment in new, nonpolluting plants must not exceed the retirement of old facilities. A series of fundamental shifts in behavioral patterns must take place. Instead of yearning for material goods, people must learn to prefer services, like education or recreation."[9]

According to the Club, the models proved that, barring massive societal reorganization, catastrophe was inevitable. Almost immediately, prominent scientists and academics began speaking out.

Yale economist Henry Wallich, in a *Newsweek* editorial, called the book "a piece of irresponsible nonsense." He criticized the models, stating, "The quantitative content of the models comes from the authors' imagination."[10] In 1972, the Editor of *Nature*, John Maddox, wrote a book called *The Doomsday Syndrome*, which countered the Club's fear-mongering. Maddox wrote, "Tiny though the earth may appear from the moon, it is in reality an enormous object. The atmosphere of the earth alone weighs more than 5,000 million-million tons, more than a million tons of air for each human being now alive…It is not entirely out of the question that human intervention could at some stage bring changes, but for the time being the vast scale on which the earth is built should be a great comfort."[11]

Limits to Growth sparked intense controversy and ongoing debate. Shortly before its publication, *The Ecologist* magazine printed a 22-page article called "Blueprint for Survival." Thirty-three distinguished U.K. scientists endorsed the article including WWF founders Thomas Huxley and Peter Scott. The article, part of thinly veiled, coordinated effort between the WWF and the Club of Rome, served to prime the public for *Limits to Growth*. Shamelessly sensationalistic, "Blueprint" warned that unrestricted economic and population growth must inevitably lead to "the breakdown of society and of the life support systems on this planet—possibly by the end of this century and certainly within the lifetime of our children."[12] The article further asserted, "Governments are either refusing to face the relevant facts or are briefing their scientists in such a way that the seriousness is played down." Consequently, "we may muddle our way to extinction."[13]

The *Ecologist* later published an expanded book version of "Blueprint." An impassioned diatribe against western lifestyles, the book contended, "The principal defect of the industrial way of life with its ethos of expansion is that it is not sustainable…By now it should be clear that the main problems of the environment do not arise from temporary and accidental malfunctions of existing economic and social systems. On the contrary, they are the warning signs of a profound incompatibility between deeply rooted beliefs in continuous growth and the dawning recognition of the earth as a space ship, limited in its resources and vulnerable to thoughtless mishandling."[14]

In 1973, Herman Daly, an acclaimed economist and former Senior Economist in the Environment Department of the World Bank, published a book called *Toward a Steady-State Economy*. In support of the *Limits to Growth* contention, Daly argued, "Environmental degradation is an iatrogenic disease induced by economic physicians who treat the basic malady of unlimited wants by prescribing unlimited economic growth...The growth paradigm has outlived its usefulness. It is a senile ideology that should be unceremoniously retired into the history of economic doctrines."[15] Daly has also declared,

- "If you have eaten poison, you must get rid of the substances that are making you ill. Let us, then, apply the stomach pump to the doctrines of economic growth that we have been force-fed for decades."[16]
- "We cannot have too many people alive simultaneously lest we destroy carrying capacity and thereby reduce the number of lives possible in all subsequent time periods."[17]

George Church, a prolific reporter for *Time*, observed, "This status quo prescription—the report calls it 'global equilibrium'—is as chilling as the doomsday prophecy."[18] Gunnar Myrdal, the former Swedish Minister of Trade and winner of the Nobel Prize in Economics, shamed the Club of Rome for their methodology:

"The use of mathematical equations and a huge computer, which registers the alternatives of abstractly conceived policies by a 'world simulation model,' may impress the innocent general public but has little, if any, scientific validity. That this 'sort of model is actually a new tool for mankind' is unfortunately not true. It represents quasi-learnedness of a type that we have, for a long time, had too much of."[19]

The debate continued, going back and forth between supporters and critics. And all the while, the Club of Rome's conniving insiders remained silent, quietly enjoying the intensifying controversy.

Shortly after the book's release, *Time* reported, "Even the club members were startled by the computer's findings but were unable to raise any important objections to them."[20] And why couldn't they raise objections? The book was a massive scam, a fear campaign targeting an unsuspecting public. The Club of Rome intentionally rigged the computer models to produce the desired outcome. For four years they maintained their position advocating heavy restrictions on economic growth. In April 1976, however, after the book had sold over three million copies, they completely *reversed* their position. *Time* reported,

"At a three-day meeting in Philadelphia sponsored mainly by the First Pennsylvania Corp., a leading bank, speaker after speaker came out for more growth. Why? The Club's founder, Italian Industrialist Aurelio Peccei, says that *Limits* was intended to jolt people from the comfortable idea that present growth trends could continue indefinitely. That done, he says, the Club could then seek ways to close the widening gap between rich and poor nations—inequities that, if they continue, could all too easily lead to famine, pollution and war. The Club's startling shift, Peccei says, is thus not so much a turnabout as part of an evolving strategy."[21]

Additionally, the *New York Times* quoted Peccei as stating that *Limits to Growth* had served its purpose of "getting the world's attention."[22]

This is a crucial point. For the Club of Rome, (a) fear is a weapon, and (b) the ends justify the means. This is classic globalist psychopathology. Was *Limits to Growth* simply an innocuous prank? Peccei admitted the project was part of "an evolving strategy." What if *Limits to Growth* was testing public response for a planned, future initiative? What if the Club of Rome is intimately connected to the current global warming hysteria? What if the most prominent global warming alarmists are all Club of Rome members? And since they admittedly lied before, what if they are lying again about global warming? The Club of Rome's "evolving strategy" was an immense social engineering initiative culminating in the current manmade global warming agenda. With *Limits to Growth* and many ensuing documents, they were executing a three-fold strategy:

1. Invent a scenario of massive environmental catastrophe.
2. Justify the catastrophe with rigged computer models, manipulated scientific studies, and endorsements by various scientists and scientific organizations.
3. Promote the catastrophe, along with its predetermined solutions, through intense media propaganda.

To understand what the Club of Rome is really working towards requires an examination of some of their other seminal publications.

Reshaping the International Order

Peccei admitted that *Limits* was part of an evolving strategy. But what was the larger agenda? At the Philadelphia meeting, Peccei also spoke about the Club's next book. They had recently commissioned Nobel Laureate professor Jan Tinbergen to explore the creation of "a new international order."[23] Peccei declared that *Limits* had "punctured the myth of exponential growth," but future progress required "puncturing a second myth—the myth of national competence."[24] In 1976, the Club of Rome published *RIO: Reshaping the International Order*. Coordinated by Tinbergen, *RIO* addressed the following fundamental question: "What new international order should be recommended to the world's statesmen and social groups so as to meet, to the extent practically and realistically possible, the urgent needs of today's population and the probable needs of future generations?"[25] The book openly called for a slow but steady dismantling of national sovereignty and the implementation of a world government characterized by "humanistic socialism."[26] So how did the Club expect to achieve this feat? "The achievement of this global planning and management system calls for the conscious transfer of power—a gradual transfer to be sure—from the nation State to the world organization. Only when this transfer takes place can the organization become effective and purposeful," they explained.[27]

To reassure the public that the New World Order would not infringe upon their cherished national sovereignties, the authors recommended shrewd semantic deception:

"The traditional concept of territorial sovereignty should be replaced by the concept of *functional sovereignty*, which distinguishes jurisdiction over specific uses from sovereignty over geographic space. This would permit the interweaving of national jurisdiction and international competences within the same territorial space and open the possibility of applying the concept of the common heritage of mankind both beyond and within the limits of national jurisdiction."[28]

In other words, people formerly known as "Americans," from the former "United States," would no longer have national sovereignty. They would instead have *functional sovereignty* whereby all rights would become privileges, granted by a world government. According to the Club of Rome, "Ultimately, we must aim for decentralized planetary sovereignty with the network of strong international institutions which will make it possible."[29]

The Club of Rome realized that word games alone would not sufficiently dupe the public. Therefore, they also recommended slick propaganda and using "experts" to portray the New World Order as both inevitable and desirable. Preferring stealth and trickery to forced implementation, they noted, "The possibility of implementing ideas of a new power structure would, in democratic societies, necessitate the acceptance of such ideas by wide sections of public opinion."[30] So how could they ensure such acceptance? The authors explained,

> "Public opinion is no phenomenon sui generis. It is in part the result of government policies and by definition politicians cannot hide behind their own creation. If some sectors of public opinion in the industrialized countries are immersed in the rhetoric and slogans associated with misunderstanding, then much of this may be inherited from their political leaders. And if these leaders are in part responsible for a situation which impedes acceptance of the need for change, then they themselves must be held responsible for changing this situation."[31]

Notice the phrase, "rhetoric and slogans associated with misunderstanding." This implies that dissenters simply *misunderstand* the altruistic ambitions of their would-be masters. And furthermore, notice how during the transitional phase, political leaders are responsible for quelling all dissent. The Club of Rome published this book in 1976. More than three decades later, is the *RIO* plan now being implemented?

What happens to those who question the theory of manmade global warming? Are they not marginalized, and oftentimes ostracized and ridiculed? The *perceived* authority of the United Nations' IPCC fuels the global warming myth. When people fail to question this perceived authority, the legitimacy of the IPCC's "science" becomes irrelevant. The *RIO* report boldly acknowledged the importance of using scientists for propaganda purposes: "One of our main weapons in this search is the vast arsenal of scientists we are potentially able to deploy. To fully utilize this resource, we must deliberately choose to focus investigation in directions we believe to be really relevant."[32]

For the Club of Rome, scientists are merely functionaries for predetermined agendas. The exploratory nature of science, far from being encouraged, is actually suppressed by New World Order architects. Scientists become "experts" — mouthpieces of the established order. The Club explained, "The 'new expert,' in actively promoting local self-reliant development, may need to subordinate his own values, even his knowledge, to those of the community he is attempting to serve."[33] The Club of Rome is sometimes blatant beyond comprehension. Here they assert that scientists, "the new experts," must abandon their values and integrity and disregard their technical knowledge, purportedly to serve the community. The "community," of course, is a euphemism for the tyrannical world state. Have the IPCC's scientists been participating, perhaps without their full comprehension, in the *RIO* plan?

RIO also called for a new monetary system, including international taxation and an international reserve currency. During the current economic crash, scores of globalist have been recommending these very same "solutions." The *RIO* authors proposed,

- "The gradual introduction of a system of international taxation which should be handled by a World Treasury, both to meet the current as well as the development needs of the poorer nations."
- "The creation of an international reserve currency by an international authority, such as an International Central Bank, which should be under international management without being dominated by the interests of one particular group of nations."[34]

1976 was a monumental year for the Club of Rome. First Peccei announced that *Limits to Growth* was an exaggerated hoax, part of an "evolving strategy." Then they released *RIO*, the next phase of this strategy, their New World Order vision. This vision involved dissolving national sovereignty and imposing global taxation and a global currency. If the Club of Rome's intentions were really benevolent, why have their plans involved so much deception? Why do they feel they must coerce scientists, and manipulate public opinion with propaganda? Do they believe that otherwise the public would simply "misunderstand" their true intentions? Of course not, they fear the opposite. They fear that people would absolutely understand, and would vehemently oppose their true intentions. Nobody wants totalitarian global government, except for these elites. Thus, only by stealth can they achieve their ends. Another pivotal Club of Rome publication was their 1974 book, *Mankind at the Turning Point: The Second Report to The Club of Rome.*

Mankind at the Turning Point

Following up on *Limits to Growth*, this book again used absurd computer modeling to predict global catastrophe. Foreshadowing the current scare about manmade global warming, which the Club of Rome helped conceive and orchestrate, the authors asserted, "The modern crises are, in fact, man-made, and differ from many of their predecessors in that they can be dealt with."[35] But *how* can they be dealt with? The Club insisted that mankind must adopt an "organic growth" model for development, another thinly veiled euphemism for totalitarian world government. The following statements encapsulated their cleverly framed argument:

- "In Nature organic growth proceeds according to a 'master plan,' a 'blueprint.'"[36]
- "Such a 'master plan' is missing from the process of growth and development of the world system."[37]
- "The concept of the 'organic growth' of mankind, as we have proposed in this report, is intended as a contribution toward achieving that end."[38]

This all seems quite rational, desirable even, yet it hinges on the meaning of "world system." The authors elaborated: "And cooperation, finally, requires that the people of all nations face up to an admission that may not come easy. Cooperation by definition connotes interdependence. Increasing interdependence between nations and regions must then translate as a decrease in independence. Nations cannot be interdependent without each of them giving up some of, or at least acknowledging limits to, its own independence."[39] This book also hinted at the globalists' planned new world religion with the following statements:

- "A world consciousness must be developed through which every individual realizes his role as a member of the world community."[40]
- "It must become part of the consciousness of every individual that the basic unit of human cooperation and hence survival is moving from the national to the global level."[41]

Notice that *survival* depends on accepting this global system. But who controls this global government? Do ordinary people have any say in the matter? No. The new system would be run by financial and governmental elites. The book also discussed the planned world economic system:

- "Now is the time to draw up a master plan for organic sustainable growth and world development based on global allocation of all finite resources and a new global economic system. Ten or twenty years from today it will probably be too late."[42]
- "The solution of these crises can be developed only in a global context with full and explicit recognition of the emerging world system and on a long-term basis. This would necessitate, among other changes, a new world economic order and a global resources allocation system."[43]

During the current financial collapse, world leaders and media pundits have been incessantly parroting calls for a new world economic order. And yet, Club of Rome elites were openly planning for this more than three decades ago. A final noteworthy point about *Mankind at the Turning Point* is its inclusion as a chapter header a particular Alan Gregg quotation. Gregg was an ardent eugenicist and former director of the Rockefeller Foundation's Division of Medical Sciences.[44] Two years before his death in 1957, Gregg declared, "The world has cancer and the cancer is man."[45] By showcasing this quotation, the Club of Rome demonstrated their utter disdain and contempt for humanity. Remember, elites consider themselves separate from the public, a superior race. Bertrand Russell's arrogant pronouncement bears repeating: "Gradually, by selective breeding the congenital differences between rulers and ruled will increase until they become almost different species. A revolt of the plebs would become as unthinkable as an organised insurrection of sheep against the practice of eating mutton."[46]

Following the 1972 release of *Limits to Growth*, George Church, writing for *Time* magazine, made an uncomfortable, yet penetrating observation about the Club of Rome. His observation is befitting to this day. Church stated, "It is hard to see how growth could be halted, or even substantially slowed, without a world dictatorship — the more so as citizens of underdeveloped countries already suspect that the no-growth argument is an elitist, aristocratic, white man's conspiracy to lock them into perpetual poverty."[47] Church was right. In the ensuing years, the Club would openly announce their New World Order plans. And while attempting to portray these plans as peaceful and humanitarian, in 1992 they fully confirmed the totalitarian nature of their agenda. With the 1992 publication of *The First Global Revolution*, the Club of Rome openly declared war against humanity. Not only that, they openly announced the creation of a fake adversary, manmade global warming, to mobilize humanity into a planned global government. The following two quotations recapitulate their contempt for democracy:

- "This is where the question must be raised — what sort of democracy is required today and for what purpose? The old democracies have functioned reasonably well over the last two hundred years, but they appear now to be in a phase of complacent stagnation with little evidence of real leadership and innovation."[48]

- "Democracy is not a panacea. It cannot organize everything and it is unaware of its own limits. These facts must be faced squarely, sacrilegious though this may sound. In its present form, democracy is no longer well suited for the tasks ahead. The complexity and the technical nature of many of today's problems do not always allow elected representatives to make competent decisions at the right time."[49]

To be clear, the Club of Rome called for governance by unelected, unaccountable bureaucrats, which, incidentally, has transpired within the European Union. Next, they announced their contrived, mythical climate crisis — the organizing factor in their final push towards world government. And most shockingly, they boldly identified their true enemy. "The real enemy," they explained, "is humanity itself."[50] The report stated,

- "The opposition of the two political ideologies [capitalism versus communism] which have dominated this century no longer exists, leaving nothing but a crass materialism."[51]

- "It would seem that men and women need a common motivation, namely a common adversary against whom they can organize themselves and act together."[52]

- "The need for enemies seems to be a common historical factor...either a real one, or else one invented for the purpose."[53]

- "New enemies have to be identified, new strategies imagined, and new weapons devised."[54]

- "In searching for a common enemy against whom we can unite, we came up with the idea that pollution, the threat of global warming, water shortages, famine and the like, would fit the bill. In their totality and their interactions these phenomena do constitute a common threat which must be confronted by everyone together. But in designating these dangers as the enemy, we fall into the trap, which we have already warned readers about, namely mistaking symptoms for causes. All these dangers are caused by *human* intervention in natural processes, and it is only through changed attitudes and behaviour that they can be overcome. The real enemy then is humanity itself."[55]

So humanity needs a collective enemy against which to organize. If no enemy exists, then, according to the Club, an enemy can be "invented for the purpose." What is this invented enemy? Is it global warming? Yes, but global warming is actually just a symptom of the real disease, the real enemy. The real enemy is *humanity*? While completely shocking, this statement is consistent with the Club of Rome's principal message. According to them, the world must always have an enemy — real or invented. As documented throughout this book, the invented enemy, al-Qaeda for example, is easier to manage and thus preferable. So the Club of Rome decided — they just "came up with the idea" — that the new "invented" enemy, the "common adversary," would be global warming. Global warming would be the phantom enemy, like the elusive al-Qaeda, whereas the real enemies of the globalists are the common people of North and South America, Europe, Asia, Africa — humanity itself.

Club of Rome Members

Notable members of the Club of Rome include Al Gore, Bill Gates,[56] Prince Philippe of Belgium, Queen Beatrix of the Netherlands, Queen Doña Sophia of Spain, former President of Georgia Eduard Shevardnadze, Prime Minister of India Manmohan Singh, former President and former Prime Minister of Portugal Mario Soares, former President of Germany Richard von Weizsäcker, former President of Mexico and Director of The Yale Centre for the Study of Globalization Ernesto Zedillo, Director-General of UNESCO Koïchiro Matsuura, former Director-General of UNESCO Federico Mayor Zaragoza, former President of Uruguay Luis Lacalle Herrera, former Prime Minister of Finland Mauno Koivisto, former Prime Minister of the Netherlands Ruud Lubbers, King of Spain Juan Carlos I, former President of Czechoslovakia Vaclav Havel, former President of the USSR Mikhail Gorbachev, former President of Hungary Arpad Göncz, former President of Columbia César Gaviria, former President of the European Commission Jacques Delors, former President of Brazil Fernando Henrique Cardoso, former President of Columbia Belisario Betancur,[57] and Secretary General of the Council of the European Union Javier Solana.[58] Throughout this book, various Club of Rome members are analyzed in depth. Part III, for example, focuses considerable attention on Al Gore. This study of the Club of Rome will analyze one prominent member, Ruud Lubbers.

The 1999 Club of Rome Annual Meeting featured a presentation by Ruud Lubbers entitled, "Governance in an era of Globalization." Lubbers was the Dutch Prime Minister from 1982 to 1994 and a United Nations High Commissioner in 2001. Lubbers is also a habitual sex offender found guilty, by an official UN investigation, of a "pattern of sexual harassment."[59] Consequently, he was forced to resign. In his paper, Lubbers observed, "The new world order is no perfect world. It is more scaring than promising. Certainly, there are opportunities, but a too optimistic view will destroy motivations and actions to improve quality of life in our world. At the same time we have to stress that a too negative, apocalyptic view will do the same. Globalization as such is a given, but the outcomes will depend on the combined effects of human action."[60] Lubbers also wrote,

- "National democracies are weakening."
- "The state is less effective in realizing societal values and therefore politics becomes less credible."
- "The sovereignty of the people as guaranteed by national parliaments is limited to national policies. The more these policies are being embedded in and dependent on the supra-national juridical-political surrounding, the less meaningful parliaments can be."
- "In short, the more the international and supra-national level gains in importance for world-governance, the less power there is for national parliament and the bigger the democratic deficit will be."[61]

As a seasoned globalist with firsthand experience in subverting national democracy, Lubbers is certainly qualified to speak about these matters. During his time as Prime Minister, according to the *International Herald Tribune*, "he was a strong defender of the [NATO] alliance, overcoming considerable domestic opposition to win approval for the stationing of intermediate-range nuclear weapons in the country in the mid-1980s."[62] Lubbers also cooperated with a massive clandestine international arms ring organized by Abdul Qadeer Khan, a Pakistani nuclear scientist. According to the *San Francisco Chronicle*,

"Former Dutch Prime Minister Ruud Lubbers is on record as saying that the government of The Netherlands knew Khan was stealing nuclear secrets but let him go on two occasions after the CIA said it wanted to continue to monitor his movements."[63]

Lubbers is just another run-of-the-mill globalist. He is corrupt to his core, arrogant, and pompous. He disdains democracy and national sovereignty. Among his peers, there is nothing exceptional about this man. He is mediocre, undistinguished, unremarkable, lackluster and ultimately forgettable. There are hundreds more like him. The Club of Rome is a *Club of Elites*. After finally understanding and accepting what this means, society can retire this gaggle of plutocrats, reclaim its inherent sovereignty and rights, and build genuine world peace and prosperity.

Club of Rome Satellites

The Club of Madrid is a Club of Rome satellite consisting of "more than 70 democratic former Heads of State and Government from 50 countries who contribute their time, experience and knowledge."[64] Its members, predominantly drawn from the ranks of the Club of Rome, include Gro Harlem Brundtland, Bill Clinton, Javier Pérez de Cuellar (former UN Secretary-General), Vicente Fox, and Mikhail Gorbachev.[65] Also Diego Hidalgo, founder and President of the European Council on Foreign Relations,[66] Jimmy Carter and Kofi Annan are members.[67]

The Global Leadership for Climate Action (GLCA), a joint initiative of Ted Turner's United Nations Fund and the Club of Madrid, is "a task force of world leaders committed to addressing climate change through international negotiations."[68] They claim, "The scientific diagnosis has been made. The time for action is now."[69] Their objective is "to mobilize political will and invigorate international negotiations toward an agreement on climate change beyond 2012."[70] Their members include the usual suspects—Gro Harlem Brundtland, Mary Robinson, George Soros, Ted Turner, Tim Wirth (President of UN Fund) and James Wolfensohn.[71]

The Club of Budapest is yet another Club of Rome satellite. According to their website, "The idea of the Club of Budapest was developed in 1978 in a discussion between Aurelio Peccei, founder and first president of the Club of Rome, and Ervin Laszlo, systems philosopher and also member of the Club of Rome at that time. They were convinced that the enormous challenges to humanity can only be dealt with through the development of a cultural and cosmopolitan consciousness. Based on these ideas, the Club of Budapest was founded by Dr. Laszlo in 1993."[72] Not surprisingly, the Club of Budapest also pushes the manmade global warming myth. In 2008 they published *State of Global Emergency*, an ultra-alarmist document declaring,

- "Leading climate scientists who have pioneered the research are now stating that global CO_2 emissions must be reduced to ZERO as soon as possible to stabilize the Earth's climate and avoid the 'worst case' climate change scenarios for humanity."[73]
- "Time-estimates of when the 'point of no return' will be reached for the global system of humanity have shrunk from the end of the century to mid-century, then to the next twenty years, and recently to the next five to twenty years."[74]
- "This is close in time to the Mayan 2012 prophecy for the end of the current world."[75]
- "Failure to implement a worldwide shift in the window of time available to us will

almost certainly lead to the breakdown of our civilization and possibly to the demise of our species."[76]

CO_2 is what humans exhale. A reduction of global CO_2 emissions to ZERO is another way of saying kill everyone, except perhaps a small group of elites. The report also pointed to "exponential growth in population" as the main driver for numerous challenges including those posed by resource depletion, food production and distribution, air quality, freshwater availability, and arable land.

The main operational body of the Club of Budapest is called the WorldShift Network. Their mission statement declares, "Maybe the last minute has come where it is possible to stop the downward trend and transform planet earth into the world, for which the most noble emotional, intellectual and spiritual abilities of humans are meant…Thus the WorldShift Network is ready to help giving birth to a new global culture, deeply imbued with spirit, fully aware of and supporting the evolution of global consciousness, inspired by love, and preserving and fostering life."[77] Honorary members of the Club of Budapest include deep ecologist Thomas Berry, author and futurist Sir Arthur C. Clarke, author Paul Coelho, The Dalai Lama, musician Peter Gabriel, primatologist Jane Goodall, Mikhail Gorbachev, Princess of the Netherlands Irene van Lippe-Biesterfeld, Robert Muller, and Desmond Tutu.[78]

The World Commission on Global Consciousness and Spirituality is another pompous Club of Rome spin-off. According to their website, "The Commission seeks to inspire consciousness of the wholeness of the human family and the sacred tapestry of all life…The Commission is action oriented and acknowledges awakening consciousness and spirituality as transformative powers for the common good…Perhaps the single most powerful event facing humanity today is a great awakening on a planetary scale that has been millennia in the making. We humans are in the midst of a profound advance as a species to a higher form of global consciousness that has been emerging across cultures, religions and worldviews through the centuries."[79] Members of the various Global Councils of the World Commission include Al Gore, Mikhail Gorbachev, Steven Rockefeller, Deepak Chopra, El Hassan bin Talal, the Dalai Lama, Hans Kung, Fritjof Capra, Andrew Weil, Bono, Jane Goodall, Peter Gabriel, and Steven Spielberg.[80]

Chapter 16

Manmade Warm-ups for Manmade Warming

The manmade global warming scam is nothing new. False environmentalists have attempted, though on a smaller scale, many such scams before. One memorable example is the "Alar Apple Scam of 1989," a fear campaign designed and executed by false environmentalists based upon manipulated and fraudulent science. Alar is a chemical compound sprayed on apple trees prior to apple formation to extend the harvest season. Apple growers had been using Alar since 1968. In the late 60s, the U.S. government approved Alar based on laboratory tests that concluded, "The amount fed to mice before any effect was noted was equivalent to an average adult eating 28,000 pounds of Alar-treated apples each year for 70 years, or a 10-pound infant eating 1,750 pounds per year."[1] The Environmental Protection Agency (EPA) reconfirmed Alar's safety in 1983.[2] In 1989, however, the National Resources Defense Council (NRDC), an organization founded with generous grants from the Rockefeller Brothers Fund, the Tides Foundation, the Ford Foundation and the Turner Foundation, claimed they had scientific evidence showing that Alar causes cancer.[3] Their alleged evidence, however, was their own private, non-peer-reviewed study. Nevertheless, NRDC requested that EPA declare Alar an "imminent hazard." This would have resulted in an immediate ban. The EPA rejected NRDC's request after an EPA Special Review Panel determined that NRDC's evidence was flawed.[4]

NRDC had been expecting this ruling. They knew their report was bogus. In fact, a favorable EPA decision would have ruined NRDC's larger scheme. Five months earlier, NRDC had hired Fenton Communications to organize a sophisticated media campaign designed to scare the public, cripple the apple industry, and subsequently funnel money back into NRDC pockets. Furthermore, they wanted to devise a model they could apply to future fear campaigns with much larger implications, like manmade global warming. David Fenton, founder of Fenton Communications, would later proudly assert, "We submit this campaign as a model for other non-profit organizations."[5]

This remark came from an internal memo written by Fenton and uncovered and published by the *Wall Street Journal* several months after NRDC successfully executed their mission. The memo revealed that NRDC and Fenton Communications had concocted and engineered the entire fraudulent scheme. Fenton bragged that a "sea change in public opinion" has "taken place because of a carefully planned media campaign, conceived and implemented by Fenton Communications with the Natural Resources Defense Council." He further admitted, "Of course, this had to be achieved with extremely limited resources. In most regards, this goal was met. A modest investment by NRDC re-paid itself many-fold in tremendous media exposure (and substantial, immediate revenue for future pesticide work)."[6] The following statements come directly from Fenton's self-congratulatory memo:

- "In October of 1988 NRDC hired Fenton Communications to undertake the media campaign for its report."
- "The campaign was based on NRDC's report 'Intolerable Risk: Pesticides in Our Children's Food.' Participation by the actress Meryl Streep was another essential element."

- "Our goal was to create so many repetitions of NRDC's message that average American consumers (not just the policy elite in Washington) could not avoid hearing it—from many different media outlets within a short period of time."

- "Media coverage included two segments on CBS *60 Minutes*, the covers of *Time* and *Newsweek* (two stories in each magazine), the *Phil Donahue Show*, multiple appearances on *Today*, *Good Morning America* and *CBS This Morning*, several stories on each of the network evening newscasts, MacNeil/Lehrer, multiple stories in the *N.Y. Times*, *Washington Post*, *L.A. Times* and newspapers around the country, three cover stories in *USA Today*, *People*, four women's magazines with a combined circulation of 17 million (*Redbook*, *Family Circle*, *Women's Day* and *New Woman*), and thousands of repeat stories in local media around the nation and the world."

- "As the report was being finalized, Fenton Communications began contacting various media. An agreement was made with *60 Minutes* to 'break' the story of the report in late February. Interviews were also arranged several months in advance."

- "It was agreed that one week after the study's release, Streep and other prominent citizens would announce the formation of NRDC's new project, *Mothers and Others for Pesticide Limits*."

- "The separation of these two events was important in ensuring that the media would have two stories, not one, about this project. Thereby, more repetition of NRDC's message was guaranteed."

- "On February 26th *CBS 60 Minutes* broke the story to an audience of 40 million viewers. The next morning, NRDC held a news conference attended by more than 70 journalists and 12 camera crews."

- "In the ensuing weeks, the controversy kept building. Articles appeared in food sections of newspapers around the country."[7]

As a result of this massive hoax, public hysteria caused apple sales to plummet, causing 20,000 American apple growers to suffer financial damages. This happened despite federal government estimates that at most 15% of U.S. apple growers were using Alar, and perhaps only 5% were using it during 1988-1989.[8]

Hey Farmer Farmer, Don't Put Away That DDT

False environmentalists are crafty. They even fooled Joni Mitchell and much of the early-1970s counterculture. One of the biggest-ever environmental scams, and certainly one of the deadliest, was the banning of DDT. The Environmental Defense Fund, the organization largely responsible for the ban, formed in 1969 thanks to Ford Foundation grants.[9] The leading researcher on the politically motivated suppression of DDT was the late J. Gordon Edwards, a former Emeritus professor at San Jose State University. Just before his death in 2004, Edwards published an article in the *Journal of American Physicians and Surgeons* called, "DDT: A Case Study in Scientific Fraud," the last of his many exposés on the subject. Edwards wrote,

"The chemical compound that has saved more human lives than any other in history, DDT, was banned by order of one man, the head of the U.S. Environmental Protection Agency (EPA). Public pressure was generated by one popular book and sustained by faulty or fraudulent research. Widely believed claims of carcinogenicity, toxicity to birds, anti-androgenic properties, and prolonged environmental persistence are false or grossly exaggerated. The worldwide effect of the U.S. ban has been millions of preventable deaths."[10]

In 1948, Swiss scientist Paul Hermann Müller won the Nobel Prize in Physiology for his work on the synthesis of DDT.[11] The chemical proved highly effective in combating insect borne diseases including typhus, yellow fever, encephalitis, and especially malaria. By 1959, the use of DDT nearly eradicated malaria in the U.S. and Europe, portions of the Soviet Union, Chile, and several Caribbean islands.[12] It was so effective that in 1970, the National Academy of Sciences declared, "To only a few chemicals does man owe as great a debt as to DDT. In little more than two decades DDT has prevented 500 million human deaths due to malaria that would have otherwise have been inevitable."[13] Yet today, largely because of the ban on DDT, malaria inflicts 300 to 500 million people per year and claims between 1 million and 2.5 million lives per year.[14] Edwards noted, "Many South American countries suffered more than 90 percent increases in malaria rates after halting DDT use, but Ecuador used DDT again and enjoyed a 61 percent reduction in malaria."[15]

The book that initiated the propaganda campaign against DDT was Rachel Carson's *Silent Spring*, first published in 1962. Carson referred to DDT as "The most alarming of all man's assaults upon the environment." She asserted, "This pollution is for the most part irrecoverable; the chain of evil it initiates not only in the world that must support life but in living tissues is for the most part irreversible."[16] During the ensuing years, anti-DDT activists began linking DDT to bird death, marine algae death, general environmental contamination and destruction, and ultimately, cancer in humans. Paul Ehrlich, for example, bolstered the hype with his 1968 article, "Eco-Catastrophe!" He predicted dead oceans, devoid of fish by 1979 due to DDT poisoning.[17]

In 1967, George Woodwell (founder of the Woods Hole Research Center, of which current White House Science Czar John P. Holdren was recently Director) wrote an unsettling article for *Science* purporting to show DDT concentrations between 13-32 pounds per acre on Long Island and elsewhere in North America. "These concentrations," Woodwell warned, "approach those in animals dying from DDT poisoning, which suggests that many natural populations in this area are now being affected, possibly limited, by DDT residues."[18] During 1972 EPA hearings, however, a USDA attorney challenged Woodwell on the veracity of these findings. Attorney O'Connor asked, "Isn't it a fact, Dr. Woodwell, that after you wrote this, or after you initially studied this salt marsh, that you continued your samplings, and that you found as a result of your continued sampling that you were getting around or less than one—an average of one pound per acre of DDT?" Woodwell replied, "No, I wouldn't agree with that." O'Connor continued,

> "Dr. Wurster, your coauthor, made the following statement at the Washington state hearings, and I'm quoting him verbatim. He testified, 'We have since sampled that marsh much more extensively and we found that the average, the overall figure on the marsh was closer to one pound per acre. The discrepancy was caused by the fact that our initial sampling was in a convenient place, and this turned out to be a convenient place for the Mosquito Commission's spray truck, too'…Doctor, have you ever published a retraction of this 13 pounds per acre or a qualification?"[19]

Woodwell replied, "I never felt that this was necessary." Despite the chicanery of Woodwell, Ehrlich and their ilk, the claims against DDT were largely unfounded.

In 1975, the EPA determined that Americans, on average, were ingesting no more than 15 *micrograms* of DDT per day. In a separate study, human volunteers ingested 35 *milligrams*, more than two thousand times the supposed daily exposure rate. They ingested this daily dosage for two years and experienced no negative effects.[20] Furthermore, workers in the Montrose Chemical Company experienced 19 years (1,300 man-years) of DDT exposure at 17 milligrams per man per day without any negative health effects.[21] Instead of carcinogenic effects, multiple studies actually demonstrated an anti-carcinogenic effect from DDT. Edwards explained, "DDT ingestion induces hepatic microsomal enzymes, which destroy carcinogenic aflatoxins and thereby inhibit tumors."[22] In 1969, the fledgling EDF, backed by Ford Foundation cash, launched an all-out assault on DDT. The *New York Times* reported,

> "Spokesmen for the Environmental Defense Fund have said they would like to see a legal ban on DDT traces in food products, together with waivers that would allow gradually diminishing trace amounts of the pesticide over the next several years. Such an arrangement would ensure withdrawal of the pesticide from use, they said."[23]

In 1972, the EPA appointed Administrative Law Judge Edmund Sweeny to convene seven months of hearings on DDT. In his final 113-page decision, Sweeney determined, "DDT is not a carcinogenic, mutagenic, or teratogenic hazard to man. The uses under regulations involved here do not have a deleterious effect on fresh water fish, estuarine organisms, wild birds, or other wildlife and there is a present need for essential uses of DDT."[24] Nevertheless, EPA Administrator William Ruckelshaus, who hadn't attended even one day of the seven months of hearings, overruled Sweeny's decision. Ruckelshaus even admitted he hadn't read any of the hearings' transcripts. And in 1979, in a letter to the President of the American Farm Bureau Federation, Ruckelshaus admitted, "Decisions by the government involving the use of toxic substances are political with a small 'p'. Science has a role to play, but the ultimate judgment remains political, (and) the power to make this judgment has been delegated to the Administrator of EPA."[25] Most significantly, Ruckelshaus was a *member* of the Environmental Defense Fund, and had even solicited donations for the organization using his own letterhead.[26]

In 2001, Gro Brundtland, then Director of the World Health Organization, paid lip service to malaria control, expressing goals of halving malaria mortality by 2010 and halving it again by 2015. Her strategy, however, ignored using DDT. The World Bank and USAID have consistently extended funds for malaria suffering countries, but always on the condition that they refrain from using DDT.[27] In 1998, the World Health Organization (WHO), the United Nations Children's Fund (UNICEF), the United Nations Development Programme (UNDP) and the World Bank combined to launch the Roll Back Malaria Partnership (RBM). Their mission was, "To work together to enable sustained delivery and use of the most effective prevention and treatment for those affected most by malaria."[28] According to the *New York Times*, "Roll Back Malaria sees its mosquito-control strategy as promoting bed nets, period."[29] And furthermore, "Global health institutions like W.H.O. and its malaria program, Roll Back Malaria, actively discourage countries from using [DDT]."[30] The *Times* also reported, "Treated bed nets are indeed a useful tool for controlling malaria. But they have significant limitations, and one reason malaria has surged is that they have essentially become the only tool promoted by Western donors."[31]

Yet in 2006, in what the *Wall Street Journal* deemed "a sign that widely used methods of fighting malaria have failed to bring the catastrophic disease under control," the World

Health Organization finally announced plans to reverse their 30-year opposition to DDT.[32] Prior to this pivotal decision, many malaria-stricken countries had been bucking the oppressive anti-DDT policies imposed by the U.S. and the UN. South Africa resumed DDT use in 2003 with great success, decreasing malaria levels by 80 percent in KwaZulu, the most affected province. By 2004, Zambia, Zimbabwe, Uganda and Kenya were actively seeking resumed DDT campaigns.[33]

South Africa's DDT campaign seized the attention of at least one conscientious onlooker within the WHO—and a very influential one at that. Arata Kochi, Director of WHO's Global Malaria Program declared, "We must take a position based on the science and the data. One of the best tools we have against malaria is indoor residual house spraying. Of the dozen insecticides WHO has approved as safe for house spraying, the most effective is DDT."[34]

The *Wall Street Journal* noted that Kochi "has been shaking up policies and personnel, hoping to repeat his success in the 1990s overseeing the WHO's comprehensive tuberculosis-control strategy."[35] In the tuberculosis effort, Kochi confronted the Rockefeller Foundation head-on. The *New York Times* reported, "He ruled the Stop TB campaign with an iron fist, colleagues say, and by his own admission, so alienated the Rockefeller Foundation and other partners that he was ultimately forced out of the job."[36] After spearheading the resurrection of DDT, Kochi candidly acknowledged, "The malaria community hates me."[37] His rebuttal, however, was even more direct: "I said, basically, 'You are stupid.' Their science is very weak. The community is small and inward-looking and fighting each other."[38]

Nevertheless, despite the WHO's decision, supposed "malaria control" groups have yet to embrace DDT. Even Roll Back Malaria, partnered by the WHO, failed to endorse DDT in its 2008 Global Malaria Action Plan. Hidden among the footnotes of the report, they admitted, "DDT remains the best choice for IRS [indoor residual spraying]." Yet they illogically concluded, "due to problems with acceptance," alternate chemicals "are equally effective."[39]

Self-declared malaria-control supporter Bill Gates, through his Bill and Melinda Gates Foundation, continues to suppress DDT. In the foundation's *2009 Letter from Bill Gates*, Gates claimed, "The world hoped in the 1950s and 1960s that [malaria] could be eliminated by killing mosquitoes with DDT, but that tactic failed when the mosquitoes evolved to be resistant to the chemical."[40]

Environmental groups including the Sierra Club and the Environmental Defense Fund condemned the WHO's announcement and continue to demonize DDT.[41] Other groups, like the WWF and the NRDC, have reluctantly withdrawn from their previously strict opposition and now endorse "emergency" DDT only when "accompanied by efforts to develop alternative approaches."[42]

According to the vehemently anti-DDT USAID, "The negative environmental effects of DDT use that led to its banning were due to massive, widespread agricultural use."[43] But DDT use today consists of spraying minute quantities once or twice per year on the walls of homes. USAID official Anne Peterson admitted that the reason her agency doesn't finance DDT is because of the chemical's unwarranted negative image, firmly entrenched in the public mind. A reversal of USAID's position, the *New York Times* reported, "would require a battle for public opinion." Peterson complained, "You'd have to explain to everybody why this is really O.K. and safe every time you do it."[44] For Peterson and

USAID, explaining why DDT treatment is safe and effective is a nuisance, an inconvenient hassle they would rather avoid. Does USAID (remembering their steadfast dedication to genocidal depopulation) lack the resources to write a simple report on DDT safety? They are welcome to post this chapter of this book, in its entirety, on their website — if they think preventing 300 to 500 million annual cases of malaria and 1 to 2.5 million annual deaths is worth the inconvenience.

The World Bank has taken a particularly Malthusian approach towards DDT. After the 1999 El Niño, Ecuador suffered large malaria outbreaks requiring emergency DDT control. The World Bank, however, *blocked* emergency assistance. A spokesman from the Pan American Health Organization, an affiliate of the WHO, reported, "The bank's environmental group told him it was fighting for the elimination of DDT and could not allow the bank to finance DDT while advocating a ban."[45] Alexander King, a co-founder of the Club of Rome, has also embraced Malthusian justifications for banning DDT. He candidly admitted, "In Guyana, within two years, [DDT] had almost eliminated malaria, but at the same time the birth rate had doubled. So my chief quarrel with DDT in hindsight is that it greatly added to the population problem."[46]

Robert S. Desowitz was an eminent researcher and teacher in the field of infectious disease. He held many government and university jobs during his accomplished career.[47] In his book *The Malaria Capers*, Desowitz quoted Edwin Cohn, a USAID official as stating, "The third world didn't require a healthy labor force because there was a surplus of workers; better some people should be sick with malaria and spread the job opportunities around." Furthermore, Cohn stated that women in poor, tropical countries are "better dead than alive and riotously reproducing."[48]

The upcoming chapters detail the biggest hoax of them all. The stakes are now much bigger. The New World Order is heavily pushing manmade global warming. The title of Al Gore's documentary is appropriate — *An Inconvenient Truth*. The truth is out. And for the New World Order, it is indeed inconvenient.

PART III: GLOBAL WARMING

"I know that most men, including those at ease with problems of the greatest complexity, can seldom accept even the simplest and most obvious truth if it be such as would oblige them to admit the falsity of conclusions which they delighted in explaining to colleagues, which they have proudly taught to others, and which they have woven, thread by thread, into the fabric of their lives." — Leo Tolstoy[1]

Historically, the environmental movement's most powerful and most influential figures have been congenital liars, control-freaks, psychopaths, and globalists. By their own admissions, these people want to eliminate the nation state model of world order and, upon its ashes, construct a New World Order global government. By their further admissions, they are using the environmental movement to mask their true ambitions. Both the crimes they've committed and deception they've engaged in should elicit immediate suspicion regarding their tenacious claims of environmental catastrophe, no matter how plausible such claims may sound. Therefore, understanding the history of the environmental movement greatly facilitates understanding the current global warming hysteria.

The first chapter of this section examines the United Nations' Intergovernmental Panel on Climate Change. The next chapter explores Earth's climate history to determine whether the current climate is abnormal and whether a warmer climate would actually be problematic or, perhaps, beneficial. The next chapter investigates carbon dioxide, what it is and why it's become the new Al-Qaeda—the ubiquitous, yet invisible enemy. The next chapter examines the relationship between the Sun and Earth's climate. Could the most powerful nuclear reactor in the Solar System be influencing the climate of Earth, Mars, Jupiter and the other planets? The last three chapters evaluate the rhetoric and tactics of those who promote the manmade global warming theory, those who incessantly chorus the mantra of scientific "consensus."

Chapter 17

United Nations Ringleaders

In 1988, the United Nations established the Intergovernmental Panel on Climate Change (IPCC). Nineteen years later, along with Al Gore, the IPCC won the Nobel Peace Prize "for their efforts to build up and disseminate greater knowledge about man-made climate change."[1] According to its mandate, the IPCC should be transparent and objective, and neutral with respect to policy:

> "Its role is to assess on a comprehensive, objective, open and transparent basis the latest scientific, technical and socio-economic literature produced worldwide relevant to the understanding of the risk of human-induced climate change. IPCC reports should be neutral with respect to policy, although they need to deal objectively with policy relevant scientific, technical and socio-economic factors."[2]

Notice the reference to *human-induced climate change*. How can the IPCC be neutral and objective when their mandate presupposes that human activity is significantly affecting climate? What if human activity is inconsequential? What if natural forces, far beyond human control, are the primary drivers of climate change? The IPCC does not care. They preemptively decided that human activity is driving climate change.

Roughly once every five years, the IPCC publishes an Assessment Report, a public document written primarily for government officials and policymakers. From their 1st Assessment Report in 1990 through their 4th Assessment Report in 2007, they have consistently warned that human CO_2 emissions, if left unchecked, will lead to catastrophic global warming. Their Assessment Reports detail the findings of the IPCC's three Working Groups. Working Group I (WG1) is the most important because it determines, supposedly from a scientific perspective, the underlying causes of climate change.[3] Based on WG1's findings, WG2 and WG3 investigate options for preventing and adapting to climate change. IPCC supporters frequently parrot the claim that 2,500 scientists contribute to the Assessment Reports. Yet WG1, the most important group, consists of only 600 scientists. During a 2008 speech, IPCC Chairman Rajendra Pachauri failed to mention this crucial point. "The Panel," he stated, "mobilizes thousands of the best scientists in the world for its assessment of various aspects of climate change. This work is carried out with complete transparency and objectivity in all the procedures followed and peer reviews carried out at each stage of the process by experts as well as governments."[4] But who are these thousands of scientists? And what role do they have within the IPCC? Do they necessarily agree with the IPCC's conclusions and recommendations? Pachauri and other IPCC officials are reluctant to address such questions because, despite his assertions, the IPCC process is neither transparent, nor objective.

Before 2007, the IPCC had always refused to disclose scientists' comments on Assessment Report drafts. The United States' Freedom of Information Act, however, forced the reversal of this policy. The now-public comments show that IPCC editors have consistently supported predetermined conclusions, while systematically ignoring opposing perspectives. In the 4th Assessment Report, for example, the critical chapter linking human activity to global warming was Chapter 9 from WG1, entitled, "Understanding and Attributing Climate Change." It contained the statement, "Greenhouse gas forcing has *very likely* caused most of the observed global warming over the last 50 years."[5] Chapter 9 had 53 authors and 55 reviewers. Of the 55 reviewers, only 4 commented favorably on the entire chapter, with an additional 3 commenting favorably on portions of the chapter. Fifty-three authors plus seven favorable reviewers makes only 60 scientists — from the celebrated 2,500 — who explicitly supported the IPCC's manmade global warming conclusions.[6]

Ross McKitrick of the University of Guelph, Canada was one of the scientists who reviewed Chapter 9. Referring to the above quotation, he stated, "A categorical summary statement like this is not supported by the evidence in the IPCC WG1 report. Evidence shown in the report suggests that other factors play a major role in climate change, and the specific effects expected from greenhouse gases have not been observed."[7] Dr. Vincent Gray of New Zealand, another scientist who reviewed Chapter 9, asserted, "The text of the IPCC report shows that this is decided by a guess from persons with a conflict of interest, not from a tested model."[8]

In January 2007, Richard Lindzen appeared on the *Larry King Live Show* to explain the reality of the IPCC. Lindzen, an Alfred P. Sloan professor of atmospheric science at Massachusetts Institute of Technology, was a Lead Author for the 3rd Assessment Report and Contributing Author for the 2nd Report. He explained,

"It's not 2,500 people offering their consensus, I participated in that. Each person who is an author writes one or two pages in conjunction with someone else. They travel around the world several times a year for several years to write it and the Summary for Policymakers has the input of about 13 of the scientists, but ultimately, it is written by representatives of governments, of environmental organizations like the Union of Concerned Scientists, and industrial organizations, each seeking their own benefit."[9]

The United Nations and governments throughout the world recognize the IPCC as the preeminent organization concerned with climate change. They are the most prominent advocates of massive CO_2 emission reductions, purportedly to stave off catastrophic global warming. Yet for a Nobel Peace Prize winning organization, the IPCC has a disturbing history of subterfuge, deception and fraud.

First Assessment Report

Following the 1990 publication of their 1st Assessment Report, massive corruption within the IPCC was already apparent. Patrick Michaels, a Senior Fellow at the Cato Institute and retired professor from the University of Virginia, participated as an Expert Reviewer for this report and as a Contributing Author for its 1992 update. In his book, *Sound and Fury*, Michaels summarized the IPCC process as follows:

"Approximately 200 scientists, bureaucrats, and administrators were invited to contribute various chapters on agriculture, ecosystems, and climate history, but the document was really crafted by a small number of lead authors (usually one to three people per subject area) who were able, with input from the 200, to tailor each chapter as they best saw fit. An additional team of approximately 100 reviewers critiqued the original document, and again the lead authors had final say about admission of a reviewer's comments...The overall IPCC document is several hundred pages long and prefaced by a brief Policymakers Summary, which is the section that has been read the most, and it is doubtful that many officials who speak so glibly about the IPCC consensus ever got beyond that summary."[10]

IPCC WG1, chaired by John Houghton, intentionally crafted the 1990 Summary to frighten the public and motivate the politicians. Houghton's representative, Christopher Folland, admitted this much during a 1991 climatologist meeting in Asheville, North Carolina. Michaels recalled, "Dr. Folland related that the problem in creating a summary was the somewhat equivocal nature of the entire report. Of what use to policymakers, he asked, would a document be if punctuated by repeated cautions and 'on the other hand' statements. So, Folland stated, the group made a conscious decision to produce the Policymakers Summary that did not equivocate as much as the overall report."[11]

The Summary for Policy makers featured decisive language, such as, "We are certain of the following...emissions resulting from human activities are substantially increasing the atmospheric concentrations of the greenhouse gases: CO_2, methane, CFCs and nitrous oxide. These increases will enhance the greenhouse effect, resulting on average in an additional warming of the Earth's surface."[12] Deeper into the report, however, such certainties were noticeably absent. The Executive Summary to Chapter 7, for example, reported,

"A global warming of larger size has almost certainly occurred at least once since the end of the last glaciation without any appreciable increase in greenhouse gases. Because we do not understand the reasons for these past warming events, it is not yet possible to attribute a specific proportion of the recent, smaller warming to an increase of greenhouse gases."[13]

The aforementioned Richard Lindzen summarized, "the body of the report is extremely ambiguous, and the caveats are numerous," yet Houghton's "summary largely ignores the uncertainty in the report and attempts to present the expectation of substantial warming as firmly based science."[14]

Second Assessment Report

In 1996, following the publication of the 2nd Assessment Report, Frederick Seitz, the former President of the U.S. National Academy of Sciences, wrote an op-ed for the *Wall Street Journal* condemning IPCC leaders for their unscrupulous behavior. According to Seitz, they had completely undermined the peer-review process by scrapping the report that IPCC scientists had authorized and instead publishing a non-reviewed, unauthorized version, which supported their alarmist agenda. Seitz wrote,

"This report is not what it appears to be—it is not the version that was approved by the contributing scientists listed on the title page. In my more than 60 years as a member of the American scientific community, including service as president of both the National Academy of Sciences and the American Physical Society, I have never witnessed a more disturbing corruption of the peer-review process than the events that led to this IPCC report."[15]

He further noted that nothing in the official IPCC Rules permitted the switch. The IPCC either changed or deleted 15 sections from Chapter 8 of the report in question, "The Science of Climate Change." This was "the key chapter setting out the scientific evidence for and against a human influence over climate." Furthermore, the changes were not merely cosmetic. Seitz explained, "Nearly all [the changes] worked to remove hints of the skepticism with which many scientists regard claims that human activities are having a major impact on climate in general and on global warming in particular."

In his article, Seitz revealed the censored statements, which clearly demonstrated the cautious skepticism of the participating scientists. They regarded manmade global warming as an unproven theory, not necessarily wrong, yet lacking sufficient supporting evidence. Nevertheless, the IPCC leaders wanted to generate fear while creating a sense of urgency. Deception and fraud were acceptable tactics towards accomplishing this end. Seitz concluded,

"Whatever the intent was of those who made these significant changes, their effect is to deceive policy makers and the public into believing that the scientific evidence shows human activities are causing global warming. If the IPCC is incapable of following its most basic procedures, it would be best to abandon the entire IPCC process, or at least that part that is concerned with the scientific evidence on climate change, and look for more sources of advice to governments on this important question."[16]

Christopher Monckton, a member of the British House of Lords and a former advisor to Margaret Thatcher, also saw the 2nd Assessment Report's infamous Chapter 8. Monckton blamed Lead Author Ben Santer for the deletions. "In comes Santer and re-writes it for them," Monckton explained, "after the scientists have sent in their finalized draft, and that finalized draft said at five different places, there is no discernable human effect on global temperature—I've seen a copy of this—Santer went through, crossed out all of those and substituted a new conclusion, and this has been the official conclusion ever since."[17] When questioned, on camera, about Monckton's allegations, Santer candidly admitted, "Lord Monckton points to deletions from the chapter, and there were deletions from the chapter, to be consistent with the other chapters we dropped the summary at the end."[18]

Third Assessment Report

Richard Lindzen was a Lead Author for the 3rd Assessment Report, and a Contributing Author for the 2nd Assessment Report. In 2001, in an article for the *Wall Street Journal*, he asserted, "I cannot stress this enough—we are not in a position to confidently attribute past climate change to carbon dioxide or to forecast what the climate will be in the future…there is no consensus, unanimous or otherwise, about long-term climate trends and what causes them."[19] In 2001, the U.S. Senate Committee on Environment and Public Works summoned Lindzen to testify regarding possible IPCC corruption. He stated,

"Throughout the drafting sessions, IPCC 'coordinators' would go around insisting that criticism of models be toned down, and that 'motherhood' statements be inserted to the effect that models might still be correct despite the cited faults. Refusals were occasionally met with ad hominem attacks. I personally witnessed coauthors forced to assert their 'green' credentials in defense of their statements."[20]

He also informed the committee that "almost all reading and coverage of the IPCC is restricted to the highly publicized Summaries for Policymakers which are written by representatives from governments, NGOs and business; the full reports, written by participating scientists, are largely ignored."[21]

Another Lead Author for the 3rd Assessment Report was John Christy, a Fellow of the American Meteorological Society and a recipient of the 1991 NASA Exceptional Scientific Achievement Medal. Christy has contributed in some capacity to all four IPCC Assessment Reports. Reflecting on his long-term involvement with the IPCC, in 2007 he penned an article for *BBC News* entitled, "No consensus on IPCC's level of ignorance."[22] Confirming that the

IPCC had become extremely politicized, he recalled, for example, a Lead Author Meeting in New Zealand during which fellow (supposedly policy-neutral) Lead Authors openly stated, "We must write this report so strongly that it will convince the U.S. to sign the Kyoto Protocol."[23] Christy also criticized the IPCC's heavy reliance on uncertain climate models, the models used to justify their statements of 'near-certainty' regarding manmade global warming. According to Christy, "The agreement displayed is just as likely to do with clever software engineering as to the first principles of science." He concluded, "Answering the question about how much warming has occurred because of increases in greenhouse gases and what we may expect in the future still holds enormous uncertainty, in my view…Fundamental knowledge is meager here, and our own research indicates that alarming changes in the key observations are not occurring"[24]

Fourth Assessment Report

The 4th Assessment Report earned the IPCC the Nobel Peace Prize. Since its 2007 publication, however, the IPCC has suffered considerable reputational damage via a steady stream of embarrassing scandals, critiques and revelations. In 2008, for example, Kiminori Itoh, an Expert Reviewer for the 4th Assessment Report, wrote *Lies and Traps in the Global Warming Affairs*, a book (written in Japanese) in which he characterized global warming as "the worst scientific scandal in history."[25] Madhav Khandekar, a distinguished scientist with fifty years experience in climate and weather science, was also an Expert Reviewer for the 4th Assessment Report. In 2008, Khandekar penned an article for the scientific journal *Energy & Environment* entitled, "Has The IPCC Exaggerated Adverse Impact Of Global Warming On Human Societies?"[26] Answering his own question, he asserted, "Yes, Certainly!" Commenting on his involvement with WG2, Khandekar explained,

> "I felt time and time again that there were areas where the chapter authors highlighted *adverse impact* of GW (Global Warming) on human societies, while downplaying possible beneficial impacts. The IPCC authors referred to several publications which projected adverse impacts while ignoring many excellent studies which have questioned these projections. Throughout the text of this important chapter of WG2, there were many instances where adverse impact was highlighted or exaggerated, while possible beneficial impacts were totally ignored. Further, IPCC authors while assessing *observed changes in natural systems* chose to highlight only those changes which support the GW hypothesis while completely ignoring other observed changes which did not conform to the human-induced GW hypothesis and change. Such cherry-picking of observed climate change to bolster claims of human-caused GW and climate change is disingenuous and does not help [us] understand the real cause of how and why the earth's climate has changed in historical and geological times."[27]

The IPCC describes its procedure as open, transparent and robust, and encompassing only *peer-reviewed* scientific literature. Khandekar and many other scientists, however, have repeatedly denounced the IPCC for making exaggerated claims based on unpublished, non-peer-reviewed reports. For example, Benny Peiser, a fellow of the Royal Astronomical Society and director of the Global Warming Policy Foundation, recently harshly criticized the IPCC's contempt for the peer-review process. "Conflicting data and evidence," he observed, "even if published in peer reviewed journals, are regularly ignored, while exaggerated claims, even if contentious or not peer-reviewed, are often highlighted in IPCC reports."[28]

In the 4th Assessment Report, the IPCC claimed that global warming had already caused increased natural disasters, including hurricanes and floods, and increased economic losses. They based these claims on unpublished, non-peer-reviewed reports while completely ignoring contradictory evidence from *peer-reviewed studies*. For example, they ignored a study by prominent U.S. climatologist Stanley Changnon published in 2003 in the journal *Natural Hazards* entitled, "Shifting Economic Impacts from Weather Extremes in the United States: A Result of Societal Change, Not Global Warming."[29] Although Khandekar publicized this IPCC subterfuge in May 2008, the mainstream press didn't react until January 2010, a time when the IPCC's credibility was rapidly imploding.

The IPCC's claim of "rapidly rising costs due to extreme weather-related events since the 1970s" came from an unpublished 2006 academic paper written by Robert Muir-Wood, the head of a London-based consultancy who later became a Contributing Author for the 4th Assessment Report.[30] In 2008, however, when Muir-Wood finally published his paper, he included the following qualifying statement: "We find insufficient evidence to claim a statistical relationship between global temperature increase and catastrophe losses."[31] The IPCC completely ignored this turnabout. They issued no retraction or statement of clarification. Furthermore, according to the *London Times*, "at least two scientific reviewers who checked drafts of the IPCC report urged greater caution in proposing a link between climate change and disaster impacts—but were ignored."[32] Roger Pielke, a Colorado University environmental studies professor who commissioned the original Muir-Wood study, commented, "All the literature published before and since the IPCC report shows that rising disaster losses can be explained entirely by social change. People have looked hard for evidence that global warming plays a part but can't find it. Muir-Wood's study actually confirmed that."[33] When this story finally broke in January 2010, IPCC Vice-Chair Jean-Pascal van Ypersele responded, "We are reassessing the evidence." Unconvincingly, he added, "Despite recent events the IPCC process is still very rigorous and scientific."[34]

"Recent events" referred to an outpouring of humiliating revelations of dramatic IPCC claims based on non-peer-reviewed studies. For example, the IPCC claimed that global warming has already caused observed reductions in mountain ice in the Andes, Alps and Africa. Of their two sources for this claim, one was Mark Bowen, a journalist for *Climbing* magazine who made the claim based on anecdotal evidence from climbers.[35] The other source was a dissertation paper by Dario-Andri Schworer, a professional mountain guide, climate change campaigner and geography student who made the claim based on interviews with about 80 mountain guides from the Bernina region of the Swiss Alps.[36] In another example, the IPCC claimed that global warming would cut rain-fed crop production in North Africa by up to 50 percent by 2020. This claim became a talking point, parroted by both Pachauri and UN Secretary General Ban Ki-moon in various speeches.[37] The source of this claim, however, as revealed by the *London Times*, was a non-peer-reviewed 2003 policy paper written by the International Institute for Sustainable Development, a Canadian think tank. Chris Field, the new lead author for the IPCC's climate impact team candidly acknowledged, "The IPCC needs to investigate a more sophisticated approach for dealing with emerging errors."[38]

The most audacious of the IPCC's "emerging errors" was their claim that, "Glaciers in the Himalaya are receding faster than in any other part of the world and, if the present rate continues, the likelihood of them disappearing by the year 2035 and perhaps sooner is very high if the Earth keeps warming at the current rate."[39] Himalayan glaciers are the largest source of fresh water for all of northern India. They provide half the water for the Ganges and are the primary source for the Indus, the Brahmaputra, the Mekong, the Irrawady and the Yellow and Yangtse rivers.[40] Consequently, the IPCC's glacier claims caused great fear and distress throughout these regions. Most disturbingly, the IPCC deliberately included these unscientific, exaggerated claims to arouse fear and motivate political action. In an interview with the *Daily Mail*, Murari Lal, a Lead Author of the chapter in question, acknowledged that no peer-reviewed research supported these claims yet, "We thought that if we can highlight it, it will impact policy-makers and politicians and encourage them to take some concrete action."[41]

The IPCC borrowed the 2035 disappearance claim from a 2005 WWF report, one of at least sixteen references to WWF reports throughout the 4th Assessment Report.[42] The WWF produces *advocacy* reports, not peer-reviewed science. They *sometimes* cite peer-reviewed studies, but not always. The WWF based the 2035 disappearance claim, for example, on speculative statements by Indian scientist Syed Iqbal Hasnain, published in a 1999 *New Scientist* article. Curiously, Hasnain now works for The Energy Research Institute (TERI), which is headed by none other than IPCC Chairman Rajendra Pachauri. Furthermore, in early January 2010, TERI received $500,000 from the Carnegie Institute specifically to study Hasnain's fallacious claim.[43] Murari Lal conceded, "We knew the WWF report with the 2035 date was 'grey literature.' But it was never picked up by any of the authors in our working group, nor by any of the more than 500 external reviewers, by the governments to which it was sent, or by the final IPCC review editors.[44] He also claimed, "Had we received information that undermined the claim, we would have included it." Numerous scientists, however, did indeed challenge these erroneous claims.

According to the Union of Concerned Scientists, the IPCC's glacier claims were "openly challenged by some researchers during the review and editing process."[45] Georg Kaser, a glacier expert from Austria and Lead Author of another IPCC chapter, for example, wrote to Lal denouncing the claim as patently false and urging its removal. Lal, however, simply claimed he never received Kaser's letter. Another reviewer, Hayley Fowler of Newcastle University, informed the IPCC that Himalayan glaciers in the Karakoram region are growing rapidly. Fowler cited a peer-reviewed paper published in *Nature*, yet the IPCC claimed they were "unable to get hold of the suggested references."[46]

Basing the 2035 claim on non-peer-reviewed reports was bad enough, yet the IPCC simply *invented* the claim that Himalayan glaciers are retreating faster than anywhere else in the world. Michael Zemp from the World Glacier Monitoring Service explained, "There are simply no observations available to make these sorts of statements...we simply do not have the glacier change measurements. The Himalayas are among those regions with the fewest available data."[47] The IPCC also ignored an exhaustive study of 150 years of data from 25 Himalayan glaciers, commissioned by India's Environment Minister Jairam Ramesh. "The health of glaciers," Ramesh remarked, "is a cause of grave concern but the IPCC's alarmist position that they would melt by 2035 was not based on an iota of scientific evidence."[48] In November 2009, IPCC Chairman Rajendra Pachauri dismissed the Indian report as "voodoo science," while accusing Ramesh of arrogance.[49] Three months later, however, in response to the glacier scandal, Pachauri reluctantly admitted, "There is absolutely no information available on the rate at which Himalayan glaciers are melting."[50] He further conceded, "We slipped up,"[51] yet he nevertheless insisted, "Our procedure is robust."[52]

Pachauri's credibility suffered yet another devastating blow when the *London Times* reported he had lied on January 22, 2010, when he stated, "I became aware of this [2035 error] when it was reported in the media about ten days ago. Before that, it was really not made known. Nobody brought it to my attention."[53] Asked whether he concealed the error to avoid embarrassment before the IPCC's December 2009 Copenhagen Climate Conference, Pachauri insisted, "That's ridiculous. It never came to my attention before the Copenhagen summit. It wasn't in the public sphere."[54] Yet in November 2009, Pallava Bagla, a prominent journalist for *Science*, informed Pachauri that a leading glaciologist had determined that the 2035 date was off by at least 300 years. In late January 2010, during an interview with Pachauri for *Science*, Bagla asked, "I pointed it out to you in several e-mails, several discussions, yet you decided to overlook it. Was that so that you did not want to destabilise what was happening in Copenhagen?"[55] Pachauri awkwardly replied, "Not at all, not at all. As it happens, we were all terribly preoccupied with a lot of events." Then, once again, he unconvincingly defended the IPCC's integrity. "This presumption," Pachauri retorted, "on your part or on the part of any others is totally wrong. We are certainly never — and I can say this categorically — ever going to do anything other than what is truthful and what upholds the veracity of science."[56]

The Hurricane Hoodwink: A Closer Look at IPCC Manipulation

In October 2004, following an intense hurricane season, Harvard University issued a promotional press release regarding an upcoming hurricane conference. They announced, "Experts to warn global warming likely to continue spurring more outbreaks of intense hurricane activity."[57] To Christopher Landsea, this announcement was rather suspicious. As one of the world's preeminent experts on hurricanes and tropical storms, Landsea participated both as Contributing Author and Expert Reviewer for both the 2nd and 3rd IPCC Assessment Reports. While the alarmist language of the press release perturbed him, Landsea was shocked to learn that Harvard had invited a grand total of *zero* hurricane experts to speak at the conference. Of the scientists they did invite, *none* had conducted research on hurricane variability. Furthermore, at the time, the IPCC's position was that global warming *does not* induce bigger or more frequent hurricanes. The 3rd Assessment Report, which Landsea had worked on, clearly stated, "Changes globally in tropical and extra-tropical storm intensity and frequency are dominated by inter-decadal and multi-decadal variations, with no significant trends evident over the 20th century."[58] Scanning the list of invited scientists, Landsea decided to contact his IPCC colleague, Kevin Trenberth.

Trenberth was a top official at the National Center for Atmospheric Research and a Lead Author for both the 3rd and 4th Assessment Reports. Since he had no expertise on hurricanes, Landsea volunteered to brief him before the upcoming conference. Trenberth flatly declined. The conference, and especially Trenberth's address, sparked frenzied headlines across the world. Furthermore, it foreshadowed the IPCC's about-face concerning hurricanes, which later appeared in the 4th Assessment Report. Hans von Storch, a climatologist at the Institute for Coastal Research in Geesthacht, remarked, "It's a demonstration of how highly politicized the IPCC process has become."[59]

Indignant, Landsea wrote to top IPCC officials, "Where is the science, the refereed publications, that substantiate these pronouncements? What studies are being alluded to that have shown a connection between observed warming trends on the earth and long-term trends in tropical cyclone activity? As far as I know, there are none."[60] Landsea had been invited to participate in the 2007 IPCC report, but the Harvard debacle made him question the IPCC's integrity. He wrote them, seeking assurances:

"[Trenberth] seems to have already come to the conclusion that global warming has altered hurricane activity and has publicly stated so. This does not reflect the consensus within the hurricane research community…Thus I would like assurance that what will be included in the IPCC report will reflect the best available information and the consensus within the scientific community most expert on the specific topic."[61]

These assurances never came. Landsea withdrew from the IPCC and subsequently published an open letter on the University of Colorado's website stating,

- "I am withdrawing because I have come to view the part of the IPCC to which my expertise is relevant as having become politicized. In addition, when I have raised my concerns to the IPCC leadership, their response was simply to dismiss my concerns."

- "Given Dr. Trenberth's role as the IPCC's Lead Author responsible for preparing the text on hurricanes, his public statements so far outside of current scientific understanding led me to concern that it would be very difficult for the IPCC process to proceed objectively with regards to the assessment on hurricane activity. My view is that when people identify themselves as being associated with the IPCC and then make pronouncements far outside current scientific understandings that this will harm the credibility of climate change science and will in the longer term diminish our role in public policy."

- "I personally cannot in good faith continue to contribute to a process that I view as both being motivated by pre-conceived agendas and being scientifically unsound."[62]

Since the Harvard conference, manmade global warming advocates have increasingly blamed tropical storms on human activity. Trenberth, for example, published a 2006 paper claiming, "About half of last year's extra [ocean] warmth was due to global warming."[63] In 2007, as expected, the IPCC reversed its position from 2001. They claimed it was "more likely than not" that humans were causing increasing hurricane activity. They also determined that, "Based on a range of models, it is [>66%] *likely* that future tropical cyclones (typhoons and hurricanes) will become more intense."[64] In another study, Judith Curry and Greg Holland determined that category 4 and 5 storms have been increasing sharply since the 1970s. *USA Today* reported, "Co-author Judith Curry of Georgia Tech said the team is confident that the measured increase in sea surface temperatures is associated with global warming."[65] Landsea harshly criticized this study noting that, although they looked at storms worldwide, "for most of the world there was no way to determine objectively what the winds were in 1970."[66] Landsea maintains that storms *seem* to have increased because advanced satellite technology now enables scientists to identify and track storms that would have gone unnoticed just decades ago. At a 2008 meeting of the American Meteorological Society, Landsea challenged Holland on this point. Landsea requested, "Can you answer the question?" Holland, like an immature child, responded, "I'm not going to answer the question because it's a stupid question."[67]

William Gray, another top hurricane expert, described by the *Associated Press* as "America's most reliable hurricane forecaster,"[68] also strongly opposes the recent hype. In 2007 he wrote, "Although global surface temperatures have increased over the last century and over the last 30 years, there is no reliable data available to indicate increased hurricane frequency or intensity in any of the globe's seven tropical cyclone basins, except for the Atlantic over the past 12 years. Meteorologists who study tropical cyclones have no valid physical theory as to why hurricane frequency or intensity would necessarily be altered significantly by small amounts of global mean temperature change."[69] At 77 years of age, Gray has seen plenty of alarmists come and go. Regarding Al Gore, he aptly concluded, "He's one of these guys that preaches the end of the world type of things. I think he's doing a great disservice and he doesn't know what he's talking about."[70]

The Spread of Malaria: A Contrived Threat?

In 2001, the U.S. State Department nominated Paul Reiter, head of the Insects and Infectious Disease Unit of the Pasteur Institute, for Lead Author of the IPCC's 4th Assessment Report. At the time, many alarmists had been suggesting that global warming would facilitate mosquito habitation and enable the potential deadly spread of malaria and other so-called "tropical diseases" far beyond the tropics. Reiter did not share this perspective. Therefore, the IPCC rejected him in favor of two candidates who had never even published a peer-reviewed article on mosquito-born diseases. Both these candidates, however, did believe that global warming could intensify the spread of these diseases. While not surprised by his rejection, Reiter protested, "I know of no major scientist with any long record in this field who agrees with the pronouncements of the alarmists at the IPCC."[71]

The 1995 2nd Assessment Report claimed, "Indirect effects of climate change include increases in the potential transmission of vector-borne infectious diseases (e.g., malaria, dengue, yellow fever and some viral encephalitis) resulting from extensions of the geographical range and season for vector organisms."[72] They concluded that malaria contraction could increase "on the order of 50-80 million additional annual cases." In written testimony to the U.K. House of Lords, Reiter blasted the IPCC and their authoritative pronouncements on subjects they had clearly failed to understand. He observed, "Not one of the lead authors had ever written a research paper on the subject!"[73]

In the same testimony, Reiter documented the facts about malaria: principally that malaria is not a tropical disease. It can and often has broken out in temperate climates. He explained that malaria was endemic in England during the Little Ice Age from mid-15th century until early 18th century, that the worst known outbreak of malaria occurred in the 1920s in Russia with a peak incidence of 13 million cases per year and 600,000 deaths, and that malaria has persisted as far north as Siberia and the Arctic. Reiter documented both the misinterpretation *and* misrepresentation of data by the IPCC. He concluded, "After careful review of the pronouncements of the Health chapter in Working Group II of the IPCC Second Assessment, it is my opinion that that they were not based on authentic science."[74]

For the 3rd Assessment Report, Reiter briefly participated as a Contributing Author on the Health chapter. He then resigned, however, because, "My colleagues and I repeatedly found ourselves at loggerheads with persons who insisted on making authoritative pronouncements, although they had little or no knowledge of our speciality."[75] Yet after resigning, his name remained on the IPCC's draft. "I requested its removal," Reiter explained, "but was told it would remain because 'I had contributed.' It was only after strong insistence that I succeeded in having it removed." Reiter concluded his House of Lords testimony with the following concise, powerful statement:

> "The recent resurgence of many of these diseases is a major cause for concern, but it is facile to attribute this resurgence to climate change, or to use models based on temperature to 'predict' future prevalence. In my opinion, the IPCC has done a disservice to society by relying on 'experts' who have little or no knowledge of the subject, and allowing them to make authoritative pronouncements that are not based on sound

science. In truth, the principal determinants of transmission of malaria and many other mosquito-borne diseases are politics, economics and human activities. A creative and organized application of resources is urgently required to control these diseases, regardless of future climate change."[76]

Further Opposition and Dissent

Vincent Gray, a cofounder of The New Zealand Climate Science Coalition, has been an Expert Reviewer on every IPCC Assessment Report. Consequently, his firsthand knowledge of the IPCC process has made him one of their harshest critics. Gray explained,

"Over the years, as I have learned more about the data and procedures of the IPCC, I have found increasing opposition by them to providing explanations, until I have been forced to the conclusion that for significant parts of the work of the IPCC, the data collection and scientific methods employed are unsound. Resistance to all efforts to try and discuss or rectify these problems has convinced me that normal scientific procedures are not only rejected by the IPCC, but that this practice is endemic, and was part of the organization from the very beginning."[77]

In 2006, the *New Zealand Herald* interviewed Gray about his experience with the IPCC. He stated, "The more I looked into it, the more suspicious I got. I've reached the stage now where I think the whole thing is almost a gigantic fraud."[78] By 2007, Gray finally *had* reached that stage. A group of academics leading an effort to reform the IPCC asked Gray to participate. He refused, stating, "The IPCC is fundamentally corrupt. The only 'reform' I could envisage would be its abolition."[79]

Yuri Izrael is the former Vice-President of the World Meteorological Organization, and former Vice-President of the Institute of Global Climate and Ecology Institute, which is part of the Russian Academy of Sciences. Izrael once held one of the highest-ranking positions within the IPCC, Vice-Chair. He broke ranks, however, after realizing that the panic surrounding global warming was greatly exaggerated. In 2003 he stated, "All the scientific evidence seems to support the same general conclusions, that the Kyoto Protocol is overly expensive, ineffective and based on bad science."[80] In a 2005 article, Izrael wrote,

"There is no proven link between human activity and global warming. According to 10,000 meteorological stations, average temperatures have increased by just 0.6 degrees in the last 100 years. But there is no scientifically sound evidence of the negative processes that allegedly begin to take place at such temperatures…The European Union has established by fiat that a two-degree rise in global temperatures would be quite dangerous. However, this data is not scientifically sound."[81]

Freeman Dyson is a legendary mathematician and physicist known for unifying three versions of quantum electrodynamics. His scientific expertise and knowledge is vast. Most recently he is known for challenging what he calls "the holy brotherhood of climate model experts and the crowd of deluded citizens who believe the numbers predicted by the computer models."[82] In his 2007 book, *A Many-Colored Glass*, Dyson wrote,

"Of course, they say, I have no degree in meteorology and I am therefore not qualified to speak. But I have studied the climate models and know what they can do. The models solve the equations of fluid dynamics, and they do a very good job of describing the fluid motions of the atmosphere and the oceans. They do a very poor job of describing the clouds, the dust, the chemistry, and the biology of fields and farms and forests. They do not begin to describe the real world that we live in."[83]

The World Federation of Scientists (WFS), founded in 1973, is a prestigious organization of more than 10,000 scientists from over 110 countries. Membership is open to all scientists sympathetic to the Federation's ideals and committed to upholding its principals. The group is primarily concerned with monitoring and averting planetary emergencies by encouraging the free flow of scientific information and the reduction of secrecy.[84] They have identified 15 Planetary Emergencies. Climate change is one of them. Their stated priorities for dealing with climate change are,

- "To encourage and support free access to data on climate change."
- "To monitor the monitoring of the global environment."
- "To stimulate the education of the public with regard to the causes and effects of climate change."[85]

They acknowledge both natural episodic influences and possible human influences as contributing factors. They oppose, however, dogmatic claims like "the science is settled" and "the debate is over." Antonio Zichichi, President of the WFS, openly lambasted the IPCC during the 2007 Vatican-sponsored Climate Change and Development Conference. He asserted that the "mathematical models used by the IPCC do not correspond to the criteria of the scientific method."[86] Zichichi informed the conference that the IPCC had used "the method of 'forcing' to arrive at their conclusions that human activity produces meteorological variations." Zichichi believes that human activity has less than a 10 percent impact, the dominant influence being fluctuations in cosmic rays.[87]

Besides being WFS President, Zichichi was formerly President of the European Physical Society, President of the Italian National Institute for Nuclear and Sub-nuclear Physics, and President of the NATO Science Committee for Disarmament Technology. As reported by Canada's *National Post*, "He is also Italy's most renowned scientist, credited with the discovery of nuclear antimatter, the discovery of the 'time-like' electromagnetic structure of the proton, the discovery of the effective energy in the forces which act between quarks and gluons, and the proof that, despite its complex structure, it is impossible to break the proton."[88]

In January 2009, the IPCC's legitimacy continued to crumble when one of the world's most prominent forecasting experts, Scott Armstrong, founder of the International Journal of Forecasting, the Journal of Forecasting, and the International Institute of Forecasters, reported to Senator James Inhofe that "the forecasting process reported on by the Intergovernmental Panel on Climate Change (IPCC) lacks a scientific basis."[89] Armstrong explained, "We conducted an audit of the procedures described in the IPCC report and found that they clearly violated 72 scientific principles of forecasting." Independent of this study, John Theon, a retired NASA chief atmospheric scientist and former supervisor of arch-alarmist James Hanson, wrote to Inhofe's committee,

"My own belief concerning anthropogenic climate change is that the models do not realistically simulate the climate system because there are many very important sub-grid scale processes that the models either replicate poorly or completely omit. Furthermore, some scientists have manipulated the observed data to justify their model results. In doing so, they neither explain what they have modified in the observations, nor explain how they did it. They have resisted making their work transparent so that it can be replicated independently by other scientists. This is clearly contrary to how science should be done. Thus there is no rational justification for using climate model forecasts to determine public policy."[90]

Political Backlash to IPCC Fraud

In 2005, the House of Lords Select Committee on Economic Affairs published a report highly critical of the obviously politicized IPCC. Their comments included the following:

- "We can see no justification for an IPCC procedure which strikes us as opening the way for climate science and economics to be determined, at least in part, by political requirements rather than by the evidence. Sound science cannot emerge from an unsound process."

- "We are concerned that there may be political interference in the nomination of scientists to the IPCC."

- "We conclude that there are weaknesses in the way the scientific community, and the IPCC in particular, treats the impacts of climate change. We call for a more balanced approach and look to the government to take an active role in securing that balance of research and appraisal."[91]

After releasing this report, Lord Nigel Lawson, former Chancellor of the Exchequer, testified before the U.S. Senate Committee on Environment and Public Works. Lawson boldly declared, "I believe the IPCC process is so flawed, and the institution, it has to be said, so closed to reason, that it would be far better to thank it for the work it has done, close it down, and transfer all future international collaboration on the issue of climate change."[92]

Senator James Inhofe has been the most active American politician exposing the inept IPCC and their prevailing myths. In December 2005, Inhofe wrote an open letter to Rajendra Pachauri, Chairman of the IPCC, stating, "The more I have researched the issue, the more convinced I have become that climate science is being co-opted by those who care more deeply about promoting doomsday scenarios to further their own, broader agendas than they do about scientific integrity."[93] He further explained,

"My primary concerns lie with how certain scientific conclusions are selected or excluded from the IPCC's consideration and presentation, and how the science has been manipulated in order to reach a predetermined conclusion. These problems must be remedied in order for the IPCC to present a fair and impartial conclusion as to the current state of climate science."[94]

In November 2009, Jairam Ramesh, India's Environment Minister, released a report demonstrating that Himalayan glaciers are not retreating due to manmade global warming, contradicting the conclusions of the IPCC. "So far," Ramesh explained, "we have been depending on research conducted by the West on what is happening to our glaciers and environment. There is an urgent need to have our own studies by our scientists."[95] India began challenging the IPCC's presumed authority in June 2008 when they announced their National Action Plan on Climate Change, which called for sustainable development through clean technologies while also challenging the orthodoxy of manmade global warming.[96]

According to the Plan, India *will not* commit to CO_2 emissions reductions and will instead *increase* coal-fired power plants five-fold by 2030, making them the third largest CO_2 producer.[97] The *Washington Post* lashed out at India, claiming they "continued to balk at the approach favored by the United States."[98] How dare they? Who authorized *India* to introduce opposing viewpoints into political and scientific forums? According to their Action Plan, "No firm link between the changes described below and warming due to anthropogenic climate change has yet been established."[99] India had openly announced its noncompliance. So why then did Rajendra Pachauri endorse the Action Plan?[100]

Not only does Pachauri head the IPCC, he also vociferously leads their fear-campaign. He even once compared Danish academic Bjørn Lomborg, a global warming "skeptic," to Hitler.[101] Pachauri has publicly complained, "India can not be held for any emission control. [Developed countries] should get off the back of India and China."[102] Has not Pachauri fully discredited himself? On the one hand, he claims "the overwhelming weight of scientific evidence" supports manmade global warming.[103] He provokes fear and ridicules anyone who dares disagree with his unsubstantiated claims. He even compares manmade global warming skeptics to people who once believed the earth was flat.[104] Yet on the other hand, Pachauri endorses India's National Action Plan, a plan that openly questions the validity of manmade global warming.

If Pachauri and the IPCC really believe their own propaganda, then why do they only support emission caps on Western nations? If their theory is true then the entire world must immediately curb CO_2 emissions to avert disaster. As previously discussed, the environmental movement is not always what it appears to be. The real agenda behind global warming is global government. They must financially cripple the West to force people into accepting the new system. It has nothing to do with climate, nor carbon.

David Henderson, former chief economist of the Organisation for Economic Co-operation and Development (OECD), blasted the IPCC in a 2007 *Wall Street Journal* op-ed entitled, "Misplaced Trust: Inconvenient truths about the UN's global warming panel." Henderson called attention to the absence of objectivity and transparency within the IPCC. He explained,

"Despite the numbers of persons involved, and the lengthy formal review procedures, the preparation of the IPCC Assessment Reports is far from being a model of rigor, inclusiveness and impartiality…A basic general weakness is the uncritical reliance on peer review as a qualifying criterion for published work to be taken into account in the assessments. Peer review is no safeguard against dubious assumptions, arguments and conclusions if the peers are largely drawn from the same restricted professional milieu. What is more, the peer-review process as such is insufficiently rigorous, since it does not guarantee due disclosure of sources, methods and procedures. Failures of disclosure, such as many journals would not tolerate, have characterized published work that the IPCC has drawn on. The Panel has failed to acknowledge this problem and take appropriate action to deal with it."[105]

Vaclav Klaus, President of the Czech Republic and then-President of the European Union, participated in the 2009 International Conference on Climate Change. He expressed frustration regarding the IPCC's political hijacking of the climate change debate. He explained,

"The politicians—hopefully—sometimes look at the very condensed versions of the IPCC's Summaries for Policymakers but these documents do not represent science, but politics and environmental activism. It is difficult to change their minds. They fully subscribe to the idea that the IPCC publications represent *the* climate science. We know that is not true."[106]

He also reflected, "It reminds me of the frustration people like me felt in the Communist era. Whatever you said—any convincing and well-prepared arguments you used, any relevant data you assembled—no reaction. It all fell into emptiness." As discussed in previous chapters, tax-exempt foundations, international bankers and corporatists have controlled and directed the environmental movement since its outset. In his speech, Klaus also described the lower rungs of this hierarchy as "a very powerful rent-seeking group" of financial opportunists. "Very much like the politicians," Klaus explained, "these people are interested neither in temperature, carbon dioxide, competing scientific hypotheses and their testing, nor in freedom nor markets. They are interested in their businesses and their profits—made with the help of politicians."

Climategate and Copenhagen

The IPCC relies on climate data from four primary sources: NASA, the University of East Anglia's Climate Research Unit (CRU), the National Oceanic and Atmospheric Administration (NOAA), and the British Met Office. These reputable organizations curiously and consistently dodge Freedom of Information requests from scientists wishing to analyze source data and thereby reproduce experimental results. In 2005, for example, CRU director Phil Jones audaciously replied to one such request, "We have 25 or so years invested in the work. Why should I make the data available to you, when your aim is to try and find something wrong with it?"[107] But why should Jones fear independent analysis? Does he care only about his reputation, or does he care about the science? Unfortunately, reputation and arrogance are oftentimes directly related. Nevertheless, reputable scientists and scientific organizations are not infallible. They do make mistakes.

In 2007, for example, independent researchers forced NASA to revise its temperature record, which had been showing 1998 as the hottest year of the century. Steve McIntyre, the editor of ClimateAudit.com, noticed significant errors with NASA's record. He informed them and, accordingly, they conceded that 1934 was actually the century's hottest year.[108] Shortly thereafter however, without explanation, NASA released new revisions, this time claiming that 1998 and 2006 were the hottest years, with 1934 slightly behind.[109] Throughout the controversy, NASA flatly refused to release its source data, despite Freedom of Information Act (FOIA) requests from Christopher Horner, a senior fellow at the Competitive Enterprise Institute. For two years, NASA continued evading these requests. Sound science requires reproducible results. Why then are reputable scientific organizations operating so secretly? *Fox News* columnist John Lott commented, "No researcher should be trusted if he or she is not willing to share their data gladly…what are global warming supporters trying to hide?"[110]

By late November 2009, the answer to Lott's question was echoing around the world via 61 megabytes of data contained in an anonymously released file labeled FOIA (Freedom of Information Act). Hackers, or perhaps insiders, had accessed the computer servers of East Anglia University's Climate Research Unit and published online the massive file, which contained 1,000 plus emails and some 3,000 documents. A *London Telegraph* headline declared,

"This is the worst scientific scandal of our generation."[111] Dubbed Climategate, the emails featured top CRU and IPCC scientists, including Phil Jones, Michael Mann, and Kevin Trenberth, discussing strategies for manipulating climate data, avoiding freedom of information requests, corrupting the peer-review process, and most significantly, acknowledging the fragility of the manmade global warming theory. Both the University and the implicated scientists confirmed the emails' authenticity. Amazingly, however, they claimed innocence. Their words, they insisted, had been taken "completely out of context."[112]

The emails, however, told a different story. In one email Phil Jones wrote to prominent IPCC scientist Michael Mann, "I can't see either of these papers being in the next IPCC report. Kevin [Trenberth] and I will keep them out somehow—even if we have to redefine what the peer-review literature is!"[113] In another email, Jones, alluding to Mann's *Nature* publications, stated, "I've just completed Mike's *Nature* trick of adding in the real temps to each series for the last 20 years (i.e. from 1981 onwards) and from 1961 for Keith's to hide the decline."[114] In another exchange, Jones and Mann discussed coercing an academic journal into excluding contrarian viewpoints. "Perhaps we should encourage our colleagues in the climate research community to no longer submit to, or cite papers in, this journal," Mann wrote. Jones replied, "I will be emailing the journal to tell them I'm having nothing more to do with it until they rid themselves of this troublesome editor."[115]

For years, independent researchers have attempted to acquire CRU data for verification purposes. Yet the Climategate emails show top CRU scientists scheming to break and otherwise avoid Freedom of Information laws. For example, Jones told Mann he would sooner destroy data than release it to climate skeptics like Steve McIntyre and Ross McKitrick. "The two MMs," Jones wrote, "have been after the CRU station data for years. If they ever hear there is a Freedom of Information Act now in the UK, I think I'll delete the file rather than send to anyone."[116] In January 2010, the U.K.'s Information Commissioner's Office ruled that East Anglia had in fact broken Freedom of Information laws. Prosecutions, however, were impossible because too much time had elapsed between the breaches and the allegations.[117]

In perhaps the most revealing email thread, Kevin Trenberth, the IPCC's hurricane "expert," frustratingly wrote, "The fact is that we can't account for the lack of warming at the moment and it is a travesty that we can't."[118] He further admitted, "The data are surely wrong. Our observing system is inadequate."[119] Later in the thread, Trenberth lamented,

"We are no where close to knowing where energy is going or whether clouds are changing to make the planet brighter. We are not close to balancing the energy budget. The fact that we can not account for what is happening in the climate system makes any consideration of geoengineering quite hopeless as we will never be able to tell if it is successful or not! It is a travesty!" [120]

The Climategate scandal forced Phil Jones to temporarily step down from his CRU directorship, pending an investigation.[121] Jones also conceded that no statistically significant global warming had occurred for 15 years.[122] Also the CRU confessed to having destroyed their original climate data. They explained, "We do not hold the original raw data but only the value-added (quality controlled and homogenised) data."[123] Yet without the original data, scientists are unable to verify the CRU's conclusions. Roger Pielke, professor of environmental studies at Colorado University, commented, "The CRU is basically saying, 'Trust us'. So much for settling questions and resolving debates with science."[124]

Defenders of the manmade global warming theory largely dismissed Climategate. For example, Thomas Karl, Director of NOAA's National Climatic Data Center, claimed, "These emails do nothing to undermine the very strong scientific consensus that the earth is warming and that human activity is largely responsible."[125] James Hansen, head of NASA's Goddard Institute for Space Studies (GISS), remarked, "The important point is that nothing was found in the East Anglia e-mails altering the reality and magnitude of global warming in the instrumental record."[126] Al Gore dismissed the emails as "taken wildly out of context," and "sound and fury signifying nothing."[127] Gordon Brown, the British Prime Minister, declared, "With only days to go before Copenhagen we mustn't be distracted by the behind the times anti-science, flat earth climate skeptics. We know the science. We know what we must do."[128] While revelations from Climategate did not alone disprove the grandiose manmade global warming theory, they certainly substantiated the growing body of evidence against it. Above all, they demonstrated the urgent need for honest, independent investigations and the immediate discontinuation of all proposed carbon taxation and cap-and-trade schemes.

No national government, however, even entertained the idea of investigating Climategate or the larger implications of unsound science guiding public policy decisions. The Obama administration, for example, completely disregarded the emails. White House Press Secretary Robert Gibbs, declared, "there's no real scientific basis for the dispute."[129] Initially, the IPCC acknowledged the seriousness of the situation. IPCC Chairman Rajendra Pachauri declared, "We will certainly go into the whole lot and then we will take a position on it...We certainly don't want to brush anything under the carpet. This is a serious issue and we will look into it in detail."[130] Furthermore, he admitted that the IPCC does not conduct climate research. And since it merely collates climate research, it values a strong peer-review process. Pachauri explained,

> "The IPCC does not carry out climate research. All it does, it brings together all the papers from round the world from peer-reviewing, etc, brings them together to make a position, but it depends of course on the legitimacy of the peer-reviewing process and that, in a sense, is what's being brought into question."[131]

Just days after these remarks, Pachauri completely backtracked, stating that the IPCC *would not* investigate Climategate, but would merely probe the incident "just to see if there are any lessons for us that we might want to take onboard."[132] He further explained, "We are determining how best to do that. But I want to clarify that this is not an investigation." Pachauri shamelessly defended the Climategate scientists while vilifying the whistleblowers. "The persons who have worked on this report," he stated, "and those who unfortunately have been victims of this terrible and illegal act, are outstanding scientists, and have contributed enormously over the 20, 21 years of the existence of the IPCC." Pachauri concluded, "The only issue that has to be dealt with as far as this occurrence is concerned is to find out who is behind it."

Despite repeated denials and rationalizations by establishment scientists, politicians, and journalists, Climategate certainly influenced the United Nations' 15th Climate Change Conference (COP15), held in Copenhagen in early December 2009. The goal of the confab, which attracted more than 20,000 participants, including delegates from 193 nations,[133] was to secure a legally binding agreement on global carbon emissions to replace the Kyoto Protocol, which expires at the end of 2012. UN Secretary General Ban Ki-moon remarked, "We need to have a very strong, robust, binding political deal that will have an immediate operational effect. This is not going to be a political declaration, just for the sake of declaration. It is going to be a binding political deal, which will lead to a legally binding treaty next year."[134] More significantly, Ki-moon admitted the world's elite, through the guise of climate change, are planning and preparing for global governance. "We will establish a global governance structure to monitor and manage the implementation of this," he told the *Los Angeles Times*.[135] Furthermore, this wasn't his first such admission. Weeks before, he told the *New York Times* that an emissions "deal must include an equitable global governance structure."[136]

Despite high expectations, most participants left Copenhagen disappointed. Trying desperately to comment favorably, President Obama concluded, "Rather than see a complete collapse in Copenhagen, in which nothing at all got done and would have been a huge backward step, at least we kind of held ground and there wasn't too much backsliding from where we were."[137] While Climategate was an unexpected and troublesome wrench for conference organizers, it was not the sole reason COP15 failed. For example, before the conference, China and the U.S. had been quarreling regarding America's demands that China submit to an international emissions monitoring system. As reported by the *New York Times*, after China announced its reduction target, "American officials privately said the target was too low and raised questions about the reliability of Beijing's reporting methods, saying that some form of international monitoring would be necessary. China protested and declared that it would not sacrifice its sovereignty to an outside verification scheme."[138]

As the conference began, the drama escalated. On day two, as reported by the *Guardian*, "developing countries reacted furiously to leaked documents that show world leaders will next week be asked to sign an agreement that hands more power to rich countries and sidelines the UN's role in all future climate change negotiations."[139] The so-called Danish Text, obtained by the *Guardian*, departed from the Kyoto Protocol whereby rich nations assume the bulk of the emissions burden. Instead, the Danish Text proposed allowing richer nations to emit double the per capita emissions of developing countries. Furthermore, the Text proposed handing "effective control of climate change finance to the World Bank,"[140] an organization notoriously known for imposing crippling loans upon poor nations and thereby allowing private interests to seize precious natural resources and infrastructure when the nations inevitably default. John Perkins, for example, author of *New York Times* bestseller *Confessions of an Economic Hit Man*, has thoroughly documented this treachery based on his first-hand experience. Perkins describes himself as "a former economic hit man—a highly paid professional who cheated countries around the globe out of trillions of dollars."[141] When asked by Amy Goodman of *Democracy Now*, "How closely did you work with the World Bank," Perkins replied, "Very, very closely with the World Bank. The World Bank provides most of the money that's used by economic hit men, it and the IMF."[142]

Also Joseph Stiglitz, the Nobel Prize-winning former Chief Economist of the World Bank, has frequently condemned the World Bank and IMF for predatory policies. When asked by *The Progressive*, "Who wins when the U.S. and the IMF impose their bailouts on countries in crisis," Stiglitz replied, "These policies protect foreign creditors…In the nineteenth century, they used gunboats. Now they use economic weapons and arm-twisting."[143] In 2009, Stiglitz penned an article for *Vanity Fair* criticizing so-called free market ideology. He observed,

"The World Bank and the I.M.F. said they were doing all this for the benefit of the developing world…Not surprisingly, people in developing countries became less and less convinced that Western help was motivated by altruism…Free-market ideology turned out to be an excuse for new forms of exploitation. 'Privatization' meant that foreigners could buy mines and oil fields in developing countries at low prices. It meant they could reap large profits from monopolies and quasi-monopolies, such as in telecommunications. 'Liberalization' meant that they could get high returns on their loans—and when loans went bad, the I.M.F. forced the socialization of the losses, meaning that the screws were put on entire populations to pay the banks back."[144]

On the heals of the Danish Text leak, billionaire globalist George Soros proposed the creation of a $100 billion IMF fund, which would dole out "green loans" to developing countries. Noting that climate change "is really an existential problem for the world," Soros added, "The fund should be invested in such a way that it could potentially provide a return."[145] But is Soros a trustworthy source for economic advice? He is, after all, a notorious market manipulator who, in 2002, was convicted of insider trading in France.[146] Also in 2009, Hungary's financial supervisory watchdog fined him for market manipulation.[147] The Danish Text authors apparently didn't realize that underhanded attempts by Soros to position the World Bank and IMF as regulators of the new green economy do not inspire confidence, especially among nations previously victimized by World Bank and IMF neocolonial policies.

Understandably, distrust and suspicion ran high in Copenhagen, especially among the developing countries. One delegate lamented, "It is being done in secret. Clearly the intention is to get Obama and the leaders of other rich countries to muscle it through when they arrive next week. It effectively is the end of the UN process."[148] Days later, negotiations deteriorated even further. Lumumba Di-Aping, chief negotiator for the G77-China bloc of 130 nations, explained, "It has become clear that the Danish presidency—in the most undemocratic fashion—is advancing the interests of the developed countries at the expense of the balance of obligations between developed and developing countries."[149]

Despite the gridlock, attending nations finally agreed to "take note" of the Copenhagen Accord, a non-binding "statement of intention."[150] Besides standard rhetoric like, "We underline that climate change is one of the greatest challenges of our time," and, "We agree that deep cuts in global emissions are required according to science, and as documented by the IPCC Fourth Assessment Report," the Accord also declared that, "New multilateral funding for adaptation will be delivered through effective and efficient fund arrangements, with a governance structure."[151] Christopher Monckton commented,

> "The one and only single aim of this entire global warming conference [was] to establish the mechanism, the structure, and above all the funding for a world government...World government is coming because the leaders of the West have given up. They no longer care about democracy. They no longer care about the truth about the climate. They are willing to go along with this world government because they see roles for themselves in that world government in exactly the same way as the leaders of the EU did."[152]

The IPCC, as demonstrated, is a political body with a political agenda. Scores of IPCC participants with firsthand experience have exposed the organization's duplicity. The IPCC claims there is a scientific consensus, yet a consensus does not even exist within the IPCC. So what about the science behind climate change? What scientific information has the IPCC been suppressing?

Chapter 18
Never-Ending Climate Change

Is global warming real? Are temperatures increasing? Are they decreasing? Yes, yes, and yes. Earth's climate is *always* changing. Global warming *is* real, but so is global cooling. The question, as to whether temperatures are increasing or decreasing, depends upon the time interval under consideration. For example, four million years ago, during the Pliocene era, average temperatures were 2-3°C warmer than today. Therefore, based on the interval from the Pliocene until the present, temperatures are decreasing (Figure 1). However, twenty-five thousand years ago, during the depths of the last ice age, average temperatures were 12-13°C colder than today. Based on the interval from twenty-five thousand years ago until the present, temperatures are increasing dramatically (Figure 2). One hundred years ago, temperatures were about 1°C cooler. Based on this interval, temperatures are increasing slightly (Figure 3). The IPCC and other alarmists narrowly focus on the past several decades, insisting that current warming trends are anomalous. Yet their theories ignore, and in some cases *reinvent*, climate history.

Climate Cycles

Fred Singer and Dennis Avery, in their book *Unstoppable Global Warming: Every 1,500 Years*, discussed Earth's two naturally occurring climate cycles. During the 100,000-year cycle, the Earth oscillates between long, deep Ice Ages and relatively shorter periods of moderate temperatures. Today, obviously, the Earth is passing through the moderate phase of this 100,000-year cycle. Within this cycle, however, there is another cycle, which spans roughly 1,000 to 1,500 years. During this cycle, the planet undergoes one warming period and one cooling period, each lasting roughly 500 to 750 years (Figure 4).

Since about 1850, Earth has been passing through a warming period. Therefore, this moderate warming trend should continue for several more centuries, followed by another cooling period beginning around the 24th or 25th century. The period that preceded the current warming period is called the Little Ice Age. It began around 1300 and lasted until 1850. The Little Ice Age was characterized by crop failures, reduced harvests, famines, colder and longer winters, and oftentimes wildly unpredictable weather. According to Singer and Avery, "The standard deviation of winter temperatures in England and the Netherlands was about 40 to 50 percent greater during the coldest centuries of the Little Ice Age than during the 20th century. It was virtually impossible to adapt and all too necessary to suffer."[1]

Before the Little Ice Age, humanity enjoyed the Medieval Warm Period, starting around 900. During this period, plentiful harvests enabled European aristocracies to build fabulous castles and cathedrals by mobilizing excess laborers away from the farms and towards their self-congratulatory construction projects. Also, during the Medieval Warm Period, Europe's population increased around 50 percent. Singer and Avery explained,

> "The food abundance of the Medieval Warming ensured the growth of population, cities, transportation, and great buildings. Even a one-degree difference in the winter climate made a significant difference in the length of the growing season, especially the absence of untimely frosts."[2]

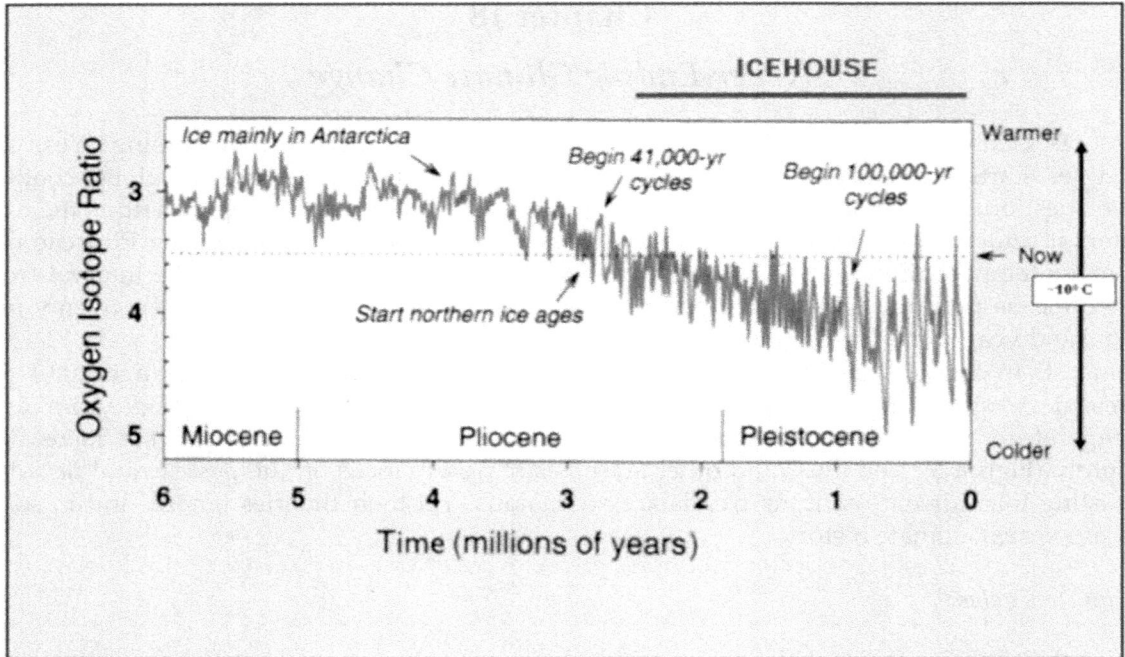

Figure 1: Composite deep-ocean temperate curve from DSDP Sites 86 and 849 (North Pacific) during past 6 million years.[3]

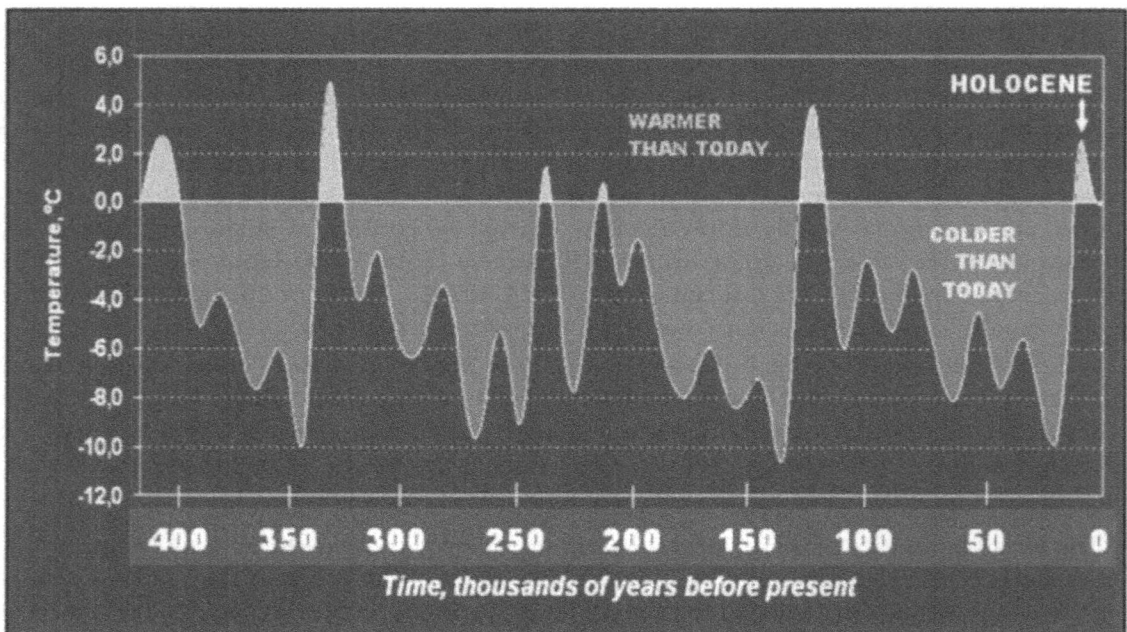

Figure 2: Surface air temperature over Antarctica's Vostok station during past 400,000 years.[4]

Global Average Near-Surface Temperatures 1850–Nov 2008

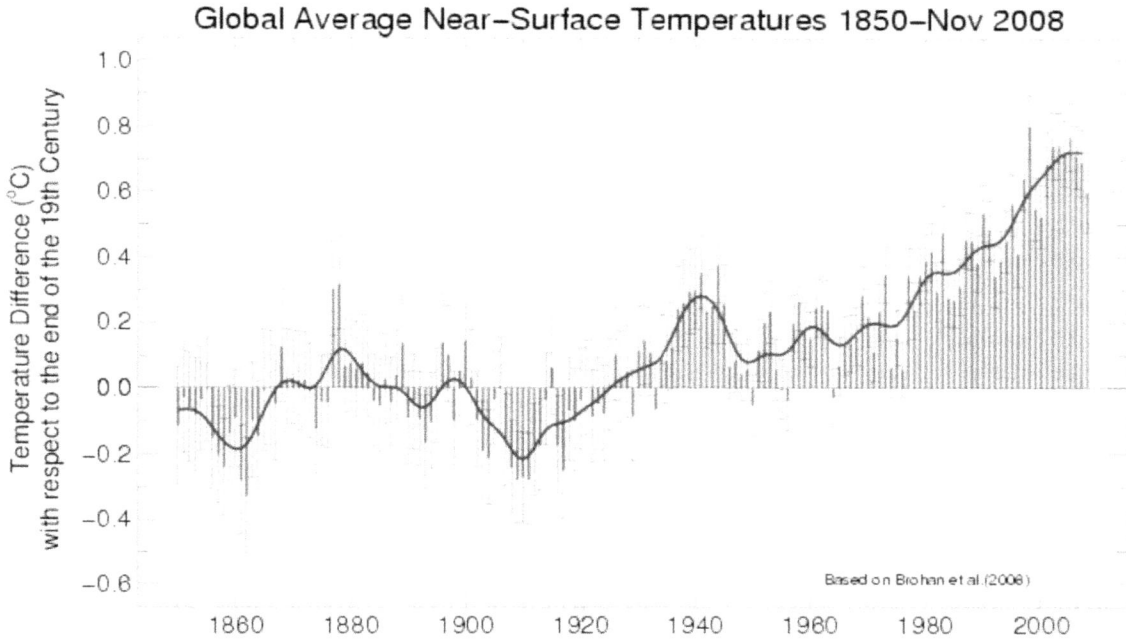

Figure 3: Ground Thermostat Records from U.K. Meteorological Office.[5]

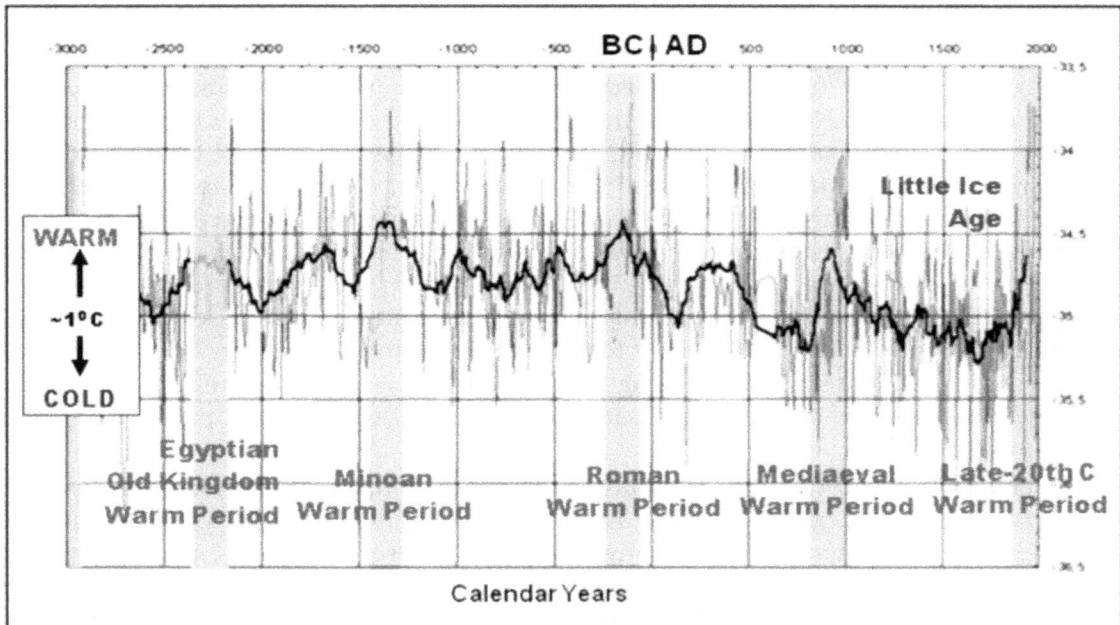

Figure 4: Surface air temperature over Greenland during past 5,000 years.[6]

During this period, people spread out and inhabited the far-northern regions of the globe. Richard Tkachuck of the Geosciences Research Institute discovered that the ancestors of today's American Eskimos were then living in northern Greenland. He explained,

> "At these locations, large dwellings made from driftwood have been found. There is also archeological evidence of large villages that were developed for whaling and fishing. These settlements eventually were forced south by climatic change until they came in contact with Viking colonies in southern Greenland. Conflict occurred, and the Viking colonies eventually died out in the 1400s."[7]

Grape vines are another key indicator of climatic conditions during the Medieval Warm Period. Tkachuck explained, "The cultivation of grapes for wine making was extensive throughout the southern portion of England from about 1100 to around 1300. This represents a northward latitude extension of about 500 km from where grapes are presently grown in France and Germany."[8]

Before the Medieval Warm Period, during the depths of the Dark Ages, humanity endured another cooling period, which started around 400 and lasted until 900. While the collapse of the Roman Empire certainly ushered in the Dark Ages, climate change also greatly affected these times. Droughts led to famines, which in turn led to plagues and massive reductions in trade. Mike Baillie, a professor from Queen's University in Belfast has confirmed this cooling period based on his expertise in tree-ring studies. The tree ring record also suggests that around 540, the earth experienced some sort of catastrophic event. He explained,

> "The trees are unequivocal that something quite terrible happened. Not only in Northern Ireland and Britain, but right across northern Siberia, North and South America—it is a global event of some kind. There seem to have been comets, meteors, earthquakes, dimmed skies and inundations and, following the famines of the late 530s, plague arrived in Europe in the window A.D. 542-545."[9]

The period before this cooling period, from 200 B.C. to A.D. 400, was the Roman Warm Period. Grape and olive farming extended further north than was previously possible. The European climate became milder, while the tropics became wetter. John Oliver, in his book *Climate and Man's Environment*, explained, "By 350 A.D., the climate had become milder in northern realms while in tropical regions it appears to have become excessively wet. Tropical rains in Africa caused high-level Nile floods and temples built earlier (1250 B.C.) were inundated. At this time too, Central America experienced heavy precipitation and tropical Yucatan was very wet."[10] Finally, before the Roman Warming, Egyptian records document a cooling period starting around 750 B.C. Cooler weather caused the beneficial Nile floods to decrease. Singer and Avery noted, "Early Roman authors wrote of a frozen Tiber River and of snow remaining on the ground for lengthy periods. Those events would be unthinkable today. We also know that European glaciers advanced during the early part of the Roman civilization."[11]

Overall, the historical record supports the contention that Earth regularly oscillates between warming and cooling periods, each lasting roughly 500 to 750 years. In 2003, the Harvard-Smithsonian Center for Astrophysics (CFA) sponsored a study, partly funded by NASA, to collate, analyze, and summarize 240 climate studies over the past 40 years. According to the CFA press release, "Their report, covering a multitude of geophysical and biological climate indicators, provides a detailed look at climate changes that occurred in different regions around the world over the last 1000 years."[12] Coming from such a prestigious institution, this study surely embarrassed the global warming orthodoxy, which at the time was attempting to rewrite history *without* the previous warming periods. Nevertheless, the CFA announced,

"The 20th century is neither the warmest century nor the century with the most extreme weather of the past 1000 years. Their review also confirmed that the Medieval Warm Period of 800 to 1300 A.D. and the Little Ice Age of 1300 to 1900 A.D. were worldwide phenomena not limited to the European and North American continents. While 20th century temperatures are much higher than in the Little Ice Age period, many parts of the world show the medieval warmth to be greater than that of the 20th century."

The CFA analyzed 112 studies with information about the Medieval Warm Period. 92 percent showed direct evidence of the warming while only 2 studies refuted the warming. They analyzed 124 studies concerning the Little Ice Age, of which 98 confirmed the cooling. They analyzed 102 studies containing information on whether the 20th century was the warmest on record. Only 3 studies supported this contention, and each of these 3 showed 20th century warming occurring mostly in the century's early decades, long before human CO_2 emissions could have factored into the equation.[13]

The Hockey Stick Scandal

Princeton University physics professor William Happer is the author of over 200 peer-reviewed papers, and the former Director of Energy Research at the U.S. Department of Energy (DOE) from 1990 to 1993, where he supervised all DOE work on climate change. In February 2009, Happer addressed the U.S. Senate Environment and Public Works Committee regarding the ongoing global warming debate. He recalled,

"When I was a schoolboy, my textbooks on earth science showed a prominent Medieval Warm Period at the time the Vikings settled Greenland, followed by a vicious Little Ice Age that drove them out. So I was very surprised when I first saw the celebrated 'hockey stick curve,' in the Third Assessment Report of the IPCC. I could hardly believe my eyes. Both the Little Ice Age and the Medieval Warm Period were gone, and the newly revised temperature of the world since the year 1000 had suddenly become absolutely flat until the last hundred years when it shot up like the blade on a hockey stick.

This was far from an obscure detail, and the hockey stick was trumpeted around the world as evidence that the end was near. We now know that the hockey stick has nothing to do with reality but was the result of incorrect handling of proxy temperature records and incorrect statistical analysis. There really was a Little Ice Age and there really was a Medieval Warm Period that was as warm or warmer than today.

I bring up the hockey stick as a particularly clear example that the IPCC summaries for policy makers are not dispassionate statements of the facts of climate change. It is a shame, because many of the IPCC chapters are quite good. The whole hockey stick episode reminds me of the motto of Orwell's Ministry of Information in the novel *1984*. 'He who controls the present, controls the past. He who controls the past, controls the future.'"[14]

A thirty-one year old climatologist named Michael Mann published the hockey stick graph in *Nature* in 1998.[15] Three years later, the IPCC prominently displayed it as striking, visual evidence of manmade global warming (Figure 5). Also Al Gore included the hockey stick in his 2006 documentary, *An Inconvenient Truth*.[16] Perhaps Mann's graph was the result of careless error, not scientific fraud. Nevertheless, for manmade global warming enthusiasts, the Medieval Warm Period was, quite simply, an inconvenient truth. Thus an audacious attempt to rewrite history — to control the past, to control the future — began in 1995.

David Deming, a geophysics professor at the University of Oklahoma, published a 1995 article in *Science* regarding borehole temperature data in North America. His conclusions seemed to support the manmade global warming theory. After publishing his article, he recalled gaining significant credibility from the manmade climate change community. "They thought I was one of them," he explained, "someone who would pervert science in the service of social and political causes. One of them let his guard down."[17] In 2006, Deming testified before the Senate Committee on Environment and Public Works about that person:

"I had another interesting experience around the time my paper in *Science* was published. I received an astonishing email from a major researcher in the area of climate change. He said, 'We have to get rid of the Medieval Warm Period.' The Medieval Warm Period (MWP) was a time of unusually warm weather that began around 1000 AD and persisted until a cold period known as the 'Little Ice Age' took hold in the 14th century. Warmer climate brought a remarkable flowering of prosperity, knowledge, and art to Europe during the High Middle Ages. The existence of the MWP had been recognized in the scientific literature for decades. But now it was a major embarrassment to those maintaining that the 20th century warming was truly anomalous. It had to be 'gotten rid of.'"[18]

Referring to Mann's elimination of the MWP via his 1998 hockey stick graph, Deming stated,

"Normally in science when you have a novel result that appears to overturn previous work, you have to demonstrate why the earlier work was wrong. But the work of Mann and his colleagues was initially accepted uncritically, even though it contradicted the results of more than 100 previous studies. Other researchers have since reaffirmed that the Medieval Warm Period was both warm and global in its extent."[19]

The 1st IPCC report, in 1990, included a graph displaying the traditionally accepted temperature construction of the past thousand years (Figure 6). Yet in 2001, the 3rd IPCC report presented a radically different view — the hockey stick graph. In 2003, Canadian statisticians Steve McIntyre and Ross McKitrick decided to investigate Mann's methodology. They requested the original study data from *Nature*. An excessively long wait-time (plus other factors) indicated that the study had not yet been peer-reviewed. When they finally received the data, McIntyre and McKitrick discovered that due to "collation errors, unjustifiable truncations of extrapolation of source data, obsolete data, geographical location errors, incorrect calculations of principal components, and other quality control defects," the hockey stick graph was bogus.[20] The duo published a peer-reviewed critique demonstrating Mann's error in the journal *Energy & Environment*.[21]

Statistician Francis Zwiers of Environment Canada (a government agency) agreed that Mann's method "preferentially produces hockey sticks when there are none in the data."[22] Because the IPCC's strong reliance on the hockey stick had apparently greatly skewed the ongoing global warming debate, certain Congressmen began asking tough questions. Representative Joe Barton from the Energy and Commerce committee requested detailed records of Mann's data and methods, and his government and private funding. The American

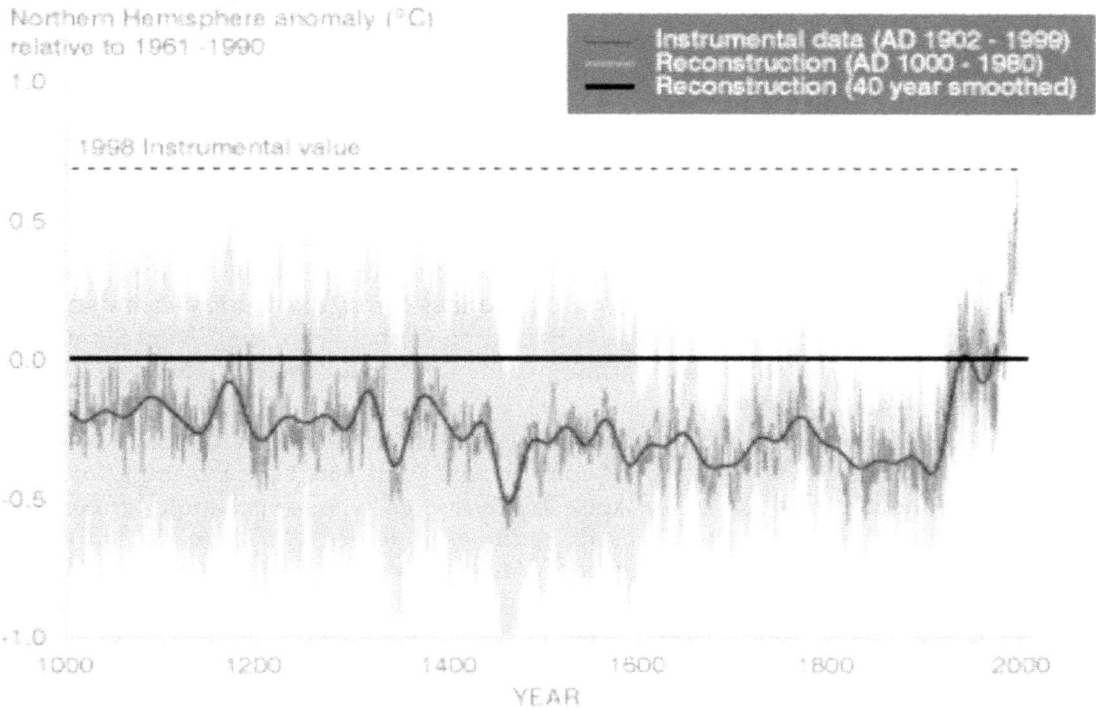

Figure 5: Graph from 2001 IPCC Report.[23]

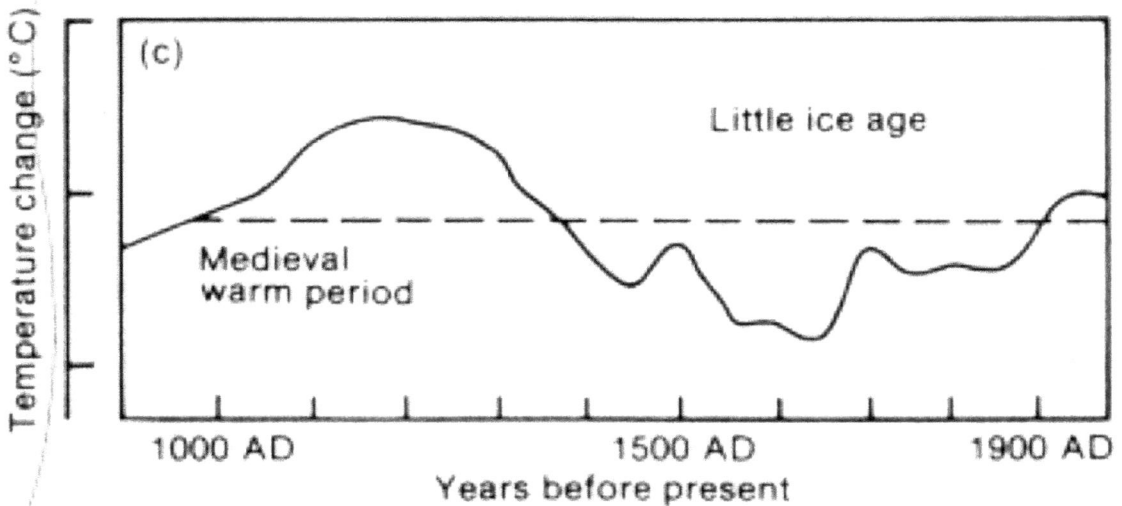

Figure 6: 1990 IPCC Graph (Figure 7.1c) showing schematic diagram of global temperature variations during past 1,100 years. The dotted line nominally represents conditions near the beginning of the twentieth century.[24]

Association for the Advancement of Science (AAAS), however, characterized Barton's request as an intimidation tactic.[25] Consequently, Congress requested two investigations regarding the hockey stick controversy, one by the National Academy of Sciences (NAS), and one by an independent team of statisticians.

NAS was *unable* to confirm Mann's results. Yet in June 2006, alarmists victoriously embraced the NAS report, boldly asserting that Mann had been vindicated. "The hockey stick is alive and well," declared Raymond Bradley, one of Mann's colleagues who worked on the original study.[26] Sherwood Boehlert, chairman of the House Science Committee insisted, "There is nothing in this report that should raise any doubts about the broad scientific consensus on global climate change."[27] Typical headlines across the country included the *New York Times'* "Panel Supports a Controversial Report on Global Warming,"[28] the *Boston Globe's* "National panel supports '98 global warming evidence,"[29] and the *San Francisco Chronicle's* "It's official: We live in hot times."[30]

It was incredible deception, yet completely transparent. While NAS could not confirm Mann's elimination of the MWP, they *did* confirm that Earth has been warming for the past 400 years since the depths of Little Ice Age. This, however, was never a point of contention. The problem with the hockey stick was its *exclusion* of the MWP. Nevertheless, the spin-doctors simply embraced the unchallenged confirmation of recent warming and declared victory. Despite their misleading headlines, the aforementioned articles readily admitted the hockey stick was indeed still broken. The *San Francisco Chronicle*, for example, remarked, "they showed less confidence in the researchers' conclusion that the climate is warmer now than it has been in 1,000 years, a conjecture the scientists said was only scientifically plausible."[31] The hockey stick was still "alive," in a *Weekend at Bernie's* sense, yet the facade was rapidly waning.

Several weeks later, the independent team of statisticians released their report, a sound refutation of Mann's methodology, to Congress. Edward Wegman, a statistics professor at George Mason University and fellow of numerous scientific societies including the American Statistical Association, had assembled a panel of expert statisticians for the investigation. When they applied proper statistical methodology, the hockey stick disappeared, confirming the original refutations made by McIntyre and McKitrick. Mann had grossly miscalculated, or perhaps worse. The infamous Climategate emails suggest the latter. In 2003, for example, Mann wrote to Jones and other scientists, "It would be nice to try to contain the putative 'MWP.'"[32] Wegman noted the obvious absurdity of the climate change community's reliance on statistics without having consulted trained statisticians. He commented, "If statistical methods are being used, then statisticians ought to be funded partners engaged in the research to insure as best we possibly can that the best quality science is being done."[33]

The press almost completely ignored the Wegman study. The *Washington Post*, however, published an article by H. Sterling Burnett, a senior fellow from the National Center for Policy Analysis. Burnett chastised the media for refusing to allow the hockey stick to die:

"[Recent] press reports typically highlighted the limited areas where the NAS supported the hockey stick research and downplayed the substantive flaws the NAS confirmed...the mainstream press has largely abdicated its role as a provider of objective, balanced reporting on global warming and has adopted the role of advocate. In doing so, the press has done the public, scientific progress and the journalistic profession a profound disservice"[34]

While the IPCC quietly retracted the hockey stick from their 2007 4th Assessment Report, Al Gore's delusions regarding the Medieval Warm Period still persist. He has only acknowledged some minor "nuances and distinctions" contained in his film.[35] Michael Mann has never acknowledged his errors either. Conversely, Mann now claims the hockey stick handle should stretch back *two* thousand years![36]

Urban Heat Islands

Climate alarmists maintain that ground temperature increases of roughly 1°C constitute a serious environmental threat. However, as previously discussed, such increases are well within the range of natural climactic fluctuation and thus no reason for concern. Furthermore, ground temperature readings are likely skewed due to the Urban Heat Island Effect. Cities are generally several degrees warmer than rural areas because concrete, glass, and steel—pervasive throughout urban environments—absorb radiant energy from the sun, thus artificially raising temperatures. Even a small village of 1,000 people can create a heat island, raising temperatures 2° to 3°C.[37] This creates a problem, however, for climate researchers relying on official temperature readings from urban locations. A 2003 article in *Nature* suggested that U.S. surface warming has been overestimated by as much as 40 percent.[38]

In 2007, meteorologist Anthony Watts launched an effort to photograph all 1,221 official surface temperature stations in the U.S. to see where changes in land use have been skewing temperature records. For example, why does a station in Marysville, California show accelerating warming over the past several decades while a station in nearby Orland, California shows no such trend? Quite simply, land use around the Orland station has remained unchanged while land use around the Marysville station has changed completely. Asphalt driveways, cell phone towers, and nearby air conditioning units have all contributed to artificially higher temperature readings in Marysville as compared to Orland.[39]

Watts assumed responsibility for the project after years of frustrating correspondence with the National Oceanic and Atmospheric Administration (NOAA), the organization responsible for maintaining the U.S. Historical Climatology Network. Despite their supposed competence, NOAA has never properly surveyed U.S. temperature stations to determine the impact and extent of the Urban Heat Island Effect. Furthermore, in 1999, the United Nations determined that ground temperature measuring equipment is "inadequate and is deteriorating worldwide."[40]

A decade later, NOAA continues to stall on upgrade initiatives. Watts commented, "Given such a massive failure of bureaucracy to perform something so simple as taking some photographs and making some measurements and notes of a few to a few dozen weather stations in each state, it seemed that a grass roots network of volunteers could easily accomplish this task."[41] As of November 2009, 865 stations had been surveyed. The results indicate widespread data contamination. 89 percent of stations fail to meet National Weather Service standards requiring stations to be at least 30 meters away from artificial heat sources. 89 percent of stations show temperatures at least 1 degree higher than has actually occurred, of which 58 percent show temperatures 2 degrees or higher than actual, and 11 percent show temperatures 5 degrees or higher than actual.[42] "The conclusion is inescapable," Watts wrote, "The U.S. temperature record is unreliable."[43]

Melting Ice, Rising Seas?

Ahead of the UN's 2009 Copenhagen Climate Conference, Maldivian President Mohamed Nasheed organized a dramatic publicity stunt. Wearing full scuba gear, he and his Cabinet held an underwater meeting, culminating with a declaration calling for worldwide carbon emission cuts. Asked what would happen to his 1,200-island nation, most of which lies just 1.5 meters above sea level, should the Copenhagen conference fail, Nasheed declared, "We are all going to die."[44]

Are the polar ice caps really melting? Are polar bear really endangered? Are sea levels rising dangerously? The estimated polar bear population has increased from 5 thousand in 1972 to 20-25 thousand today.[45] Furthermore, having existed for 150 thousand (and possibly 1 million) years, polar bears have endured extreme fluctuations in both temperatures and Arctic ice coverage.[46] Nevertheless, Gore continues to bemoan, "Their habitat is melting...beautiful animals, literally being forced off the planet. They're in trouble, got nowhere else to go."[47] Regarding sea levels, Gore presents computer-generated graphics of submerged coastal cities and disappearing Pacific islands. He suggests that sea levels might rise 20 feet "in the near future." In their 4th Assessment Report, however, the IPCC predicted 59cm by 2100, ten times lower than Gore's doomsday scenario. In 2007, a London High Court ruled that Gore's film contained *nine* significant errors, including his claims regarding polar bears and sea levels. Judge Justice Burton dismissed Gore's sea level claims as "distinctly alarmist."[48]

Nevertheless, Antarctica, the world's most massive block of ice, *is* changing, especially its northern peninsula, which extends upwards toward Chile. In 1995, for example, the 1,600 sq km 'Larson A' ice shelf broke away from the continent. In 1998, the 1,100 sq km 'Wilkins' ice shelf broke away, and in 2002, the 13,500 sq km 'Larson B' ice shelf broke away.[49] These dramatic, transformational events provided powerful, visual 'evidence' for climate alarmists. But what do these events mean? Changes are not necessarily crises, nor even causes for concern. At least that's the contention of Duncan Wingham, the European Space Agency's principal scientist and professor of Climate Physics at University College, London. Using satellite data, Wingham tracks fluctuations in the mass of Antarctica. In 2005, he and three colleagues published an article in the *Journal of the Royal Society* entitled, "Mass Balance of the Antarctic Ice Sheet." Based on data collected from 1992-2003, they determined that 72% of Antarctica is actually growing and creating "a sink of ocean mass sufficient to lower global sea levels by 0.08 mm per year."[50]

In 2009, the Scientific Committee on Antarctic Research confirmed that east Antarctica, which is four times the size of west Antarctica, has shown "significant cooling in recent decades."[51] Ian Allison, head of the Antarctic Division of the Australian government's Glaciology Program also agreed, stating, "Sea ice conditions have remained stable in Antarctica generally."[52] These findings dovetail with temperature measurements from the South Pole station, established by the United States in 1957. Since then, local temperatures have been *falling*. In 2010, the European Space Agency will launch CryoSat-2, a sophisticated satellite, capable of measuring the whole of the continent.[53]

The Arctic icecap has been the focus of even greater alarmism. For example, in 2008, *National Geographic* reported that the Arctic was on the verge of completely melting away. David Barber, a University of Manitoba scientist, proclaimed, "We're actually projecting this year that the North Pole may be free of ice for the first time."[54] Sheldon Drobot, a University of Colorado climatologist, added, "I would say the ice in the vicinity of the North Pole is primed for melting, and an ice-free North Pole is a good possibility."[55] However, at the time of these predictions, data from NASA's Marshall Space Flight Center showed that 2008's Arctic ice was nearly identical to that of 2002, 2005, and 2006.[56] In fact, according to the National Snow and Ice Data Center (NSIDC) in Boulder, Colorado, the 2008 sea ice minimum was up 9.1% over 2007's minimum, and during 2009 it grew another 12.9% as compared to 2008.[57]

Nevertheless, alarmist organizations like Greenpeace continue to fear-monger about ice-free Arctic summers. "We are looking at ice-free summers in the Arctic as early as 2030," Greenpeace reported in 2009.[58] Shortly thereafter, the BBC's Stephen Sackur challenged Greenpeace Executive Director Gerd Leipold on his organization's claim. Sackur accused Greenpeace of using "scare tactics" and making "misleading" and "preposterous" claims. When pressed about these claims, Leipold conceded, "I don't think it will be melting by 2030. That may have been a mistake."[59] Greenpeace and other alarmists never mention that Arctic ice coverage depends largely on variable oceanic and atmospheric currents. Even NASA, an organization staunchly supportive of the manmade global warming theory, has conceded, "not all the large changes seen in Arctic climate in recent years are a result of long-term trends associated with global warming."[60] James Morrison, the lead scientist of a 2007 NASA study explained, "Our study confirms many changes seen in upper Arctic Ocean circulation in the 1990s were mostly decadal in nature, rather than trends caused by global warming."[61]

According to sea level expert Nils-Axel Mörner, sea levels, like the entire climate system, fluctuate naturally. Before his 2005 retirement, Mörner was head of Geodynamics at Stockholm University. He was also the Chairman of INQUA (an International Commission on Sea Level Change) and a former Lead Reviewer for the IPCC. He famously exposed the IPCC for falsely claiming that sea levels were rising 2.3 millimeters per year. As reported by the *London Telegraph*,

> "IPCC's favored experts had drawn on the finding of a single tide-gauge in Hong Kong harbor showing a 2.3mm rise. The entire global sea-level projection was then adjusted upwards by a 'corrective factor' of 2.3mm, because, as the IPCC scientists admitted, they 'needed to show a trend.'"[62]

Regarding the IPCC's so-called experts, Mörner remarked, "I was astonished to find that not one of their 22 contributing authors on sea levels was a sea level specialist: not one."[63] According to his research, the sea "hasn't risen in 50 years," and should any rising occur this century, it will "not be more than 10cm, with an uncertainty of plus or minus 10cm."[64]

After the Maldivian underwater meeting, Mörner published an open letter in the *Spectator* addressed to Mohamed Nasheed (his forth attempt to contact the Maldivian President). Mörner explained that, based on his own hands-on research and detailed documentation of 4,000 years of sea level changes around the Maldives, nothing extraordinary is happening today. For the past 30 years, sea levels around the Maldives have been stable, and during the 1970s, levels actually *dropped* 20cm in the region. Furthermore, during the 17th century, sea levels around the Maldives were 50cm *higher* than present. "This bodes well for their prospects of surviving the next change," Mörner observed.[65] He concluded, "Mr. President, you and your ministers in the Maldives really don't need to worry about a future life beneath the waves. You should pass on this message to the people of the Maldives. It is high time to release them from this terrible psychological burden."[66]

Chapter 19
Deadly Carbon Dioxide

Two oxygen atoms bonded to one carbon atom: despite its recent demonization, carbon dioxide is essential to life on Earth. Even Al Gore acknowledges this. In his book *Earth in the Balance*, Gore stated, "The human lung inhales oxygen and exhales carbon dioxide…Trees and other plants pull CO_2 out of the atmosphere and replace it with oxygen, transforming the carbon into wood, among other things."[1] Nevertheless, Gore and the global warming alarmists have reinvented CO_2 as a noxious, deadly pollutant. For example, NOAA senior scientist Susan Solomon recently commented about carbon dioxide, "I think you have to think about this stuff as more like nuclear waste."[2] In his book, Gore also declared, "Unfortunately, the most common technologies for converting energy into usable forms of power happen to release enormous quantities of pollutants, including most prominently the growing concentrations of carbon dioxide (CO_2) now circling the earth."[3]

An 'inconvenient truth' for Gore and his ilk is that carbon dioxide is harmless in concentrations anywhere near current or projected atmospheric concentrations. Currently, the atmosphere consists of 386 parts per million CO_2. Global warming alarmists are horrified by the prospects of 700 to 800 parts per million. But according to the United States Environmental Protection Agency (EPA), CO_2 poses no human health risks below 10,000-15,000 parts per million, more than thirty times current atmospheric levels. According to Gore's logic, water should also be classified as a pollutant. After all, consuming thirty times more water than normal is indeed dangerous. So just how did CO_2 become so demonized? In the words of Richard Lindzen, "Controlling carbon is a bureaucrat's dream. If you control carbon, you control life."[4]

Wanting desperately to impose a global carbon tax, the New World Order must convince the public that breathing (exhaling CO_2) is actually *harmful* to the environment. As discussed in previous chapters, the environmental movement is intimately linked to the eugenics movement. And eugenicists incessantly monitor population levels and periodically "cull the herd" according to their preferences. Even before the advent of global warming hysteria, eugenicists were cleverly advocating population reduction under the guise of saving humanity from overconsumption of food, minerals, water, and other resources. To these standby excuses, they are now impregnating society with the idea, in the words of the Club of Rome, that "the real enemy is humanity itself." The very existence of humans, with their ceaseless exhalation of carbon dioxide, is detrimental to the environment. Therefore, each person must atone for his existence through various draconian carbon allowance and carbon taxation schemes.

Western governments are rapidly mobilizing to implement such programs. For example, the British Environment Agency has proposed the implementation of a nationwide "carbon allowance" program. As reported by the *London Telegraph*, the plan "would involve people being issued with a unique number which they would hand over when purchasing products that contribute to their carbon footprint, such as fuel, airline tickets and electricity."[5] Similarly, physicist Hans Joachim Schellnhuber, an IPCC member and the German government's top climate advisor, has called for *worldwide* carbon allowances. As reported by *Der Spiegel*, Schellnhuber "argues that drastic measures must be taken in order to prevent a catastrophe. He is proposing the creation of a CO_2 budget for every person on the planet, regardless whether they live in Berlin or Beijing."[6]

In 2009, the Swedish National Food Administration issued new food guidelines giving equal weight to climate and health. Accordingly, Swedes should now favor carrots over tomatoes or cucumbers because, supposedly, carrots have a lower carbon footprint. The Swedish government is also implementing new labeling standards requiring listings of CO_2 emissions associated with food products.[7]

In May 2009, House Speaker Nancy Pelosi declared that because of climate change, "Every aspect of our lives must be subjected to an inventory...of how we are taking responsibility."[8] Finally, and most outrageously, in December 2009, the EPA declared CO_2 a deadly toxin. As reported by the *Los Angeles Times*, "Under the so-called 'endangerment finding,' the Environmental Protection Agency asserts the power to regulate carbon dioxide and other gases that scientists blame for global warming."[9] An anonymous Obama administration official told the *Wall Street Journal* that should Congress fail to pass acceptable climate legislation, the EPA would "regulate [carbon] in a command-and-control way."[10] So what, if any, evidence do these alarmists have that elevated CO_2 concentrations are destroying the planet?

The crux of their argument is that increasing CO_2 levels lead to increasing temperatures, which trigger drastically elevated sea levels, the northward spread of malaria, droughts, crop failures, more intense storms, and many other doomsday scenarios. There are many problems, however, with this theory. First, as demonstrated via the historical record, higher temperatures have always made living conditions *easier* on Earth. Second, the alarmists have inverted the temperature/CO_2 relationship. In his film, *An Inconvenient Truth*, Al Gore cleverly displayed two graphs, one above the other. The first showed atmospheric CO_2 levels stretching back hundreds of thousands of years. The second showed global average temperatures for the same period (Figure 7). Gore condescendingly explained, "The relationship is actually very complicated. But there is one relationship that is far more powerful than all the others, and it is this: when there is more CO_2, the temperature gets warmer, because it traps more heat from the sun."[11] However, Gore intentionally kept the two graphs separate. This sleight of hand creates the illusion that CO_2 is driving temperature. Nevertheless, superimposing one graph on top of the other effectively shatters this illusion. It shows that *temperature* drives CO_2. When temperatures increase, CO_2 *follows* (Figure 8). The lag time is roughly 800 years. Gore borrowed the split-graph concept from United Nations Environment Program (UNEP), which published a similar version in *Nature* in 1999. Christopher Monckton, a former policy advisor to Margaret Thatcher, referring to the UNEP graph in an article for the *London Telegraph*, wrote,

"It displays two 450,000-year graphs: a sawtooth curve of temperature and a sawtooth of airborne CO_2 that's scaled to look similar. Usually, similar curves are superimposed for comparison. The UN didn't do that. If it had, the truth would have shown: the changes in temperature preceded the changes in CO_2 levels."[12]

Laurie David, the ex-wife of *Seinfeld* creator Larry David, is another global warming alarmist who has also been caught pushing deceptive graphs. Besides being an Executive Producer for Al Gore's film, she has also been a trustee for the Natural Resources Defense Council, an organization whose false environmentalism was detailed in Chapter 16. In 2007, she published *The Down-to-Earth Guide to Global Warming*, a book targeting children with global warming propaganda. David included a single graph with CO_2 and temperature together, one superimposed on top of the other. She explained to her young readers, "The more carbon dioxide in the atmosphere, the higher the temperature climbed. The less carbon

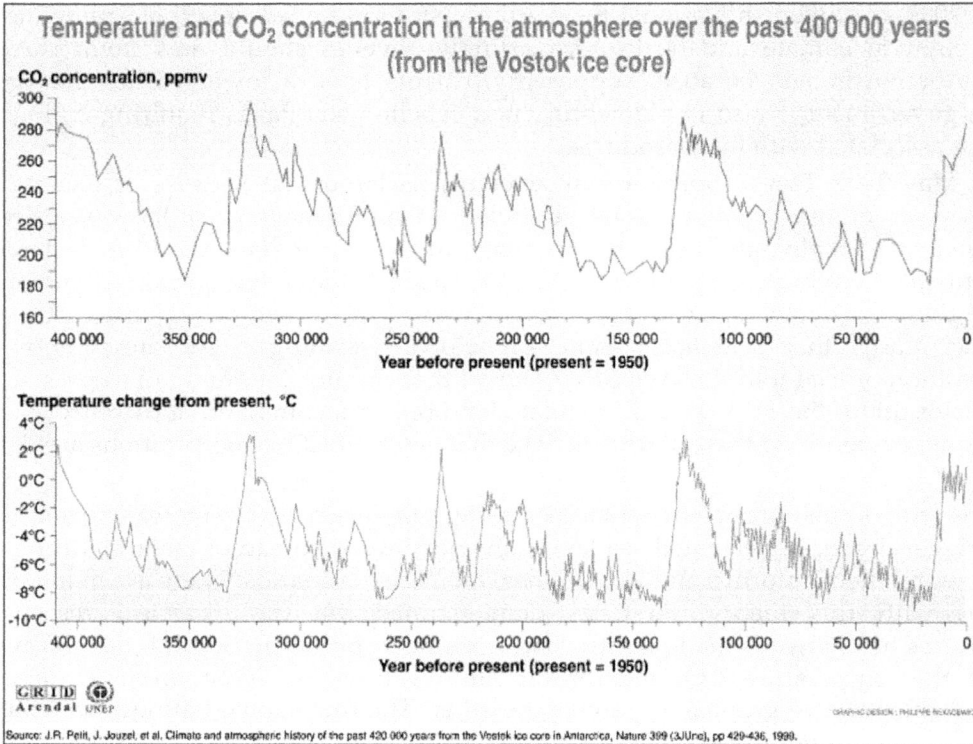

Figure 7: UNEP graph used by the IPCC, Al Gore and others to suggest that CO_2 drives temperature. Superimposing one graph upon the other shows that temperature drives CO_2. [13]

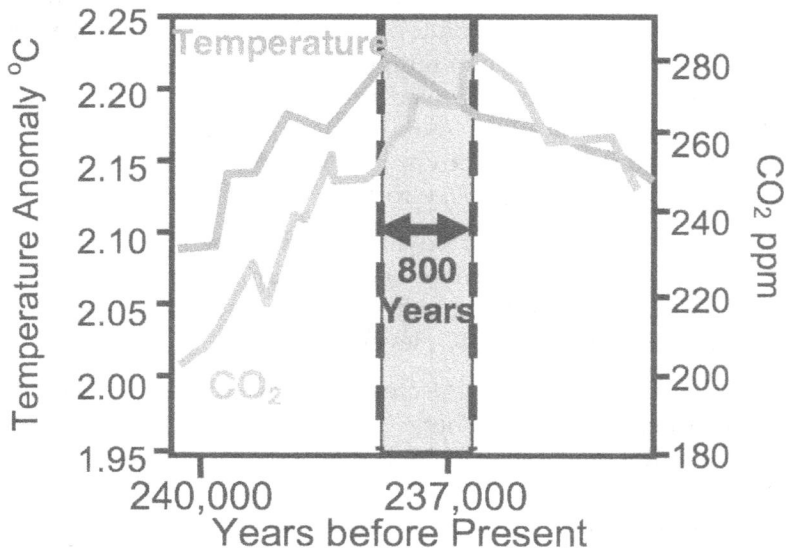

Figure 8: Ice core data shows that CO_2 generally lags 800 years behind temperature. Temperature drives CO_2, despite the claims of Al Gore, Laurie David and other alarmists. [14]

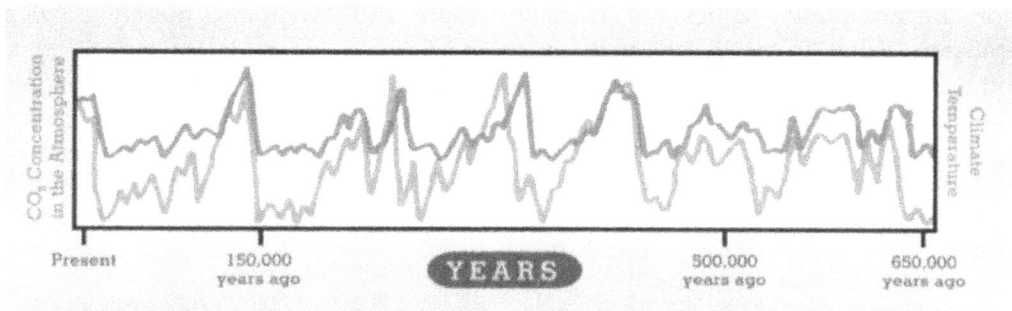

Figure 9: Graph from Laurie David's book *The Down-to-earth Guide to Global Warming* mislabels the upper line as CO_2 and the lower line as temperature, suggesting that CO_2 leads temperature (graph reads right to left).[15]

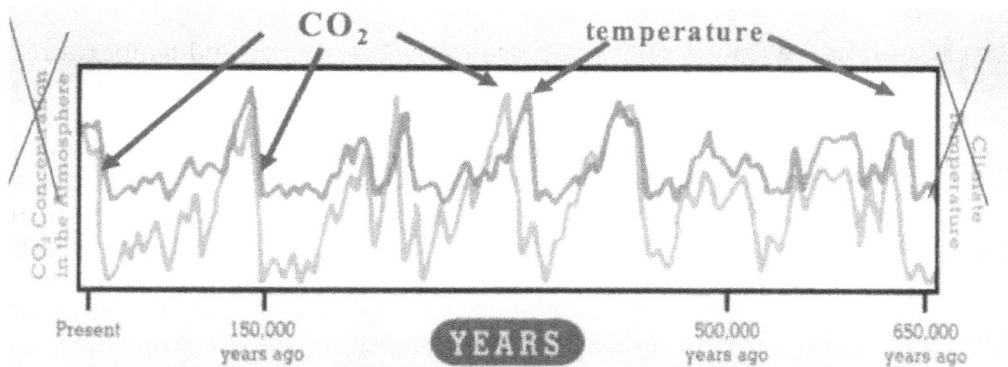

Figure 10: Corrected version of Laurie David's graph showing how temperature actually leads CO_2 (graph reads right to left).[16]

dioxide, the more the temperature fell. You can see this relationship for yourself by looking at the graph. What makes this graph so amazing is that by connecting rising CO_2 to rising temperatures, scientists have discovered the link between greenhouse-gas pollution and global warming."[17]

As noted by the Science and Public Policy Institute (SPPI), however, "What really makes their graph 'amazing' is that it's dead wrong." SPPI discovered that David had done the unthinkable. She had reversed the graphs, labeling CO_2 as temperature, and temperature as CO_2. The entire premise of her book was that CO_2 drives temperature, yet her "amazing" proof was a fraudulent graph! David cited a 2005 *Science* article by Urs Siegenthaler as her source for the graph. Yet as SPPI reported, Siegenthaler's article clearly stated, "The lags of CO_2 with respect to the Antarctic temperature over glacial terminations V to VII are 800, 1600, and 2800 years, respectively, which are consistent with earlier observations during the last four glacial cycles."[18] The hubris of Gore and David is astonishing. Had their errors been genuine mistakes, they could have simply apologized. But they can't. They have completely invested themselves in this myth. Even fraud and deception are acceptable tactics for them. But Gore and David are merely advocates, policy-pushers. What do actual scientists say about CO_2?

When temperature goes up, CO_2 follows. Hubertus Fischer and his team from the Scripps Institute of Oceanography analyzed ice core records from Antarctica's Vostok Station stretching back 250,000 years. Additionally, they cross-correlated this analysis with CO_2 records from Antarctica's Taylor Dome, covering 35,000 years, to construct a temperature-CO_2 history, broken down by decade. They reported in *Science*, "The time lag of the rise in CO_2 concentrations with respect to temperature is on the order of 400 to 1,000 years during all three glacial-interglacial transitions."[19]

Nicolas Caillon, a scientist from the French Atomic Energy Commission, also measured Antarctic ice cores, but with a different method (argon isotopes), which he thought was more accurate. While he calculated the lag to be 200-800 years, he nevertheless reached the same conclusion—temperature drives CO_2. Publishing his results in *Science*, Caillon stated, "This confirms that CO_2 is *not* the forcing that initially drives the climatic system during a deglaciation."[20]

The Ocean-Iron Feedback loop is one reason why CO_2 lags behind temperature. Singer and Avery explained, "During cold, dry ice ages, huge tracts of forest and prairie turn to desert. Iron dust from Patagonia, for instance, would blow into the iron-starved waters of the Southern Ocean. Then a surge in phytoplankton growth would pull large amounts of CO_2 from the air. This feedback loop amplifies the global cooling trend. The oceans hold more than 70 times as much CO_2 as the atmosphere, and warm water holds less of any gas than cold. Thus, when the sun warms the oceans, they release more CO_2 to the air."[21] The CO_2 interchange between the atmosphere and the oceans is natural, desirable and unending. The sun is the primary driver of global temperature. CO_2 follows, lagging hundreds of years behind. The IPCC, however, claims that CO_2 *drives* temperature. Their greenhouse warming models suggest that CO_2 released from the oceans, part of the natural, eternal exchange with the atmosphere, is actually *accelerating* global warming. Tom Segalstad disagrees.

Head of the Geological Museum of the University of Oslo, Segalstad flatly rejects the IPCC's CO_2-driven, greenhouse warming models. An Expert Reviewer for the 3rd Assessment Report, Segalstad also rejects the IPCC's alarmism and questions their understanding of basic geological processes. He remarked, "The IPCC needs a lesson in geology to avoid making fundamental mistakes...Most leading geologists, throughout the world, know that the IPCC's view of Earth processes are implausible if not impossible."[22] He explained that real world measurements of atmospheric CO_2 retention directly refute CO_2-driven warming models, which depend on CO_2 staying in the atmosphere for long periods of time.

Geologists contend that CO_2 cannot remain longer than 10 years because of the magnificent ability of the oceans to reabsorb CO_2. Nevertheless, IPCC modeling presupposes much longer atmospheric CO_2 retention than scientists have *ever* observed. While the physical evidence actually *refutes* their modeling, the IPCC is remarkably nonchalant. Segalstad explained, "They simply dismiss evidence that is, for all intents and purposes, irrefutable. Instead, they substitute their faith, constructing a kind of science fiction or fantasy world in the process."[23]

The Earth is constantly balancing atmospheric and oceanic CO_2 levels to an approximate 1 to 50 ratio. Highlighting the absurdity of the IPCC's position, Segalstad noted, "The IPCC postulates an atmospheric doubling of CO_2, meaning that the oceans would need to receive 50 times more CO_2 to obtain chemical equilibrium...This total of 51 times the present amount of carbon in atmospheric CO_2 exceeds the known reserves of fossil carbon—it represents more carbon than exists in all the coal, gas, and oil that we can exploit anywhere in the world."[24] Because atmospheric CO_2 measurements don't agree with IPCC models, the IPCC postulates the existence of an undiscovered "missing sink" which is holding the extra CO_2 demanded by their models. Segalstad concluded, "It is a search for a mythical CO_2 sink to explain an immeasurable CO_2 lifetime to fit a hypothetical CO_2 computer model that purports to show that an impossible amount of fossil fuel burning is heating the atmosphere."[25] The "missing sink" theory is just one of many problems with the CO_2-driven, greenhouse warming models.

The Greenhouse Effect

The greenhouse effect is real. Without it, the world would be uninhabitable due to extremely cold nighttime temperatures. However, there is a difference between the greenhouse effect and greenhouse warming theory. While the warming theory is based on the effect, real world observations of the greenhouse effect actually disprove the warming theory. According to the warming theory, greenhouse gases accumulate in the atmosphere, trapping more and more of Earth's radiant heat. In turn, the lower atmosphere slowly warms, causing surface temperatures to rise. Therefore, if human-emitted CO_2 is causing global warming, the lower atmosphere must warm *first*. However, both satellite and high-altitude weather balloon data confirms the lower atmosphere is *not* trapping excess heat. Singer and Avery explained that surface temperature "is warming faster than the lower atmosphere where the CO_2 is accumulating. This is strong evidence that CO_2 is not the primary climate factor."[26]

The greenhouse theory also predicts that CO_2-driven warming will start, and be strongest, at the northern and southern poles. This is not happening. The thin finger extending from Antarctica upwards towards Argentina has been warming, but temperatures over the remaining 98 percent of the continent have been steadily declining since the 1960s according to both satellite and surface station measurements.[27] The University of Chicago's Peter Doran explained, "Although previous reports suggest slight recent continental warming, our spatial analysis of Antarctic meteorological data demonstrates a net cooling on the Antarctic continent between 1966 and 2000."[28] Regarding the North Pole, meteorological stations broadly scattered across the region show that temperatures in the Arctic, Greenland, and the surrounding seas are *colder* today than in the 1930s.[29]

Much Ado About Nothing

Patrick Michaels, a climatologist and Expert Reviewer for the IPCC's 1st Assessment Report, contends that current CO_2 levels are not unusually high. If anything they are unusually low. Michaels explained, "Throughout most of the past billion years, the CO_2 concentration of the atmosphere has been greater than it is today. The same is also true for most of the past 100 million years, which is the period during which most of our food and fiber crops evolved. Only since the beginning of the ice ages, some 5 million years ago, have temperatures and atmospheric CO_2 fallen to current levels...Thus, from the perspective of both geological and evolutionary history, the atmosphere is currently impoverished in CO_2. An additional historical peculiarity is that gas bubbles trapped in Antarctic ice tell us that the temperature dropped before the CO_2 concentration changed, not after."[30]

Based on geological history, Yale geology professor Robert Berner has shown that warmer temperatures and higher CO_2 levels are *better* for life. For example, during the period from 300 million years ago until 30 million years ago, average CO_2 concentrations were *five times* greater than today.[31] Earth was then a tropical paradise, teeming with life including dinosaurs, large marine reptiles, and the primitive ancestors of many modern plants, mammals and fish. For most of Earth's history since the Cambrian era, beginning roughly 600 million years ago, the planet has been warmer than today and the poles have had little to no ice. The exceptions were the two major ice ages, both of which coincided with unusually low atmospheric CO_2 levels.[32]

Michaels explained that CO_2 is a life-giving gas and certainly nothing to fear: "Plants take in CO_2 and fix it in the form of carbohydrates in their roots, stems, and leaves. CO_2 in current concentrations is what is known as a 'limiting nutrient': there is currently so little of it in the atmosphere that plants cannot get enough. Increasing the concentration increases the growth of almost all plant species, and both laboratory and field experiments have demonstrated that plants flourish as CO_2 concentration goes up."[33] So as CO_2 levels increase, trees and plants increasingly absorb and convert CO_2 into stored energy. Michaels further explained, "As leaves are shed and ultimately decompose, an increment of CO_2 is released back into the atmosphere, only to be recaptured during the next growing season. That annual cycle represents the breathing of the photosynthesizing biosphere, which requires CO_2 for life. The breathing of the earth has begun to deepen as a result of increasing concentrations of CO_2; one interpretation is that the planet is becoming greener...The finding, which indicates that plants are taking more CO_2 than they did, should not be surprising, because the atmosphere is merely returning to CO_2 levels that were characteristic during the evolutionary history of almost all terrestrial plants."[34]

Sherwood Idso, a former research physicist with the U.S. Department of Agriculture and current President of the Center for the Study of Carbon Dioxide and Global Change, agrees that CO_2 *benefits* life. Idso and his two sons have published extensively about the benefits of CO_2 for both plants and animals. CO_2 is a natural fertilizer for plants, especially in warmer regions of the world. Increasing levels of CO_2 makes the process of photorespiration more efficient, thus saving plants' energy and enabling them to grow more vigorously.[35] A doubling of CO_2 raises the productivity of herbaceous plants by 30 to 50 percent and trees and woody plants by 50 to 80 percent.[36] Higher temperatures also benefit both plants and animals. As temperatures increase, plant and animal species can effectively broaden their inhabitable range. Idso explained, "Consequently, individual animal species, like individual plant species, have measurably increased the areas of the planet's surface that they occupy, creating more overlapping of ranges, greater local species richness, and an improved ability to avoid extinction." [37]

This of course favorably impacts agricultural prospects. Danish agricultural expert Jorgen Olesen of the Danish Institute of Agricultural Sciences has predicted that despite aridity, which causes crop reductions in some parts of southern Europe, global warming will have an overall positive effect in terms of agricultural production throughout the whole of Europe.[38] Furthermore, in 1997 the U.S. Department of Agriculture demonstrated that an additional 100 parts per million of CO_2 would increase wheat production by 72 percent under sufficiently watered conditions, and by 48 percent under semi-drought conditions, for an average crop yield gain of 60 percent.[39] Singer and Avery commented, "These results are consistent with a wide range of CO_2 enrichment studies done in more than a dozen countries on many different crops."[40]

Chapter 20

Here Comes the Sun King

Long before the current 6.5 billion human population, before the industrial revolution, before the factories and the SUVs, Earth was undergoing regular warming and cooling periods driven not by carbon dioxide, but by the giant, raging inferno high in the sky—the Sun. Despite the assertions of Al Gore and the IPCC, serious scientists know, both through common sense and meticulous scientific research, that the Sun is the most influential and dominant driver of climate change, not just on Earth, but throughout the solar system. In 1996, Lloyd Keigwin, a Senior Scientist from the Massachusetts Woods Hole Oceanographic Institution, summarized the current scientific understanding regarding the solar-climate link. In a study published in *Science*, Keigwin wrote, "The solar-climate links implied by our record are so dominant over the last 12,000 years that it seems almost certain that the well-documented connection between the Maunder Solar Minimum and cold decades of the Little Ice Age could not have been a coincidence."[1]

Five years later, Gerard Bond, a geologist from New York's Lamont-Doherty Earth Observatory, published a study in *Science* proving this striking correlation extends even further back, at least *32,000* years.[2] Commenting on the paper, glaciologist Richard Alley from Pennsylvania State University remarked, "The Bond et al., data are sufficiently convincing that [solar variability] is now the leading hypothesis to explain the roughly 1,500-year oscillation of the climate seen since the last Ice Age, including the Little Ice Age of the 17th century."[3]

Back in 1991, Eigil Friis-Christensen, director of the Danish National Space Center, coauthored the first peer-reviewed paper suggesting a correlation between solar activity and global climate change. Friis-Christensen and Knud Lassen, both scientists at the Danish Meteorological Institute in Copenhagen, published their article in *Science*. The *New York Times* reported,

> "Dr. Friis-Christensen and Dr. Lassen analyzed the lengths of sunspot cycles, which vary from 9 to 13 years with an average of 11 years. Shorter cycles, they said, are associated statistically with more intense solar activity. The variations in cycle length, the scientists found, almost precisely match the fluctuations in Northern Hemisphere temperatures from the mid-1850's to the mid-1980's."[4]

Their compelling article generated considerable interest within the scientific community. "While the correlation established by Dr. Friis-Christensen and Dr. Lassen falls short of definite proof, a number of scientists nevertheless called it remarkable in its close fit between the solar and temperature trends. The findings were termed 'a major advance' by Dr. Robert Jastrow, a professor of earth sciences at Dartmouth College who is also the president of the George C. Marshall Institute in Washington," the *Times* reported.

In January 1992, Friis-Christensen traveled to Guangzhou, China as part of the Danish delegation for an IPCC Climate Convention. Naturally, he expected the IPCC to give serious consideration to his work. Yet as Canada's *National Post* reported,

> "To his astonishment, and despite the recent publication of his *Science* article, the IPCC refused to consider the sun's influence on Earth's climate as a topic worthy of investigation. The scientists at the IPCC had decided that man-made causes and man-

made causes alone deserved their attention. But ignoring the potential role of the sun didn't make it go away, especially since Dr. Friis-Christensen and other solar scientists refused to abandon their research."[5]

Today, the Danish National Space Center, the European Organization for Nuclear Research in Geneva, the Max Planck Institute for Solar System Research in Germany, and the Pulkovo Astronomical Observatory in Russia are among the many space-related research institutions studying the Sun's role in climate change.

Sami Solanki, the director of Germany's world-renowned Max Planck Institute for Solar System Research, studies the solar-climate link fastidiously. In 2004, he led a team of scientists that quantitatively reconstructed the past 11,400 years of solar activity. Their study, published in *Nature*, concluded that since 1940, solar activity has been higher than during any time in the past 1,000 years, and two and a half times higher than the long-term average.[6] Solar activity refers to sunspots, flares, and eruptions, which project huge solar winds toward Earth and the other planets. How do changes in solar activity influence Earth's climate? Singer and Avery explained, "The key amplifier is cosmic rays. The sun sends out a 'solar wind' that protects the Earth from some of the cosmic rays bombarding the rest of the universe. When the sun is weak, however, more of the cosmic rays get through to the Earth's atmosphere. There, they ionize air molecules and create cloud nuclei. These nuclei then produce low, wet clouds that reflect solar radiation back into outer space. This cools the Earth."[7] This relationship is further explored below.

According to Solanki, 8,000 years ago the sun went through several short periods of intense activity similar to that of recent years. *Nature News* explained, "This suggests that the current high is part of the Sun's normal activity and that it will probably calm down again, returning to normal levels within the next few decades."[8] Solanki's study is also consistent with the historical astronomical record. Since the early 17th century, astronomers have regularly observed the sun by telescope. Their records, confirmed by Solanki, show that solar activity was extremely low from 1650 to 1700, a period known as the Maunder Minimum. *Nature News* explained, "It coincided with the coldest part of a chilly period on Earth called the Little Ice Age, and some scientists have speculated that this provides evidence that the Sun's activity significantly affects our climate."[9]

Prominent environmentalist David Bellamy, founder of the Conservation Foundation applauded Solanki's efforts, noting,

> "Global warming—at least the modern nightmare version—is a myth. I am sure of it and so are a growing number of scientists. But what is really worrying is that the world's politicians and policy-makers are not. Instead, they have an unshakeable faith in what has, unfortunately, become one of the central credos of the environmental movement: humans burn fossil fuels, which release increased levels of carbon dioxide—the principal so-called greenhouse gas—into the atmosphere, causing the atmosphere to heat up. They say this is global warming: I say this is poppycock."[10]

A victim of the rampant politicization of the environmental movement, after making these and other such comments, Bellamy lost his presidencies of both Plantlife International and the Royal Society of Wildlife Trusts. The *London Times* reported, "They have been acutely embarrassed to discover that while they have been campaigning to raise awareness, their president has been leading seminars and writing articles in science magazines declaring that man-made warming is a myth."[11]

Cosmic Influence

The Sun is not the only star influencing Earth's climate; distant stars are also highly significant. Both exploding supernovas and collisions of star remnants with nebula create massive amounts of cosmic rays. These rays then dissipate throughout the galaxy with some inevitably reaching Earth. The magnetic field of the Sun, however, deflects many cosmic rays back into space. When the Sun is active, strong solar winds shield the Earth from cosmic radiation. When the Sun is less active, more rays penetrate Earth's atmosphere. While observing satellite data in 1997, Henrik Svensmark, the head of Sun-climate research at the Danish National Space Center, noticed a positive correlation between cloud cover and the amount of incoming cosmic rays. When cosmic rays increased, Earth became cloudier. More clouds led to cooler temperatures, whereas fewer clouds led to warmer temperatures.[12]

Svensmark reasoned that both the Sun *and* distant stars profoundly influence global temperatures on Earth. During the Little Ice Age, the Earth was cooling because the Sun was less active. Therefore, more cosmic rays were reaching Earth. Thus more clouds formed, making the planet cooler. During the past seventy years, the Earth has been warming because the Sun has been more active. Fewer cosmic rays have been reaching the Earth, fewer clouds have been forming, and thus Earth has been warming. Svensmark explained,

> "I think the Sun is the major driver of climate change, and the reason I'm saying that is that if you look at historical temperature data and then solar activity and cosmic ray activity, it actually fits very beautifully. If CO_2 is a very important climate driver then you would expect to see its effect on all timescales; and for example when you look at the last 500 million years, or the last 10,000 years, the correlation between changes in CO_2 and climate are very poor."[13]

Although Svensmark's theory was viable, it was somewhat vulnerable because meteorologists and other scientists never before thought that cosmic rays could influence cloud formation. The *National Post* commented, "Exactly how cosmic rays might create clouds was a mystery—unprovable theory, many said. Some even claimed that it was inconceivable for cosmic rays to influence cloud cover."[14] In 2005, however, everything changed. With a team from the Danish National Space Center, Svensmark conducted experiments strongly supportive of his hypothesis. Singer and Avery explained, "They put realistic mixtures of earth's atmospheric gases into a large reaction chamber, with ultraviolet light as a stand-in for the sun. When they turned on the UV, microscopic droplets of water and sulphuric acid— cloud seeds—started floating through the chamber."[15]

In 2007, Svensmark published his orthodox-shattering study in the peer-reviewed journal *Astronomy & Geophysics*. He remarked, "Variations in the cosmic-ray influx due to solar magnetic activity account well for climatic fluctuations on decadal, centennial and millennial timescales. Over longer intervals, the changing galactic environment of the solar system has had dramatic consequences, including Snowball Earth episodes."[16] He concluded,

> "The physics of the Sun and the heliosphere runs through the story on all timescales from the early Earth to the present day. Whatever the verdict may be about the relative importance of cosmic rays and greenhouse gases in current and future climate change, there is an obvious need to predict future solar behaviour better, by clearer observations of the magnetic field at the Sun's poles."

CERN, the massive European particle-physics facility based in Geneva, has initiated a project called CLOUD to test and possibly confirm Svensmark's hypothesis. The BBC noted, "If that happens, he can expect to see a Nobel Prize and thousands of red-faced former IPCC members queuing up to hand back the one they have just received."[17]

Warm Solar Neighborhood

Another major inconvenient truth for manmade global warming enthusiasts is the warming trends currently happening on other planets. Are human CO_2 emissions causing Mars and Jupiter to warm? Is humanity's carbon footprint really so vast? Or does the Sun have something to do with climate change? Despite the alarmists' efforts to downplay this embarrassing information, the news has nevertheless spread worldwide. In 2001, for example, a coalition of Australian MPs noted, "Another problem with the view that it is anthropogenic greenhouse gases that have caused warming is that warming has also been observed on Mars, Jupiter, Triton, Pluto, Neptune and others."[18] In 2007, the *Taipei Times* reported, "This is a cosmic conspiracy — for if Jupiter and Triton are heating up, then the cause cannot be human. Our polluting influence does not extend to the solar system's perimeter. Only the sun has that kind of impact. So forget carbon emissions. Blame rising solar radiation for all global warming."[19] In 2009, the *Jerusalem Post* reported, "Mars, Jupiter and Pluto are currently experiencing warming, presumably without any contribution of human-produced fluorocarbons."[20] And in 2009, the *Denver Post* remarked, "While the climate was warming on Earth, 'global warming' was also occurring on Mars, Jupiter, Neptune's largest moon and even on Pluto. Obviously, there are far-reaching effects from CO_2 levels on Earth."[21] The secret is out. So what have the alarmists to say about this?

In defense of their dogma, some alarmists have resorted to absurdly delusional claims. David Mayer de Rothschild, from the notorious Rothschild banking dynasty, is a prominent environmentalist and author of *The Live Earth Global Warming Survival Handbook*. In 2007, Rothschild appeared on the nationally syndicated radio program *The Alex Jones Show*. Jones asked Rothschild, "The polar ice caps of Mars are receding several miles per year, much faster than ours. And the moons of Saturn and Jupiter are melting. In fact, several of their moons were ice and are now liquid seas. Now, how are SUVs causing that David Rothschild?" Rothschild confidently chuckled, "That's because those planets are closer to the Sun, my friend."[22] Rothschild epitomizes the environmental leadership. Drunk on arrogance and deceit, they presume the public will believe whatever they're told to believe. Jupiter is closer to the Sun? Okay Rothschild, whatever you say. Unchecked CO_2 emissions will destroy the Earth? Okay, you're the expert. But why do so many other experts disagree?

2005 data from NASA's Mars Global Surveyor and Odyssey missions revealed that Mars' carbon dioxide "ice caps" are steadily diminishing. Habibullo Abdussamatov, head of the International Space Station's Astrometry project and head of space research for the Russian Academies of Sciences' Pulkovo Observatory, believes the Sun is causing both the Martian warming *and* our warming here on Earth. In an article for *National Geographic News*, he explained, "The long-term increase in solar irradiance is heating both Earth and Mars. Man-made greenhouse warming has made a small contribution to the warming seen on Earth in recent years, but it cannot compete with the increase in solar irradiance."[23] Abdussamatov estimates that solar irradiance will bottom out by approximately 2040, causing steep cooling on Earth. The Russian and Ukrainian space agencies, under Abdussamatov's leadership, have launched a joint project to study the duration and extent of this anticipated cooling.

Chapter 21
What Consensus?

In 1992, while the global warming debate was only just beginning, Al Gore boldly declared it over. "Only an insignificant fraction of scientists," Gore claimed, "deny the global warming crisis. The time for debate is over. The science is settled."[1] Gore's mantra has come to symbolize an intense, politically charged, multi-decadal fear and propaganda campaign targeting naive and unsuspecting populations across the planet. Since 1992, Gore has not dampened his rhetoric. If anything, his attacks and unsubstantiated claims have greatly intensified. In 2007, for example, he discussed the "*very* strong scientific consensus that is as strong a consensus as you'll ever see in science that the climate crisis is real [and] human beings are responsible for it."[2] As strong as you'll ever see in science, he claims? That's some consensus, as irrefutable, apparently, as the laws of gravity, motion, and thermodynamics!

Did Gore's early 1990s statements have any merit at the time? Do they now? The short answer is a resounding *no*. At the time of his original declaration of consensus, a *Gallup* poll of climate scientists from the American Meteorological Society and the American Geophysical Union showed the vast majority was skeptical about manmade global warming claims. 49% refuted these claims, 33% were unsure, and only 18% believed some manmade global warming had occurred.[3] Furthermore, a Greenpeace poll revealed that 47% of climatologists didn't think runaway greenhouse warming was imminent, 36% thought it was possible, and only 13% thought it was probable.[4]

While thousands of scientists agree with manmade global warming theory, thousands also disagree. Among those who agree, however, only a minute cadre clings to the ridiculously indefensible notion of consensus. They are a small, outspoken group of zealots, scientific minions of the financial and political elite. So how do they manage to justify and proliferate this lofty claim of consensus?

The Oreskes Study

Many alarmists point to Naomi Oreskes' 2004 metanalysis, published by *Science*, as unequivocal proof of the scientific consensus. Oreskes analyzed "928 abstracts, published in refereed scientific journals between 1993 and 2003, and listed in the ISI database with the keyword 'climate change.'"[5] In his film, Al Gore referred to Oreskes' study when he said, "There was a massive study of every scientific article in a peer reviewed journal written on global warming for the last ten years. They took a big sample of ten percent, 928 articles. And you know the number of those that disagreed with the scientific consensus that we're causing global warming and that it's a serious problem? Out of the 928 — zero."[6] However, what Gore neglected to mention is that two scientists, independent of each other, both investigated Oreskes' study, both found evidence of massive deceit, both wrote rebuttal letters to the editor of *Science*, and both had their letters flatly rejected.

Notice the clever wording of Gore's statement: "the number of those that disagreed with the scientific consensus." He leads his audience to believe that just because someone didn't disagree, then they must have agreed. But what about those studies where manmade global warming was never even mentioned? What about studies where the author expressed neither agreement nor disagreement? By Gore's clever wording, any scientist who published any study including both (a) the term 'climate change,' and (b) no explicit disagreement with the hypothesis of manmade global warming, suddenly became a supporter of the manmade global warming theory.

Oreskes' study raised many eyebrows in the scientific community. For example, professor Dennis Bray of the GKSS National Research Centre in Geesthacht (Germany), skeptical of her results, conducted his own metanalysis using similar criteria. He found that fewer than 10 percent of climate scientists believe climate change is principally caused by human activity. Yet when Gray sent a rebuttal letter to *Science*, they rejected it for reasons best described as anti-scientific. Bray told the *London Telegraph*, "They said it didn't fit with what they were intending to publish."[7]

Dr. Benny Peiser, a professor at Liverpool John Moores University and a fellow of the Royal Astronomical Society, also questioned Oreskes' results. Using the same input parameters, he recreated Oreskes' study and found that the vast majority of articles didn't even mention manmade global warming. Only one third *implicitly* backed the consensus view, and only one percent did so *explicitly*.[8] The *Telegraph* reported, "Dr. Peiser submitted his findings to *Science* in January, and was asked to edit his paper for publication—but has now been told that his results have been rejected on the grounds that the points he makes had been 'widely dispersed on the internet.'"[9] Furthermore, a spokesperson for *Science* told the *Telegraph* that Peiser's rebuttal was rejected "for a variety of reasons" and that "the information in the letter was not perceived to be novel."[10] With such ridiculous excuses, *Science* was not even attempting to conceal their obvious bias and censorship.

Professor Roy Spencer of the University of Alabama, a leading authority on satellite measurements of global temperatures, also experienced firsthand *Science's* overt bias. He stated, "It's pretty clear that the editorial board of *Science* is more interested in promoting papers that are pro-global warming. It's the news value that is most important."[11] Even Donald Kennedy, then the Editor-in-Chief of *Science*, wrote an extremely slanted editorial just six months before Oreskes' article. Kennedy argued that manmade global warming was certain; the only question was how severe was the danger. He also accused many publications of downplaying the seriousness of the situation. Kennedy declared, "The results are everywhere, except in popular accounts of what's going on. Those, unfortunately, often emphasize distant possibilities rather than probable outcomes."[12]

Institutional Endorsements

Several vocal front men, Al Gore and James Hansen for example, actively promote the myth of scientific consensus. Organizations, institutions, and academies, however, are the functional backbone of this myth. By projecting their perceived authority, they elicit passivity and submission from the public. Such reactions, however, actually *invite* further manipulation. As demonstrated in Part I, hierarchy and compartmentalization create conditions whereby great lies can flourish. Within most organizations, for example, small boards or directorships preside over policy. Therefore, the leadership can use *official* declarations and endorsements to advance the myth of scientific consensus. They can declare, unambiguously, that manmade global warming is real, that all scientists agree, and that immediate action is necessary. With such declarations, they create the perception that *individual* members within these organizations also support the *official* declarations. Meanwhile, dissenting members are coerced into silence because speaking out oftentimes means losing their research funding. The IPCC is so committed to the myth of manmade global warming that they refuse to acknowledge the scores of scientists (from within their own organization even) who have publicly questioned the myth. Furthermore, they make every effort possible to conceal from the public this intellectual mutiny. The IPCC is the unrivaled leader of this agenda, yet many other organizations play supporting roles.

The American Association for the Advancement of Science (AAAS) describes itself as an "international non-profit organization dedicated to advancing science around the world by serving as an educator, leader, spokesperson and professional association."[13] Their 10 million "members" are actually just customers. Membership is open to anyone who purchases a subscription to *Science*, the organization's publication. The AAAS is certainly not democratic. The leadership does not poll the membership before making policy decisions. So what does it mean when AAAS declares global warming an imminent threat to humanity? It simply means that a small group of politically compromised board members has taken a position. In December 2006, the AAAS board declared, "The scientific evidence is clear: global climate change caused by human activities is occurring now…The conclusions in this statement reflect the scientific consensus."[14] The board of directors consisted of just fourteen people. Was the board representing the ethos of the organization, or simply their own self-interests? At the time, John Holdren was AAAS Chairman of the Board. As a member of the Council on Foreign Relations, Holdren has long been cozy with the ruling class.[15] From 1993-2001, he served on President Clinton's Committee of Advisors on Science and Technology. Today he is Obama's top Science and Technology advisor, a post from which he aggressively pushes the UN's climate change agenda. Is Holdren a scientist? Is he a politician? Or is he merely a worker-bee for the elite?

If nothing else, Holdren is certainly a eugenicist. In his 1977 book *Ecoscience*, coauthored with Paul Ehrlich, Holdren called for forced abortions, coercive mass sterilization, and a "Planetary Regime" responsible for determining optimal world population levels and enforcing population control, as well as controlling "the development, administration, conservation, and distribution of *all* natural resources, renewable or nonrenewable."[16] According to Holdren, the U.S. Constitution permits forced abortions. He explained, "Indeed, it has been concluded that compulsory population-control laws, even including laws requiring compulsory abortion, could be sustained under the existing Constitution if the population crisis became sufficiently severe to endanger the society."[17] Most shocking though was Holdren's advocacy of "A program of sterilizing women after their second or third child." He explained, "The development of a long-term sterilizing capsule that could be implanted under the skin and removed when pregnancy is desired opens additional possibilities for coercive fertility control. The capsule could be implanted at puberty and might be removable, with official permission, for a limited number of births."[18] He even discussed the possibility of "Adding a sterilant to drinking water or staple foods," so long as it could be "uniformly effective, despite widely varying doses received by individuals, and despite varying degrees of fertility and sensitivity among individuals."[19]

A longtime colleague of Paul Ehrlich, Holdren has been fear-mongering at least since the publication of his 1972 Sierra Club book, *Energy: A Crisis in Power*. Parroting the Club of Rome's *Limits to Growth* doctrine, Holdren asserted, "It is fair to conclude that under almost any assumptions, the supplies of crude petroleum and natural gas are severely limited. The bulk of the energy likely to flow from these sources may have been tapped within the lifetime of many of the present population."[20] In 2006, Holdren suggested that global sea levels might rise thirteen feet by the end of the century. "We are experiencing dangerous human disruption of the global climate and we're going to experience more," Holdren warned.[21] In a 2008 article for the *Boston Globe*, he showcased his arrogance by labeling anyone who dares contradict, or even question, his scientific authority as "denier fringe." He stated, "Members of the public who are tempted to be swayed by the denier fringe should ask themselves how it is possible, if human-caused climate change is just a hoax, that: The leaderships of the national academies of sciences of the United States, United Kingdom, France, Italy, Germany, Japan, Russia, China, and India, among others, are on record saying that global climate change is real, caused mainly by humans, and reason for early, concerted action."[22] His question answers itself. The "leaderships" of these academies include only a handful of politically compromised people.

To understand how a scientific organization can become corrupted and politicized, it is helpful to hear from an insider. Richard Lindzen, an M.I.T. professor and member of the National Academy of Sciences (NAS), often discusses the political burden imposed upon climate scientists. In a 2008 paper, for example, Lindzen wrote about the contamination of the otherwise rigorous nomination process for NAS membership:

"The Academy is divided into many disciplinary sections whose primary task is the nomination of candidates for membership in the Academy. Typically, support by more than 85% of the membership of any section is needed for nomination. However, once a candidate is elected, the candidate is free to affiliate with any section. The vetting procedure is generally rigorous, but for over 20 years, there was a Temporary Nominating Group for the Global Environment to provide a back door for the election of candidates who were environmental activists, bypassing the conventional vetting procedure. Members, so elected, proceeded to join existing sections where they hold a veto power over the election of any scientists unsympathetic to their position. Moreover, they are almost immediately appointed to positions on the executive council, and other influential bodies within the Academy.

One of the members elected via the Temporary Nominating Group, Ralph Cicerone, is now president of the National Academy. Prior to that, he was on the nominating committee for the presidency. It should be added that there is generally only a single candidate for president. Others elected to the NAS via this route include Paul Ehrlich, James Hansen, Steven Schneider, John Holdren and Susan Solomon."[23]

Each of these people is now leveraging the reputation of NAS to aggressively push global warming propaganda. NAS officially endorsed the consensus myth in 2008 with its *Expert Consensus Report*.[24] Other scientific organizations pushing the consensus myth include the American Meteorological Society (AMS) and the Joint Science Academies (JSA).[25]

The AMS consensus endorsement is particularly interesting considering the results of a January 2010 poll of AMS broadcast meteorologists. The study, funded by the National Environmental Education Foundation, showed that only 24 percent of respondents agree with the IPCC's assertion that, "Most of the warming since 1950 is very likely human-induced."[26] Also only 19 percent agree that, "Global climate models are reliable in their projection for a warming of the planet." Joe D'Aleo, Executive Director of the International Climate and Environmental Change Assessment Project and first Director of Meteorology at the Weather Channel, observed,

"AMS has tried very hard to brainwash broadcast meteorologists by forcing them to attend conferences and teleconferences with one-sided presentations where global warming evangelism is preached. Broadcasters send me notifications they get from AMS telling them they must attend these conferences where only the alarmist point of view is preached. This survey shows that broadcast meteorologists are not swayed by these one-sided presentations."[27]

The JSA is a network of national science academies from various nations. For many years, the Joint Science Academies' Statements (JSAS) strongly endorsed the consensus myth. However, in recent years they have backpedalled, somewhat. Sixteen national science academy presidents, mostly European, signed the first JSAS in 2001. The statement proclaimed, "The work of the Intergovernmental Panel on Climate Change (IPCC) represents the consensus of the international scientific community on climate change science."[28] The next JSAS, from 2005, again recognized an "international scientific consensus," however, this time only eleven academy presidents signed.[29] Curiously, the next JSAS, from 2007, dropped all language concerning consensus.[30] Again in 2008 and 2009, they refrained from even mentioning the scientific consensus.[31] Why the change? Were they quietly acknowledging that the scientific community is actually rife with dissenting and skeptical voices? How else can one explain this subtle retraction?

There is no consensus. There never was. A declared or implied consensus without supporting evidence is completely unreasonable, speculative at best, and intentionally deceitful at worst. If Gore insists upon trumpeting his declaration of scientific consensus, why won't he prove its existence? Clearly he has no proof. His position is so tenuous that he consistently refuses invitations for public debate. The fact remains that thousands of scientists have already spoken out, effectively shattering his asinine claims of consensus.

Petitions and Appeals

The myth pushers want to create the illusion of scientific consensus. They substantiate their claims by implication, associating the members of scientific organizations with the views expressed through official statements by boards and directorships. However, thousands of scientists throughout the world have shattered this illusion by using the opposite strategy. Instead of relying on unseen, unnamed scientists, these scientists make their opinions readily known by signing their names to various petitions and appeals, thus proving, unequivocally, that *there is no consensus*. Petitions are genuine benchmarks of opinions within the scientific community. They are explicit and open rather than implicit and discreet. The following petitions prove that scientific consensus, by definition, cannot exist concerning manmade global warming. Furthermore, they suggest that perhaps more scientists than not question man's impact on climate.

In 1998, Arthur Robinson, an accomplished biochemist and founder of the Oregon Institute of Science and Medicine, started the Oregon Petition for scientists opposing the Kyoto Protocol and the overall notion of manmade global warming. Frederick Seitz, the former president of the U.S. National Academy of Sciences, participated by writing the petition's official cover letter. Seitz wrote, "This treaty is, in our opinion, based upon flawed ideas. Research data on climate change do not show that human use of hydrocarbons is harmful. To the contrary, there is good evidence that increased atmospheric carbon dioxide is environmentally helpful."[32]

Since 1998, the petition has spread like wildfire throughout the scientific community, attracting thousands of signatories. In May 2008, the *London Telegraph* reported, "More 31,000 scientists have signed a petition denying that man is responsible for global warming. The academics, including 9,000 with PhDs, claim that greenhouse gases such as carbon dioxide and methane are actually beneficial for the environment."[33] Against 31,000 scientists, Al Gore remains relentlessly committed to his myth of consensus. In 2007 on MSNBC, he declared,

> "The reason the IPCC was awarded the Nobel Peace Prize [is because] the thousands of scientists that make up that group have for almost 20 years now created a *very* strong scientific consensus that is as strong a consensus as you'll ever see in science that the climate crisis is real, human beings are responsible for it, the results would be very bad for the United States and for the entire world community—all human beings—unless we do something about it, and there is still time to solve it."[34]

As strong a consensus as you'll ever see in science? The Oregon Petition is just one of many such petitions. But do any petitions actually support the notion of consensus?

The largest signed statement by scientists urging the reduction of greenhouse gas emissions came in May 2008 from the Union of Concerned Scientists (UCS). Despite its alarmist rhetoric, this petition *did not* declare a scientific consensus. 1,733 scientists and economists endorsed the statement, including 6 Nobel Prize winners, 30 members of the National Academy of Sciences, 10 members of the National Academy of Engineering, and 100 members of the IPCC.[35] This statement, however, is more interesting for who didn't sign. The National Academy of Sciences has 2,100 members and 350 foreign associates.[36]

At the same time the UCS was distributing its petition, the NAS was officially endorsing the myth of consensus. If their endorsement actually represented the members' views, why then did only 30, less than 2% of members, sign the UCS petition?[37] The reason, as discussed above, is because institutional endorsements are poor indicators of the opinions and perspectives of members. They merely reflect the political affiliations of the leadership. The Union of Concerned Scientists applauded itself for attracting 100 members of the IPCC, yet IPCC cheerleaders routinely brag that 2,500 scientists contribute to IPCC reports. Thus only 4% of IPCC scientists felt compelled to sign the UCS petition.

Shortly after Climategate, the British Met Office scrambled to circulate a petition among U.K. scientists "to defend our profession against this unprecedented attack to discredit us and the science of climate change."[38] While this petition also included no claims of consensus, it did insist, "Warming of the climate system is unequivocal" and "Most of the observed increase in global average temperatures since the mid-20th century is very likely due to the observed increase in anthropogenic greenhouse gas concentrations."[39] Although 1,700 scientists agreed to sign, the Met Office may have implicitly pressured them to do so. One scientist anonymously told the *Times*, "The Met Office is a major employer of scientists and has long had a policy of only appointing and working with those who subscribe to their views on man-made global warming."[40]

The Oregon Petition was not the first to question manmade global warming. Following the 1992 Earth Summit in Rio, 425 scientists signed the Heidelberg Appeal, a document opposing environmental hysteria based on unsubstantiated claims. The Heidelberg Appeal included the following statements:

- "We fully subscribe to the objectives of a scientific ecology for a universe whose resources must be taken stock of, monitored and preserved. But we herewith demand that this stock-taking, monitoring and preservation be founded on scientific criteria and not on irrational pre-conceptions."

- "We intend to assert science's responsibility and duty towards society as a whole. We do however forewarn the authorities in charge of our planet's destiny against decisions which are supported by pseudo-scientific arguments or false and non-relevant data."

At the time, the media largely ignored this appeal. Today, more than 4,000 scientists from 106 countries, including 72 Nobel Prize winners, have signed.[41] Nevertheless, the media blackout continues.

In December 2007, 103 distinguished scientists sent a letter to UN Secretary General Ban Ki-moon criticizing the IPCC and concluding, "Attempts to prevent global climate change from occurring are ultimately futile, and constitute a tragic misallocation of resources that would be better spent on humanity's real and pressing problems."[42] In December 2008, the U.S. Senate Environment and Public Works Committee released a 231 page report entitled, "More Than 650 International Scientists Dissent Over Man-Made Global Warming Claims, Scientists Continue to Debunk 'Consensus' in 2008."[43] The report, an update from the 2007 version featuring 400 dissenting scientists, stated, "The over 650 dissenting scientists are more than 12 times the number of UN scientists (52) who authored the media-hyped IPCC 2007 Summary for Policymakers. The chorus of skeptical scientific voices grew louder in 2008 as a steady stream of peer-reviewed studies, analyses, real world data and inconvenient developments challenged the UN's and former Vice President Al Gore's claims that the 'science is settled' and there is a 'consensus.'"

In 2008, the Association of Professional Engineers, Geologists and Geophysicists of Alberta (APEGGA) surveyed their 51,000 members concerning human-influenced climate change. Their results further illustrated the clear non-consensus within the scientific community. The *Edmonton Journal* reported, "A 99-per-cent majority believes the climate is changing." However, "The expert jury is divided, with 26 per cent attributing global warming to human activity." The survey also found that "68 per cent disagree with the popular statement that 'the debate on the scientific causes of recent climate change is settled.'"[44]

On March 4, 2008, the International Conference on Climate Change (subtitled *Global Warming: Truth or Swindle?*) convened in New York City. The *Washington Times* editorialized, "This will be one of the rare opportunities for the media to hear the other side of the story—for those old-fashioned journalists who still believe that their job is to inform the public, rather than promote an agenda."[45] At the conclusion of the conference, 114 scientists and researchers signed The Manhattan Declaration on Climate Change (650 climate scientists or scientists in closely related fields have since signed), which asserted,

- "That current plans to restrict anthropogenic CO_2 emissions are a dangerous misallocation of intellectual capital and resources that should be dedicated to solving humanity's real and serious problems."

- "That there is no convincing evidence that CO_2 emissions from modern industrial activity has in the past, is now, or will in the future cause catastrophic climate change."

- "That all taxes, regulations, and other interventions intended to reduce emissions of CO_2 be abandoned forthwith."[46]

Consensus Busters

More than twenty years ago, Al Gore was already condemning scientists who dared challenge the false consensus. In *Time* magazine in 1989, he wrote, "The fact that we face an ecological crisis without any precedent in historic times is no longer a matter of any dispute worthy of recognition. And those who, for the purpose of maintaining balance in debate, take the contrarian view that there is significant uncertainty about whether it's real are hurting our ability to respond."[47] Why must the alarmists constantly reassert their claims of consensus? Why can't the science speak for itself? If there really were a consensus, wouldn't it be obvious? And wouldn't speaking about the consensus be redundant and awkward? For example, imagine Gore lecturing from his pulpit, "There is an extremely strong scientific consensus that water consists of two molecules of hydrogen and one molecule of oxygen. The debate is over." Such a statement would be superfluous. A consensus indicates an extremely high degree of certainty attained through scientific research. Any declared consensus absent such certainty is simply the rhetoric of alarmists.

Michael Crichton, the award-winning science fiction writer and filmmaker, was also an M.D. (Harvard Medical School), an anthropology professor at Cambridge University, a visiting writer at Massachusetts Institute of Technology, and a courageous voice of reason against the corrupting influence of politics on science. During an historic 2003 speech at California Institute of Technology, Crichton concisely addressed the supposed scientific consensus on global warming:

> "I want to pause here and talk about this notion of consensus, and the rise of what has been called consensus science. I regard consensus science as an extremely pernicious development that ought to be stopped cold in its tracks. Historically, the claim of consensus has been the first refuge of scoundrels; it is a way to avoid debate by claiming that the matter is already settled. Whenever you hear the consensus of scientists agrees on something or other, reach for your wallet, because you're being had. Let's be clear: the work of science has nothing whatever to do with consensus. Consensus is the business of politics. Science, on the contrary, requires only one investigator who happens to be right, which means that he or she has results that are verifiable by reference to the real world. In science consensus is irrelevant. What *is* relevant is reproducible results. The greatest scientists in history are great precisely because they broke with the consensus."[48]

He went on to describe several of the seemingly endless examples where the scientific establishment declared consensus on a particular issue only to later be proven wrong. Crichton concluded, "Finally, I would remind you to notice where the claim of consensus is invoked. Consensus is invoked only in situations where the science is not solid enough. Nobody says the consensus of scientists agrees that $E=mc^2$. Nobody says the consensus is that the sun is 93 million miles away. It would never occur to anyone to speak that way."

Chapter 22

Who Is Al Gore?

Meet the hero of humanity. Most people don't realize that Al Gore, besides courageously saving the planet from deadly carbon dioxide, also created the Internet. Outrageous? Yes! But according Gore, it's true. In 1999, while seeking the Democratic presidential nomination, Gore appeared on CNN's *Late Edition* with Wolf Blitzer. Blitzer asked Gore what set him apart from his opponents. Gore replied, "During my service in the United States Congress, I took the initiative in creating the Internet."[1] Never mind that DARPA, an agency of the Department of Defense, started working on the Internet during the 1960s, at least a decade before Gore's first Congressional term.[2]

Distortion and deceit have come to characterize Gore's political career. A faithful New World Order minion, he has pushed the global warming agenda for more than two decades. For his efforts, he has reaped obscene profits. According to the *New York Times*, "Mr. Gore is poised to become the world's first 'carbon billionaire,' profiteering from government policies he supports that would direct billions of dollars to the business ventures he has invested in."[3]

Gore is the co-founder, along with former Goldman Sachs Asset Management CEO David Blood, of Generation Investment Management (GIM), which sells carbon offset opportunities. Appropriately nicknamed Blood and Gore, GIM is a top owner of the Chicago Climate Exchange (CCX),[4] which describes itself as "North America's only cap and trade system for all six greenhouse gases, with global affiliates and projects worldwide."[5] Additionally, the notoriously corrupt Goldman Sachs owns a 10 percent stake in CCX. As reported by *Rolling Stone*, "From tech stocks to high gas prices, Goldman Sachs has engineered every major market manipulation since the Great Depression—and they're about to do it again."[6]

CCX owes its existence to generous grants donated in 2000 and 2001 by the Joyce Foundation. During these years, Barack Obama enabled these grants while serving on the Joyce Foundation board of directors. In March 2009, *Fox News* revealed, "Barack Obama helped fund a carbon trading exchange that will likely play a critical role in the cap-and-trade carbon reduction program he is now trying to push through Congress as president."[7] Incidentally, Maurice Strong, a key architect of the modern day false environmental movement, is a member of the CCX board of directors.[8] Gore claims that "every penny" he earns from his climate investments goes to his charity, Alliance for Climate Protection.[9] Nevertheless, for decades he has fronted for the biggest scientific hoax of modern times, which, if successful, will detrimentally affect the lives of most of the world's population.

Decades of Duplicity

In 1993, William Happer, Director of Energy Research for the United States Department of Energy, was fired for expressing his professional opinion on climate change. He presented evidence that ultraviolet radiation had actually been decreasing when, according to the ozone depletion theory, it should have been increasing. Testifying before the House Energy and Water Development Subcommittee on Appropriations, Happer stated,

"I think that there probably has been some exaggeration of the dangers of ozone and global climate change."[10] These words effectively sealed his fate. Happer recalled, "After a few months, Secretary O'Leary called me in to say that I was unacceptable to Al Gore and his environmental advisers, and that I would have to be replaced. She was apologetic and gracious during this discussion, and she did not elaborate on the exact reasons for Gore's instructions."[11]

The *Wall Street Journal* quoted a top staff member as saying that Happer was "philosophically out of tune" with the new administration.[12] On Gore's repressive, suffocating approach towards science, Happer stated, "Many atmospheric scientists are afraid for their funding, which is why they don't challenge Al Gore and his colleagues. They have a pretty clear idea of what the answer they're supposed to get is. The attitude in the administration is, 'If you get a wrong result, we don't want to hear about it'...I was told that science was not going to intrude on policy."[13]

In his 1992 book *Earth in the Balance*, Gore claimed that 98 percent of scientists share the view that human CO_2 emissions are detrimentally warming the planet. "The skeptics," he announced, "are vastly outnumbered by former skeptics who now accept the overwhelming weight of the accumulated evidence."[14] He offered no proof for his bold proclamations, but such is the faith-based Gospel of Gore. Gore's past raises serious doubts regarding his trustworthiness. The Roger Revelle story is the quintessential example.

The *Boston Globe* has called Roger Revelle "the grandfather of the greenhouse effect," and "the godfather of global warming."[15] He was also Al Gore's first mentor on climate change. Gore and Revelle met at Harvard University in 1967. Ten years earlier, Revelle had co-authored a seminal paper demonstrating how fossil fuel consumption contributes to increased CO_2 levels.[16] In 1965, under Revelle's leadership, the President's Science Advisory Committee Panel on Environmental Pollution published the first authoritative report identifying CO_2 emissions as a potential global problem.[17] Al Gore, in his book *Earth in the Balance*, recalled,

"I was introduced to the idea of a global environmental threat as a young student when one of my college professors was the first person in the world to monitor carbon dioxide (CO_2) in the atmosphere...Professor Revelle explained that higher levels of CO_2 would create what he called the greenhouse effect, which would cause the earth to grow warmer. The implications of his words were startling: we were looking at only eight years of information, but if this trend continued, human civilization would be forcing a profound and disruptive change in the entire global climate."[18]

Although he presents himself as a humble Revelle devotee, Gore's claims on greenhouse warming, scientific consensus, and the need for draconian legislation each directly contradict Revelle's beliefs prior to his death in 1991. Three months before he died, Revelle co-authored a paper for *Cosmos* entitled, "What To Do About Greenhouse Warming: Look Before You Leap." His co-authors were Fred Singer, an accomplished atmospheric physicist, and Chauncey Starr, a visionary electrical engineer. The authors wrote:

- "The scientific basis for a greenhouse warming is too uncertain to justify drastic action at this time."

- "Even if significant warming were to occur in the next century, the net impact to the entire planet may well be beneficial."
- "It would be prudent to complete the ongoing and recently expanded research so that we will know what we are doing before we act. 'Look before you leap' may still be good advice."[19]

For Gore, busily pushing his climate crusade, this public rebuttal from his former mentor was a harrowing setback. Singer reflected, "The contradiction between what Senator Gore wrote about what he learned from Dr. Revelle and what Dr. Revelle had written in the *Cosmos* article embarrassed Senator Gore, who had become the leading candidate for the vice presidential slot of the Democratic Party."[20] Gore was mute on the issue until the Vice-Presidential debates when James Stockdale, the running mate of Ross Perot, questioned Gore regarding Revelle's paper. Gore claimed, "He died last year and just before he died he co-authored an article which had statements taken completely out of context."[21] Did Revelle take *his own* statements out of context? Or was Gore preying on the public's naivety by distorting his deceased mentor's message?

Before the 1991 *Cosmos* article, Revelle made numerous statements consistent with his "look before you leap" advice. In 1988, for example, he wrote several letters to Congress confirming both his respect for the scientific method and his prudence. In July 1988, Revelle sent a letter to Senator Tim Wirth advising, "We should be careful not to arouse too much alarm until the rate and amount of warming becomes clearer. It is not yet obvious that this summer's hot weather and drought are the result of a global climatic change or simply an example of the uncertainties of climate variability. My own feeling is that we had better wait another ten years before making confident predictions."[22] He also sent a letter to Representative Jim Bates stating,

> "Most scientists familiar with the subject are not yet willing to bet that the climate this year is the result of 'greenhouse warming.' As you very well know, climate is highly variable from year to year, and the causes of these variations are not at all well understood. My own personal belief is that we should wait another ten or twenty years to really be convinced that the greenhouse effect is going to be important for human beings, in both positive and negative ways."[23]

Also in 1984, *Omni* magazine interviewed Revelle about the implications of global warming. His position was decidedly non-alarmist. Revelle remarked, "There may be lots of effects. Increased CO_2 in the air acts like a fertilizer for plants…you get more plant growth. Increasing CO_2 levels also affect water transpiration, causing plants to close their pores and sweat less. That means plants will be able to grow in drier climates."[24] He concluded, "I estimate that the total increase [in CO_2] over the past hundred years has been about 21 percent. But whether the increase will lead to a significant rise in global temperature, we can't absolutely say."

Revelle was no alarmist. To the contrary, he consistently approached climate concerns with prudence and judiciousness. After his death, however, Gore and his associates attempted to portray Revelle in a manner more aligned with their political ambitions. In the months preceding the Vice-Presidential debates, several prominent journalists highlighted Revelle's *Cosmos* article as compared to Gore's alarmism, mendacity, and hypocrisy. These included George Will, a regular columnist for the *Washington Post* and contributing editor to *Newsweek*, and Gregg Easterbrook, a frequent contributor to numerous publications including the *New York Times*, *Washington Post* and *Los Angeles Times*.[25]

After these articles, Justin Lancaster, one of Revelle's former students, contacted Singer demanding he retract Revelle's name from the *Cosmos* article. When Singer refused, Lancaster began publicly denouncing Singer, claiming he had deceived Revelle and coerced him into lending his name to the article. Singer recalled, "He later added the charge that I had pressured an aging and sick colleague, suggesting that Dr. Revelle's mental capacities were failing at the time."[26] Senior Gore staff member Anthony Socci also directed similar accusations towards Singer.

Facing intense defamation, Singer filed suit against Lancaster. According to Lancaster's official deposition, Gore had telephoned Lancaster following the publication of Easterbrook's article to ask Lancaster about Revelle's mental condition. Lancaster also wrote in his deposition that Revelle *agreed* with the content of the *Cosmos* article. Prior to publication, Revelle even showed Lancaster the manuscript of the article and remarked that he "felt it was honest to admit the uncertainties about greenhouse warming, including the idea that our ignorance could be hiding benefits as well as catastrophes."[27]

Lancaster's association with Gore did not stop with Gore's initial phone call. The *Boston Globe* reported, "Gore was concerned enough to ask Lancaster, a Lexington oceanographer, to write a rebuttal to one of the articles suggesting Revelle had changed his views on global warming."[28] Lancaster prepared seven drafts, faxing each one to Gore's office. On one fax he asked, "Is this close to what the Senator had in mind?"[29] The *Globe* also revealed that Lancaster was receiving financial support from the Environmental Defense Fund and the Natural Resources Defense Fund, both of which supported Gore's global warming theory. Gore then attempted to rescue Lancaster from media scrutiny by contacting prominent news programs and urging them to discredit Singer and investigate other scientists who were skeptical about manmade global warming. The *Washington Times* reported,

> "Concurrent with Mr. Lancaster's attack on Mr. Singer, Mr. Gore himself led a similar effort to discredit the respected scientist. Mr. Gore reportedly contacted 60 Minutes and Nightline to do stories on Mr. Singer and other opponents of Mr. Gore's environmental policies. The stories were designed to undermine the opposition by suggesting that only raving ideologues and corporate mouthpieces could challenge Mr. Gore's green gospel. The strategy backfired. When Nightline did the story, it exposed the vice president's machinations and compared his activities to Lysenkoism: The Stalinist politicization of science in the former Soviet Union."[30]

On the February 24, 1994 edition of *Nightline*, Ted Koppel chastised Gore:

> "There is some irony in the fact that Vice President Gore, one of the most scientifically literate men to sit in the White House in this century, that he is resorting to political means to achieve what should ultimately be resolved on a purely scientific basis...The issues of global warming and ozone depletion are undeniably important. The future of mankind may depend on how this generation deals with them. But the issues have to be debated and settled on scientific grounds, not politics. There is nothing new about major institutions seeking to influence science to their own ends. The church did it, ruling families have done it, the communists did it, and so have others, in the name of anti-communism. But it has always been a corrupting influence, and it

always will be. The measure of good science is neither the politics of the scientist nor the people with whom the scientist associates. It is the immersion of hypotheses into the acid of truth. That's the hard way to do it, but it's the only way that works."[31]

With his back to the wall, Lancaster conceded defeat. On April 29, 1994, he signed the following statement: "I retract as being unwarranted any and all statements, oral or written, I have made which state or imply that Professor Revelle was not a true and voluntary coauthor...I apologize to Professor Singer for the pain my conduct has caused him and for any damage that I may have caused to his reputation."[32] Gore was exposed and discredited. Nevertheless, people are still swallowing his propaganda. In 2007 alone, Gore won the Nobel Peace Prize, his film won an Academy Award, the audio-book version won a Grammy, he won a Primetime Emmy Award and he was *Time* magazine's runner-up for Person Of The Year. In 2009, with a straight face, feigning dignity, Gore accepted the inaugural Roger Revelle Award from the Scripps Institution of Oceanography. "And now that we have arrived at the moment when we have to choose, I invoke Roger Revelle's name in asking all of you to go to work and do what's necessary to pass the legislation and ratify the treaty and solve global warming," Gore declared.[33]

In 2007, Gore suffered an embarrassing setback when British courts blocked the distribution of his film to schools, except under certain conditions. The British government had wanted every school in the country to show the film, but High Court Judge Justice Burton ruled that the film could not be show without an accompanying notice documenting nine significant errors. Burton ruled that the "apocalyptic vision" of the film has many false claims arising in "the context of alarmism and exaggeration."[34] Included in the nine errors was Gore's claims regarding extreme sea level rise. Burton commented, "The Armageddon scenario he predicts, insofar as it suggests that sea level rises of seven meters might occur in the immediate future, is not in line with the scientific consensus." Burton also shamed Gore's attempt to connect the drying of Lake Chad, the melting of snow on Mount Kilimanjaro, and Hurricane Katrina all to global warming.

Another person who dared challenge Gore was John Coleman, the founder of the Weather Channel. In 2008, Coleman filed suit against Gore for fraud, hoping the lawsuit would generate genuine debate, since Gore has always flatly refused requests and challenges to debate. "Since we can't get a debate, I thought perhaps if we had a legal challenge and went into a court of law, where it was our scientists and their scientists, and all the legal proceedings with the discovery and all their documents from both sides and scientific testimony from both sides, we could finally get a good solid debate on the issue," Coleman explained.[35] While many people are now aware of Gore's duplicity, a critical evaluation of the environmental leadership would certainly greatly benefit both the environment and humanity.

Chapter 23
The Rhetoric of Alarmists

Pigeonholes and Fear Traps

Pigeonholing is a common tactic of global warming propagandists. By directing extreme language and labels towards those who contradict their views, they effectively discourage neutral and undecided people from even questioning their dogma. For example, they introduce the idea that global warming skeptics are the modern version of those who once believed the Earth was flat. They repeat this mantra over and over again. They use the media, advertising, marketing, television, cinema, and music for full-fledged propaganda campaigns.

Since the 2006 release of Al Gore's film, such propaganda has become ubiquitous. Rupert Murdoch's News Corporation, for example, now openly admits to inserting manipulative messages into their programs. One of the company's promotional videos, for example, opens with the caption, "In 2006, News Corp. embarked upon a company wide initiative to reduce the size of its carbon footprint."[1] Fox Television chairman Gary Newman then explained, "The biggest thing we've done is inserting messages about the environment into some of our content." The video scrolled through several examples of global warming propaganda woven into programs like *24* and *The Simpsons*.

Why are these shows called *programs*? Who are they programming? They suggest to viewers, directly or indirectly, that global warming skeptics are delusional or crazy. The subconscious absorbs these messages and associations. Consequently, people adapt 'herd mentality,' conforming to beliefs they think they are *supposed* to hold. This tactic also provides cover for the propagandists' refusal to publicly debate. They simply imply, "We can't waste our time debating these imbeciles. Their position is so absurd, it doesn't warrant debate."

Pigeonholing is the last vestige of the desperate while they struggle to defend an indefensible position. For example, Pulitzer Prize winning reporter Ellen Goodman wrote for the *Boston Globe*, "I would like to say we're at a point where global warming is impossible to deny. Let's just say that global warming deniers are now on a par with Holocaust deniers, though one denies the past and the other denies the present and future."[2] Such embarrassingly distasteful associations are commonplace within the alarmist community. A completely transparent tactic, they attempt to elicit intense guilt by comparing, for example, climate 'skeptics' to the scores of apathetic onlookers during the Holocaust. The incessantly parroted term 'climate change denier,' for example, conjures up immediate associations with the dreaded 'Holocaust denier.' Al Gore, for example, frequently warns about an impending "ecological holocaust," while James Hanson, Gore's chief scientific advisor, has testified, "If we cannot stop the building of more coal-fired power plants, those coal trains will be death trains."[3]

Yvo de Boer, the UN's top climate official, is another rabid alarmist who warns policymakers that global warming will "threaten the very survival" of some people and "failing to recognize the urgency of this message and act on it would be nothing less that criminally irresponsible."[4] And of course, IPCC Chairman Rajendra Pachauri regularly injects his trademark ridicule and contempt. In 2008, he told the *Chicago Tribune*, "There is, even today, a Flat Earth Society that meets every year to say the Earth is flat. The science about climate change is very clear. There really is no room for doubt at this point."[5] If logic and science were on the alarmists' side, would such pigeonholing really be necessary?

Another common tactic is the exploitation of fear, linking global warming to unrelated threats. In 2003, John Houghton, a notorious alarmist, penned an article for the *London Guardian* declaring, "As a climate scientist who has worked on this issue for several decades, first as head of the Met Office, and then as co-chair of scientific assessment for the UN Intergovernmental Panel on Climate Change, the impacts of global warming are such that I have no hesitation in describing it as a 'weapon of mass destruction.'"[6] Houghton, a Co-Chair of the IPCC and Lead Editor of its first three reports, cannot substantiate his global warming claims with science. Therefore, he associates his doomsday climate scenario with terrorism. He paralyzes people with fear so they believe what they are told to believe. For Houghton, the exploitation of fear to mobilize humanity is completely justifiable. In the 1994 edition of his book *Global Warming: The Complete Briefing*, he asserted, "Unless we announce disasters, no one will listen."[7] What do you suppose people would do if you simply told the truth, Mr. Houghton?

In 2008, New York City Mayor Michael Bloomberg also pulled the terrorism card. After addressing the UN General Assembly, he told reporters, "Terrorists kill people. Weapons of mass destruction have the potential to kill an enormous amount of people, but global warming in the long term has the potential to kill everybody."[8] Bloomberg then renewed his calls for carbon taxes. "So long as there's no penalty or cost involved in producing greenhouse gases, there will be no incentive. For that reason, I believe the U.S. should enact a tax on carbon emissions," Bloomberg asserted.[9]

Alarmists also use fear to create urgency and panic. In 1988, for example, on the hottest day of the summer, Al Gore and James Hansen launched the modern global warming fear campaign with an elaborate stunt. Hansen was testifying before a U.S. Senate Committee chaired by Gore. According to Tim Wirth (then a U.S. Senator), "What we did is went in the night before and opened all the windows, I will admit, right? So that the air conditioning wasn't working inside the room…So it was sort of a perfect collection of events that happened that day, with the wonderful Jim Hansen, who was wiping his brow at the witness table and giving this remarkable testimony."[10]

Today Hansen still frequently warns about Earth's "imminent peril."[11] In 2009, for example, he wrote, "We have only four years left for Obama to set an example to the rest of the world,"[12] and in 2006 he claimed, "We have at most ten years—not ten years to decide upon action, but ten years to alter fundamentally the trajectory of global greenhouse emissions."[13] And in 2010, Hansen endorsed *Time's Up!: An Uncivilized Solution to a Global Crisis*, an ultra-radical book in which author Keith Farnish suggests, "The only way to prevent global ecological collapse and thus ensure the survival of humanity is to rid the world of Industrial Civilization."[14] Farnish also called for "razing cities to the ground, blowing up dams and switching off the greenhouse gas emissions machine." According to Hansen, "Keith Farnish has it right: time has practically run out, and the 'system' is the problem."[15]

Hansen's deceptions and radicalism have not gone unnoticed within the scientific community. In 2009, for example, his former supervisor at NASA, atmospheric scientist John Theon, publicly reprimanded him. In a letter to the U.S. Senate Environment and Public Works Committee, Theon wrote:

- "As Chief of several NASA Headquarter Programs (1982-94), an SES position, I was responsible for all weather and climate research in the entire agency, including the research work by James Hansen, Roy Spencer, Joanne Simpson, and several hundred other scientists at NASA field centers, in academia, and in the private sector who worked on climate research."

- "Hansen was never muzzled even though he violated NASA's official agency position on climate forecasting (i.e., we did not know enough to forecast climate change or mankind's effect on it). Hansen thus embarrassed NASA by coming out with his claims of global warming in 1988 in his testimony before Congress."

- "I appreciate the opportunity to add my name to those who disagree that global warming is man-made."[16]

Science, Irrelevant; Honesty, Optional

Since the 1980s, Stanford professor Stephen Schneider has continually warned about catastrophic manmade global warming. Yet, ironically, during the 1970s Schneider was an outspoken alarmist regarding *global cooling* and the coming *ice age*. In a 1971 *Science* article, Schneider predicted, "An increase by only a factor of 4 in global aerosol background concentration may be sufficient to reduce the surface temperature by as much as 3.5° K. If sustained over a period of several years, such a temperature decrease over the whole globe is believed to be sufficient to trigger an ice age."[17] The media and scores of self-righteous environmentalists parroted Schneider's alarmism for years.

By 1978, he was predicting radical and potentially catastrophic changes within decades: "By the year 2000, these changes could be as large as any documented since the end of the last ice age when the earth's surface temperature apparently varied by about five degrees centigrade between glacial and interglacial extremes."[18] In the same year, however, a Defense Department survey of 24 leading atmospheric scientists (including Schneider) concluded, "The most likely event will be a climate that resembles the average of the past 30 years."[19] The *Los Angeles Times* portrayed Schneider as representative of "one pole of that community and a minority of the 24 who were surveyed." Schneider nevertheless defended his position while declaring himself "very leery of science by consensus." Today, however, he is rather fond of "science by consensus." In 2008, for example, he told Australia's *ABC Radio National*, "Unfortunately, there are some self-professed climate contrarians who try to discredit the strong scientific consensus, by citing what they believe to be data that disproves the consensus."[20]

In 1992, Schneider even condemned journalists who dare report on so-called contrarian viewpoints. "It is journalistically irresponsible," Schneider declared, "to present both sides as if it were a question of balance. Given the distribution of views, with groups like the National Academy of Science expressing strong scientific concern, it is irresponsible to give equal time to a few people standing out in left field."[21] Since 1994, he has participated in all four IPCC reports, either as a Lead or Contributing Author. As an IPCC member, he shares the Nobel Prize with Al Gore. In a 1989 interview with *Discover* magazine, Schneider showcased the dangerous mentality of the climate crusaders:

"On the one hand, as scientists we are ethically bound to the scientific method, in effect promising to tell the truth, the whole truth, and nothing but—which means that we must include all doubts, the caveats, the ifs, ands and buts. On the other hand, we are not just scientists but human beings as well. And like most people we'd like to see the world a better place, which in this context translates into our working to reduce the risk of potentially disastrous climate change. To do that we need to get some broad based support, to capture the public's imagination. That, of course, means getting loads of media coverage. So we have to offer up scary scenarios, make simplified, dramatic statements, and make little mention of any doubts we might have. This 'double ethical bind' we frequently find ourselves in cannot be solved by any formula. Each of us has to decide what the right balance is between being effective and being honest. I hope that means being both."[22]

In other words, honesty is optional. This ideological thread connects the IPCC, Al Gore, and scores of other propagandists. In 1998, for example, Canada's *National Post* quoted then Canadian Environment Minister Christine Stewart as stating, "No matter if the science is all phony, there are collateral environmental benefits...Climate change [provides] the greatest chance to bring about justice and equality in the world."[23] Her shocking statement reveals the mentality of the global warming leadership. For them, scientific evidence is irrelevant. Here was a prominent environmental leader admitting that science is merely a political tool, a steering device. Terence Corcoran, a distinguished Canadian journalist and editor of the *Post*, remarked, "Consolidating Ms. Stewart's statements, we reach some horrific conclusions. Whether global warming actually exists is irrelevant. It is, in the hands of government and environmental activists, a convenient front for the introduction of programs and economic policies that Canadians—and most citizens of the world—would not otherwise accept."[24]

Tim Wirth, another high priest of the environmental movement, also admittedly shares this anti-scientific attitude. As Under Secretary of State for Global Affairs during the Clinton Administration, Wirth worked side by side with Al Gore to advance the global warming agenda. In 1992, the *Washington Post* quoted Wirth as saying, "We've got to ride the global-warming issue. Even if the theory of global warming is wrong, we will be doing the right thing anyway in terms of economic policy and environmental policy."[25]

At the 1996 UN climate conference in Geneva, Wirth presented his ideas for "riding the global-warming issue." He delivered an impassioned rant, claiming, "Scientific consensus does not mean unanimity. There are a handful of scientists who remain in a very different position from the overwhelming consensus of the international scientific community...Human health is at risk from projected increases in the spread of diseases like malaria, yellow fever and cholera; food security is threatened in certain regions of the world; water resources are expected to be increasingly stressed, with substantial economic, social and environmental costs in regions that are already water-limited, and perhaps even political costs where there is already conflict over limited resources...This problem cannot be wished away. The science cannot be ignored and is increasingly compelling. The obligation of policymakers is to respond with the same thoughtfulness that has characterized the work of the world's scientific community."[26]

The following year Wirth, the man who claimed that erroneous science is an acceptable basis for economic and environmental policy, quit his job in the State Department because President Clinton refused to follow his radical recommendations. The *New York Times* reported, "Mr. Wirth advocated an aggressive plan for significant cutbacks in the nation's emissions of greenhouse gases—one that could have had significant costs for the economy. Bit by bit, that plan was scaled back as Mr. Clinton listened, and listened again, to the warnings of senior economic advisers."[27]

Wirth then accepted an invitation from Ted Turner to head the newly formed United Nations Foundation. One of the foundation's core objectives is "Climate and Energy." According to their website, "The time for action is now. Addressing climate change is one of humanity's greatest and most pressing challenges—and one that requires an urgent and cooperative response."[28] According to Turner, failure to immediately impose draconian legislation will result in catastrophe. In 2008, on the *Charlie Rose Show*, Turner declared, "We'll be eight degrees hotter in 10—not 10, but in 30 or 40 years. And basically none of the crops will grow. Most of the people will have died, and the rest of us will be cannibals. Civilization will have broken down. The few people left will be living in a failed state like Somalia or Sudan, and living conditions will be intolerable. The droughts will be so bad there will be no more corn growing. It will—not doing it is suicide."[29] Is Ted Turner sane?

Turner exemplifies the danger to society posed by psychopathic elites. Not only does he support crippling the economy with harsh carbon taxes, he also advocates mass genocide. In the same interview Turner declared, "The simplest answer is that the world's population should be about two billion, and we've got about six billion now."[30] Not surprisingly, so-called "population stabilization" is another core objective of the United Nations Foundation.[31] As discussed in previous chapters, this is what global warming is really about, population reduction—genocide.

To be clear, most people within the environmental movement are compassionate and sincere in their beliefs. They truly believe that global warming and overpopulation threaten the very existence of humanity. Nevertheless, they have been co-opted by ruthless psychopaths, eugenicists, and monopoly men. These criminals appeal to popular sentiment then misdirect this sentiment according to their own agenda. For centuries elites have been brooding about overpopulation, preferring smaller populations because they are easier to manage. Their concerns have nothing to do with the environment. If they can convince humanity that the world is overpopulated, they can convince the people to participate in their own demise.

Currently, these elites are openly merging the depopulation agenda with the global warming campaign. Turner, for example, proclaimed, "We're too many people. That's why we have global warming. We have global warming because too many people are using too much stuff. If there were less people, they'd be using less stuff."[32] He even advocated Chinese-style one-child policies. Despite having five children of his own, Turner proposed, "On a voluntary basis, everybody in the world has got to pledge to themselves that one or two children is it."[33] Of course it begins *voluntarily*, just like it did China.

In 2008, Turner announced another *voluntary* initiative to save the world from global warming. The Global Sustainable Tourism Criteria (GSTC) is an effort by Turner and the United Nations to impose "voluntary standards [that] help travel suppliers around the world meet increasing consumer demand for products and services that will have positive effects on communities and the environment."[34] This is precisely how you domesticate wild animals. First, you appeal to that which seems reasonable to them. Then you coax them into altering their normal behavior, ever so slightly. After some time, they come to accept a new standard of normality. Then you repeat the process. Eventually, these once wild animals become domesticated cage-dwellers, behaviorally and psychologically completely transformed.

Climate Skeptics Funded by Oil Companies?

Global warming believers commonly accuse transnational oil companies of funding and encouraging contrarian viewpoints. For example, in 2007, prominent environmentalist David Mayer de Rothschild childishly called radio talk show host Alex Jones an "oil boy." "Get back onto your little rodeo bull," Rothschild sneered, "go out there and deny it all, protect your oil shares and make sure that your kids don't have any planet to live on."[35] The fallacy of such allegations is that transnational oil companies firmly support the global warming orthodoxy and its proposed solutions. In 2009, for example, both Shell and British Petroleum participated in the Prince of Wales' Corporate Leaders Group on Climate Change. Both companies signed the Copenhagen Communiqué, which proposed absurd, economically devastating, greenhouse gas emission reductions of 50-85 percent by 2050. The Communiqué declared, "Developed countries need to take on immediate and deep emission reduction commitments that are much higher than the global average."[36]

The big oil companies, including Exxon, British Petroleum and Shell, all accept the IPCC's global warming conclusions. Furthermore, each company supports government-mandated carbon reduction schemes. British Petroleum, for example, describes itself as "an advocate of legislation and co-operation to address climate change."[37] They accept the IPCC's conclusions and support mandatory carbon reduction policies: "The scale of change required to address the issue means that it can only be achieved through government policy to drive emissions reduction and stimulate investments in low-carbon technologies."[38] Likewise, Dutch Royal Shell's website contends, "Recent scientific reviews have confirmed the reality of global warming, effectively shifting the emphasis from a debate about the evidence to the formulation of a response."[39] In 2007, Shell CEO Jeroen van der Veer declared, "Governments must urgently provide the rules that can foster lower carbon dioxide emissions."[40] He advocated a *worldwide* cap-and-trade system: "Policymakers the world over should make wider use of flexible market mechanisms such as the capping of carbon emissions and trading of credits that has occurred under the European emissions trading scheme since 2005. While Europe's CO_2 market is a good start, trading needs to become global in order to be truly effective."[41]

Oil giant Exxon not only supports the IPCC's position, two of their in-house scientists are actually IPCC contributing authors.[42] Exxon also confirms its globalist allegiances with the admission that, "We contribute to an array of public policy organizations that research and promote discussion on climate change and other domestic and international issues, including…the Council on Foreign Relations."[43] Rather than a cap-and-trade system, however, Exxon supports direct carbon taxation. During an October 2009 speech before the Economic Club of Washington, DC, CEO Rex Tillerson acknowledged the prevailing attitude of New World Order elites that "it is easier and more politically expedient to support a cap-and-trade approach, because the public will never figure out where it is hitting them. They will just know they hurt somewhere in their pocketbook."[44] Tillerson and Exxon, however, favor direct carbon taxation because, "a carbon tax may be a more viable framework for engaging participation by other nations." [45] The oil companies are hardly concerned about losing profits because, ultimately, consumers will bear the brunt of any new climate legislation. Furthermore, the executives and principal owners of these corporations are intimately involved with the New World Order agenda and its various steering committees including Bilderberg and the CFR. They have no reason to fund contrary viewpoints.

The Precautionary Principle

The UN's Rio Declaration on Environment and Development included 27 seemingly altruistic principals for the purpose of "Recognizing the integral and interdependent nature of the Earth, our home." Principal number 15 deserves special attention. It reads,

> "In order to protect the environment, the precautionary approach shall be widely applied by States according to their capabilities. Where there are threats of serious or irreversible damage, lack of full scientific certainty shall not be used as a reason for postponing cost-effective measures to prevent environmental degradation."[46]

This proclamation enables manmade global warming advocates to disavow themselves of scientific integrity and responsibility. Where else but from scientific evidence should such "threats of serious or irreversible damage" arise?

The precautionary principle, often characterized by the phrase *better safe than sorry*, is the Bush doctrine of preemptive war, applied to environmental law. The Bush doctrine presumes that "we can't wait for proof." The danger is too extreme. As Bush so eloquently declared, "We cannot wait for the final proof — the smoking gun — that could come in the form of a mushroom cloud."[47] In other words, the precautionary principle is the tool of tyrants.

Vaclav Klaus, President of the Czech Republic and then-President of the European Union, in his speech to the 2009 International Conference on Climate Change, addressed the underlying machinations of the precautionary principle. He asserted, "It is evident that the environmentalists don't want to change the climate. They want to change us and our behavior. Their ambition is to control and manipulate us. Therefore, it should not be surprising that they recommend 'preventive,' not 'adaptive' policies. Adaptation would be our voluntary behavior, which is not what they aim at. They do not want to recognize that — to quote Nigel Lawson — 'the capacity to adapt is arguably the most fundamental characteristic of mankind' and that our 'adaptive capacity is increasing all the time with the development of technology.' The environmentalists speak about 'saving the planet.' From what? And from whom? One thing I know for sure: we have to save it — and us — from them."[48]

PART IV: LAPDOG MEDIA

"The Corporate grip on opinion in the United States is one of the wonders of the Western world. No First World country has ever managed to eliminate so entirely from its media all objectivity — much less dissent."[1] — Gore Vidal

In 1785, referring to the British government, Thomas Jefferson wrote, "You well know, that that government always kept a kind of standing army of news-writers, who, without any regard to truth, or to what should be like truth, invented, and put into papers whatever might serve the ministers. This suffices with the mass of the people, who have no means of distinguishing the false, from the true paragraphs of a newspaper."[2] Upon learning about the New World Order, many people ask, "Why hasn't the media said anything about this?" This question, however, begs several questions in response: Why should you *expect* the media to say something? What is the *purpose* of the media? Who *controls* the media? According to Walter Annenberg, billionaire publishing magnate and former U.S. ambassador to the United Kingdom, the media has "at least as much power in determining the course of the republic as the executive, legislative, and judicial branches set forth in the Constitution."[3] The ruling class has always known this. Therefore, just as in Jefferson's day, they have always attempted to control the media, either through direct ownership or through various manipulative tactics.

News organizations are corporations, intent on selling products. Their products, however, are not merely magazines and newspapers. Their "products" are conceptions of reality. They aim to "sell" certain worldviews, while discouraging others. Media conglomerates have vested interests in protecting and perpetuating the ruling class. Therefore, they must sell the worldview that government is generally benevolent and trustworthy. To do so, they must disseminate distorted, illusory and otherwise fabricated information, while filtering out any information that challenges the legitimacy of the ruling class. They can imply incompetence, they can report periodic scandals, but they can never expose systemic corruption emanating from the highest levels of business and government. In his book *Necessary Illusions*, M.I.T. professor Noam Chomsky observed,

> "In short, the major media—particularly, the elite media that set the agenda that others generally follow—are corporations 'selling' privileged audiences to other businesses. It would hardly come as a surprise if the picture of the world they present were to reflect the perspectives and interests of the sellers, the buyers and the product. Concentration of ownership of the media is high and increasing. Furthermore, those who occupy managerial positions in the media, or gain status within them as commentators, belong to the same privileged elites, and might be expected to share the perceptions, aspirations, and attitudes of their associates, reflecting their own class interests as well. Journalists entering the system are unlikely to make their way unless they conform to these ideological pressures, generally by internalizing the values; it is not easy to say one thing and believe another, and those who fail to conform will tend to be weeded out by familiar mechanisms."[4]

Hofstra University media professor Dennis Mazzocco knows first-hand about the indoctrination of media employees. He spent 28 years working for networks including

ABC, NBC, CBS, Fox, TNT and ESPN with jobs spanning from management executive to producer, director, and writer for hundreds of programs. Furthermore, he earned nine Emmy awards and twenty career Emmy nominations.[5] In his book *Networks of Power: Corporate TV's Threat to Democracy*, Mazzocco reflected,

"Like most U.S. media workers, I did not learn about the full extent of corporate control over global media production and distribution until after I resigned from ABC in 1988. To succeed in the competitive world of U.S. broadcasting (in which corporate structures increasingly resemble the top-down, tightly ordered hierarchy of a military organization), I enthusiastically endorsed the basic tenets of the system. I claimed that it was democratic, open to all citizens, and under the basic control of the audience through the ratings...U.S. media workers are paid to subtly (and at times not so subtly) influence public opinion. They learn to adopt the owners' views in order to succeed, even when their paychecks or political and social connections may be those of ordinary citizens. No media worker who wants to keep his or her job will ever admit this publicly. Nor will anyone who wants to succeed in U.S. broadcasting publicly confirm that management investment decisions are made to protect the firm's political-economic power and inevitably affect their company's on-air programming."[6]

Other media insiders have acknowledged that media conglomerates intentionally "dumb down" the news. For example, Tom Fenton, a multi-award winning, 38-year correspondent for CBS News, reported in his 2005 book *Bad News*, "The gatekeepers of the news—the executives, editors, and producers who decide what information will make it to the public—will tell you that the average reader simply cannot absorb that much hard news, especially about events abroad. On one level, they're merely restating the problem: Americans are too broadly under informed to digest nuggets of information that seem to contradict what they know of the world. Yet whose fault is that, and whose responsibility is it to correct? Instead, news channels prefer to feed Americans a constant stream of simplified information, all of which fits what they already know. That way they don't have to devote more air time or newsprint space to explanations or further investigation."[7] Fenton further admitted, "Politicians and the media have conspired to infantilize, to dumb down, the American public. At heart, politicians don't believe that Americans can handle complex truths, and the news media, especially television news, basically agrees."[8]

On the journalistic integrity of television reporting, *Time* magazine columnist Margaret Carlson explained, "What's good TV and what's thoughtful analysis are two different things. That's been conceded by most producers and bookers. They're not looking for the most learned person; they're looking for the person who can sound learned without confusing the matter with too much knowledge."[9] Consequently, sports, celebrity gossip and other trivial matters replace "thoughtful analysis." In his 1996 book *Hot Air*, longtime *Washington Post* writer Howard Kurtz observed, "The culture of news, once the straight-laced, buttoned-down preserve of Walter Cronkite and Huntley-Brinkley, has merged with the relentlessly glitzy world of entertainment, producing one great, roaring Oprahfied ooze of headlines and hype."[10] In 1976, CBS News President Richard Salant wrote, "We in broadcast journalism cannot, should not, and will not base our judgments on what we think the viewers and listeners are 'most interested' in...Our job is to give people not what they want but what they ought to have."[11] According to ruling-class elites, what the people "ought to have" is anything and everything except the truth.

Chapter 24
Ruling-Class Control

According to his 1995 *New York Times* obituary, Ferdinand Lundberg was "a prolific iconoclast whose books of social criticism pilloried the wealthy and what he perceived as their financial grip on the nation's economic life."[1] Lundberg was a journalist for the *Chicago Daily News, United Press International*, and, from 1927 to 1934, the *New York Herald Tribune*. He was also an adjunct professor of social philosophy at New York University from 1952 to 1968.[2] In his 1937 book *America's 60 Families*, Lundberg documented the inordinate media control wielded by J.P. Morgan and the Rockefeller family. The Rockefellers dominated not through direct ownership, but through their control of advertising revenue. However, "When advertising alone is not sufficient to insure the fealty of a newspaper, the Rockefeller companies have been known to make direct payments in return for a friendly editorial attitude."[3]

The Rockefellers also gained inordinate influence and indirect control through their numerous business associations with press-owning families, such as the Whitney, Harkness, Mellon, Astor, DuPont, McCormick, Hearst, and Morgan families. Like the Rockefellers, J.P. Morgan also dominated the press through advertising. Lundberg noted, "More advertising is controlled by the J.P. Morgan junta than by any other single financial group, a factor which immediately gives the banking house the respectful attention of all alert independent publishers."[4] Nevertheless, "Through various [sic] of its corporations J.P. Morgan and Company has maintained a direct hold over many newspapers, irrespective of any influence which exists by reason of the advertising at its disposal."[5]

Council on Foreign Relations historian Carroll Quigley confirmed the ruling elites' control over the press, both in England and America. He explained, "The power and influence of this Rhodes-Milner group in British imperial affairs and in foreign policy since 1889, although not widely recognized, can hardly be exaggerated. We might mention as an example that this group dominated *The Times* from 1890 to 1912, and has controlled it completely since 1912 (except for the years 1919-1922)."[6] Regarding the American press, Quigley remarked,

"The American branch of this 'English Establishment' exerted much of its influence through five American newspapers (*The New York Times, New York Herald Tribune, Christian Science Monitor*, the *Washington Post*, and the lamented *Boston Evening Transcript*). In fact, the editor of the *Christian Science Monitor* was the chief American correspondent (anonymously) of *The Round Table*, and Lord Lothian, the original editor of *The Round Table* and later secretary of the Rhodes Trust (1925-1939) and ambassador to Washington, was a frequent writer in the *Monitor*. It might be mentioned that the existence of this Wall Street, Anglo-American axis is quite obvious once it is pointed out."[7]

Lundberg's penetrating and insightful analysis is still relevant today. He further observed,

"The journalism of the United States, from top to bottom, is the personal affair — bought and paid for — of the wealthy families. There is little in American journalism today, good or bad, which does not emanate from the family dynasties. The press lords of America are actually to be found among the multimillionaire families.

Newspapers all over the world exist, and have existed, in the service of economic and political power rather than in that of truth and noble ideals. This distinction should clarify at the outset a highly complicated field of human activity, and should indicate at once that America's millionaires have not been guilty, in their journalistic preoccupation, of perpetrating some unusual crime against the ostensible sanctity and purity of the press. Yet, in saying that newspapers do not exist to serve truth, it does not follow that they never serve truth and are never actuated by idealism, although truth in established journalism is always secondary, and very often is incidental, even accidental."[8]

The Constitution guarantees *freedom*, not truth, of the press. Thus, implicitly, the Constitution protects the freedom to slant, spin, and distort truth. Therefore, the root of the problem rests not with the media corporations, but with the people who naively accept as truth that which the media calls "news." Healthy skepticism and common sense analysis would greatly neutralize manipulative reporting. For example, why does the media continually cover approved topics — celebrities, sports, drugs, and blue-collar crime — while continually avoiding taboo topics — the erosion of civil liberties, the 9/11 truth movement, the illegality of the Iraq war, and white-collar crime? Whose interests are served when the public knows very little about foreign policy, and about elite organizations like the Council on Foreign Relations? Lundberg further observed,

"Because so many fields of editorial investigation and exposition are taboo, the press as a whole must confine itself to a relatively restricted 'safe' area. This accounts for the undue measure of attention given to the underworld; to petty scandals involving actresses, baseball players, and minor politicians; to sporting affairs and the activities of the quasi-wealthy. The press, in short, must compensate for enforced lack of vitality in dynamic fields by artificial enthusiasm in static fields. In place of an evenhanded, vital, varied daily news report, the American press as a whole is obliged to present a lopsided news report that is of doubtful reader interest. And in order to recapture the constantly waning attention of readers it must rely upon comic strips, inane 'features,' contests, gossip columns, fiction, cooking recipes, instruction columns in golf, chess, bridge, and stamp collecting, and similar nonsense. American newspapers, in short, are paradoxically and with few exceptions, not newspapers at all."[9]

Callaway Warns America

In February 1917, Representative Oscar Callaway attempted to warn America that certain notorious finance oligarchs had seized control of the press. Although Congressional leaders blocked him from presenting his case verbally, he nevertheless entered a written statement into the Congressional Record. Representative Hampton Moore protested that Callaway's allegations had been "buried in the Record."[10] Nevertheless, the controversy forced the *New York Times* and the *Washington Post* to print portions of his incredible statement. Callaway wrote,

"In March, 1915, the J.P. Morgan interests, the steel, shipbuilding and powder interests, and their subsidiary organizations got together twelve men high up in the newspaper world and employed them to select from the most influential newspapers in the United States, a sufficient number of them to control generally the policies of the daily press of the United States. These twelve men worked the problem out by selecting 179 newspapers and then began, by an elimination process, to retain only those necessary for the purpose of controlling the general policy of the daily press throughout the country.

They found it was only necessary to purchase the control of twenty-five of the greatest newspapers. The twenty-five papers were agreed upon; emissaries were sent to purchase the policy, national and international, of these papers; an agreement was reached, the policy of the papers was bought, to be paid for by the month; an editor was furnished for each paper to properly supervise and edit information regarding the questions of preparedness, militarism, financial policies and other things of national and international nature considered vital to the interests of the purchasers.

The contract is in existence at the present time and it accounts for the news columns of the daily press of the country being filled with all sorts of preparedness arguments and misrepresentations as to the present condition of the United States Army and Navy, and the possibility and probability of the United States being attacked by foreign foes. The policy also included the suppression of everything in opposition to the wishes of the interests served. The effectiveness of this scheme has been conclusively demonstrated by the character of the stuff carried in the daily press throughout the country since March 1915. They have resorted to anything necessary to commercialize public sentiment and sandbag the national Congress into making extravagant and wasteful appropriations for the army and navy, under false pretense that it was necessary. Their stock argument is 'patriotism.' They are playing on every prejudice and passion of the American people."[11]

Callaway further alleged that Morgan and his cohorts were buying influence, paying publishers "for the service they are rendering in the promotion of the war spirit and of a false patriotism in the United States, misleading the common people of this land into the belief that this war in Europe is an American war."[12] These stunning allegations should have sparked massive outrage and inquiry, possibly dramatically altering the course of world history. But the subservient press dutifully misdirected the public's attention. The *New York Times* printed only two short articles while the *Washington Times* printed six. What little coverage they did provide served mostly to ridicule Moore and Callaway and distort their allegations. They also ignored Moore's demands for a formal investigation. They were too busy hyping the war and ignoring legitimate domestic threats—like those emanating from Wall Street. Moore proclaimed, "I question whether a single newspaper in the United States has taken cognizance of the remarks of the gentleman from Texas."[13]

While the *Times* did print Callaway's remarks, they certainly didn't emphasize them. After their initial article on February 14, they ran just one follow-up—a hit-piece penned by Douglas Johnson, the Chairman of the American Rights League. Johnson declared, "Hampton Moore, Representative from Pennsylvania, delivered in Congress a sensational attack against that portion of the American press and public which has patriotically stood for an unwavering defense of American rights and American honor."[14] Just as Callaway had warned, the media was frantically exploiting and manipulating the country's strong sense of patriotism. They were demonizing Germany while ignoring the country's domestic enemies. Regarding Moore's attempt to investigate these criminals, Johnson proclaimed, "His condemnation was reserved for real Americans who would rather see the country involved in an honorable war for a righteous cause than to see it seduced by unpatriotic pacifists into purchasing peace with dishonor."[15]

The *Washington Post's* coverage of the Callaway-Moore effort was similarly deceitful. With six articles over seven days, they continually degraded Moore while pushing decidedly pro-war sentiments. For his patriotism, they labeled Moore a "pacifist." Meanwhile, they praised Representative Gardner for his hawkish war speech, which attacked these so-called pacifists and accused them of "appealing to the cowardice which lurks in every man's breast, a cowardice which is the result of the strongest of human instincts, the instinct of self-preservation. This cowardice they glorify by calling it good will toward men, and timidity they have rechristened 'service to humanity.'"[16]

Guided by an insatiable hunger for war, the press and their masters had thrown caution to the wind. On several occasions, the *Post* challenged Moore to present evidence of Morgan's plot. The *Post* editorialized, "Until and unless you produce something to support your statements it is you and not the newspapers that are in danger of public contempt."[17] But Moore had repeatedly requested this very opportunity. He was calling for formal investigations to fully divulge the truth. This would have involved subpoenaing witnesses and gathering sworn testimony. Nevertheless, the *Post* childishly demanded evidence upfront. Would a prosecuting attorney reveal evidence outside the courtroom? This, of course, would give the defense a huge advantage. Another Representative asked, "Does not the gentleman believe that this House has already wasted enough of the peoples' money in making investigations of charges that have no foundations whatever?"[18] Yet Congressional leaders were blocking the investigation and the press was facilitating the matter. The March 1915 timeframe that Callaway referenced was just before the callous false flag attack on the Lusitania, the pretext for America's entrance into the war.[19]

Ickes Battles the Press Lords

Speaking before the American Civil Liberties Union on December 8, 1937, Secretary of the Interior Harold Ickes candidly remarked, "Our ancestors fought to prevent a state censorship of news and ideas. Our ancestors did not fight for the right of a few lords of the press to have almost exclusive control of and censorship over the dissemination of news and ideas. Yet under the stress of economic forces our press and news agencies are coming more and more under the domination of a handful of corporate publishers who may print such news as they wish to print and omit such news as they do not wish to print."[20] Later in January 1939, appearing on the popular radio program *America's Town Meeting of the Air*, Ickes condemned the press for kowtowing to financial interests. The *New York Times* reported, "According to Secretary Ickes, the press is not free because it is limited by financial affiliations, is subject to the influence of advertisers and is unfair to certain groups, especially workers whose interests conflict with those of its advertisers and financial backers."[21]

On this radio program, Ickes debated Frank Gannett, President of Gannett Newspapers. He challenged Gannett concerning his refusal to cover a 1937 bakers' strike in Rochester. Desperate for press coverage, the bakers' union attempted to purchase advertising space to report their story. Gannett management, however, flatly rejected their ads. Ickes reminded Gannett that "just before this, during the auto workers' strike in Detroit, both Gannett papers in Rochester ran full page ads giving the General Motors version of that strike."[22] Ickes concluded, "There are several great newspapers in the United States for which I have a profound admiration, but even they are not wholly free. Their vast financial investment, running high into the millions, binds them closely to the business world from which they draw their sustenance. Freedom is impossible, even for a publisher with a sense of civic responsibility or an editor with noble ideals, when the counting office holds the whip hand."[23]

CFR Influence

According to the Council on Foreign Relations' website, 3 of the top 5 media conglomerates are CFR Corporate Members. These are Time Warner, News Corporation (Fox), and General Electric (NBC).[24] In their 2008 Annual Report, the CFR also disclosed that 398 of their 4,338 members are journalists, correspondents or editors.[25] Rear Admiral Chester Ward, a Judge Advocate General of the U.S. Navy from 1956 to 1960, and a 16-year member of the CFR, observed, "Previous attempts to document CFR's influence have been ignored or smeared by the liberal press as 'exaggerated.' This is to be expected, considering the beachheads that key CFR members hold in all parts of the media, and especially because any attempt to tell the truth about the power and activities of CFR members is bound to sound exaggerated. Actually, however, all the published accounts thus far have understated CFR's influence."[26] Nevertheless, Richard Harwood, in a 1993 *Washington Post* article admitted about CFR journalists, "They do not merely analyze and interpret foreign policy for the United States; they help make it."[27] Harwood further commented, "The editorial page editor, deputy editorial page editor, executive editor, managing editor, foreign editor, national affairs editor, business and financial editor and various writers as well as Katharine Graham, the paper's principal owner, represent the *Washington Post* in the council's membership. The executive editor, managing editor and foreign editor of the *New York Times* are members, along with executives of such other large newspapers as the *Wall Street Journal* and *Los Angeles Times*, the weekly newsmagazines, network television executives and celebrities—Dan Rather, Tom Brokaw and Jim Lehrer, for example—and various columnists, among them Charles Krauthammer, William Buckley, George Will and Jim Hoagland."[28]

Chapter 25

Covert Government Propaganda

CIA Media Manipulation

According to William Schaap, "Governments around the world have secretly used the media for their purposes for many hundreds of years, probably thousands."[1] Schaap is a lawyer, professor, publisher, author and expert on government disinformation and propaganda. During the 1970s and 1980s, he was Staff Council for New York's Center for Constitutional Rights, and during the late 1980s, he taught courses on disinformation and propaganda at New York's City University. For more than 20 years, he has published *Covert Action Quarterly*, a magazine reporting on U.S. and foreign intelligence agencies. He has contributed to numerous books and written numerous articles for the *New York Times* and *Washington Post*. As an expert witness, Schaap has testified before U.S. state and federal courts, the United Nations, and the U.S. Congress. On November 30, 1999, he testified as an expert witness during the Martin Luther King murder trial.

During the King trial, Schaap discussed media manipulation by the U.S. government since 1917, the year the U.S. entered World War I. To influence public opinion towards supporting U.S. intervention, President Wilson established the Committee for Public Information. He appointed George Creel, a public relations executive, as Chairman, and Edward Bernays, the "Father of Spin," and nephew of Sigmund Freud, as chief advisor. In his 1928 book *Propaganda*, Bernays later bragged,

> "The conscious and intelligent manipulation of the organized habits and opinions of the masses is an important element in democratic society. Those who manipulate this unseen mechanism of society constitute an invisible government which is the true ruling power of our country. We are governed, our minds are molded, our tastes formed, our ideas suggested, largely by men we have never heard of. This is a logical result of the way in which our democratic society is organized. Vast numbers of human beings must cooperate in this manner if they are to live together as a smoothly functioning society…In almost every act of our daily lives, whether in the sphere of politics or business, in our social conduct or our ethical thinking, we are dominated by the relatively small number of persons—a trifling fraction of our hundred and twenty million—who understand the mental processes and social patterns of the masses. It is they who pull the wires which control the public mind."[2]

Regarding the Committee for Public Information, Schaap explained, "It was the policy of this committee to have no compunctions about falsifying the news whenever it was felt that that was necessary to help the war effort. They purported very often to release documents, supposedly genuine documents, to the press in order to substantiate whatever particular position the Wilson government might have been taking at the time."[3] An example was the 1917 "disinformation campaign to suggest that the Russian revolutionaries, Lenin in particular and Trotsky, were actually German agents being paid by the Kaiser."[4] Schaap continued, "The Government and Creel's committee made up the story. They created phony documents. They passed it all to friends in the major newspapers. And almost immediately this was front page news around the United States and around the world." Between the two

World Wars, government manipulation of media decreased, but with the advent of World War II, it greatly intensified.

In 1942, William Donovan headed up the newly formed Office of Strategic Services (OSS), the precursor of the CIA. Donovan, known as Wild Bill, "believed the saying that George Creel had—his philosophy from World War I—which was that you should lie to the people whenever it's necessary, whenever you think lying will help maintain morale and win the war."[5] After the war, the OSS morphed into the CIA and consequently, "The acceptability of lying to the public became very widespread and acceptable, even in time of peace, and it spread to other agencies, including the FBI which also began to engage in media manipulation in a very, very large way...the infrastructure that had been set up to spread disinformation, to be able to lie, became institutionalized and began operating at a greater and greater level."[6] This manipulation has succeeded, Schaap contends, because "Americans generally tend to believe what their government tells them, to believe that government officials on all levels generally tell the truth. And if you have that, that absence of skepticism, it's a major plus for the disinformationists."[7] Conversely, for example, "If the Italian government issues a statement, the average Italian on the street will say it's probably a lie until you can prove to me otherwise that it's not a lie...The average American would hear something from the government or hear the news on television and assume that what they're hearing is the truth unless they're shown otherwise. They assume that almost nothing is ever a conspiracy."[8]

The American public has endured covert government propaganda for many decades. Media manipulation endures because, "The longer that you, whoever you are, can control the spin on a story, the more that spin becomes accepted as the absolute truth."[9] Based on consultations with Harvard Medical School professors, Schaap reported, "When you associate one thing with another over time, just the mention of the one brings the association of the other. What this will sometimes mean is that even when something is later exposed as a lie, if it was accepted as a truth for a long time, the exposure of it as a lie is not believed."[10] For example, during the mid-1970s, Cuba sent troops to Angola to help the Angolans fight against South Africa. Wanting desperately to discredit the Cubans, the CIA invented a story about Cuban soldiers raping Angolan women. "With one phone call and one visit," Schaap explained, "it went over the wire services, it went into Europe, it went into the United States, it went around the world. And for about a six-month period there were all these stories about the horrible Cuban rapes in Angola. And what that does is, when you hear—when the average person hears Angola or Cuban, they'll think rape of the women...these patterns build up so that that becomes the truth embedded in your mind."[11] John Stockwell, a CIA agent based in the Congo at the time, quit the CIA and later wrote a book exposing this terrible lie. The *New York Times* published an extensive story about the hoax. Nevertheless, the conditioning was so strong that people continued believing the lie.

Operation Mockingbird

CIA agent E. Howard Hunt, the man from whom Tom Cruise's *Mission Impossible* character Ethan Hunt derived his name,[12] was an integral part of Operation Mockingbird, an enormous, covert CIA operation to infiltrate and control the media. In his book *American Spy*, Hunt admitted, "Much of what I worked on was exposed in revelations about Operation Mockingbird by a Frank Church-headed Senate investigation in 1975 that divulged a lot of agency dirty laundry about our infiltration of the U.S. and international media."[13]

The Church Committee, as well as subsequent congressional and independent hearings and investigations, revealed crucially important information regarding the cozy relationship between the media and the shadow government. According to the Church Committee's final report, published in 1976, "The CIA currently maintains a network of several hundred foreign individuals around the world who provide intelligence for the CIA and at times attempt to influence opinion through the use of covert propaganda. These individuals provide the CIA with direct access to a large number of newspapers and periodicals, scores of press services and news agencies, radio and television stations, commercial book publishers, and other foreign media outlets."[14] The report also stated, "The Committee is concerned that the use of American journalists and media organizations for clandestine operations is a threat to the integrity of the press."[15]

Representative Otis Pike chaired a concurrent investigation for the House Intelligence Committee. The White House, however, intervened, forcing the House to block publication of the Pike Committee's final report. Nevertheless, CBS News reporter Daniel Schorr acquired a copy and leaked it to the *Village Voice* for publication. The report highlighted the CIA's proclivity for "frequent manipulation of Reuters wire service dispatches,"[16] and revealed that,

> "Some 29 percent of forty Committee-approved covert actions were for media and propaganda projects. This number is probably not representative. Staff has determined the existence of a large number of CIA internally-approved operations of this type, apparently deemed not politically sensitive. It is believed that if the correct number of all media and propaganda projects could be determined, it would exceed Election Support as the largest single category of covert action projects undertaken by the CIA."[17]

According to Frank Church, the CIA's media manipulation had been costing American taxpayers roughly $265 million a year.[18] Economist Sean Gervasi observed that the CIA's propaganda budget was larger than the budget of any news agency. "In fact," he determined, "the CIA propaganda budget is as large as the combined budgets of Reuters, United Press International and the Associated Press."[19]

Unsatisfied by these congressional investigations, *Washington Post* reporter Carl Bernstein quit his job in 1977 to begin his own independent, six-month investigation of the CIA. At the time, Bernstein was one the most prolific journalists in the country. In 1972, for example, he and Bob Woodward broke the story on the Watergate burglaries. Their subsequent investigations led to Nixon's resignation and earned them the 1973 Pulitzer Prize for Public Service. In October 1977, Bernstein published his findings as "The CIA and the Media," a 25,000-word *Rolling Stone* cover story.

Through interviews with Church Committee members, Bernstein confirmed that both the Pike and Church Committee reports were indeed significantly watered down. "William Colby and George Bush," he wrote, "persuaded the committee to restrict its inquiry into the matter and to deliberately misrepresent the actual scope of the activities in its final report."[20] Committee member Senator Gary Hart confirmed, "It hardly reflects what we found. There was a prolonged and elaborate negotiation [with the CIA] over what would be said."[21] Bernstein concluded, "The CIA's use of the American news media has been much more extensive than Agency officials have acknowledged publicly or in closed sessions with members of Congress...There is ample evidence that America's leading publishers allowed themselves and their news organizations to become handmaidens to the intelligence services."[22]

Although the CIA was heavily influencing the media, they were not outright *controlling* it—they didn't need to. Media executives share the same ruling-class objectives as high-level CIA and government officials. Therefore, they enthusiastically cooperated with the agency. For example, Katherine Graham, a prominent CFR member[23] and longtime director of the *Washington Post*, was unashamed, boastful even, of this relationship. In a well-publicized 1988 speech at CIA headquarters, she declared, "We live in a dirty and dangerous world. There are some things the general public does not need to know and shouldn't. I believe democracy flourishes when the government can take legitimate steps to keep its secrets, and when the press can decide whether to print what it knows."[24] Graham was one of many media executives who collaborated directly with the CIA. Others included William S. Paley of CBS, Arthur Hays Sulzberger of the *New York Times*, and Henry Luce of *Time*. ABC, NBC, the Associated Press, United Press International, Reuters, Hearst Newspapers, Scripps-Howard, *Newsweek*, the Mutual Broadcasting System, the *Miami Herald*, the *Saturday Evening Post*, and the *New York Herald-Tribune* also cooperated.[25]

In December 1977, after a three-month investigation, the *New York Times* published a series of four articles. The *Times* further confirmed and expanded upon Bernstein's conclusions. They determined, "The CIA has at various times owned or subsidized more than 50 newspapers, news services, radio stations, periodicals and other communications entities...At one time, according to agency sources, there were as many as 800 such 'propaganda assets,' mostly foreign journalists."[26] When asked if the CIA told these "assets" what to write, former CIA Director William Colby boasted, "Oh, sure, all the time."[27] The CIA also propagandized through book publishing: "Some of the thousand or so books published by the CIA or on its behalf have contained propaganda ranging from tiny fictions to outright deceptions."[28] Although the *Times* investigation focused on CIA's foreign propaganda, they acknowledged that these efforts inevitably polluted the news read by Americans. One CIA official admitted, "We would not tell U.P.I. or A.P. headquarters in the U.S. when something was planted abroad."

Summarizing the CIA's treachery, Schaap explained, "What we have learned from these reports is that—the first thing was that about a third of the whole CIA budget went to media propaganda operations...close to a billion dollars is [now] being spent every year by the United States on secret propaganda...When the Church Committee reported on the CIA media operations, for example, beyond friends in the press, beyond having people who just generally thought along similar lines, it turned out that they had thousands of journalists in their employ. Not merely friendly, not merely agents, not merely someone you could pass a story to, but people, who might have appeared to the outside world to be a reporter for CBS, was in fact a CIA employee getting a salary from the CIA. And that was repeated thousands of times all around the world. They also *owned outright*, the CIA, at that time, about 250 or more media organizations. That's wire services, newspapers, magazines, radio, TV stations— all around the world that they owned outright. The actual shareholder of the company turned out to be some CIA front."[29]

Dirty Laundry

By definition, terrorism is the use of violence or intimidation to advance a political agenda. Therefore, strictly speaking, the CIA is a terrorist organization. A closer look at CIA history shows why effective tyranny requires an effectively controlled media. For without Operation Mockingbird, the CIA's terrorist agenda would have been politically impossible. When Americans finally discover the CIA's dirty laundry, they will promptly demand the agency's abolishment. In January 2010, venerable Texas Congressman Ron Paul called for such action. In a devastating critique, Paul stated,

"There's been a coup, have you heard? It's the CIA coup. The CIA runs everything. They run the military. They're the ones who are over there lobbing missiles and bombs on these countries. It's not even the military that does that. The CIA runs this. And of course the CIA is every bit as secret as the Federal Reserve. And yet think of the harm they have done since they were established, since World War II. They are a government unto themselves. They're in businesses—in drug businesses—they take out dictators. We need to take out the CIA."[30]

Some former CIA agents have also called for terminating the agency. Ralph McGehee, for example, worked for the CIA for 25 years. Upon his retirement in 1971, he received the agency's Career Intelligence Medal.[31] Despite his honorable service, McGehee later harshly criticized the CIA. In 1981, he blew the whistle on CIA involvement in Indonesia's 1965 military coup that ousted Indonesian President Sukarno and the subsequent slaughter of more than 300,000 people. McGehee attempted to publicize these and other allegations in *The Nation* until CIA officials intervened demanding censorship. The American Civil Liberties Union filed a lawsuit on behalf of McGehee and *The Nation* but the CIA prevailed.[32] Nevertheless, McGehee had the final word in 1983 when he published *Deadly Deceits: My 25 Years in the CIA*, a scathing exposé calling for abolishment of the agency. He wrote, "Efforts to create a workable intelligence service must begin by abolishing the CIA. For a host of reasons I believe the CIA as it now exists cannot be salvaged."[33] McGehee documented decades of CIA crimes and confirmed the CIA's massive propaganda campaign against the American people:

"The CIA is not now nor has it ever been a central intelligence agency. It is the covert action arm of the President's foreign policy advisers. In that capacity it overthrows or supports foreign governments while reporting 'intelligence' justifying those activities. It shapes its intelligence, even in such critical areas as Soviet nuclear weapon capability, to support presidential policy. Disinformation is a large part of its covert action responsibility, and the American people are the primary target audience of its lies."[34]

He continued, citing numerous examples:

"The Vietnam War was the Agency's greatest and longest disinformation operation. From 1954 until we were ejected in 1975, the Agency lied in its intelligence while propagandizing the American people. It planted a weapons shipment, forged documents, deceived everyone about the Tonkin Gulf incident, and lied continually about the composition and motivation of the South Vietnamese communists. Even now Agency historians and ex-employees try to perpetuate the propaganda themes through which it tried first to win and later to maintain American support for the war. As recently as April 22, 1981, former CIA director William Colby wrote an article for the *Washington Post*, portraying the Vietnam War—even in light of the Pentagon Papers disclosures—as the altruistic U.S. coming to the assistance of the South Vietnamese people. He had the audacity to recommend the period from 1968 to 1972—the era of CIA assassination teams—as a model for use in El Salvador.

Not much has changed since I left the Agency. It follows all the same patterns and uses the same techniques. We have seen this in relation to El Salvador, where it fabricated evidence for a White Paper, the same way it did in Vietnam in 1961 and 1965. We have seen it in Iran, where it cut itself off from all contact with potential revolutionary groups to support the Shah. We have seen it in the recruitment ads seeking ex-military personnel to man its paramilitary programs. We have seen it in relation to Nicaragua, where it arms Miskito Indians in an attempt to overthrow the Nicaraguan government. In this case it again exploits a naive minority people who will be discarded as soon as their usefulness ends, as happened with the Hmong in Laos. We have seen it in its attempts to rewrite and censor the truth. I personally have experienced this kind of Agency effort recently when it censored an article I wrote about its successful operation to overthrow the government of Achmed Sukarno of Indonesia."[35]

Despite overwhelming incriminating evidence as revealed by the Church and Pike Committees and various independent investigations and personal accounts, the CIA continues to function, unabated. Far from reforming the CIA or increasing CIA oversight, the Reagan Administration, with former CIA Director George Bush as Vice-President, actually *increased* CIA power and *decreased* oversight. Reagan signed Executive Order 12333 on December 4, 1981, enabling the CIA to conduct covert operations within the United States and Executive Order 12356 on April 2, 1982, restricting public access from CIA documents.[36] McGehee lamented, "The President wants the Agency free of the constraints of public exposure so that it can gather and fabricate its disinformation unharried by criticisms and so that it can overthrow governments without the knowledge of the American people."[37] Regarding CIA media manipulation, E. Howard Hunt candidly reflected, "Most interestingly, however, while employed by the CIA, I have to say I *never heard* about a specific Operation Mockingbird, although my projects would certainly have come under its umbrella. But even if there was no official project by that name, we did carry out the basic operations, and I cannot contradict the findings."[38]

Roots of CIA Power

The National Security Act of 1947 established the CIA and effectively undermined the Constitution. It was a tyrant's dream. Any national security risk, whether real or illusory, could be used to justify outrageous usurpations of power. Consider the following questions: Why for more than forty-six years have pertinent JFK assassination documents remained classified?[39] Why for more than eight years have 84 videos of the Pentagon during the 9/11 attacks remained classified?[40] How did Bush justify domestic spying?[41] What was the backdrop for Cheney's unwavering resolve to legalize torture?[42] In each case, the answer is *national security*. According to the U.S. Department of State, "The National Security Act of 1947 mandated a major reorganization of the foreign policy and military establishments of the U.S. Government. The act created many of the institutions that Presidents found useful when formulating and implementing foreign policy, including the National Security Council." And most notably, "The act also established the Central Intelligence Agency (CIA), which grew out of World War II era Office of Strategic Services and small post-war intelligence organizations."[43] Yet not surprisingly, this modest description completely ignores the far-reaching implications of the Act. Indeed, the National Security Act gave the CIA unrestrained, carte blanche power to conduct any covert actions deemed necessary, *legal or illegal*. Although the President must report illegal CIA activities to Congress, the Act also enabled the President to postpone this reporting *indefinitely*.

The power to conduct *any* activity comes from the following provision: "Nothing in this subchapter shall be construed as requiring the approval of the congressional intelligence committees as a condition precedent to the initiation of any significant anticipated intelligence activity."[44] In other words, U.S. intelligence agencies can do whatever they want, whenever they want, wherever they want. Regarding illegal activities, the Act states, "The President shall ensure that any illegal intelligence activity is reported promptly to the congressional intelligence committees, as well as any corrective action that has been taken or is planned in connection with such illegal activity."[45] In other words, the President (assuming he has been informed) *shall* (not must) report illegal activities to Congress. However, such reporting "may be postponed beyond the time provided," indefinitely, if such reporting "will impede the work of officers or employees of the intelligence community in a manner that will be detrimental to the national security of the United States."[46] The Act also explicitly permits "activities of the United States Government to influence political, economic, or military conditions abroad, where it is intended that the role of the United States Government will not be apparent or acknowledged publicly."[47] In other words, with the National Security Act, the U.S. government granted itself the right to meddle in the affairs of sovereign nations.

The National Security Act demonstrates obvious contempt for foreign nations and peoples, but what about its domestic implications? The law states, "No covert action may be conducted which is intended to influence United States political processes, public opinion, policies, or media."[48] However, *intended* and *influence* are subjective terms allowing for liberal interpretations of the law. Furthermore, the United States is not insulated from and unaffected by international events. How could CIA operations abroad *not* influence American "political processes, public opinion, policies, or media?" Remember, the CIA has absurdly and arrogantly declared the right to overthrow foreign governments and assassinate foreign leaders. How could such activities *not* influence American policies? How could covert CIA operations *not* distort and corrupt America's media? Consider the following examples.

Chile 1973

In September 1973, the CIA orchestrated the violent overthrow of democratically elected Chilean president Salvador Allende and the subsequent imposition of military dictator General Augusto Pinochet. Declassified documents from the National Security Archives reveal that then-National Security Advisor Henry Kissinger, in conjunction with the CIA, began plotting this coup in September 1970. Speaking with Secretary of State William Rogers, Kissinger stated, "I think what we have to do is make a cold-blooded assessment, get a course of action this week some time and then get it done."[49] In another conversation between Kissinger and Nixon, Nixon stated, "We don't want a big story leaking out that we are trying to overthrow the Govt."[50] Several days after these conversations, the CIA produced multiple planning documents. One stated, "President Nixon had decided that an Allende regime in Chile was not acceptable to the United States. The President asked the Agency to prevent Allende from coming to power or to unseat him. The President authorized ten million dollars for this purpose, if needed."[51] Another document revealed,

"It is firm and continuing policy that Allende be overthrown by a coup. It would be much preferable to have this transpire prior to 24 October but efforts in this regard will continue vigorously beyond this date. We are to continue to generate maximum pressure toward this end utilizing every appropriate resource. It is imperative that these actions be implemented clandestinely and securely so that the USG [U.S. government] and American hand be well hidden."[52]

In his *Rolling Stone* article, Bernstein explained, "In the Sixties, reporters were used extensively in the CIA offensive against Salvador Allende in Chile; they provided funds to Allende's opponents and wrote anti-Allende propaganda for CIA proprietary publications that were distributed in Chile."[53] Much of this propaganda trickled into American publications. Enabled by the National Security Act of 1947, the CIA heavily manipulated and propagandized Americans and the world. On September 11 (the New World Order's sacred day), 1973, the Chilean military overthrew Allende. The CIA claimed no responsibility.

Iran 1953

Another heinous CIA crime was the 1953 overthrow of Iran's democratically elected Prime Minister Mohammed Mossadegh. Western elites despised Mossadegh for his audacious presumption that Iran's wealth and natural resources should belong not to Britain, but to the Iranian people. During the late 1940s and early 1950s, public contempt for British control was escalating. The British, having occupied Iran during WWII, maintained control of Iranian oil after the war through the Anglo-Iranian Oil Company (AIOC), known today as British Petroleum. Purportedly a jointly owned company by the British and Iranian governments, the British thoroughly dominated AIOC. In 1947, for example, AIOC reported after-tax profits of £40 million, of which a mere £7 million went to Iran.[54] In 1951, the Iranian Parliament voted to nationalize the oil industry. Mossadegh, the leading advocate of the idea, became Prime Minister. For his sudden rise in popularity, *Time* magazine named him "Man of the Year."[55] Through the eyes of British oil tycoons, however, Mossadegh's radical ideas were simply unacceptable. Along with the CIA, they conspired to overthrow him and install a pro-Western dictatorship under Shah Mohammad Reza Pahlavi, whose reign extended 27 years.

Kermit Roosevelt, Jr., a CIA agent and grandson of former-President Theodore Roosevelt, directed the mission known as Operation Ajax. According to CIA documents, the CIA staged vicious terror attacks, blamed them on unknown communists, and having already infiltrated the Iranian press, planted editorials questioning Mossadegh's competence, and eventually calling for his removal from office. As reported by the *New York Times*, "The history says agency officers orchestrating the Iran coup worked directly with royalist Iranian military officers, handpicked the prime minister's replacement, sent a stream of envoys to bolster the shah's courage, directed a campaign of bombings by Iranians posing as members of the Communist Party, and planted articles and editorial cartoons in newspapers."[56]

Regarding the CIA's media infiltration and its inevitable manipulation of Americans' perceptions of the world, the *Times* reported, "The Iran desk of the State Department, the document says, was able to place a CIA study in *Newsweek*, 'using the normal channel of desk officer to journalist.' The article was one of several planted press reports that, when reprinted in Tehran, fed the 'war of nerves' against Iran's prime minister, Mohammed Mossadegh."[57] The *Times* continued,

"Western correspondents in Iran and Washington never reported that some of the unrest had been stage-managed by CIA agents posing as Communists. And they gave little emphasis to accurate contemporaneous reports in Iranian newspapers and on the Moscow radio asserting that Western powers were secretly arranging the shah's return to power."[58]

Instead the U.S. media "prominently mentioned the role of Iran's Communists in street violence leading up to the coup."[59] The press was completely subservient to the CIA. Kenneth Love, the *New York Times* correspondent in Tehran during the coup, later conceded, "The only instance since I joined the *Times* in which I have allowed policy to influence a strict news approach was in failing to report the role our own agents played in the overthrow of Mossadegh...I knew there had been U.S. involvement in the coup."[60] And of course, because of the National Security Act of 1947, the CIA could legally justify this operation.

Today, as Washington grows increasingly hostile towards Iran, the CIA still defends its 1953 terror campaign. In 2007, CIA director Michael Hayden addressed the Council on Foreign Relations in New York. Journalist Lucy Komisar asked him, "I wonder whether looking back a little bit in history you think there are any lessons to learn from the fact that the CIA, having overthrown Mossadeq, set the stage for the problems that we are facing now." Hayden blithely replied, "All I can tell you is that the Central Intelligence Agency, within a framework of law, carries out the foreign policy of the United States that is constructed by the people that we elect in both the executive and legislative branch."[61] In other words, because CIA-sponsored terrorism is "legal," morality is irrelevant.

During his June 4, 2009 speech in Cairo, President Obama acknowledged America's involvement in 1953 coup, yet by no means apologized for nor condemned the CIA's terrorism. "The United States," Obama remarked, "played a role in the overthrow of a democratically elected Iranian government...This history is well known. Rather than remain trapped in the past, I've made it clear to Iran's leaders and people that my country is prepared to move forward."[62] Three weeks later, during particularly turbulent post-election riots in Iran, Obama scoffed at allegations of CIA involvement. "There are reports," Obama stated, "suggesting that the CIA is behind all this. All of which is patently false. But it gives you a sense of the narrative that the Iranian government would love to play into."[63]

Despite Obama's assurances, two years earlier, President Bush officially authorized the CIA to once again overthrow the Iranian government! As reported by the *London Telegraph*, "President George W. Bush has given the CIA approval to launch covert 'black' operations to achieve regime change in Iran, intelligence sources have revealed. Mr. Bush has signed an official document endorsing CIA plans for a propaganda and disinformation campaign intended to destabilise, and eventually topple, the theocratic rule of the mullah."[64] These plans included "manipulating the country's currency and international financial transactions," and "giving arms-length support, supplying money and weapons, to an Iranian militant group, Jundullah, which has conducted raids into Iran from bases in Pakistan."[65] Since then, Jundullah has conducted numerous bombings against Iranian government targets including a particularly deadly attack in October 2009, which killed 42 people including seven senior officers of Iran's Revolutionary Guard.[66] One might reasonably ask oneself how a nation can arm and support known terrorists while simultaneously fighting a War on Terror.

Cuba 1961

A third example of CIA manipulation of American media and public opinion is the 1961 Bay of Pigs fiasco. According to declassified CIA documents, CIA leadership had been considering assassinating Fidel Castro as early as December 1959. A memorandum from CIA director Allen Dulles to Colonel J.C. King recommended "thorough consideration be given to the elimination of Fidel Castro. None of those close to Fidel such as his brother Raúl or his companion, Che Guevara, have the same mesmeric appeal to the masses. Many informed people believe that the disappearance of Fidel would greatly accelerate the fall of the present government."[67] The CIA preferred the corrupt, brutal dictatorship of Fulgencio Batista, against whom Castro led the Cuban Revolution. Batista was friendly to U.S. corporate interests; Castro was not. Under Batista, American corporations controlled 40 percent of the Cuban sugar industry, 80 percent of Cuba's utilities, and 90 percent of its mining.[68] Batista also allowed the American mafia to control casinos, hotels and prostitution in Havana.[69] All this changed under Castro. Like Mossadegh had done in Iran, Castro initiated policies to return his country's wealth and resources back to its people. To the U.S. State Department, however, such policies were outright theft. In January 1960, *Time* magazine reported, "The U.S. said that in seizing a third of the $850 million U.S. investment in Cuba, Dictator Fidel Castro is violating both 'Cuban law and generally accepted international law.'"[70] Despite *Time's* obviously biased reporting, "Dictator Castro" was wildly popular among Cubans. The CIA, of course, knew this well.

The CIA's Cuban Task Force began planning the Bay of Pigs invasion in March 1960. The minutes from their first meeting read, "Colonel King said that the first problem is to reduce the base of support of the Castro government with the masses (reportedly the regime now enjoys the support of between 60 and 70 percent of the population)."[71] Therefore, the CIA's primary objective was to undermine Castro's popular support. Their multifaceted agenda included establishing *Radio Swan* on Great Swan Island, near Cuba, and subsequently pumping Cuba with continuous propaganda,[72] flying planes high above Havana and dropping anti-Castro pamphlets calling for violent revolution,[73] and exploding bombs on numerous occasions, including one in the Nobel Academy in Havana's Vibora district, which wounded seven students and one professor.[74]

The CIA didn't only propagandize Cubans; they also sought to manipulate the American public by portraying Castro as an imminent threat to democracy. Citing further declassified documents, the *New York Times* reported, "The agency's headquarters had 'the capability of placing items directly on the wire service tickers' as part of its 'regular propaganda apparatus.'"[75] The *Times* observed, "The newly declassified document says flatly that the intelligence agency could essentially dictate articles and have them sent around the world."[76] Despite such vast efforts to manipulate the public, in February 1961, Arthur Schlesinger, Jr., wrote a memorandum to President Kennedy acknowledging that Americans, and people throughout the world weren't swallowing the propaganda. "However well disguised any action might be," Schlesinger wrote, "it will be ascribed to the United States. The result would be a wave of massive protest, agitation and sabotage throughout Latin America, Europe, Asia and Africa (not to speak of Canada and of certain quarters in the United States)."[77] He instead suggested a more patient approach to manufacturing fear and generating support:

"One can conceive a black operation in, say, Haiti which might in time lure Castro into sending a few boatloads of men on to a Haitian beach in what could be portrayed as an effort to overthrow the Haitian regime. If only Castro could be induced to commit an offensive act, then the moral issue would be clouded, and the anti-US campaign would be hobbled from the start."[78]

Nevertheless, the CIA carried out their April 1961 Bay of Pigs invasion, an unsuccessful attempt by CIA-trained paramilitary forces to infiltrate and overthrow the Castro government. The Cuban army soundly defeated the invaders after just three days.

Although the CIA and American establishment were thoroughly embarrassed, they continued to push for military intervention. A declassified document dated March 13, 1962 demonstrates the Machiavellian ruthlessness of these elites and their utter disdain for both the American and Cuban populations. Operation Northwoods, signed by Lyman Louis Lemnitzer, Chairman of the Joint Chiefs of Staff, proposed staging terror attacks *against the American people* to justify military intervention in Cuba. The report read,

"This plan, incorporating projects selected from the attached suggestions, or from other sources, should be developed to focus all efforts on a specific ultimate objective which would provide adequate justification for U.S. military intervention. Such a plan would enable a logical build-up of incidents to be combined with other seemingly unrelated events to camouflage the ultimate objective and create the necessary impression of Cuban rashness and irresponsibility on a large scale, directed at other countries as well as the United States. The plan would also properly integrate and time phase the courses of action to be pursued. The desired resultant from the execution of this plan would be to place the United States in the apparent position of suffering defensible grievances from a rash and irresponsible government of Cuba and to develop an international image of a Cuban threat to peace in the Western Hemisphere."[79]

Among the many proposals, described in disturbing detail, were the following:

- "We could develop a Communist Cuban terror campaign in the Miami area, in other Florida cities and even in Washington. The terror campaign could be pointed at Cuban refuges seeking haven in the United States. We could sink a boatload of Cubans enroute to Florida (real or stimulated)."
- "We could blow up a US ship in Guantanamo Bay and blame Cuba."
- "Hijacking attempts against civil air and surface craft should appear to continue as harassing measures condoned by the government of Cuba."
- "It is possible to create an incident which will demonstrate convincingly that a Cuban aircraft has attacked and shot down a chartered civil airliner enroute to Jamaica, Guatemala, Panama or Venezuela. The destination would be chosen only to cause the flight plan route to cross Cuba. The passengers could be a group of college students off on a holiday or any group of persons with a common interest to support chartering a non-scheduled flight."[80]

Anyone who still believes that al-Qaeda carried out the 9/11 attacks should visit the National Security Archive's website and download and read the complete document.[81] They should then ask themselves, "Was 9/11 was simply an updated, more sophisticated version of Operation Northwoods?"

Rather than safeguarding the country from propaganda, and protecting it from undue influence, the National Security Act of 1947 has dramatically compromised "United States political processes, public opinion, policies, *and* media."[82] Anthropologist and researcher Francisco Gil-White asked, "What do you have when you give the president the power to corrupt the U.S. media and make it so that, in order to be discovered, the president must tell on himself?" He answered, "The opposite of a mechanism to prevent the corruption of the U.S. media. *This is an invitation to do it.*"[83] And when congress does investigate such activities, as with the Church Committee, their reports tend to filter more truths than they expose.

Help Wanted: Nazis Preferred

Having established the CIA just after the Nazi abomination of Word War II, the U.S. government presumably realized that government power, if left unchecked, quickly spirals out of control. Therefore, presumably, they decided to staff the CIA with only the highest caliber, most trusted men and women. Presumably they screened out any corrupt, power-hungry applicants, especially those with psychopathic or sadistic tendencies. But they didn't. Unfathomably, the U.S. government consciously *recruited and hired* the same degenerate Nazis who served Hitler's death-machine. Most Americans never learned this important lesson in history class, yet even the *Washington Post* has candidly reported,

> "It is no longer necessary—or possible—to deny the fact: the U.S. government systematically and deliberately recruited active Nazis by the thousands, rescued them, hired them and relied upon them to serve American interests and purposes in postwar Europe…the American public was told that U.S. policy was to round up and punish (not entice and reward) those who had been instrumental in Hitler's crimes against humanity."[84]

The government secretly hired thousands of Nazis, concealed this obscenity, and lied to the public. But what the *Post* neglected to mention, and what Gil-White so astutely recognized, is that by hiring these Nazis and lying about it, the government was not breaking any laws. He explained, "The U.S. government had *permission* to lie, because this permission was granted by the same National Security Act of 1947 that also created the CIA out of these Nazi thugs."[85] Hitler did the same thing. With the Enabling Act of 1933 and similar legislation, he preemptively rewrote his country's laws, safeguarding himself from future crimes.[86]

With the National Security Act, the U.S. had taken a page from Hitler's book. And their fledgling CIA was teeming with Nazis owing no allegiance to the United States. Therefore, that the CIA continues to intentionally subvert the media and mislead the public is hardly surprising. The *Toronto Star* reported,

> "The nature and extent of the post-war threat posed by the Soviet Union was grossly misrepresented by former Nazis and Nazi collaborators sitting comfortably in America on U.S. intelligence agencies' payroll. Indeed, many of the perceptions of the Soviet Union concocted by those twisted sources have remained the conventional wisdom of Americans until only recently."[87]

The *Star* further commented on alliances forged between these Nazis and certain American media moguls: "Their cooked-up anti-Communist propaganda was systematically (and illegally) disseminated throughout the U.S. by the CIA during the 1950s, significantly hardening public opinion towards the Cold War."[88]

It is difficult to overstate the CIA's enduring influence on the media. They have repeatedly been caught planting propaganda and terrorizing the public, both in America and abroad. Following the Church hearings, Edward Morgan penned an article for The *New York Times* questioning the effectiveness, and perhaps the intentions, of the committee's investigation. He stated,

> "What bothers me is the calm *after* the storm. The press has been had by the Central Intelligence Agency...Yet where is the loud collective outrage from what I like to consider the honorable trade of journalism, the only one I have ever plied? Can we minions of the news media be so busy righteously defending freedom of the press under the First Amendment that we have no time to discover (or admit) that we have been subverted?"[89]

Morgan was outraged by the lax restrictions proposed by the committee and the nonchalant attitude and apathy of the press. He griped that some 50 American journalists were linked to the CIA and yet, "Even under new restrictive guidelines, half of these relationships will be continued."[90] Of course the actual number was much higher than 50, as Bernstein and others have documented, yet the point remains — the Church Committee offered few assurances of meaningful change regarding the CIA's infiltration of the press. Morgan asked, "Even if the new Senate Intelligence Committee maintains a tough and skeptical overview and a tight leash on the CIA budget (an iffy prospect), how can we be sure how or whether the agency actually removes its honeycomb of activities from the catacombs of official secrecy, where the skeletons of power abuse still twitch?" He concluded, "The CIA's penetration of the Fourth Estate has created a treacherous and intolerable situation. It could well undermine what respect and integrity the press has left with a skeptical public." As Morgan and many others have insisted, the cloak of secrecy must be removed — a free press cannot exist under the influence of covert manipulation.

Further Government Propaganda

The CIA is not the only center of government-sponsored propaganda. The White House, Pentagon, and other politicians also heavily influence or directly control the press. For at least a century, esteemed journalists have depended upon "official sources" within the government for political news. This dependency creates subservience. These journalists are therefore highly unlikely to challenge the established order. Over the decades, countless media professionals have commented on this lamentable relationship:

- Henry Fowles Pringle, writing for *Harper's* in 1928, observed, "Most of the men who write the political news have become mouthpieces (without being paid by them, of course) for politicians...Probably seventy-five percent of the news that comes from Washington is news that some politicians desires to see in print. Whether it is true or false, propaganda or not, is seldom even a matter for idle reflection."[91] Why don't journalists harshly criticize or print disparaging remarks about politicians? Describing one particular instance, Pringle explained, "Had we attempted to give our papers the truth (a course in which we should probably have failed) it would have been impossible for us to have obtained news from this source in the future. More than that, we should have become objects of suspicion and resentment among other politicians, and our usefulness as political writers would have been dissipated."[92]

- Joseph Kraft, in a 1965 article for *Harper's* entitled, "Politics of the Washington Press Corps," remarked, "In the typical Washington situation, news is not nosed out by keen reporters and then purveyed to the public. It is manufactured inside the government, by various interested parties for purposes of their own, and then put out to the press in ways and at times that suit the source."[93]

- Leonard Downie Jr., longtime editor of the *Washington Post*, observed in 1976, "To examine critically the institutions and mores of government might mean breaking friendships with trusted government contacts, missing the consensus front-page stories everyone else is after, of failing to be followed down a new path of inquiry."[94] In other words, since nobody else critically examines government, we can't afford to be the only ones to do so.

- According to Pulitzer Prize winning historian Doris Kearns Goodwin, former President Lyndon Johnson stated, "Reporters are puppets. They simply respond to the pull of the most powerful strings. Every reporter has a constituency in mind when he writes his stories. Sometimes it is simply his editor or his publisher, but often it is some political group he wants to please or some intellectual society he wants to court. The point is there is always someone. Every story is always slanted to win the favor of someone who sits somewhere higher up. There is no such thing as an objective news story."[95]

- Wes Gallagher, former *Associated Press* President, admitted in the mid-70s, "The Washington politician's view of what is going on in the United States has been substituted for what is actually happening in the country."[96]

- Stephen Hess of the Brookings Institute, a powerful Washington nongovernmental organization wrote in 1981, "This is a well-known 'secret' in the press corps: *Washington news is funneled through Capitol Hill.*"[97]

- Stanford Communications professor Theodore Glasser wrote in 1984, "Sources supply the sense and substance of the day's news. Sources provide the arguments, the rebuttals, the explanations, the criticism."[98]

- Walter Karp, in a 1989 article for *Harper's* entitled, "All The Congressmen's Men: How Capitol Hill Controls The Press," explained, "For to say that the press *does* things conceals the fundamental truth that the press, strictly speaking, can scarcely be said to do anything. It does not act, it is acted upon. This immediately becomes clear when one considers how and where reporters find the news. Very few newspaper stories are the result of reporters digging in files; poring over documents; or interviewing experts, dissenters, or ordinary people. The overwhelming majority of stories are based on official sources—on information provided by members of Congress, presidential aides, and politicians…It is a bitter irony of source journalism that the most esteemed journalists are precisely the most servile. For it is by making themselves useful to the powerful that they gain access to the 'best' sources."[99]

White House Control

Whether through payoffs, threats, intimidation, or outright lies, the White House has been controlling the press for many decades. While examples are plentiful, the following analysis examines just four cases—one from the 1970s, two from the 1980s and one from the recent Bush administration:

Example 1: On October 27, 1972, CBS News broke the media's silence on Watergate. Walter Cronkite detailed "charges of a high-level campaign of political sabotage and espionage apparently unparalleled in American history."[100] The following day, White House staffer Charles Colson kicked off a thuggish intimidation campaign against CBS, targeting their highest executives. Colson demanded that CBS substantially weaken their scheduled follow-up report on the scandal. Furthermore, he threatened that should CBS continue broadcasting unfavorable news about the president, the White House would use Wall Street and Madison Avenue to financially destroy the company. To CBS Vice Chairman Frank Stanton, Colson avowed, "We'll break your network."[101] Stanton, rather than exposing this criminal blackmail, cowardly succumbed. *Harper's* reflected, "What else is a free press for if not to help a free people hold the powerful to account? Yet here was a president grossly abusing the power of his office (which was newsworthy in itself) in order to censor the news (which was doubly newsworthy) so that the electorate might not hold him accountable at the polls — which was newsworthy three times over."[102] Only years later, when the Administration was fully unraveling, did Stanton reveal Colson's tactics.[103]

Example 2: As reported by the *Washington Post*, the Reagan Administration, starting in August 1986, launched a massive disinformation campaign "designed to convince Libyan leader Moammar Gadhafi that he was about to be attacked again by U.S. bombers and perhaps be ousted in a coup."[104] The campaign involved planting false and greatly exaggerated stories in U.S. and foreign newspapers. Having acquired leaked government memos, the *Post* reported in October 1986 that, "Senior administration officials have been frustrated that Gadhafi has been able to remain in power despite a presidentially authorized, year-long CIA effort to oust him." The *Post* further explained how the U.S. government terrorized Americans by falsely suggesting that Gadhafi was planning terror attacks against America: "Beginning with an Aug. 25 report in the *Wall Street Journal*, the American media — including the *Washington Post* — reported as fact much of the false information generated by the new plan. Published articles described renewed Libyan backing for terrorism and a looming, new US-Libyan confrontation." In other words, the U.S. government engaged in psychological warfare against the American people. Perhaps Gadhafi was their ultimate target, but they were completely indifferent to the inevitable side effect — the terrorization of American citizens. The press insisted they had no prior knowledge of this disinformation campaign. At the reporter level, perhaps this was true. However, considering the incestuous CFR-CIA-government-media relationship, complicity and cooperation at higher levels was almost certain.

Example 3: Concurrent with the Libya deception, the Reagan administration, for three years, was manipulating the media to deceive the American public about Nicaragua. According to the *Miami Herald*, "[The strategy] was to gradually destroy the favorable image Nicaragua enjoyed following the Sandinista revolutionary triumph and neutralize the once intense opposition in Congress to Reagan's policies of providing military aid to El Salvador and contra rebels."[105] A typical example of this deception occurred on election night 1984 when the White House leaked information to the press that a Soviet ship was transporting MiG combat jets to Nicaragua. The *Herald* later admitted, "No MiGs were aboard the ship, but the leak enhanced administration efforts to portray Nicaragua as a potential military threat."[106]

Example 4: During the George W. Bush administration, the White House wasted millions of taxpayer dollars paying off reporters in return for favorable news coverage and buying fake news segments, called Video News Releases (VNRs). VNRs looked like genuine reports, yet they were paid advertisements deceptively inserted into news programming without informing audiences. During Bush's first term alone, the administration spent $250 *million* on public relations, of which at least $97 million went to Ketchum Communications.[107] Through a Freedom of Information Act request, *USA Today* obtained documents showing that Armstrong Williams, a nationally syndicated television show host, received $240,000 to promote Bush's No Child Left Behind (NCLB) program. Williams' contract with Ketchum required him "to regularly comment on NCLB during the course of his broadcast," and to invite Education Secretary Rod Paige onto his show several times for interviews.[108] Williams was also required to use his contacts with America's Black Forum, a group of black broadcast journalists, "to encourage the producers to periodically address" NCLB.[109] *New York Times* columnist Frank Rich observed, "It is a brilliant strategy. When the Bush administration isn't using taxpayers' money to buy its own fake news, it does everything it can to shut out and pillory real reporters who might tell Americans what is happening in what is, at least in theory, their own government."[110]

In 2005, Bush asserted, "There needs to be a nice independent relationship between the White House and the press."[111] Yet concurrent with this statement, the White House clarified that, "The president's prohibition did not apply to government-made television news segments."[112] In three separate decisions, the Government Accountability Office (GAO), the investigative arm of Congress that studies the government and its expenditures, ruled that VNRs were illegal "covert propaganda" because they were "designed and executed to be indistinguishable from news stories produced by private sector television news organizations."[113] Nevertheless, the administration's response was simply to disregard these rulings. Furthermore, the Justice Department circulated a memo to all executive branch agencies ordering them to ignore the GAO ruling.[114] When asked about the GAO ruling, Bush declared, "There is a Justice Department opinion that says these—these pieces are within the law so long as they're based upon facts, not advocacy."[115] The Bush administration, however, frequently blurred the lines between facts and advocacy—weapons of mass destruction in Iraq, for example.

Pentagon Newsmakers

Similar to the White House program to *buy* the news, the Pentagon had a program to *make* the news. To mold public opinion regarding the Iraq war, the Pentagon recruited more than 75 retired military officers to masquerade on TV as "independent" military analysts. They provided these analysts with talking points, which the analysts subsequently shoveled upon the unsuspecting viewers of Fox, NBC, CNN, CBS, and ABC. These analysts fetched handsome fees from the networks—$500 to $1,000 per appearance—and also profited from their financial ties to military contractors and military lobbying groups. The *New York Times* reported, "The group was heavily represented by men involved in the business of helping companies win military contracts. Several held senior positions with contractors that gave them direct responsibility for winning new Pentagon business...Still others held board positions with military firms that gave them responsibility for government business...Several were defense industry lobbyists."[116]

The *Times* originally broke this story after successfully suing the Defense Department to acquire 8,000 pages of emails, transcripts and records fully documenting an organized program to groom these analysts for effective propagandizing. They reported, "These records reveal a symbiotic relationship where the usual dividing lines between government and journalism have been obliterated. Internal Pentagon documents repeatedly refer to the military analysts as 'message force multipliers' or 'surrogates' who could be counted on to deliver the administration's 'themes and messages' to millions of Americans 'in the form of their own opinions'…Again and again, records show, the administration has enlisted analysts as a rapid reaction force to rebut what it viewed as critical news coverage."[117]

The networks fully complied with the arrangement. For example, "they did not hold their analysts to the same ethical standards as their news employees regarding outside financial interests." While regular employees were required to disclose such interests, the networks maintained a different policy—don't ask, don't tell—for military analysts. The program was largely successful. For example, the documents show Brent Krueger, a senior aide involved with the project, bragging that on some days, "We were able to click on every single station and every one of our folks were up there delivering our message."[118] Some analysts, however, were angered when the military asked them to lie. Robert S. Bevelacqua, a Fox News analyst and retired Green Beret, attended a special briefing for select analysts in early 2003 regarding Iraq's weapons of mass destruction. He recalled asking the briefer if the United States had "smoking gun" proof. "We don't have any hard evidence," was the reply. Bevelacqua said he and other analysts were alarmed by this concession: "We are looking at ourselves saying, 'What are we doing?'"[119]

In February 2002, the Pentagon created a new office with a multimillion-dollar budget to propagandize the War. The *Times* reported, "The military has long engaged in information warfare against hostile nations—for instance, by dropping leaflets and broadcasting messages into Afghanistan when it was still under Taliban rule. But it recently created the Office of Strategic Influence, which is proposing to broaden that mission into allied nations in the Middle East, Asia and even Western Europe."[120] The head of the new office, General Simon Worden, "envisions a broad mission ranging from 'black' campaigns that use disinformation and other covert activities to 'white' public affairs that rely on truthful news releases, Pentagon officials said." Joint Chiefs chairman General Meyers even acknowledged Pentagon plans to target civilians with psychological warfare operations: "Perhaps the most challenging piece of this is putting together what we call a strategic influence campaign quickly and with the right emphasis. That's everything from psychological operations to the public affairs piece to coordinating partners in this effort with us."[121] Furthermore, the *Times* acquired a classified Pentagon briefing stating, "the office should find ways to 'coerce' foreign journalists and opinion makers and 'punish' those who convey the wrong message."[122]

Facing a heavy backlash of criticism, the Pentagon closed the office days after its existence surfaced in the press. Donald Rumsfeld announced, "The office has clearly been so damaged that it is pretty clear to me that it could not function effectively."[123] Of course he didn't say it was a bad idea, just that too many people found out. Nevertheless, the Pentagon simply enacted plan B—they outsourced the project to private companies. In December 2003, the *Times* acquired an internal Pentagon document called, "Winning the War of Ideas." According to the document, a major Department of Defense goal was to establish a "road map for creating an effective D.O.D. capability to design and conduct effective strategic influence and operational and tactical perception-management campaigns."[124] It also revealed a $300,000 contract involving SAIC, a major defense consultant. The Pentagon hired SAIC to develop an "effective strategic influence" campaign to combat global terror.[125] Senior Pentagon officials claimed a low-level staffer had written the document and his "poor choice of words" had simply caused a misunderstanding.

Chapter 26
Big Media and Government Collusion

Whereas the last chapter documented government manipulation of the media, the media also frequently manipulates the government. Or more accurately, government and media are simply two different heads of the same beast. Neither truly manipulates the other. Rather, both work together to manipulate the public. This is especially apparent regarding America's governmental media regulatory body, the Federal Communications Commission (FCC). This chapter shows how media conglomerates and the FCC have colluded to misinform the pubic. The United States, however, is certainly not the only country where such collusion occurs. For example,

- In Italy, Prime Minister Silvio Berlusconi outright owns and controls much of the Italian media. Veteran CBS News correspondent Tom Fenton explained, "As head of the government, he already ran state-owned RAI television. Only in the old days, the political parties that used to make up the governing coalitions shared between them the control of the different RAI channels. Berlusconi did away with the old arrangement and now virtually dictates what runs on RAI. What's more, he has used his influence and personal fortune to buy leading newspapers and build an empire of private television channels that feed the public a diet of his right-wing views and trashy game and girlie shows." [1]

- French weapons conglomerates Lagardere and Serge Dassault own and control more than 70 percent of the French press.[2] For example, Serge Dassault, the dominant French arms manufacturer, owns *Le Figaro*, the leading conservative daily periodical. In August 2004, Dassault met with *Le Figaro*, journalists and told them, "There are times when it's necessary to take many precautions. There are articles that talk about contracts under negotiation. There is some information that is more bad than good. And this puts at risk the commercial and industrial interests of our country." [3]

- After the fall on the Soviet Union, the Russian media attained some degree of autonomy. Television and independent newspapers began exposing the brutality of the Russian army and government. "But after Vladimir Putin was elected president," explained Fenton, "things began to change…he reasserted tight control over television—the main means of propaganda in a vast country where few people can still afford a daily newspaper. Even television news magazines and talk programs, which had briefly flourished in the 1990s, were curbed." [4]

- Most Arab nations also have state-controlled media, with only minor competition from satellite channels such as Al Jazeera, Al Arabiya, and Al Hura (funded by the U.S.). Fenton noted, "The proliferation of channels has taken the pressure off state-controlled channels to make any internal change." [5]

- China strictly censors all Internet sites deemed politically hostile, especially those concerning Tibetan and the 1989 Tiananmen massacre.

- In 2008, under the guise of fighting child pornography, Australia also announced plans to censor the net. Communications minister Stephen Conroy declared, "We are talking about mandatory blocking, where possible, of illegal material. The *Melbourne Herald Sun* noted, "The government has declared it will not let Internet users opt out of the proposed national Internet filter." [6]

Federal Communications Commission

According to U.S. law, the radio spectrum (the medium for radio and television broadcasting) is public property. Radio and television companies are therefore legally obligated to serve the public's interest. The Federal Communications Commission (FCC) is the government agency responsible for protecting public interest by monitoring and regulating these companies. Created by the Communications Act of 1934, the FCC grants licenses to commercial broadcasters for commercial use. In return, broadcasters agree to serve the public's interest by offering relevant news programming. The bill clearly states, "Nothing in the foregoing sentence shall be construed as relieving broadcasters, in connection with the presentation of newscasts, news interviews, news documentaries, and on-the-spot coverage of news events, from the obligation imposed upon them under this Act to operate in the public interest and to afford reasonable opportunity for the discussion of conflicting views on issues of public importance."[7] However, the FCC's "revolving door" has since heavily compromised the Communications Act of 1934.

The revolving door is a euphemism for the back and forth movement of media executives between media corporations and the FCC. University of Illinois professor Robert McChesney observed, "The regulators for the Federal Communications Commission almost always upon leaving office, go to work for the people they were supposedly regulating in the public interest."[8] Veteran media expert Dennis Mazzocco elaborated,

> "Currently, the Federal Communications Commission, like other federal regulatory bodies, is primarily dominated by ex-officials of media companies, all of whom have been appointed on the basis of political, economic, or social ties to the upper echelon of corporate America. Although the commission is mandated by law to include a balanced representation from both the Democratic and Republican parties, the vast majority of past FCC commissioners have far more links to power than the average citizen. These ties are often the only reason why one is placed on the FCC, or other federal regulatory agencies."[9]

With a truly independent FCC, the unprecedented media deregulation of the past two decades would have been politically impossible. Nevertheless, by the mid-1990s, the leading media corporations had successfully infiltrated the FCC, and via the Telecommunications Act of 1996, they started heavily consolidating media ownership.

According to the FCC, "The Telecommunications Act of 1996 is the first major overhaul of telecommunications law in almost 62 years. The goal of this new law is to let anyone enter any communications business — to let any communications business compete in any market against any other. The Telecommunications Act of 1996 has the potential to change the way we work, live and learn. It will affect telephone service — local and long distance, cable programming and other video services, broadcast services and services provided to schools."[10] The Act has certainly changed "the way we work, live and learn," yet not necessarily for the better.

Supporters of the Telecommunications Act claimed it would foster competition. Instead, the opposite has occurred. In 2003, media mogul Ted Turner acknowledged, "There's really five companies that control 90 percent of what we read, see and hear."[11] In 2005, Representative Maurice Hinchey introduced the *Media Ownership Reform Act of 2005*, noting that,

"Despite the increase in information outlets, ownership and control of those is shrinking. A handful of companies control a large portion of both programming and distribution. Five companies now own the broadcast networks, 90 percent of the top 50 cable networks, produce three-quarters of all prime time programming, and control 70 percent of the prime time television market share. The same companies that own the nation's most popular newspapers and networks also own over 85 percent of the top 20 Internet news sites."[12]

Hinchey's resolution died in committee. On December 31, 2005, Viacom and CBS Corporation split. Therefore, the following six companies now control the vast majority of American media:

- News Corporation (FOX, HarperCollins, *New York Post*, *Wall Street Journal*, *The Times (London)*, 30 newspapers and 35 TV stations)
- General Electric (NBC, Bravo, Universal Pictures and 28 TV stations)
- Time Warner (AOL, CNN, HBO, Turner Broadcasting System, Time Incorporated, New Line Cinema and 130-plus magazines)
- Disney (ABC, Disney Channel, ESPN, 10 TV and 72 radio stations)
- CBS Corporation (CBS, Simon & Schuster)
- Viacom (Comedy Central, BET, Nickelodeon, MTV, VH1, CMT, Paramount Pictures and DreamWorks)

The Telecommunications Act of 1996 exacerbated an already prevalent consolidation trend, a trend Pulitzer Prize-winning reporter Ben Bagdikian has been studying since the early 1980s. The former Editor of the *Saturday Evening Post*, former Assistant Managing Editor for the *Washington Post* and current Dean Emeritus of the University of California at Berkeley's Graduate School of Journalism, Bagdikian is a recognized media expert.[13] In his 1983 book *The Media Monopoly*, he warned that only 50 companies controlled the production and distribution of the vast majority of American media including magazines, radio, TV programs, books, motion pictures, cable, and music. Critics dismissed him as an alarmist. Nevertheless, by 1992, ownership had shrunk to only 20 companies,[14] and by 1997, to only 10.[15] In his 2004 book, *The New Media Monopoly*, Bagdikian lamented,

- "Five global-dimension firms, operating with many of the characteristics of a cartel, own most of the newspapers, magazines, book publishers, motion picture studios, and radio and television stations in the United States."
- "Today, none of the dominant media companies bother with dominance merely in a single medium. Their strategy has been to have major holdings in all the media, from newspapers to movie studios. This gives each of the five corporations and their leaders more communications power than was exercised by any despot or dictatorship in history."
- "No imperial ruler in past history had multiple media channels that included television and satellite channels that can permeate entire societies with controlled sights and sounds. The leaders of the Big Five are not Hitlers and Stalins. They are American and foreign entrepreneurs whose corporate empires control every means by which the population learns of its society. And like any close-knit hierarchy, they find ways to cooperate so that all five can work together to expand their power, a power that has become a major force in shaping contemporary American life. The Big Five have similar

boards of directors, they jointly invest in the same ventures, and they even go through motions that, in effect, lend each other money and swap properties when it is mutually advantageous."[16]

According to the *Columbia Journalism Review*, in 2003, News Corporation, Disney, Viacom, and Time Warner had 45 interlocking directorships.[17]

Bagdikian is certainly not the only prominent critic of media consolidation. The aforementioned professor Dennis Mazzocco observed, "In helping to 'manufacture consent,' concentrated media ownership continues to provide elite business interests with an awesome state propaganda apparatus for citizen thought control. Corporate media systems, even when they promise profits, prosperity, and freedom have remained steadfast in their opposition to citizen control of media."[18] During his years at ABC, Mazzocco realized, "The company's prime endeavor has been to argue extensively in Washington before the FCC, the Justice Department, and the federal courts, for even wider broadcast deregulation which would allow it (and its U.S. media cartel partners) to grow even larger. The inevitable result will be a media system that sells, and rarely criticizes, the government or the corporate structure that protects its continued existence."[19]

On February 8, 1996, President Bill Clinton quietly signed the Telecommunications Act into law. In his book *Corporate Media and the Threat to Democracy*, University of Illinois professor Robert McChesney quoted a career lobbyist as saying, "I have never seen anything like the Telecommunications Bill. The silence of public debate is deafening. A bill with such astonishing impact on all of us is not even being discussed."[20] It wasn't being discussed because the major networks weren't discussing it! During *9 months* of congressional debate, the major networks devoted a combined total of *19 minutes* of coverage.[21] Charles Lewis, director of the Center for Public Integrity, observed, "The media did not want to discuss what it was doing."[22] Furthermore, unbeknownst to the public, the Telecommunications Act gifted the major conglomerates the exclusive broadcasting rights to the emerging digital spectrum, worth an estimated $70 billion.[23]

The Act also eliminated restrictions preventing companies from owning more than 40 radio stations. Consequently, major conglomerates started voraciously acquiring smaller stations. Ownership dropped nearly 30% in the first five years, a reduction of 1,100 owners.[24] By late 2006, Clear Channel owned more than 1,150 stations nationwide, a *2,275 percent increase* over the previously permitted ownership level. They also owned as many as eight stations per individual market.[25] In 2002, the Future of Music Coalition published *Radio Deregulation: Has It Served Citizens and Musicians?* They concluded, "Oligopolies control almost every geographic market. Virtually every geographic market is dominated by four firms controlling 70 percent of market share or greater. In smaller markets, consolidation is more extreme. The largest four firms in most small markets control 90 percent of market share or more."[26]

In 2004, McChesney concluded, "We grant these monopoly rights to TV stations, they don't pay a penny to the people for getting these monopoly rights to TV frequencies. Then they turn around and sell time on the public property and make billions of dollars and destroy our political system. All the while they pay off the politicians themselves to keep getting these monopoly rights. The corruption here is dreadful."[27] Also in 2004, Senator Bernie Sanders observed,

"The truth of the matter is, that increasingly, what we see, what we hear, and what we read is being controlled by fewer and fewer large, multinational corporations...People don't appreciate this, you know in the last days of the Soviet Union you had dozens of newspapers, dozens of magazines, all kinds of radio and television stations, the only problem is that all of the media was controlled by the Communist Party of the Soviet Union or the government of the Soviet Union. We are moving in that direction."[28]

In 2005, Common Cause, a nonprofit public interest group headed by former Congressman Bob Edgar, concluded,

"In many ways, the Telecom Act failed to serve the public and did not deliver on its promise of more competition, more diversity, lower prices, more jobs and a booming economy. Instead, the public got more media concentration, less diversity, and higher prices...And study after study has documented that profit-driven media conglomerates are investing less in news and information, and that local news in particular is failing to provide viewers with the information they need to participate in their democracy."[29]

Further Deregulation

In June 2003, the FCC voted 3-2 to relax media ownership rules even further. The new laws allowed single companies to own television stations reaching a combined 45% of a market's audience share, up from the previous 35% limitation. The new laws also enabled companies to own a broadcasting outlet *and* a newspaper in the same city, ending restrictions against cross-ownership in all but the smallest markets. FCC Chairman Michael Powell disregarded concerns about further consolidation. He stated, "I have heard the concerns expressed by the public about excessive consolidation. Though generalized worries do not clearly suggest specific answers to the specific issues the commission must address, they have introduced a note of caution in the choices we have made."[30] However, FCC Commissioners Michael Copps and Jonathan Adelstein, the two nay votes, sharply disagreed. Copps announced, "This path surrenders to a handful of corporations awesome powers over our news, information, and entertainment. On this path, we endanger time-honored safeguards and time-proven values that have strengthened the country, as well as the media."[31] Adelstein noted, "Judging from our record, public opposition is nearly unanimous, from ultra-conservatives to ultra-liberals and virtually everyone in between. We have received about three-quarters of a million comments from the public in opposition to relaxing our ownership rules — a new record — and only a handful in support."[32]

PBS correspondent Terence Smith asked Powell, "In the hearing today, there was mention of some 750,000 comments that the commission received on this and Commissioner [Adelstein] said that 99.9 percent of those were opposed to it. What does that say to you?" Powell responded, "Well, I don't know how accurate that is. I don't think anybody has ever conducted a truly meaningful survey of what we received but I will concede the point that we had a strong amount of e-mail and news that expressed concern.[33] However, three weeks earlier, *The Nation* had examined roughly half of the 18,000 public statements then filed with the FCC and found that 97 percent opposed permitting further media concentration.[34]

Despite stunning opposition, Powell seemed utterly determined to do the media giants' bidding. He claimed he had "research" justifying further relaxation of media ownership rules. This "research," however, he refused to release to the public. Consequently, 300 academics signed a letter to the FCC protesting Powell's secrecy.[35] Also

Chicago's City Council passed a resolution declaring, "Unchecked media consolidation benefits a small number of corporate interests at the expense of the public interest."[36] Also Senator Olympia Snowe and fourteen other senators sent a letter to the FCC declaring, "We believe it is virtually impossible to serve the public interest in this extremely important and highly complex proceeding without letting the public know about and comment on the changes you intend to make to these critical rules."[37] Furthermore, as reported by *The Nation*, "Nearly 100 House Democrats signed a letter by Representatives Bernie Sanders, Maurice Hinchey and Sherrod Brown calling on Powell to delay the June 2 vote on the rules, open the process to public comment and demonstrate how his proposed changes in ownership limits will serve the public interest by promoting diversity, competition and localism."[38]

Powell scheduled just one public hearing and refused to delay the vote. *The Nation* commented, "Powell's contempt for public opinion, evidenced by his scheduling of only one official hearing on the proposed rule changes, is so great that he refused invitations to nine semiofficial hearings at which other commissioners were present. The hearings drew thousands of citizens and close to universal condemnation of the rule changes."[39]

On July 24, 2003 The House of Representatives voted overwhelmingly (400-21) to reverse the FCC's decision. The *New York Times* commented, "The vote, a clear repudiation of Mr. Powell, suggested that he miscalculated the widespread opposition to the new rules."[40] Representative David Obey of Wisconsin announced, "The House has now repudiated the FCC's attempted giveaway of the public airways to national media giants based in New York and LA...I hope the administration is listening and will fix its flawed policy, so citizens can get accurate, free-flowing information—the lifeblood of democracy."[41] On September 17, the Senate followed suit, voting 55-40 on a resolution to repeal the new media regulations. Speaking out against Chairman Powell, Senator Byron Dorgan, the chief sponsor of the resolution, commented, "I think he has made a horrible mistake. His leadership at the commission has led the commission to cave in to the special interests as quickly and as thoroughly as I've ever seen."[42]

On September 3, the U.S. Third Circuit Court of Appeals issued a decision temporarily blocking the enforcement of the new rules, pending the outcome of litigation brought against the FCC by various citizen groups in the case of *Prometheus Radio Project v. FCC*. On June 24, 2004, the Court ruled 2-1 against the FCC, blocking the new rules unless and until the FCC could provide detailed justification for their position. *Democracy Now* reported, "The Court ordered the commission to revisit the issue, saying it should focus on protecting, rather than undermining, the public interest in diverse ownership of local and national media."[43] The people had spoken and finally their elected representatives had listened—but not for long.

In September 2006, Senator Barbara Boxer publicized a sensitive FCC study on media concentration obtained, according to Boxer's office, "indirectly from someone within the FCC who believed the information should be made public."[44] The *Associated Press* reported, "The Federal Communications Commission ordered its staff to destroy all copies of a draft study that suggested greater concentration of media ownership would hurt local TV news coverage."[45] Adam Candeub, a former FCC lawyer, admitted that senior managers ordered "every last piece" of the report destroyed. He explained, "The whole project was just stopped—end of discussion." In a letter to new FCC Chairman Kevin Martin, Boxer said she was "dismayed that this report, which was done at taxpayer expense more than two years ago, and which concluded that localism is beneficial to the public, was shoved in a drawer."

She also wondered rhetorically whether the report was "shelved because the outcome was not to the liking of some of the commissioners and/or any outside powerful interests?"[46]

The FCC seemed to harbor unlimited contempt for public opinion. In November 2007, the FCC published revised proposals for media ownership rules. Not surprisingly, the new proposals continued to push for relaxing the cross-ownership ban. The FCC's new strategy was to appeal for public sympathy by portraying the newspapers as "struggling" businesses. An FCC press release declared, "According to almost every measure newspapers are struggling. At least 300 daily papers have stopped publishing over the past thirty years. Their circulation is down, their advertising revenue is shrinking and their stock prices are falling. Permitting cross-ownership can preserve the viability of newspapers by allowing them to share their operational costs across multiple media platforms."[47] FCC Chairman Kevin Martin reiterated these desperate pleas in a *New York Times* editorial. He declared, "If we don't act to improve the health of the newspaper industry, we will see newspapers wither and die. Without newspapers, we would be less informed about our communities and have fewer outlets for the expression of independent thinking and a diversity of viewpoints."[48]

Were the newspapers really facing such dire straits? FCC commissioner Michael Copps retorted, "Merrill Lynch put the average return on newspapers at 17 or 18 percent. I wish I had some investments that were making 17 or 18 percent."[49] Copps later observed, "We shed crocodile tears for the financial plight of newspapers—yet the truth is that newspaper profits are about double the S&P 500 average."[50] Even if the newspapers were to fail, where is the crisis? Tired of the lies and propaganda, people are increasingly rejecting these corporate news outlets and instead embracing the alternative media. If the newspaper giants were to fail, the ruling class, which the newspapers safeguard, would weaken and society would thus be *better* informed.

Another of the FCC's shameless tactics was their calculated suppression of public opinion. After the 2003 ordeal, they knew the vast majority of Americans would not support further consolidation. Therefore, they offered only one public hearing, held on just seven day's notice. On the day of the hearing, November 9, 2007, Representative Byron Dorgan protested, "There will be, it appears to me, perhaps a month, maximum, for the American people to weigh in on a new rule that will be proposed for final action on December 18. That doesn't meet any test of reasonableness or any standard that I know that makes any sense."[51]

More than 200 irritated citizens attended the sole public hearing, held in Seattle. Chairman Martin patronized the audience, stating, "You're asking why the rush? And why no notice?" The audience roared, acknowledging the obvious. Martin continued, "Throughout this process, I've been as transparent as I could be."[52] Prolonged sighs and boos filled the auditorium, prompting headlines the following day such as, "Seattle crowd blasts FCC on big media" by the *Seattle Times*,[53] and, "Seattle Opens Can 'o Whoop Ass on FCC Chairman" by the *Huffington Post*.[54] Nevertheless, Martin was aloof and indifferent. He knew he had the majority vote (of the FCC's five-person board).

The two dissenting Commissioners spoke candidly about the situation. Commissioner Jonathan Adelstein told attendees, "If you see a proposal for more consolidation made quickly after this final hearing, you'll know your input was dismissed."[55] Sure enough, as soon as Martin arrived back in Washington, he published his aforementioned *New York Times* editorial. In other words, the Seattle hearing was simply a charade. Martin and the FCC majority were working for the media conglomerates, and public opinion was not going to foil their agenda. Despite continued protests, the FCC voted 3-2 on December 18, 2007 to approve the new rules. Michael Copps, the other dissenting Commissioner, lamented, "I had an opportunity to read a little bit of George Orwell the other day, and it was good preparation for getting ready to deal with this particular item."[56]

Fox News: Consequences of FCC and Big Media Collusion

Rupert Murdoch's News Corporation is the world's second largest media conglomerate. Its major holdings include Fox Broadcasting Company, the *Wall Street Journal*, Times Newspapers Ltd. (U.K.), HarperCollins publishing, and scores of other newspapers, magazines, television stations, Internet companies and film studios. In 1986, Murdoch launched Fox after paying $1.55 billion to acquire independent television stations in New York, Washington, Los Angeles, Houston, Chicago and Dallas. In Washington, he bought the already successful station WTTG. Three years later, according to former WTTG producer Frank O'Donnell, Murdoch began dictating content. "We were stunned," O'Donnell recalled, "because up until that point we were allowed to do legitimate news. And suddenly we were ordered, from the top, to carry propaganda…It was made very clear to us that our activities were being monitored and if someone wasn't watching it live they were at least recording it and they would review it after the fact to see what we did."[57]

In 1996, Murdoch launched the Fox News Channel. Despite their slogans, "Fair and Balanced," and "We Report, You Decide," Fox News exists to protect the ruling class. Former Fox anchor Jon Du Pre admitted, "We weren't necessarily, as it was told to us, a news gathering organization so much as we were a proponent of a point of view."[58] Jeff Cohen, a former MSNBC/Fox news contributor further explained, "It's very hard on Fox News to separate news from commentary because it all blends together. That's what makes it so ridiculous, that slogan 'We report, you decide,' because there's no TV news channel in history that's ever reported less."[59] The company actually dictates (or at least has in the past) daily talking points to its reporters. Du Pre explained, "When Headquarters sent the memo every morning and said, 'we want to touch on the following issues, we want to cover the following stories, we want to do them in this particular way,' our job and our objective then was to execute the plan."[60] During the 9/11 Commission's official investigation, for example, Fox ordered reporters to bury any leads pointing towards government duplicity. An Internal Fox Memo stated, "The so-called 9/11 commission has already been meeting. In fact, this is its eighth session. The fact that former Clinton and both former and current Bush administration officials are testifying gives it a certain tension, but this is not, 'what did he know and when did he know it' stuff. Don't turn this into Watergate."[61]

How far does Fox News go to protect the establishment? Biotech leviathan Monsanto is the quintessential establishment corporation. They are intimately tied to the FDA, the military, the Obama administration, and the past three presidential administrations. So what happened when Jane Akre and Steve Wilson, two award-winning reporters from Tampa Bay's local Fox News affiliate WTVT conducted an investigation of Monsanto and their drug Posilac, a genetically engineered bovine growth hormone designed to increase milk production in dairy cows? The team produced a powerful report documenting (a) Monsanto's failure to properly test Posilac before receiving FDA approval,[62] (b) the FDA's failure to adhere to their own approval standards, (c) the drug's potential cancer risks according to peer-reviewed studies,[63] and (d) Monsanto's attempts to bribe Canadian government officials.[64]

WTVT management approved the four-part series and scheduled the first airing for February 24, 1997. Then Monsanto found out. Shortly thereafter a letter passed from John Walsh, a lawyer representing Monsanto, to Roger Ailes, Chairman and CEO of Fox News. Walsh viciously attacked the report, claiming it contained "recklessly made accusations" and "unsupported speculation."[65] Consequently, WTVT management agreed to postpone the series to allow for more thorough accuracy checks. However, after one week, the station's editors and lawyers reapproved the report. Monsanto redoubled with another letter, this time threatening "dire consequences for Fox News" should they decide to air the series.[66]

Fox acquiesced. Akre and Wilson, however, were not willing to compromise their journalistic integrity. After 73 rewrites and revisions, David Boylan, General Manager of WVTV, flatly ordered them to falsify their report. Akre and Wilson refused, instead threatening to report Fox to the FCC for news distortion.[67] Boylan replied, "We paid $3 billion for these television stations. We will decide what the news is. The news is what we tell you it is."[68] Fox then attempted to silence the reporters with a $125,000 cash settlement. They again refused. Finally in December 1997, the station fired the duo for insubordination and refusing to be objective.[69] Akre sued Fox. Two and a half years later, a Florida jury awarded her $425,000 in damages. The *Village Voice* reported, "The reporters say the verdict proves that Fox deliberately slanted the news."[70] Nevertheless, Fox appealed the case and won. Florida's Second District Court of Appeals determined that distortion and falsification of the news does not violate any law—it merely violates FCC *policy*. The Court ruled, "We agree with WTVT that the FCC's policy against the intentional falsification of the news— which the FCC has called its 'news distortion policy'—does not qualify as the required 'law, rule, or regulation' under section 448.102."[71]

This opinion set a dangerous precedent. It confirmed Boylan's arrogant insistence that, irrespective of truth, "The news is what we say it is." Furthermore, the opinion contradicted a 1968 FCC ruling against CBS regarding an intentionally distorted report entitled, "Hunger in America." In that case, the FCC ruled, "As a public trustee, the broadcaster may not engage in intentional and deliberate falsification (distorting, slanting, rigging, staging) of the news...As the Commission stated in its ruling in Hunger in America, 'Rigging or slanting the news is a most heinous act against the public interest—indeed, there is no act more harmful to the public's ability to handle its affairs.'"[72] The Florida Court's opinion also defied a 1969 U.S. Supreme Court decision stating, "It is the purpose of the First Amendment to preserve an uninhibited market-place of ideas in which truth will ultimately prevail, rather than to countenance monopolization of that market, whether it be by the Government itself or a private licensee."[73]

Incestuous relationships between the government and media giants like Fox have fully discredited the mainstream media. David Brock, President and CEO of Media Matters for America, concluded about Rupert Murdoch, "He doesn't believe in objectivity. He has contempt for journalism. I mean they wanted all news to be a matter of opinion because opinion can't be proven false. And I think that's very dangerous because if people don't have a set of facts that they can agree on, I think it's difficult to reach a consensus on what's correct public policy."[74]

Chapter 27
Media Duplicity: Case Studies

German philosopher Oswald Spengler observed in his 1918 book *The Decline of the West*, "It is permitted to everyone to say what he pleases, but the press is free to take notice of what he says or not. It can condemn 'truth' to death simply by not undertaking its communication to the world—a terrible censorship of silence, which is all the more potent in that the masses of newspaper readers are absolutely unaware that it exists."[1] This chapter analyzes various examples of this "censorship of silence." Surprisingly, as evidenced throughout this book, much of the New World Agenda *does* appear in the media, however, not congruously. It appears scattered such that casual observers cannot piece together the big puzzle.

If the media were truly free and independent, both of the first two examples of this chapter—the 2000 presidential election, and the modern slave trade—would be considered enormous failures of the media to inform the public. However, if the media is indeed a controlled asset of the shadow government, then both these examples have been successes because in each case, the media has insulated the establishment from potentially explosive information. In the third example, a candid *New York Times* article about the New World Order, the media relinquished surprising amounts of truth. Yet like exploding supernovas, this momentary admission quickly fizzled out. In the final example, an article by the *London Times*, the media boldly and bluntly acknowledged the New World Order agenda. This was not an exposé, however, it was a ringing endorsement with a contemptuous "whether you like it or not" tone.

2000 Election

Florida denies the right to vote to any citizen convicted of felonies *in Florida*. Both national and state laws, however, protect voting rights for ex-cons from other states who relocate to Florida. Nevertheless, during the 2000 presidential election season, Florida Governor Jeb Bush and Secretary of State Katherine Harris illegally denied the right to vote to thousands of Florida citizens who were previously convicted of crimes in other states, and who, after serving their sentences, had their voting rights restored. This contrived voter disenfranchisement operation certainly impacted the election. Had he lost Florida, Bush would have lost the presidency. With the help of brother Jeb, however, he won Florida—by only 537 votes. Based on overwhelming evidence unearthed and made public by BBC reporter Greg Palast, the NAACP filed suit on January 10, 2001, against Harris, elections unit chief Clay Roberts, and private database contractor Database Technologies. All the while, the mainstream media shamefully ignored the biggest election heist in U.S. history.

The disenfranchisement operation began six months before Florida's 1998 gubernatorial election, won by Jeb Bush. In May 1998, the Florida legislature passed a law designed to block ineligible voters from voting. This seemed reasonable, especially considering that Xavier Suarez had recently won the 1997 Miami mayoral election through widespread vote fraud, including through absentee ballots cast in the names of dead people.[2] However, *The Nation* later obtained an internal memo, dated August 1998, from the Florida State Association of Supervisors of Elections addressed to then-Secretary of State Sandra Mortham's office. The letter warned Mortham that her rush "to capriciously take names off the rolls" was wrongly removing many eligible voters.[3] Nevertheless, the Supervisors of Elections decided not to press the issue because "entering a public fight with [state officials] would be counterproductive."[4] Consequently, Jeb Bush enjoyed an unexpectedly easy victory in the November election.

The law that past in May 1998 also included an extraordinary provision enabling the government to award list-management responsibilities to private companies. In November 1998, the Republican-controlled office of the Secretary of the State handed the contract to Database Technologies (DBT).[5] Leading up the 2000 presidential election, DBT identified tens of thousands of supposedly ineligible voters. Palast reported, "Florida Secretary of State Katherine Harris, in coordination with Governor Jeb Bush, ordered local election supervisors to purge 57,700 voters from the registries, supposedly ex-cons not allowed to vote in Florida. At least 90.2 percent of those on this 'scrub' list, targeted to lose their civil rights, are innocent. Notably, more than half—about 54 percent—are black or Hispanic. You can argue all night about the number ultimately purged, but there's no argument that this electoral racial pogrom ordered by Jeb Bush's operatives gave the White House to his older brother."[6]

Among those targeted were 3,000 citizens who committed felonies in states that automatically restore voting rights after convicts serve their sentences. Therefore, under Florida state and national laws, these citizens retained their rights to vote when moving to Florida.[7] The June 1998 case *Schlenther v. Florida Department of State* specifically addressed this situation. The Florida Court of Appeals ruled unanimously that a man convicted twenty-five years earlier in Connecticut was not required "to ask [Florida] to restore his civil rights...he arrived as any other citizen, with full rights of citizenship."[8] Nevertheless, Harris's election division chiefs still ordered local officials to purge all out-of-state felons identified by DBT.

Hillsborough County officials were so indignant about this order that they demanded the state put it writing. The Governor's Office of Executive Clemency issued a letter, dated September 18, 2000, ordering the county to inform any ex-felons attempting to register to vote that they would still be "required to make application for restoration of civil rights in the state of Florida," even if they entered Florida with full rights restored by another state.[9] This was an illegal order, trumping both state and federal laws. Furthermore, an August 10, 2000 letter from Harris's office to Bush's office, unearthed by a Freedom of Information Act request, indicated that the Florida State Association of Supervisors of Elections disputed the illegal purge of ex-cons whose rights were restored by other states. Palast explained, "The supervisors' group received the same response as Hillsborough: Strike them from the voter rolls, and if they complain, make them ask Bush for clemency."[10] Those seeking clemency in Florida were asked to produce papers proving that another state had restored their rights. However, such papers didn't exist because other states restore voting rights *automatically*, by law.

Why was the mainstream media completely ignoring such bombshell information? CBS News contacted Palast, claiming they wanted to report the story. He freely offered them his research, yet the following day the CBS producer told Palast, "I'm sorry but your story didn't hold up."[11] How did they reach this conclusion? They called Jeb Bush's office! The magnitude of CBS's intellectual dishonesty is enormous. They halted their investigation based on statements of denial by representatives of Governor Jeb Bush, the very target of the allegations.

The Modern Slave Trade

Slavery epitomizes the New World Order. An exercise in unbridled power, slavery involves humiliation, manipulation, and dominance. Such behavior arises from extreme fear, which cripples empathy, eventually leading to psychopathic and sociopathic tendencies. Among New World Order elites, extreme fear is pervasive. Yet besides extreme psychopathology, the modern slave trade also requires a sophisticated, interlocking network of corporate, military, governmental and media power. Quite simply, if modern slavery indeed exists, then global elites must be involved. Only they have both the means and motives to carry out such depravity. While most people are completely unaware of the modern slave trade, it nevertheless does exist. In the aftermath of war it is shockingly common. Private contracting firms regularly exploit slave laborers, and also facilitate sex slavery. Governments hire these firms, military forces support them, and media censorship diverts public attention elsewhere. Do all these institutions act independently, creating optimal slavery conditions merely by coincidence? Or are skilled puppet masters behind the curtains pulling the strings? The following examples strongly suggest the latter.

DynCorp Sex Slavery

Insight on the News, a sister publication of the *Washington Times*, published a disturbing article in 2002, which examined sex slavery in postwar Bosnia. The victims were women and young girls from neighboring countries. The antagonist was DynCorp, a contracting company dependent upon the U.S. government for 95 percent of its business.[12] The hero was Ben Johnson, a DynCorp employee who witnessed his coworkers committing unthinkable human rights violations. Johnson reported these activities to DynCorp officials and later to the U.S. Army Criminal Investigation Division (CID). What did he get for his courageous whistle blowing? DynCorp fired him for bringing "discredit to the company and the U.S. Army while working in Tuzla, Bosnia-Herzegovina."[13]

Subsequently, Johnson filed a lawsuit proclaiming, "Johnston witnessed co-workers and supervisors literally buying and selling women for their own personal enjoyment, and employees would brag about the various ages and talents of the individual slaves they had purchased."[14] He explained, "None of the girls were from Bosnia. They were from Russia, Romania and other places, and they were imported in by DynCorp and the Serbian mafia...DynCorp leadership was 100 percent in bed with the mafia over there."[15] DynCorp didn't deny sex slavery was happening, nor did they deny their employees were involved. They simply denied responsibility. According to Kevin Glasheen, Johnson's attorney, "DynCorp says that whatever these guys were doing isn't corporate activity and they're not responsible for it. But this problem permeated their business and management and they made business decisions to further the scheme and to cover it up."[16]

The military also denied responsibility. Air Force public affairs told *Salon.com*, "The contractors themselves are responsible for their employees...the Air Force is concerned with whether the contract is carried out, not with the behavior of individual employees."[17] So the military blamed the contractors, and the contractors blamed the employees. DynCorp denied responsibility, yet the problem was endemic throughout the company, including management. Nobody cared. Peter Singer, a Senior Fellow at the Brookings Institute, explained,

"Typically when someone commits a crime, the legal system of a state is responsible for prosecuting that crime. But these companies are operating in areas where the local legal system is either unable or unwilling to hold these guys accountable. So, what you have happening is that these guys are American, but they're not being held under American law because they're inside Bosnia or Kosovo or wherever. And then there's no capacity to carry out enforcement by the locals. And if the locals do attempt to do something, the company pulls them out [of the country] because they'd rather not see their own employees get prosecuted. It's bad for business."[18]

Yet contrary to popular belief, business is not only about profit. From a New World Order perspective, business entails following an *agenda*. The globalists follow an incredibly complex, long-term business plan. They embrace slavery both for pragmatic and philosophical reasons.

The CID investigation turned up plenty of evidence, including a video featuring a DynCorp supervisor raping two women. CID, however, closed the case prematurely because neither Bosnian nor U.S. federal laws applied to the perpetrators. DynCorp subsequently conducted its own investigation. They fired just seven employees for what a DynCorp spokesperson described as "unacceptable behavior."[19] Nobody ever faced criminal charges. Widney Brown, an advocate for Human Rights Watch, stated, "Our government has an obligation to tell these companies that this behavior is wrong and they will be held accountable. They should be sending a clear message that it won't be tolerated. One would hope that these people wouldn't need to be told that they can't buy women, but you have to start off by laying the ground rules. Rape is a crime in any jurisdiction and there should not be impunity for anyone. Firing someone is not sufficient punishment. This is a very distressing story — especially when you think that these people and organizations are going into these countries to try and make it better, to restore a rule of law and some civility."[20]

During and after the scandal, Dyncorp *retained* its U.S. government contracts. Christine Dolan, founder of the International Humanitarian Campaign Against the Exploitation of Children, a Washington-based nonprofit organization, commented, "What is surprising to me is that Dyncorp has kept this contract. The U.S. says it wants to eradicate trafficking of people, has established an office in the State Department for this purpose, and yet neither State nor the government-contracting authorities have stepped in and done an investigation of this matter. It's not just Americans who are participating in these illegal acts. But what makes this more egregious for the U.S. is that our purpose in those regions is to restore some sense of civility. Now you've got employees of U.S. contractors in bed with the local mafia and buying kids for sex! That these guys have some kind of immunity from prosecution is morally outrageous. How can men be allowed to get away with rape simply because of location?"[21]

During the scandal, another DynCorp employee, Kathryn Bolkovac, also came forward with similar allegations. Predictably, DynCorp also fired her. Bolkovac also filed suit against DynCorp and later won. With Ben Johnson, DynCorp settled out of court.[22] However, these were just small slaps on the wrist for DynCorp. Nothing changed. DynCorp kept its contracts. They even bragged in internal emails about their evasion of justice. Johnson's lawyer commented, "There was a real corporate culture with a deep commitment to a cover-up. And it's outrageous that DynCorp still is being paid by the government on this contract. The worst thing I've seen is a DynCorp e-mail after this first came up where they're saying how they have turned this thing into a marketing success, that they have convinced the government that they could handle something like this."[23]

United Nations Sex Slavery

DynCorp was not the only organization involved with Bosnian sex trafficking. In late 2001, the *Washington Post* published an article entitled, "UN Halted Probe of Officers' Alleged Role in Sex Trafficking." The article began, "The United Nations quashed an investigation earlier this year into whether UN police were directly involved in the enslavement of Eastern European women in Bosnian brothels, according to UN officials and internal documents."[24] David Lamb, a former Philadelphia police officer who became a UN human rights investigator in Bosnia, investigated allegations that UN police officers had "recruited Romanian women, purchased false documents for them and then sold the women to Bosnian brothel owners."[25] The *Post* reported, "Within weeks, Lamb said, his preliminary inquiry found more than enough evidence to justify a full-scale criminal investigation. But Lamb and his colleagues said they also faced physical threats and were repeatedly stymied in their inquiries by their superiors."[26]

The UN oversight team dismissed Lamb's evidence and rashly denied systematic police involvement in sex trafficking. This reaction, though appalling, was not unexpected. Just five months earlier, The *Washington Post* also reported, "In the five years since international police officers were sent to help restore order in Bosnia, the UN police mission has faced numerous charges of misconduct, corruption and sexual impropriety. But in nearly every case, UN officials handled the allegations quietly by sending the officers home, often without a full investigation."[27]

In Kosovo, starting in 1999, the UN's role in sex slavery was so conspicuous that the organization eventually had to admit involvement. In 2004, The *London Guardian* reported, "Western troops, policemen, and civilians are largely to blame for the rapid growth of the sex slavery industry in Kosovo over the past five years, a mushrooming trade in which hundreds of women, many of them under-age girls, are tortured, raped, abused and then criminalised, Amnesty International said yesterday. In a report on the rapid growth of sex-trafficking and forced prostitution rackets since NATO troops and UN administrators took over the Balkan province in 1999, Amnesty said NATO soldiers, UN police, and western aid workers operated with near impunity in exploiting the victims of the sex traffickers."[28] The Amnesty report stated, "US, French, German and Italian soldiers were known to have been involved in the rackets...Women were bought and sold for up to £2,000 and then kept in appalling conditions as slaves by their 'owners'...They were routinely raped 'as a means of control and coercion,' beaten, held at gunpoint, robbed, and kept in darkened rooms unable to go out."[29] The UN admitted involvement. They were caught. Amnesty stated, "The UN admission in March that its peacekeepers were part of the problem was welcome."[30]

The UN wants the world to embrace its supposedly humanitarian mission. Yet time after time they discredit themselves. In an article published by *Fox News*, Wendy McElroy commented, "As the United Nations pushes for jurisdiction over the globe, it is important to remember how it has acted in Bosnia. The character of an institution, no less than of an individual, is revealed through actions, not words. It is revealed in the small behaviors. Such as the willingness to watch or participate in the selling of young girls into the living hell of Bosnian brothels."[31]

The KBR Slave-Labor Scandal

In October 2005, the *Chicago Tribune* published a two-part series documenting U.S. defense contractor involvement in human trafficking. Specifically, KBR subcontractors organized an "illicit system delivering cheap labor from impoverished countries to U.S. military bases throughout Iraq."[32] KBR is a subsidiary of Halliburton, the largest defense contractor in Iraq. Since the beginning of the Iraq war, Halliburton has pocketed billions of dollars through no-bid, U.S. government contracts. *Tribune* correspondent Cam Simpson documented the somber story of 12 Nepalese men offered jobs working in five-star hotels in Jordan. He explained, "They learned Iraq was their real destination only after their families went deeply into debt to pay huge sums demanded by the brokers who sent these sons and brothers to the Middle East."[33]

The *New York Times* also reported on the duplicity of KBR subcontractors. They related the stories of Indian men offered jobs working as butchers on military bases in Kuwait for $385 per month. These men would sign six-month contracts, yet when they arrived in Kuwait, they were abruptly transferred to Iraq and their pay was slashed to $150 per month. Furthermore, "They said their supervisor, who had taken their passports in Kuwait, told them they were obligated to work on the base for six months and could not leave."[34] The Indian media described the Iraq labor camps as "U.S. Slave Camps." Other headlines read, "Indians Abused in Iraq."[35] On April 15, 2004, the Indian government officially banned Indian workers from going to Iraq, despite the fact that hundreds, if not thousands, were already there.

Facing accusations of serious human rights violations, both Halliburton and the U.S. military denied responsibility. The *Tribune* reported, "At the time, Halliburton said it was not responsible for the recruitment or hiring practices of its subcontractors, and the U.S. Army, which oversees the privatization contract, said questions about alleged misconduct 'by subcontractor firms should be addressed to those firms, as these are not Army issues.'"[36]

How could this happen, and once revealed, how could it continue to happen? One big clue is that former Vice President Dick Cheney was Halliburton's CEO from 1995 until 2000. By 2003, Cheney still retained over 433,000 company stock options and was still collecting a deferred salary.[37] On September 21, 2003, during a nationally televised interview, Cheney stated, "I've severed all my ties with the company, gotten rid of all my financial interest. I have no financial interest in Halliburton of any kind and haven't had, now, for over three years."[38] However, the Congressional Research Committee reported that unexercised stock options and deferred salary "are among those benefits described by the Office of Government Ethics as 'retained ties' or 'linkages' to one's former employer."[39] Senator Frank Lautenberg bluntly stated, "I ask the vice president to stop dodging the issue with legalese."[40]

Cheney's Halliburton ties are extremely relevant. Besides the slave labor scandal, Halliburton also has an incredible history of fraud and misappropriation of government funds. For example, on March 28, 2006, Representative Henry Waxman, Chairman of the Committee on Oversight and Government Reform, released an analysis of Halliburton's performance on government-contracted work. Regarding Halliburton's, "'obstructive' corporate attitude toward oversight," the analysis concluded, "Two years ago, despite warnings from auditors not to enter into further contracts with Halliburton, the Defense

Department awarded Halliburton a new oil infrastructure contract, RIO 2. Internal government documents show that Halliburton's performance under RIO 2 has been deeply flawed. Among the serious and persistent problems identified in the documents are repeated examples of apparently intentional overcharging, exorbitant costs, poor cost reporting, slipping schedules, and a refusal to cooperate with the government."[41]

Furthermore, in 2004, KBR's CEO was indicted on bribery charges. In September 2008, the *New York Times* reported, "Albert J. Stanley, a former executive with a Halliburton subsidiary, pleaded guilty on Wednesday to charges that he conspired to pay $182 million in bribes to Nigerian officials in return for contracts to build a $6 billion liquefied natural gas complex."[42] In 2007, Halliburton sold KBR, for a $933 million gain.[43] And just months earlier, they announced plans to move Halliburton headquarters to Dubai. Senator Patrick Leahy decried, "This is an insult to the U.S. soldiers and taxpayers who paid the tab for their no-bid contracts and endured their overcharges for all these years."[44] Jim Donahue, co-director of Halliburton Watch, a project of the public interest organization Center for Corporate Policy, offered a more accurate observation, "With various ongoing investigations, Halliburton's sale of KBR and the move to UAE are tantamount to fleeing the scene of a crime."[45]

Could Cheney have played a role in Halliburton's contract renewals, against warnings from auditors? Could Cheney also have influenced the administration's shockingly nonchalant attitude towards human trafficking? In 2002 Bush mockingly declared, "zero tolerance" for human trafficking. In 2005, Congress passed legislation purportedly blocking trafficking, yet nevertheless perpetuating this atrocity. The *Tribune* reported, "A bill reauthorizing the nation's efforts against trafficking for the next two years was overwhelmingly passed by the House this month, but only after a provision creating a trafficking watchdog at the Pentagon was stripped from the measure at the insistence of defense-friendly lawmakers, according to congressional records and officials."[46]

Was the Bush administration quietly dedicated to *allowing* human trafficking to continue? On March 11, 2005, Representative Cynthia McKinney questioned then Secretary of Defense Donald Rumsfeld before the House Armed Services Committee. McKinney raised numerous sensitive issues including the ongoing slavery scandal involving U.S. contractors. This historic exchange encapsulates the extreme arrogance of New World Order elites, while also demonstrating their total disregard for the law and human suffering. McKinney bravely asked direct and prodding questions. Not surprisingly, the media's response was nonresponsive, a complete blackout. The information was far too explosive. On the Internet, however, clips from the C-SPAN coverage rapidly spread:

> *McKinney:* Thank you Mr. Chairman. Mr. Secretary, I watched President Bush deliver a moving speech at the United Nations in September 2003, in which he mentioned the crisis of the sex trade. The President called for the punishment of those involved in this horrible business. But at the very moment of that speech, DynCorp was exposed for having been involved in the buying and selling of young women and children. While all of this was going on, DynCorp kept the Pentagon contract to administer the smallpox and anthrax vaccines, and is now working on a plague vaccine through the Joint Vaccine Acquisition Program. Mr. Secretary, is it policy of the U.S. Government to reward companies that traffic in women and little girls?

Rumsfeld: Thank you, Representative. First, the answer to your first question is, no, absolutely not, the policy of the United States Government is clear, unambiguous, and opposed to the activities that you described.

McKinney: Well how do you explain the fact that DynCorp and its successor companies have received and continue to receive government contracts?

Rumsfeld: I would have to go and find the facts, but there are laws and rules and regulations with respect to government contracts, and there are times that corporations do things they should not do, in which case they tend to be suspended for some period; there are times then that, under the laws and the rules and regulations passed by the Congress and implemented by the Executive branch, that corporations can get out of the penalty box, if you will, and be permitted to engage in contracts with the government. They're generally not barred in perpetuity.

McKinney: This company was never *in* the penalty box.[47]

Flirting with Truth

On May 5, 1998, A.M. Rosenthal published in the *New York Times* a brave article exposing, to some degree, the true face of the New World Order. Rosenthal won the Pulitzer Prize in 1960 for international reporting and was the former Executive Editor of the *Times*. His article portrayed the New World Order for what it is—a corrupt racket, run by international criminals posing as statesmen, designed to consolidate power and wealth. Rosenthal wrote,

> "The U.S., its democratic allies and major dictatorships are rapidly building a new world order—not quite finished yet but already a central part of international life and values. Its ideology, powers, rewards and punishments are supplanting those that prevailed internationally until 1994, when President Clinton joined the new order. If it continues, it will be the most important new international concept since the end of World War II. The order was created without formal parliamentary approval by its sponsors, or any treaty. But every week, sometimes every day, the underlying tenets are revealed, in action."[48]

Of course the New World Order long-preceded Clinton. Yet while Rosenthal failed to grasp its scope, he certainly captured its tyrannical intent. He continued, "The following description of objectives and goals of the new order is so different from principles recently assumed in the West, though not always followed, that it may read as satire. It is not."[49]

Rosenthal described an arrangement whereby supposedly righteous democracies support brutal dictatorships. The democracies project benign criticism towards the dictatorships regarding certain internal practices, yet never enough to threaten underlying economic commitments. When the dictatorships inevitably collapse under the weight of their own corruption, the International Monetary Fund comes to the rescue. "The explanation given," explained Rosenthal, "is that otherwise the dictatorships' economies would disintegrate, bringing revolution. Now, the people of the dictatorships may long for revolution. Obviously that cannot be allowed to overcome saving the dictatorship and thus rescuing the money invested by nationals of democracies."[50]

Rosenthal was essentially describing an intermediate phase of the New World Order, which began after World War II. The next phase—the controlled implosion of the Western democracies and the implementation of global tyranny—however, has arrived. Brzezinski, Obama's puppet-master, seeks to catalyze this next phase through global war setting Russia and China against the West. Rosenthal foreshadowed this impending doom, observing that on May 1, 1998, the CIA had assessed that "China has nuclear missiles targeted at U.S." However, just two days later, on May 3, Washington announced it would soon permit U.S. companies to sell nuclear reactors to China. And furthermore, President Clinton would ceremoniously visit China to advance this agenda. Rosenthal lamented, "The U.S. gets to sell strategic material to China, offering as an extra a visit by the U.S. President to honor the Communist leaders and expand their power and political life span."[51] Rosenthal poignantly concluded, "Americans and Europeans may come to object for political or moral reasons, or because the new world order may after all cost them their jobs. But they will never be able to say they never knew; see above."[52]

And Now For A World Government

On December 8, 2008, the *Financial Times*, one of the most respected newspapers in the English language, openly admitted that international elites are attempting to implement nondemocratic world government, justified by global warming, the global financial crisis, and the global war on terror. Gideon Rachman, the paper's chief foreign affairs commentator, wrote, "For the first time since homo sapiens began to doodle on cave walls, there is an argument, an opportunity and a means to make serious steps towards a world government."[53] The globalists, Rachman admitted, are using deceptive language to hoodwink the public. For example, instead of speaking about "shared sovereignty," a term sometimes used in Europe, American globalists speak about "responsible sovereignty." He also explained that "global governance" really means "global government." Rachman quoted prominent globalist Jacques Attali, an advisor to French President Nicolas Sarkozy, as stating, "Global governance is just a euphemism for global government."

Rachman acknowledged that populations the world over are decidedly against global government, thus "any push for 'global governance' in the here and now will be a painful, slow process."[54] He congratulated the European Union for eroding state sovereignty, yet conceded that European populations have consistently opposed further integration. Therefore, explained Rachman, regional and world government will have to be implemented by stealth: "The EU has suffered a series of humiliating defeats in referendums, when plans for 'ever closer union' have been referred to the voters. In general, the Union has progressed fastest when far-reaching deals have been agreed by technocrats and politicians—and then pushed through without direct reference to the voters. International governance tends to be effective, only when it is anti-democratic."[55]

Despite the globalists' overwhelming control of mainstream media, there are numerous alternative news outlets providing genuine, honest reporting and penetrating, insightful analysis. The unrivaled leader is the Alex Jones news organization, operating through *Infowars.com* and *Prisonplanet.com*, a daily, syndicated radio program, and numerous documentary films. *Globalresearch.ca* is another reputable online source for news. As the New World Order agenda advances, the globalists are attempting to stamp out citizen journalism, limit free speech and regulate and control the Internet. An educated and informed public, however, can defeat them. May the most conscious triumph, and may the truth prevail.

CONCLUSION

"In the long history of the world, only a few generations have been granted the role of defending freedom in its hour of maximum danger. I do not shrink from this responsibility. I welcome it."[1] — *John F. Kennedy*

She was paralyzed, emotionally and intellectually. For several months, my friend Sofia had been passively observing my New World Order analysis. She was curious though skeptical, an always-healthy attitude. Her curiosity prevailed. She slowly began immersing herself in books and films on the subject. I anticipated her reaction; mine had been the same. My awakening engendered feelings of anger, sadness, confusion, betrayal, anxiety, fear, and embarrassment. But mostly I had been disappointed—in myself. How could I have been so naive? I knew what she was going through. She was passing a "red pill hangover."

In the film *The Matrix*, Morpheus offers Neo the chance to see reality, to know the truth. "Unfortunately," states Morpheus, "no one can be told what the Matrix is. You have to see it for yourself. This is your last chance. After this, there is no turning back."[2] He offers Neo the choice between two pills, one blue and the other red. "You take the blue pill," says Morpheus, "and the story ends. You wake in your bed and believe whatever you want to believe. You take the red pill and you stay in Wonderland and I show you how deep the rabbit hole goes. Remember, all I am offering is the truth, nothing more." Neo bravely swallows the red pill, triggering an intense paradigm-shifting journey, which would ultimately challenge his core beliefs and assumptions about reality. Truth is a hard pill to swallow. However, once the initial discomfort subsides, it reaps unfathomable rewards. It catalyzes the expansion of consciousness and, in turn, the augmentation of freedom. "And you will know the truth," said Jesus, "and the truth will set you free."

One late-summer afternoon, I invited Sophia to my favorite beach, an ideal place for serious contemplation. Taken in unison, the unending sea, the jagged, intricate landscape, the distant mountains, the serene twin-islands, and the sun's warm caress allowed for focused introspection. We sat there silently. Her distant stare was revealing. She was mourning the death of comfortable falsehoods. "I know it's true," she lamented, "but still I'm hoping I'll wake up and realize it was just a dream. Anyway, there's no turning back now. But I must say, I feel totally helpless and afraid. How can we stop this wretched New World Order?"

"Well first of all," I replied, "we must address our underlying fears. What are we so afraid of? Are we afraid of dying? Or are we, perhaps, afraid of *living*? What does it mean *to live*? Do you know the motto of the State of New Hampshire? 'Live free or die.' Do you know what Patrick Henry, the American revolutionary and United States Founding Father said in 1775? 'Give me liberty, or give me death.' Do you know what Spanish revolutionary Dolores Ibárruri said in 1936? 'It's better to die on your feet than live on your knees.' That pretty much says it all. Are we going to *live*, or are we going to capitulate to tyranny? If we choose the former, then we must first address our fears. According to the Buddha, 'The whole secret of existence is to have no fear.' Take some time and meditate on that. Meditate on your *existence*. Who *are* you? *What* are you? What are you doing here? According to most spiritual traditions, you are much, much more than just your body and your mind. When you finally realize this, your fears will greatly subside."

We sat there enjoying the gently crashing waves. After several minutes, Sofia remarked, "I suppose you're right. After all, the Bible says the same thing: 'Yea, though I walk through the valley of the shadow of death, I will fear no evil.' But after addressing our fears, what's next? I mean, once you get yourself straight with God, you can never be destroyed, spiritually speaking, no matter what. But still, can't we *do* something to counterbalance all this evil?"

"Of course," I replied, "there are thousands of things we can do. But it's important to realize that you need not save the world all by yourself. The vast majority of humanity wants the same things—peace, security, and freedom. The globalists, though influential, are extremely small in numbers. Their agenda, when properly explained, has zero popular support. Therefore, their power is extremely fragile. Just doing *something*, even something small and seemingly insignificant, *does* make a difference. Our actions have enormous ripple effects. That's why Gandhi advised, '*Be* the change you wish to see in the world.' In other words, practice what you preach. Stop philosophizing and start living."

"But good intentions aren't good enough," I continued. "Good intentions are impotent without consciousness and intelligence. The road to hell, they say, is paved with good intentions. For example, protecting the environment is certainly desirable and beneficial. Yet giving money to Greenpeace or the WWF *isn't* protecting the environment. It's *supporting* eugenics, depopulation, fraudulent climate science, and global government. Most prominent foundations and environmental advocacy organizations are simply New World Order fronts. Therefore, we must be extremely careful about which causes and organizations we support."

"The same goes for which companies we support," Sofia added. "Absolutely," I replied, "for example, the big media conglomerates continually misinform and deceive us. Why should we support them financially? Why not instead boycott their newspapers, magazines, television programs, and films? Financial decisions made hastily and unconsciously carry enormous consequences. For example, is buying slave-labor goods from Wal-Mart really worth saving a few dollars? On the other hand, supporting local businesses, farmers' markets and food co-ops, *strengthens* local communities and economies."

"That makes sense," Sofia replied, "allowing money to circulate locally, as much as possible. It's more efficient, and less wasteful." "Yes," I responded, "and furthermore, why do we deposit our money with the same banks that fund the military-industrial war machine? Catherine Austin Fitts, for example, has developed a pragmatic model for sustainable community-based investing.[3] Fitts was formerly the Managing Director of Wall Street investment bank Dillon, Read & Co., and also Assistant Secretary of Housing-Federal Housing Commissioner during the H.W. Bush Administration. In 2004, she posed the question, 'Where would Jesus bank?' She concluded that banking with megabanks is like paying for your own prison. In a highly leveraged banking system, she explained, ordinary people have enormous power to impact the economy by shifting their deposits to local banks and credit unions.

More to the point, we must educate ourselves, and each other, about the pitfalls of debt-based money. Intentionally obscured from most history books, the relentless pursuit of private banking interests to control the U.S. monetary system has been the paramount theme of American history. Some invaluable educational resources are the documentary films of Bill Still, a former newspaper editor and publisher. Nobel Prize winning economist Milton Friedman both endorsed and helped edit Still's landmark documentary *The Money Masters: How International Bankers Gained Control of America*. Still's latest documentary, *The Secret of Oz*, clearly and coherently explains monetary history while demonstrating how L. Frank Baum wrote his famous 1900 children's novel, *The Wonderful Wizard of Oz*, as an allegory criticizing monetary manipulation and advocating monetary reform."[4]

Serenely, Sofia stared into the vast, unending sea. Pausing several minutes, we quietly enjoyed the magnificent view. I then continued, "Another effective solution is taking control of your health. It should come as no surprise that big pharmaceutical companies and the FDA are key New World Order players. Think about it. If you're planning to impose massive societal changes, you want people physically and mentally weak. Cancer and autoimmune disease rates are not exploding simply by chance. Stop taking vaccines. Stop drinking fluoridated water. Stop eating genetically modified and otherwise poisonous foods. Think of these not as medicine, water, and food, but as soft-kill, biological *weapons*. Remember Rumsfeld and Cheney's PNAC club? They actually bragged that *race specific* biological weapons would soon become 'politically useful tools.' They didn't elaborate, except to say that 'advanced forms of biological warfare that can 'target' specific genotypes may transform biological warfare from the realm of terror to a politically useful tool.'[5] What do you suppose they meant by that? Study the history of Monsanto and you'll soon find out.

During the 2009 swine flu hoax, the World Health Organization declared a Level 6 Pandemic Alert. Do you know what that means? Under the State Emergency Health Powers Act and the various Patriot Acts, the CDC can now mandate *forced* vaccinations. In other words, take the shots or go to jail. That's preposterous. We have an inherent right to our own bodies, to refuse forced medication. Neither the UN nor the U.S. nor any other government can take that right away. The modern medical system, shaped largely by Rockefeller interests since the early 1900s, was never intended to cure or prevent disease. People should read 'Death by Medicine,' a groundbreaking 2003 study by five doctors, who concluded, 'It is now evident that the American medical system is the leading cause of death and injury in the U.S.'[6] The U.S. medical system kills over 783,000 people per year. In comparison, heart disease kills 700,000 and cancer kills 553,000. Information about natural, healthy foods and nontoxic, plant-based medicine is readily available. By taking control of your health, you raise your consciousness and increase your freedom."

"So I guess we need a revolution," Sofia remarked. "Let's be clear," I replied, "in America, we already had a revolution. The year was 1776. The globalists are the ones who are now fomenting revolution. They're the ones trying to overthrow *our* government. We don't need a revolution. We need to quell *their* revolution and reestablish *our* Constitution. They're the radicals. They're the revolutionaries, not us. The strength of the American political system, originally, was its separation of powers. The founders intentionally distributed power between the federal government and the states, with the states having *more* power. Within the federal government, they distributed power between the legislative, judicial and executive branches. But over the past hundred years, and especially over the past decade, the globalists have stealthily inverted this power structure. The federal government has usurped power from the states, and within the federal government, the executive has now claimed dictatorial power. When the Bush administration drafted NSPD 51 — the continuity of government plan — they told congress, 'you can't see the plan.' That's outrageous. All presidents swear an oath to defend the Constitution against all enemies, foreign and domestic. We must demand accountability and transparency. The excuse of national security is no longer acceptable. Restricting our freedoms to protect ourselves from phantom terrorists who supposedly hate our freedoms is absurd. Benjamin Franklin's words never rang more true: 'Those who would give up essential liberty, to purchase a little temporary safety, deserve neither liberty nor safety.'[7]

Against the New World Order, the solutions are many. Yet they all involve increasing our consciousness and awareness. Although the globalists are clever, we're certainly not facing an ultra-conscious enemy. They are utterly predictable and furthermore, they've *published* their plans and constantly brag about their crimes. That's like writing a book on how to rob a bank, then announcing you're intentions to do it, then doing it, then bragging that you did it. Are we not conscious enough to subdue these cocksure gangsters? We simply have to wise-up and start taking responsibility. We've grown accustomed to perpetual childhood — playing, partying, watching sports, watching TV and thousands more distractions. The media tells us, 'All is well.' The government says, 'We're protecting you.' The Buddha, however, wisely advised, 'Believe nothing, no matter where you read it, or who said it, no matter if I have said it, unless it agrees with your own reason and your own common sense.'

Let's stop being suckers. Kill your television. It's the greatest mind-control devise ever invented. Read books. Study history. Support alternative media and documentary filmmakers. Support the 9/11 Truth Movement. Tell people about Building 7. Show them the video! Get involved with politics, especially local politics. Support independent candidates. Break the false left-right paradigm. Love your children. Love your friends. And if you know what love is, love your enemies. If you want solutions, go look in the mirror. The solution is you. Stop looking for leaders. Be your own leader. Be your own guru. Be your own savior."

Without saying a word, Sofia collected her things, stood up and began walking away. I was surprised, though intrigued. I called out, "Where are you going?" She stopped, turned around, and explained, "I also heard some good advice from the Buddha: 'There are only two mistakes one can make along the road to truth; not going all the way, and not starting.'"

CITATIONS

INTRODUCTION

[1] Steven Lawson, Charles Payne and James Patterson, *Debating the Civil Rights Movement, 1945-1968*, published by Rowman & Littlefield, 2006, p.106
[2] Andrew Clark, "Bankers and academics at top of donor list," *London Guardian*, November 8, 2008; also see Jim Kuhnhenn, "Obama Taps Wall Street for Dollars," *Fox News* via *Associated Press*, July 16, 2007
[3] Karlyn Bowman, "Low Support For The Stimulus," *Forbes.com*, February 9, 2009; also see "Valley congressman: Phone calls running 300 to 2 against bailout," *Los Angeles Times (Blog)*, September 25, 2008 and Les Blumenthal, "Inslee joins state's three Republicans in voting against bailout package," *McClatchy Newspapers*, September 29, 2008
[4] Russell Berman, "Despite Criticism, Obama Stands By Adviser Brzezinski," *New York Sun*, September 13, 2007
[5] "Hijack 'suspects' alive and well," *BBC News*, September 23, 2001; Dan Eggen, George Lardner Jr. and Susan Schmidt, "Some Hijackers' Identities Uncertain," *Washington Post*, September 20, 2001; Lisa Getter, Elizabeth Mehren and Eric Slater, "AFTER THE ATTACK; THE INVESTIGATION; FBI Chief Raises New Doubts Over Hijackers' Identities; Inquiry: Mueller says several names 'are still in question.' Some of them match those of Saudi citizens who are alive," *Los Angeles Times*, September 21, 2001
[6] Federal Bureau of Investigation website, http://www.fbi.gov/wanted/terrorists/terbinladen.htm
[7] Ed Haas, "FBI says, 'No Hard Evidence Connecting Bin Laden to 9/11,'" *Ithaca Journal* and *The Muckraker Report*, June 6, 2006; also see Peter Phillips and Project Censored, *Censored 2008: The Top 25 Censored Stories*, published by Seven Stories Press, 2007, p.93
[8] http://www.journalof911studies.com/
[9] Architects and Engineers for 9/11 Truth, http://www.ae911truth.org
[10] Won-Young Kim, L. R. Sykes et al., "Seismic Waves Generated by Aircraft Impacts and Building Collapses at World Trade Center, New York City," available at http://www.ldeo.columbia.edu/LCSN/Eq/20010911_WTC/WTC_LDEO_KIM.pdf
[11] Mike Rudin, Controversy and conspiracy III," *BBC News (Blog)*, July 2, 2008
[12] http://www.patriotsquestion911.org
[13] Architects and Engineers for 9/11 Truth, http://www.ae911truth.org
[14] Laurence Arnold, "9/11 panel to get access to withheld data," *Associated Press* via *Boston Globe*, November 13, 2003
[15] Eric Boehlert, "The president ought to be ashamed," *Salon.com*, November 21, 2003
[16] Amy Goodman, "The White House Has Played Cover-Up," *Democracy Now*, March 23, 2004
[17] Ibid. Arnold
[18] "9/11 panel distrusted Pentagon testimony," *CNN.com*, August 2, 2006
[19] Farhad Manjoo, "The 9/11 deniers," *Salon.com*, June 27, 20006
[20] "Nightly News," *NBC*, February 2, 2008; also see Nick Langewis and David Edwards, "9/11 Commissioner: We had to go through Karl Rove," *RawStory.com*, February 3, 2008
[21] Thomas Kean and Lee Hamilton, "Stonewalled by the CIA" *New York Times*, January 2, 2008
[22] Thomas Kean and Lee Hamilton, *Without Precedent: The Inside Story of the 9/11 Commission*, published by Vintage, 2007, p.14
[23] Dan Eggen, "9/11 Panel Suspected Deception by Pentagon," *Washington Post*, August 2, 2006
[24] "Rebuilding America's Defenses: Strategy, Forces and Resources For a New Century," published by The Project for the New American Century, September 2000
[25] See Chapter 1
[26] Robert Stinnett, *Day Of Deceit: The Truth About FDR and Pearl Harbor*, published by Simon & Schuster, 2001
[27] Scott Shane, "Vietnam War Intelligence 'Deliberately Skewed,' Secret Study Says," *New York Times*, December 2, 2005

PART I: TYRANNICAL GLOBAL GOVERNMENT

[1] Thomas Jefferson, *Memoirs, Correspondence and Private Papers of Thomas Jefferson*, published by Henry Colburn and Richard Bentley, 1829, p.110
[2] Interview with Dan Itse, *The Alex Jones Show*, Genesis Communications Network, March 3, 2009
[3] George H.W. Bush, "Address Before a Joint Session of the Congress on the Persian Gulf Crisis and the Federal Budget Deficit," September 11, 1990; George Bush Presidential Library and Museum, archives available online at http://bushlibrary.tamu.edu
[4] Ibid. Bush
[5] Ibid. Bush
[6] Michael Gorbachev, *On My Country And The World*, published by Colombia University Press, 2000, p.221
[7] Ibid. Gorbachev, p.67-68
[8] Ibid. Gorbachev, p.269
[9] Senator Jesse Helms, Congressional Record, December 15, 1987, page S18146; As cited by Daniel Estulin, *The True Story of the Bilderberg Group*, published by TrineDay LLC, 2007, p.150
[10] Ibid. Helms
[11] David Rockefeller, *Memoirs*, published by Random House, 2002, p.405
[12] Charles Kershaw, *The Encyclopedia of Public Choice*, published by Springer, 2004, p.536
[13] David Rockefeller, "From a China Traveler," *New York Times*, August 10, 1973
[14] Gary Hart, speaking before the Council on Foreign Relations' U.S. Commission on National Security/21st Century, broadcast live on C-SPAN, September 12, 2001; video from author's personal collection
[15] Henry Kissinger, interviewed by CNBC, January 5, 2009; video from author's personal collection

CHAPTER 1: The New World Order Agenda

[1] Carroll Quigley, *Tragedy and Hope: A History of the World in Our Time*, published by Macmillan, 1966, p.324
[2] Obituary for Carroll Quigley, *Washington Star*, January 6, 1977
[3] Bill Clinton, Acceptance Speech for Democratic Party Presidential Candidate, Democratic National Convention, New York City, New York, July 16, 1992; See *Congressional Quarterly Weekly*, Vol.50, 1992, p.2130
[4] Carroll Quigley, *Tragedy and Hope: A History of the World in Our Time*, published by Macmillan, 1966, p.950

[5] Carroll Quigley, *Tragedy and Hope: A History of the World in Our Time*, published by Macmillan, 1966, p.950
[6] Ibid. Quigley, p.955
[7] Ibid. Quigley, p.953
[8] Barry M. Goldwater, *With No Apologies*, published by William Morrow and Company, 1979, p.281
[9] Ibid. Quigley, p.61
[10] Mark Pittman and Bob Ivry, "Financial Rescue Nears GDP as Pledges Top $12.8 Trillion," *Bloomberg.com*, March 31, 2009; also see Dawn Kopecki and Catherine Dodge, "U.S. Rescue May Reach $23.7 Trillion, Barofsky Says," *Bloomberg.com*, July 20, 2009
[11] Ibid. Quigley, p.62
[12] Ibid. Quigley, p.324
[13] John L. Lewis v. United States of America, 680 F.2d 1239, Amended June 24, 1982
[14] Louis McFadden, Testimony before Committee on Rules, House of Representatives, United States Congressional Record, June 10, 1932, p.12595-12596
[15] Ibid. McFadden
[16] U.S. Congressional Record, *Government Ownership of the Twelve Federal Reserve Banks: Hearings Before the Committee on Banking and Currency*, House of Representatives, Seventy-fifth Congress, third session, published by U.S. Government Printing Office, 1938, p.162
[17] Edwin Williams, *Statesman's Manual: Presidents' Messages from 1789 to 1846*, published by Edward Walker, 1847, p.877
[18] Robert Eric Wright and David Jack Cowen, *Financial Founding Fathers: The Men Who Made America Rich*, published by University of Chicago Press, 2006, p.169
[19] Allen C. Guelzo, *Abraham Lincoln: Redeemer President*, published by Eerdmans Publishing, 2003, p.381
[20] United States Congress, House Committee on Banking and Currency, Government Ownership of the Twelve Federal Reserve Banks, published by Government Printing Office, 1938, p.10
[21] George Nathan and Henry Louis Mencken, *The American Mercury*, Vol.86, 1958, p.103
[22] Milford Wriarson Howard, *The American Plutocracy*, published by Holland Publishing Company, 1895, p.156
[23] Benson John Lossing and Woodrow Wilson, *Harper's encyclopedia of United States history from 458 A.D. to 1909*, published by Harper, 1905, p.427
[24] Eldon J. Eisenach, *The Social and Political Thought of American Progressivism*, published by Hackett Publishing, 2006, p.274
[25] "Hylan Takes Stand on National Issues," *New York Times*, March 26, 1922
[26] Franklin Delano Roosevelt and Elliot Roosevelt, *FDR: His Personal Letters, 1928-1945*, published by Duell, Sloan and Pearce, 1947, p.373
[27] *Grassroots Hearings on the Economy: Hearings Before the Committee on Banking, Finance, and Urban Affairs*, House of Representatives, Ninety-seventh Congress, first session, published by U.S. Government Printing Office, 1981, p.126
[28] Barry M. Goldwater, *With No Apologies*, published by William Morrow and Company, 1979, p.282
[29] Ron Paul, "Abolish the Federal Reserve," speech before U.S. House of Representative, September 10, 2002; http://www.house.gov/paul/congrec/congrec2002/cr091002b.htm
[30] Carroll Quigley, *Tragedy and Hope: A History of the World in Our Time*, published by Macmillan, 1966, p.530
[31] Ibid. Quigley, p.337-338
[32] J. Lawrence Broz, *The International Origins of the Federal Reserve System*, published by Cornell University Press, 1997, p.175
[33] James Perloff, "Our Monetary Mayhem Began With the Fed," *The New American*, April 3, 2009
[34] Carroll Quigley, *Tragedy and Hope: A History of the World in Our Time*, published by Macmillan, 1966, p.357
[35] Murray Rothbard, *Wall Street, Banks, and American Foreign Policy*, published by Center for Libertarian Studies, 1996, p.15-16
[36] Burton J. Hendrick, *The Life and Letters of Walter H. Page Vol. II (1922)*, published by BiblioBazaar, 2007, p.240
[37] Charles Willis Thompson, "Col. House's Real Place in the Wilson Circle," *New York Times*, January 9, 1916
[38] Edward Mandell House and Charles Seymour, *The Intimate Papers of Colonel House*, published by Houghton Mifflin, 1926, p.114
[39] Edward Robb Ellis, *Echoes of Distant Thunder: Life in the United States, 1914-1918*, published by Kodansha America, 1996, p.204
[40] George Sylvester Viereck, *The Strangest Friendship in History*, published by Greenwood Press, 1976, p.106
[41] Carroll Quigley, *Tragedy and Hope: A History of the World in Our Time*, published by Macmillan, 1966, p.238
[42] Ibid. Quigley, p.239
[43] Ibid. Quigley, p.239
[44] Ibid. Quigley, p.239
[45] Thomas A. Bailey, "The Sinking of the Lusitania," *American Historical Review*, October 1935
[46] Thomas A. Bailey, "German Documents Relating To The Lusitania," *Journal of Modern History*, September 1936
[47] Edward Robb Ellis, *Echoes of Distant Thunder: Life in the United States, 1914-1918*, published by Kodansha America, 1996, p.196
[48] Ibid. Ellis, p.195
[49] Ibid. Ellis, p.195
[50] Ibid. Ellis, p.208
[51] "Disaster Bears Out Embassy's Warning," *New York Times*, May 8, 1915
[52] Ibid. Ellis, p.198
[53] Ibid. Ellis, p.198
[54] Charles E. Lauriat, *The Lusitania's Last Voyage*, published by Houghton Mifflin, 1915, p.10
[55] Thomas Andrew Bailey, Alexander DeConde and Armin Rappaport, *Essays Diplomatic and Undiplomatic of Thomas A. Bailey*, published by Appleton-Century-Crofts, 1969, p.201
[56] Dudley Field Malone, "Cargo of the Lusitania: An Official Statement," *The Nation*, January 3, 1923
[57] Horace Cornelius Peterson, *Propaganda for War: The Campaign Against American Neutrality, 1914-1917*, published by University of Oklahoma Press, 1939, p.123
[58] Ibid. Malone
[59] Ibid. Peterson, p.120
[60] Ibid. Peterson, p.129
[61] William Jennings Bryan, *Memoirs of William Jennings Bryan (1925)*, published by Kessinger Publishing, 2003, p.398-399
[62] Edward Robb Ellis, *Echoes of Distant Thunder: Life in the United States, 1914-1918*, published by Kodansha America, 1996, p.214
[63] Carl R. Burgchardt, *Robert M. La Follette, Sr.: The Voice of Conscience*, published by Greenwood Press, 1992 p.90
[64] Jules Archer, *The Plot to Seize the White House: The Shocking True Story of the Conspiracy to Overthrow FDR*, published by Skyhorse Publishing, 2007, p.160
[65] Ibid. Archer, p.139
[66] Ibid. Archer, p.26
[67] "Plot Without Plotters," *Time*, December 3, 1934
[68] Jules Archer, *The Plot to Seize the White House: The Shocking True Story of the Conspiracy to Overthrow FDR*, published by Skyhorse Publishing, 2007, p.192-193
[69] Ibid. Archer, p.209

[70] Ibid. Archer, p.207
[71] Ibid. Archer, p.212
[72] Ibid. Archer, p.206
[73] Ibid. Archer, p.214
[74] Ibid. Archer, p.x
[75] Carroll Quigley, *Tragedy and Hope: A History of the World in Our Time*, published by Macmillan, 1966, p.130
[76] Cecil Rhodes and William Thomas Stead, *The Last Will and Testament of Cecil John Rhodes*, published by William Clowes and Sons, 1902, p.96-97
[77] Ibid. Rhodes and Stead, p.58
[78] Ibid. Rhodes and Stead, p.59; Also see Antony Thomas and British Broadcasting Corporation, *Rhodes: The Race for Africa*, published by BBC Books, 1996, p.112
[79] Carroll Quigley, *Tragedy and Hope: A History of the World in Our Time*, published by Macmillan, 1966, p.131
[80] Ibid. Quigley, p.132
[81] Ibid. Quigley, p.951-952
[82] Ibid. Quigley, p.952
[83] Phyllis Schlafly and Chester Ward, *Kissinger on the Couch*, published by Arlington House Publishers, 1975, p.150
[84] Phyllis Schlafly and Chester Ward, *Kissinger on the Couch*, published by Arlington House Publishers, 1975, p.129
[85] Carroll Quigley, *The Anglo-American Establishment: From Rhodes to Cliveden*, published by GSG & Associates, 1981, p.10
[86] Ibid. Quigley, p.10
[87] Ibid. Quigley, p.161
[88] Arnold J. Toynbee, "World Sovereignty and World Culture," *Pacific Affairs*, Vol.4, No.9, September, 1931, p.758- 759
[89] Arnold J. Toynbee, "World Sovereignty and World Culture," *Pacific Affairs*, Vol.4, No.9, September, 1931, p.759-760
[90] Arnold J. Toynbee, "World Sovereignty and World Culture," *Pacific Affairs*, Vol.4, No.9, September, 1931, p.760
[91] Arnold J. Toynbee, "World Sovereignty and World Culture," *Pacific Affairs*, Vol.4, No.9, September, 1931, p.770

CHAPTER 2: *Elite Foundations*

[1] René Albert Wormser, *Foundations: Their Power and Influence*, published by The Devin-Adair Company, 1958, p.200
[2] Carroll Quigley, *Tragedy and Hope: A History of the World in Our Time*, published by Macmillan, 1966, p.955
[3] G. Edward Griffin, *The Hidden Agenda for World Government: Merging America into World Government, as Told by Norman Dodd, Congressional Investigator of Tax-Exempt Foundations*, American Media, 1990; video distributed by Quantum Communications, available at http://www.realityzone.com
[4] Ibid. Griffin
[5] Ibid. Griffin
[6] Peter Thompson, "Bilderberg And The West," chapter from Holly Sklar, *Trilateralism: The Trilateral Commission and Elite Planning for World Management*, published by South End Press, 1980, p.168
[7] Ibid. Griffin
[8] Ibid. Griffin
[9] Ford Foundation website, http://www.fordfound.org/archives/item/0196/text/005
[10] William Hoffman, *Queen Juliana: The Story of the Richest Woman in the World*, published by Harcourt Brace Jovanovich, 1979, p.133
[11] James William Fulbright and Center for the Study of Democratic Institutions, *The Elite and the Electorate: Is Government by the People Possible?* Published by Center for the Study of Democratic Institutions, 1963
[12] "Around A Motionless Center," *The Lima News* (Ohio), June 27, 1963
[13] Arthur Selwyn Miller, *The Secret Constitution and the Need for Constitutional Change*, published by Greenwood Press, 1987, p.ix
[14] Ibid. Miller, p.10
[15] Ibid. Miller, p.3
[16] Ibid. Miller, p.135
[17] Ibid. Miller, p.69
[18] Ibid. Miller, p.81
[19] Ibid. Miller, p.86
[20] Ibid. Miller, p.153
[21] Ibid. Miller, p.152
[22] Ibid. Miller, p.152
[23] "Thirty-two Million Rockefeller Gift: General Education Board to Receive Enormous Sum," *New York Times*, February 8, 1907; also see Raymond Blaine Fosdick, *The Story of the Rockefeller Foundation*, published by Harper & Row, 1954, p.9
[24] "Thirty-two Million Rockefeller Gift: General Education Board to Receive Enormous Sum," *New York Times*, February 8, 1907
[25] John Taylor Gatto, *The Underground History of American Education*, published by Oxford Village Press, 2001, p.45
[26] Charlotte Iserbyt, The Deliberate Dumbing Down of America, published by Conscious Press, 1999, p.10
[27] "Gifts Now Half a Billion; The $50,000,000 Provided Will Raise College Salaries; Equal Amount For Health," *New York Times*, December 25, 1919
[28] John Ensor Harr and Peter J. Johnson, *The Rockefeller Century, published by Scribner, 1988, p.195*
[29] John Taylor Gatto, "I may be a teacher, but I'm not an educator," *Wall Street Journal*, July 25, 1991
[30] Office of Education, Department of Health, Education & Welfare (DHEW), *Feasibility Study: Behavioral Science Teacher Education Program*, published by Government Printing Office, December 1969, p.5
[31] Ibid. DHEW, p.6
[32] Ibid. DHEW, p.233
[33] Ibid. DHEW, p.237
[34] Ibid. DHEW, p.248
[35] Ibid. DHEW, p.259
[36] Ibid. DHEW, p.255
[37] Ibid. DHEW, p.255
[38] Ibid. DHEW, p.255
[39] Ibid. DHEW, p.261
[40] Ibid. DHEW, p.251

CHAPTER 3: *The Fabian Society*

[1] John Micklethwait and Adrian Wooldridge, *The Right Nation: Conservative Power in America*, published by Penguin, 2004, p.151
[2] Adam Lowther, Donald M Snow, *Americans and Asymmetric Conflict: Lebanon, Somalia, and Afghanistan*, published by Greenwood Publishing Group, 2007, p.18
[3] Robert Charles Kirkwood Ensor, *Modern Socialism: As Set Forth by Socialists in Their Speeches, Writings and Programmes (1906)*, published by BiblioBazaar, 2008, p.359
[4] Ibid. Ensor, p.359
[5] Jim Marrs, *Rule by Secrecy*, published by HarperCollins, 2000, p.99
[6] Ibid. Micklethwait and Wooldridge, p.152
[7] Ibid. Micklethwait and Wooldridge, p.152
[8] A. M. McBriar, *Fabian Socialism and English Politics*, published by University Press, 1962, p.343
[9] "The Fabian Society: a brief history," *London Guardian*, August 13, 2001
[10] See Chapter 9
[11] Edwin Black, *War Against The Weak: Eugenics And America's Campaign To Create A Master Race*, published by Thunder's Mouth Press, 2003, p.248
[12] Bernard Shaw, *The Complete Prefaces*, published by Hamlyn, 1965, p.175
[13] Richard Lynn, *Eugenics: A Reassessment*, published by Greenwood Publishing Group, 2001, p.19
[14] A. D. Irvine, "Bertrand Russell," *The Stanford Encyclopedia of Philosophy*, Fall 2008 Edition, http://plato.stanford.edu/archives/fall2008/entries/russell/
[15] Bertrand Russell, *The Impact of Science on Society (1951)*, published by Routledge, 1985, p.117
[16] Bertrand Russell, *The Impact of Science on Society (1951)*, published by Routledge, 1985, p.125
[17] Ibid. Russell, p.115-116
[18] Ibid. Russell, p.50
[19] Ibid. Russell, p.37
[20] Ibid. Russell, p.106-107
[21] Ibid. Russell, p.61
[22] Ibid. Russell, p.56
[23] Ibid. Russell, p.40-41
[24] Ibid. Russell, p.41
[25] Ibid. Russell, p.116
[26] Ibid. Russell, p.61-62
[27] Bertrand Russell, *Marriage and Morals (1929)*, published by Taylor and Francis, 2009, p.159-160
[28] Bertrand Russell, *Marriage and Morals (1929)*, published by Taylor and Francis, 2009, p.163
[29] Bertrand Russell, *The Impact of Science on Society (1951)*, published by Routledge, 1985, p.62-63
[30] Adam Charles Roberts, *Science Fiction*, published by Routledge, 2000, p.48
[31] H.G. Wells, *The New World Order (1940)*, published by Filiquarian Publishing, 2007, p.127
[32] H.G. Wells, *A Modern Utopia*, published by Forgotten Books, 1908, p.197
[33] H.G. Wells, *A Modern Utopia*, published by Forgotten Books, 1908, p.197
[34] H.G. Wells, *Anticipations: Of the Reaction of Mechanical and Scientific Progress Upon Human Life and Thought (1902)*, published by BiblioBazaar, 2007, p.193
[35] H.G. Wells, *Anticipations: Of the Reaction of Mechanical and Scientific Progress Upon Human Life and Thought (1902)*, published by BiblioBazaar, 2007, p.194
[36] H.G. Wells, *A Modern Utopia*, published by Forgotten Books, 1908, p.201
[37] Fabian Society website, http://www.fabians.org.uk/about-the-fabian-society
[38] Fabian Society website, http://www.fabians.org.uk/about-the-fabian-society

CHAPTER 4: *The Council on Foreign Relations*

[1] Council on Foreign Relations, *2008 Annual Report*; http://www.cfr.org/about/annual_report/
[2] David Rockefeller, *Memoirs*, published by Random House, 2002, p.407
[3] Ibid. Rockefeller, p.405
[4] Richard Haass, "State sovereignty must be altered in globalized era," *Taipei Times*, February 21, 2006
[5] Robert W. Tucker and David C. Hendrickson, *The Imperial Temptation: The New World Order and America's Purpose*, published by Council on Foreign Relations, 1992, p. 198
[6] Council on Foreign Relations, *2008 Annual Report*; http://www.cfr.org/about/annual_report/
[7] Council on Foreign Relations website, http://www.cfr.org/about/corporate/roster.html
[8] Phyllis Schlafly and Chester Ward, *Kissinger on the Couch*, published by Arlington House Publishers, 1975, p.150
[9] Barry M. Goldwater, *With No Apologies*, published by William Morrow and Company, 1979, p.277
[10] William F. Jasper, "Obama Picks Come From Same Old CFR Roster," *The New American*, November 26, 2008
[11] Ibid. Jasper
[12] Ibid. Jasper
[13] Ibid. Jasper
[14] Daniel Estulin, *The True Story of the Bilderberg Group*, published by TrineDay LLC, 2007, p.81-82
[15] Ibid. Estulin, p.81-82
[16] Ibid. Estulin, p.81-82
[17] Ibid. Estulin, p.81-82
[18] Alex Jones, *The Obama Deception*, Alex Jones Film Productions, 2009
[19] Council on Foreign Relations website, http://www.cfr.org/about/faqs.html
[20] William Bundy, "The History of Foreign Affairs," Council on Foreign Relations website, http://www.cfr.org/about/history/foreign_affairs.html
[21] David Rockefeller, *Memoirs*, published by Random House, 2002, p.407
[22] Phyllis Schlafly and Chester Ward, *Kissinger on the Couch*, published by Arlington House Publishers, 1975, p.150-151
[23] Ibid. Schlafly and Ward, p.151
[24] Council on Foreign Relations website, http://www.cfr.org/about/faqs.html
[25] Gary Shapiro, "What Angelina Jolie Could Encounter at the Council on Foreign Relations," *New York Sun*, February 28, 2007

[26] Phyllis Schlafly and Chester Ward, *Kissinger on the Couch*, published by Arlington House Publishers, 1975, p.146

[27] Ibid. Schlafly and Ward, p.139

[28] Ibid. Schlafly and Ward, p.139

[29] Ibid. Schlafly and Ward, p.151

[30] Dan Smoot, *Invisible Government*, published by BiblioBazaar, 2008, p.49

[31] Barry M. Goldwater, *With No Apologies*, published by William Morrow and Company, 1979, p.278

[32] Barry M. Goldwater, *With No Apologies*, published by William Morrow and Company, 1979, p.279

[33] Curtis B. Dall, *Franklin Delano Roosevelt: My Exploited Father-in-Law*, published by Christian Crusade Publications, 1967, p.92

[34] Ibid. Dall, p.59

[35] Pat Buchanan, Tom Braden and Larry McDonald, *Crossfire*, Cable News Network (CNN), May 1983; video from author's personal collection

[36] Foreign Affairs website, http://www.foreignaffairs.com/about-us

[37] Phyllis Schlafly and Chester Ward, *Kissinger on the Couch*, published by Arlington House Publishers, 1975, p.136

[38] Kingman Brewster, Jr. "Reflections On Our National Purpose," *Foreign Affairs*, Vol.50, No.3, April 1972, p.409

[39] Ibid. Brewster, Jr., p.407

[40] Ibid. Brewster, Jr., p.414

[41] Ibid. Brewster, Jr., p.409-410

[42] Ibid. Brewster, Jr., p.415

[43] Richard Gardner, "The Hard Road to World Order," *Foreign Affairs*, Vol.52, No.3, April 1974, p.556

[44] Richard Gardner, "The Hard Road to World Order," *Foreign Affairs*, Vol.52, No.3, April 1974, p.558

[45] Ibid. Gardner, p.559

[46] Ibid. Gardner, p.559

[47] Ibid. Gardner, p.559

[48] Ibid. Gardner, p.559-560

[49] Ibid. Gardner, p.560

[50] Ibid. Gardner, p.560

[51] Ibid. Gardner, p.560

[52] Ibid. Gardner, p.558

[53] Arthur Schlesinger, Jr., "New Isolationists Weaken America," *New York Times*, June 11, 1995

[54] Ibid. Schlesinger, Jr.,

[55] Ibid. Schlesinger, Jr.,

[56] Ibid. Schlesinger, Jr.,

[57] Kent Courtney and Phoebe Courtney, *America's Unelected Rulers*, published by Conservative Society of America, 1964, p.1

[58] Benjamin Franklin and Edmund S. Morgan, *Not Your Usual Founding Father: Selected Readings from Benjamin Franklin*, published by Yale University Press, 2007, p.272

[59] Council on Foreign Relations website, http://www.foreignaffairs.com/author/james-macgregor-burns

[60] Gaddis Smith, "The Power To Lead: The Crisis of the American Presidency," *Foreign Affairs*, Vol.62, No.5, Summer 1984

[61] James MacGregor Burns, *The Power to Lead: The Crisis of the American Presidency*, published by Simon and Schuster, 1984, p.190

[62] Ibid. Burns, p.189

[63] World Affairs Councils of America website, http://www.worldaffairscouncils.org

[64] Dan Smoot, *Invisible Government*, published by BiblioBazaar, 2008, p.46

[65] "Dulles And Harriman Argue Foreign Policy, " *New York Times*, October 31, 1952

[66] R.W. Apple Jr., "Rockefeller Says U.S. Policy Lags," *New York Times*, May 1, 1968

[67] Henry Kissinger, *The Troubled Partnership: A Re-appraisal of the Atlantic Alliance*, published by Doubleday, 1966, p.36

[68] Richard C. Thornton, *The Carter Years: Toward a New Global Order*, published by Paragon House, 1991 p.78

[69] James J. Kilpatrick, "Good Intentions and Bad Judgment," *Syracuse Herald-American*, February 8, 1976

[70] Ibid. Kilpatrick

[71] Harlan Cleveland, "Foreign Policy's Missing Link: Right at Home," *New York Times*, February 29, 1976

[72] "DAR Honors State Regent At Meeting," *The Times-Bulletin* (Van Wert, Ohio), April 14, 1976

[73] Associated Press, "Byron removes name from 'declaration,'" *The Post* (Frederick, Maryland), February 23, 1976

[74] Richard Haass, "State sovereignty must be altered in globalized era," *Taipei Times*, February 21, 2006

[75] Ibid. Haass

[76] Ibid. Haass

[77] Council on Foreign Relation, *International Institutions and Global Governance Program*, document available at http://www.cfr.org/project/1369/international_institutions_and_global_governance.html

[78] Ibid. Council on Foreign Relation

[79] Ibid. Council on Foreign Relation

[80] Ibid. Council on Foreign Relation

[81] Dan Ackman, "Bring Us Osama Bin Laden," *Forbes*, September 21, 2001

[82] Bob Barr, "International Organizations Continue Threats to our Republic," press release, October 12, 2002; As cited in Thomas P. Kilgannon, *Diplomatic Divorce: Why America Should End Its Love Affair with the United Nations*, published by Stroud & Hall Publishers, 2006, p.x

[83] Michael Gorbachev, *On My Country And The World*, published by Colombia University Press, 2000, p.227

[84] United Nations website, http://www.un.org/geninfo/ir/index.asp?id=110

[85] United Nations website, http://www.un.org/aboutun/unhistory/

[86] René Albert Wormser, *Foundations: Their Power and Influence*, published by The Devin-Adair Company, 1958, p.209

[87] Jim Marrs, *Rule by Secrecy*, published by Harper Collins, 2000, p.140

[88] "Dulles Outlines World Peace Plan," *New York Times*, October 29, 1939

[89] Peter Grose, *Continuing the inquiry: the Council on Foreign Relations from 1921 to 1996*, published by University of California, 2008, p.23; Also see Council on Foreign Relations website, http://www.cfr.org/about/history/cfr/war_peace.html

[90] Ibid. Grose, p.23

[91] Curtis B. Dall, *Franklin Delano Roosevelt: My Exploited Father-in-Law*, published by Christian Crusade Publications, 1967, p.170

[92] Council on Foreign Relations website, http://www.cfr.org/about/history/cfr/war_peace.html

[93] David Rockefeller, *Memoirs*, published by Random House, 2002, p.407

[94] Neil Smith, *American Empire: Roosevelt's Geographer and the Prelude to Globalization*, published by University of California Press, 2004, p.330

[95] Michael Wala, *The Council on Foreign Relations and American Foreign Policy in the Early Cold War*, published by Berghahn Books, 1994, p.34

[96] Ibid. Wala, p.35

[97] United Nations website, http://www.un.org/aboutun/charter/history/moscowteheran.shtml

[98] Ibid. Wala, p.36

[99] John Loftus website, http://www.john-loftus.com/bio.asp

[100] John Luftus and Mark Aarons, *The Secret War Against the Jews: How Western Espionage Betrayed The Jewish People*, published by St. Martin's Press, 1994, p.165-171

[101] F. William Engdahl, *Seeds of Destruction: The Hidden Agenda of Genetic Manipulation*, published by Global Research, 2007, p.111

[102] Dan Smoot, *The Invisible Government*, published by BiblioBazaar, 2008, p.185

[103] Peter Thompson, "Bilderberg And The West," from Holly Sklar, *Trilateralism: The Trilateral Commission and Elite Planning for World Management*, published by South End Press, 1980, p.166

[104] Ibid. Smoot, p.228

[105] Institute for Defense Analyses Special Studies Group, *Study Memorandum No.7*, "A World Effectively Controlled by the United Nations," published by Special Studies Group, Institute for Defense Analyses, 1961, p.v

[106] UNESCO, *Toward World Understanding*, published by UNESCO, 1949, p.56

[107] UNESCO, *Toward World Understanding*, published by UNESCO, 1949, p.58

[108] William Jasper, "UNESCO's rotten track record," *The New American*, May 19, 2003

[109] William Preston, Edward S. Herman, and Herbert I. Schiller, *Hope & Folly: The United States and Unesco, 1945-1985*, published by University of Minnesota Press, 1989, p.229

[110] United Press International, "Senate Gets Resolution to Resist Unesco Efforts to Regulate Press," *New York Times*, June 10, 1981

[111] Ibid. Preston, Herman and Schiller, p.229

[112] Owen Bowcott, "After 18 years away America rejoins Unesco in surprise announcement," *London Guardian*, September 13, 2002

[113] Jon Henley, "Leaked documents reveal extent of Unesco corruption," *London Guardian*, October 18, 1999

[114] Ibid. Henley

[115] Phil Brennan, "UNESCO: Strangle This Monster in Its Crib," *NewsMax*, June 18, 2003; http://archive.newsmax.com/archives/articles/2003/6/17/184606.shtml

[116] Julian Huxley, *UNESCO: Its Purpose And Its Philosophy*, published by Public Affairs Press, 1948, p.61

[117] Ibid. Huxley, p.13

[118] Ibid. Huxley, p.34

[119] Ibid. Huxley, p.60

[120] Ibid. Huxley, p.55

[121] Ibid. Huxley, p.51

[122] Ibid. Huxley, p.40-41

[123] Ibid. Huxley, p.21

[124] Ibid. Huxley, p.21

[125] Ibid. Huxley, p.15

[126] Dan Smoot, *The Invisible Government*, published by BiblioBazaar, 2008, p.140

[127] Hugh Wilford, *The Mighty Wurlitzer: How the CIA Played America*, published by Harvard University Press, 2008, p.150

[128] Mortimer Jerome Adler, *How to Think about War and Peace*, published by Fordham University Press, 1995, back cover

[129] Lawrence S. Wittner, *One World Or None: A History of the World Nuclear Disarmament Movement Through 1953*, published by Stanford University Press, 1993, p.71

[130] Ibid. Wittner, p.71

[131] Ibid. Wittner, p.71

[132] Ibid. Wittner, p.70

[133] Cord Meyer, Jr., "A Progress Report on World Federation," *Bulletin of the Atomic Scientists*, Vol.5, No.10, October 1949, p.282

[134] C.P. Trussell, "UNO Bill in Senate Sparks a Debate," *New York Times*, November 27, 1945

[135] Harold B. Hinton, "Senate Unit Hears New Plan for UN," *New York Times*, February 10, 1950

[136] Ibid. Meyer, Jr., p.281

[137] Ibid. Meyer, Jr., p.281

[138] Everette Howard Hunt, *American Spy: My Secret History in the CIA, Watergate, and Beyond*, published by John Wiley and Sons, 2007, p.133

[139] Paul Joseph Watson, "Landmark E. Howard Hunt JFK Confession Video Tape Ignored," *PrisonPlanet.com*, October 24, 2008

[140] Douglas Valentine, *The Strength of the Wolf: The Secret History of America's War on Drugs*, published by Verso, 2004, p.383

[141] Dan Smoot, *The Invisible Government*, published by BiblioBazaar, 2008, p.140

[142] James P. Warburg, *The West in Crisis*, published by Doubleday, 1959, p.30

[143] United States Senate, Eighty-first Congress, second session, *Revision of the United Nations Charter: Hearings before a subcommittee of the Committee on Foreign Relations*, published by Government Printing Office, 1950, p.494; Also see Associated Press, "Views Clash On World Government," *The Troy Record (New York)*, February 18, 1950

[144] Dan Smoot, *The Invisible Government*, published by BiblioBazaar, 2008, p.20

[145] Nelson A. Rockefeller, *The Future of Federalism*, published by Harvard Press, 1962, p.64

[146] Ibid. Rockefeller, p.66

[147] Ibid. Rockefeller, p.67

[148] Ibid. Rockefeller, p.74

[149] "The Future of Federalism," *Time* magazine, June 15, 1962

[150] "The Future of Federalism," *Time* magazine, June 15, 1962

[151] Nelson A. Rockefeller, *The Future of Federalism*, published by Harvard Press, 1962, p.64

[152] Associated Press, "New World Order Pledged by Rocky," *Moberly Monitor-Index* (Missouri), July 26, 1968

[153] World Federalist Movement website, http://www.wfm.org/site/index.php/pages/2

[154] World Federalist Movement website, http://www.wfm.org/site/index.php/articles/7

[155] Obituary: Dr. John J. Logue, *The Pennsylvania Gazette*, Vol.103, No.2, November/December 2004

[156] Mortimer Jerome Adler, *How to Think about War and Peace*, published by Fordham University Press, 1995, p.xxxi

[157] "Strobe Talbott to head Center for Study of Globalization," *Yale Bulletin & Calendar*, Vol.29, No.11, November 17, 2000

[158] "Strobe Talbott to head Center for Study of Globalization," *Yale Bulletin & Calendar*, Vol.29, No.11, November 17, 2000

[159] Strobe Talbott, "America Abroad: The Birth of the Global Nation," *Time*, July 20, 1992

[160] William F. Jasper, "Target: World Government," *The New American*, September 16, 1993

[161] Stephen L. Vaughn, Encyclopedia of American Journalism, published by CRC Press, 2007, p.127

[162] Walter Cronkite, *A Reporter's Life*, published by Random House, 1997, p.79

[163] Ibid. Cronkite, p.128

[164] Walter Cronkite, Norman Cousins Global Governance Award acceptance speech, October 19, 1999, video from author's personal collection; Also see Joseph Farah, *Stop the Presses! The Inside Story of the New Media Revolution*, published by WND Books, 2007, p.32

CHAPTER 5: *The Trilateral Commission*

[1] David Rockefeller, *Memoirs*, published by Random House, 2002, p.416

[2] Zbigniew Brzezinski, Q&A session at Colombia University, New York City, March 25, 2008; video footage courtesy of wearechange.org

[3] Zbigniew Brzezinski, *Between Two Ages: America's Role in the Technetronic Era*, published by Viking Press, 1970, p.258-259

[4] Ibid. Brzezinski, p.259

[5] Trilateral Commission website, http://www.trilateral.org/moreinfo/faqs.htm

[6] David Rockefeller, *Memoirs*, published by Random House, 2002, p.405

[7] Ibid. Rockefeller, p.405

[8] Gary Benoit, "McCain Would Put Democrats in His Cabinet," *The New American*, September 11, 2008

[9] Trilateral Commission website, http://www.trilateral.org/moreinfo/faqs.htm

[10] Ibid. Rockefeller, p.416-417

[11] Barry M. Goldwater, *With No Apologies*, published by William Morrow and Company, 1979, p.286

[12] Ibid. Goldwater, p.280

[13] Ibid. Goldwater, p.281

[14] Anthony C. Sutton, *Trilaterals Over America*, published by CPA Book Publisher, 1994, p.11

[15] Zbigniew Brzezinski, *Between Two Ages: America's Role in the Technetronic Era*, published by Viking Press, 1970, p.56

[16] Michel Crozier, Samuel Huntington, and Joji Watanuki, *The Crisis of Democracy: Report on the Governability of Democracies to the Trilateral Commission*, published by New York University Press, 1975, p.2

[17] Ibid Crozier, Huntington and Watanuki, p.105

[18] Ibid Crozier, Huntington and Watanuki, p.76

[19] Declaration of Independence, Library of Congress website, http://www.loc.gov/rr/program/bib/ourdocs/DeclarInd.html

[20] Ibid Crozier, Huntington and Watanuki, p.76-78

[21] Ibid Crozier, Huntington and Watanuki, p.76

[22] Ibid Crozier, Huntington and Watanuki, p.106

[23] Ibid Crozier, Huntington and Watanuki, p.114-115

[24] Noam Chomsky and Carlos Otero, *Radical Priorities*, published by AK Press, 2003, p.140

[25] Ibid Crozier et al., p.114

[26] Andrew Napolitano, *The Constitution in Exile*, published by Thomas Nelson, 2007, p.x

[27] Ibid Crozier et al., p.113

[28] Ibid Crozier et al., p.7

[29] Missouri Information Analysis Center Strategic Report, "The Modern Militia Movement," February 20, 2009

[30] Ibid Crozier et al., p.114

[31] William Greider, "But What Are They?" *Washington Post*, January 16, 1977

[32] Ibid. Greider

[33] Barry M. Goldwater, *With No Apologies*, published by William Morrow and Company, 1979, p.287

[34] Ibid. Goldwater, p.288-289

[35] "Jimmy Carter: Not Just Peanuts," *Time*, March 8, 1976

[36] Ibid. Goldwater, p.290; Also see Aaron Latham, "Carter's Little Kissingers," *New York Magazine*, December 13, 1976

[37] Ibid. Goldwater, p.286

[38] David Rockefeller, *Memoirs*, published by Random House, 2002, p.417-418

[39] William Greider, "But What Are They?" *Washington Post*, January 16, 1977

[40] Daniel Estulin, *The True Story of the Bilderberg Group*, published by TrineDay LLC, 2007, p.161

[41] Aaron Latham, "Carter's Little Kissingers," *New York Magazine*, December 13, 1976

[42] Ibid. Goldwater, p.292

[43] See Chapter 8

[44] Kathleen Klenetsky and Herbert Quinde, "FEMA's structure for fascist rule," *Executive Intelligence Review*, November 23, 1990

[45] Ibid. Goldwater, p.293

[46] Leslie H. Gelb, "The Secretary of State sweepstakes," *New York Times*, May 23, 1976

[47] Barack Obama, *The Audacity of Hope: Thoughts on Reclaiming the American Dream*, published by Crown Publishers, 2006, p.11

[48] Janny Scott, "Obama's Account of New York Years Often Differs From What Others Say," *New York Times*, October 30, 2007

[49] Ibid. Scott

[50] Adam Goldman and Robert Tanner (Associated Press), "Old friends recall Obama's years in LA, NY," *USA Today*, May 15, 2008

[51] Ibid. Scott

[52] Webster Griffin Tarpley, *Barack H. Obama: The Unauthorized Biography*, published by Progressive Press, 2008, p.9-10

[53] Philip Elliot (Associated Press), "Obama tries to allay Jewish concerns," *USA Today*, March 14, 2008

[54] Katja Gloger, "Brzezinski about Obama: 'Very different from most American politicians,'" *Stern* (Germany), November 14, 2008; http://www.stern.de/politik/ausland/:Brzezinski-Obama-Very-American/645635.html

[55] Zbigniew Brzezinski, *The Grand Chessboard: American Primacy and Its Geostrategic Imperatives*, published by Basic Books, 1998, p.215

[56] Ibid. Brzezinski, p.57

[57] Ibid. Brzezinski, p.35-36

[58] Ibid. Brzezinski, p.210

[59] Ibid. Brzezinski, p.211

[60] Cambodian Genocide Program website, http://www.yale.edu/cgp/

[61] John Pilger, "How Thatcher gave Pol Pot a hand," *New Statesman*, April 17, 2000

[62] Strobe Talbott, "America Abroad: Defanging the Beast," *Time*, February 6, 1989

[63] Ibid. Pilger

[64] Carl Boggs, *Masters of War: Militarism and Blowback in the Era of American Empire*, published by Routledge, 2003, p.114

[65] Megan K. Stack, "The other Afghan war," *Los Angeles Times*, November 23, 2008

[66] Zbigniew Brzezinski, *Between Two Ages: America's Role in the Technetronic Era*, published by Viking Press, 1970, p.9

[67] Ibid. Brzezinski, p.253

[68] Ibid. Brzezinski, p.308-309

[69] Ibid. Brzezinski, p.70

[70] Ibid. Brzezinski, p.283

[71] Ibid. Brzezinski, p.29

[72] Ibid. Brzezinski, p.276
[73] Ibid. Brzezinski, p.13
[74] Ibid. Brzezinski, p.15
[75] Ibid. Brzezinski, p.57
[76] Ibid. Brzezinski, p.72-75
[77] Ibid. Brzezinski, p.61
[78] Ibid. Brzezinski, p.291
[79] Ibid. Brzezinski, p.300
[80] Ibid. Brzezinski, p.253

CHAPTER 6: The Bilderberg Group

[1] David Rockefeller, *Memoirs*, published by Random House, 2002, p.411
[2] Alden Hatch, *H.R.H. Prince Bernhard of the Netherlands: An Authorized Biography*, published by Harrap & Co., 1962, p.216
[3] Ibid. Hatch, p.212
[4] Peter Thompson, "Bilderberg And The West," from Holly Sklar, *Trilateralism: The Trilateral Commission and Elite Planning for World Management*, published by South End Press, 1980, p.158
[5] Ibid. Thompson, p.168
[6] Ibid. Hatch, p.224
[7] Ibid. Hatch, p.226
[8] Ibid. Hatch, p.225
[9] Ibid. Hatch, p.225
[10] Ibid. Hatch, p.223
[11] Andrew Rettman, "'Jury's out' on future of Europe, EU doyen says," *EU Observer*, March 16, 2009
[12] Ibid. Hatch, p.46-47
[13] Ibid. Hatch, p.48
[14] Ibid. Hatch, p.49
[15] Diarmuid Jeffreys, *Hell's Cartel: IG Farben and the Making of Hitler's War Machine*, published by Macmillan, 2008
[16] Ibid. Hatch, p.50
[17] Ibid. Hatch, p.43
[18] Kevin Dowling, "WWF—An Unnatural History part 4," *Noseweek*, No.23, July 1998
[19] Obituary, "HRH Prince Bernhard of the Netherlands," *London Telegraph*, December 4, 2004
[20] Obituary, "HRH Prince Bernhard of the Netherlands," *London Telegraph*, December 4, 2004
[21] Holly Sklar, *Trilateralism: The Trilateral Commission and Elite Planning for World Management*, published by South End Press, 1980, p.182
[22] Scott M. Cutlip, *The Unseen Power: Public Relations, a History*, published by Lawrence Erlbaum Associates, 1994, p.151
[23] Kevin Dowling, "WWF—An Unnatural History part 5," *Noseweek*, No.24, October 1998; Also see Anthony Sutton, *Wall Street and the Rise of Hitler*, published by Bloomfield Books, 1976, p.29
[24] Mark E. Spicka, "The Devil's Chemists on Trial: The American Prosecution of I.G. Farben at Nuremberg," *The Historian*, Vol.61, 1999
[25] Kevin Dowling, "WWF—An Unnatural History part 5," *Noseweek*, No.24, October 1998
[26] Ibid. Sklar, p.183
[27] "Dead prince admits Dutch bribery scandal," *London Times*, December 3, 2004
[28] "Dead prince admits Dutch bribery scandal," *London Times*, December 3, 2004
[29] Anthony Browne, "From beyond the grave, Prince finally admits taking $1m bribe," *London Times*, December 4, 2004
[30] James P. Tucker, "Big Surprises at Bilderberg," *American Free Press*, June 19, 2006
[31] Alex Jones, *Endgame 1.5*, Alex Jones Film Productions, 2007; Also see Alex Jones, *The Obama Deception*, Alex Jones Film Productions, 2009
[32] Ibid. Jones
[33] Alex Jones, *Endgame 1.5*, Alex Jones Film Productions, 2007; Also see Alex Jones, *The Obama Deception*, Alex Jones Film Productions, 2009

CHAPTER 7: Regional Government

[1] Walter LaFeber, *The Deadly Bet: LBJ, Vietnam, and the 1968 Election*, published by Rowman & Littlefield, p.56
[2] David Johnston, "Tall Orders; For Washington, Ideas to Secure America's Future," *New York Times*, January 24, 1988
[3] Ambrose Evans-Pritchard, "Euro-federalists financed by US spy chiefs," *London Telegraph*, June 19, 2001
[4] Ibid. Evans-Pritchard
[5] European Union website, http://europa.eu/abc/history/index_en.htm
[6] Paul Beliën, "Former Soviet Dissident Warns For EU Dictatorship," *Brussels Journal*, February 27, 2006
[7] Ibid. Beliën
[8] Ibid. Beliën
[9] David Rennie, "Keep up the pressure for a No vote, Left warned," *London Telegraph*, May 26, 2005
[10] "What the EU constitution says," *BBC News*, June 22, 2004
[11] "What the EU constitution says," *BBC News*, June 22, 2004
[12] "The Cleavage Between the People and Their Governments," *Brussels Journal*, November 12, 2007; also see Andrew Neil, "Trading Lisbon Treaty lists," *Straight Talk and Daily Politics (BBC)*, June 1, 2009
[13] Nigel Farage, speech before European Parliament, Strasbourg, February 20, 2008; See European Parliament website, http://www.europarl.europa.eu; video archive from author's personal collection
[14] Dominc Lawson, "Pessimism is Brown's enemy on Europe," *The Independent* (London), October 16, 2007
[15] William Rees-Mogg, "Alex Salmond and the stinking fish," *London Times*, October 22, 2007
[16] "The Cleavage Between the People and Their Governments," *Brussels Journal*, November 12, 2007; also see Paul Stephenson and Lorraine Mullally, "Brownie No.2—The Lisbon Treaty," *The Spectator*, April 4, 2008
[17] Fionnan Sheahan, "Greens forced to support vote on EU treaty," *Irish Independent*, June 25, 2007
[18] Andrew Stuttaford, "Angela's Ashes?" *National Review Magazine*, June 16, 2008
[19] Ibid. Stuttaford
[20] Bruno Waterfield, "EU polls would be lost, says Nicolas Sarkozy," *London Telegraph*, November 15, 2007
[21] Ibid. Waterfield

22 Ibid. Waterfield
23 Ibid. Lawson
24 Steve Bird, "Challenge over EU referendum given go-ahead by High Court," *London Times*, May 3, 2008
25 Stephen Castle and Judy Dempsey, "Split emerges in EU after Ireland rejects treaty," *International Herald Tribune*, June 15, 2008
26 Bruno Waterfield, "EU Constitution author says referendums can be ignored," *London Telegraph*, June 26, 2008
27 Bruno Waterfield, "EU Constitution author says referendums can be ignored," *London Telegraph*, June 26, 2008
28 Ibid. Lawson
29 *End of Nations: EU Takeover & the Lisbon Treaty*, produced and distributed by WiseUpJournal.com, 2008
30 Ibid. *End of Nations*
31 Ibid. *End of Nations*
32 Bruno Waterfield, "EU Treaty likely to be imposed by stealth despite Irish no vote," *London Telegraph*, June 15, 2008
33 Cristina Galindo, "Entrevista: Václav Klaus Presidente de la República Checa," *El País* (Spain), June 25, 2008; Also see Leigh Phillips, "UK millionaire's Lisbon Treaty challenge defeated," *EU Observer*, June 26, 2008
34 Ibid. Waterfield
35 "Martin hails poll showing 'steady support' for Lisbon," *Irish Times*, September 12, 2009
36 Daniel Hannan, "Herman Van Rompuy: today the EU, tomorrow the world!" *London Telegraph (Blog)*, November 21, 2009
37 Michelle Cini, "The European Commission: An Unelected Legislator?" chapter from Rinus Schendelen and Roger Scully, *The Unseen Hand: Unelected EU Legislators*, published by Routledge, 2003, p.15
38 Ibid. Cini, p.20
39 Ibid. Cini, p.24
40 Rinus Schendelen and Roger Scully, *The Unseen Hand: Unelected EU Legislators*, published by Routledge, 2003, p.118
41 Ibid. *End of Nations*
42 European Parliament website, http://www.europarl.europa.eu/parliament/public/staticDisplay.do?id=46&pageRank=2&language=EN
43 Ibid. *End of Nations*
44 Pascal Lamy, "Global Governance: Learning From The European Model," abridged speech released by WTO Press Office, November 9, 2009, http://www.egovmonitor.com/node/30634
45 Pascal Lamy, "Global Governance: Learning From The European Model," abridged speech released by WTO Press Office, November 9, 2009, http://www.egovmonitor.com/node/30634
46 Damien McElroy, "Berlin Wall: Angela Merkel challenges US power," *London Telegraph*, November 10, 2009
47 Brian Farmer, "An Amero for Your Thoughts," *The New American*, Vol.23, No.21, October 15, 2007
48 Samuel Francis, "NAFTA battle: nationalism vs. globalism—controversy over the North American Free Trade Agreement," *Insight on the News*, September 27, 1993
49 Ibid. Francis
50 Henry Kissinger, "With NAFTA, U.S. finally creates a new world order," *Los Angeles Times*, July 18, 1993
51 Ibid. Kissinger
52 William A. Orme, "NAFTA: Myths versus Facts," *Foreign Affairs*, Vol.72, No.5, November/December 1993
53 William A. Orme, "A Fistful of Trade; NAFTA Is Just One Facet of a Growing Economic Cohesion," *Washington Post*, November 14, 1993
54 Ibid. Orme
55 Robert E. Scott, Carlos Salas, Bruce Campbell, and Jeff Faux, "Revisiting NAFTA: Still Not Working For North America's Workers," *Economic Policy Institute*, Briefing Paper No.173, September 28, 2006, p.2
56 Ibid. Scott, Salas, Campbell, and Faux, p.2
57 Ibid. Scott, Salas, Campbell, and Faux, p.2
58 Ibid. Scott, Salas, Campbell, and Faux, p.2
59 Interview with Catherine Austin Fitts, *The Alex Jones Show*, Genesis Communications Network, November 25, 2009
60 Doug Palmer, "No need to renegotiate NAFTA to improve it: USTR," *Reuters*, April 20, 2009
61 Mike Blanchfield, "Harper delighted Obama won't open NAFTA," *National Post (Canada)* via *Canwest News Service*, April 22, 2009
62 John Authers, "US, Canada, Mexico pledge new alliance," *Financial Times*, March 23, 2005; Also see Lori Scott Fogleman, "Baylor Hosts President Bush, Mexican President Fox And Canadian Prime Minister Martin For Historic Meeting," *Baylor University News*, March 23, 2005
63 SPP website, http://www.spp.gov/myths_vs_facts.asp
64 Marcy Kaptur, "Ford Plan in Mexico," *Testimony before the U.S. House of Representatives*, June 21, 2006; http://www.kaptur.house.gov
65 Ibid. Kaptur
66 Judicial Watch website, http://www.judicialwatch.org/mission.shtml
67 Press Release, "Newly Uncovered Commerce Department Documents Detail 'Security and Prosperity Partnership of North America,'" September 26, 2006; See "Section IV—Launch of the North American Competitiveness Council (NACC), May 26, 2006," http://www.judicialwatch.org/5979.shtml
68 Jerome R. Corsi, "Bush sneaking North American super-state without oversight?" *World Net Daily*, June 13, 2006
69 Press Release, "Judicial Watch Releases Pentagon Records from 'North American Forum' Meetings," January 29, 2007; See "Bolton Agenda and Notes," at http://www.judicialwatch.org/6123.shtml
70 Ibid.
71 Ibid.
72 Ibid.
73 Jerome R. Corsi, "Plan for superhighway ripped as 'urban legend,'" *WorldNetDaily*, January 26, 2007
74 Ibid. Corsi
75 Ibid. Kaptur
76 William Jasper, "Continental Merger," *The New American*, Vol.23, No.21, October 15, 2007; Also see "The Situation Room," Cable News Network (CNN), August 21, 2007
77 Barack Obama, Q&A following campaign speech in Lancaster, Pennsylvania, March 31, 2008; video from author's personal collection
78 "Poll results: SPP plans are 'treason'," *WorldNetDaily*, April 25, 2008
79 Lou Dobbs, *Lou Dobbs Tonight*, Cable News Network (CNN), January 18, 2007; http://transcripts.cnn.com/TRANSCRIPTS/0701/18/ldt.01.html
80 House Concurrent Resolution 40, 110th Congress 2007-2008, 1st Session, January 22, 2007; www.govtrack.us
81 Chris Cobb, "Committee of MPs to study secret 'Three Amigos' scheme," *The Ottawa Citizen*, April 26, 2007
82 Ron Paul, "CNN/YouTube Republican Presidential Debate," *Cable News Network (CNN)*, November 28, 2007; http://edition.cnn.com/2007/POLITICS/11/28/debate.transcript/index.html
83 CNN Election Center 2008 Debate Scorecard, CNN/YouTube Debate, November 28, 2007; http://edition.cnn.com/ELECTION/2008/debates/scorecard/youtube.debate.112807/results.html
84 Andrew Malcolm, "News shocker: Ron Paul was biggest GOP fundraiser last quarter," *Los Angeles Times (Blog)*, February 1, 2008

[85] Jake Tapper, "Ron Paul, a Republican outsider, sets fund-raising record," *ABC News (Blog)*, February 4, 2008
[86] Associated Press, "Ron Paul, a Republican outsider, sets fund-raising record," *International Herald Tribune*, December 18, 2007
[87] Jerome R. Corsi, "Hoffa: Bush creating North American Union," *WorldNetDaily*, September 14, 2007
[88] Lou Dobbs, *Lou Dobbs Tonight*, Cable News Network (CNN), January 18, 2007; http://transcripts.cnn.com/TRANSCRIPTS/0701/18/ldt.01.html
[89] Ibid. Dobbs
[90] Ibid. Dobbs
[91] Rafael Fernandez de Castro and Rossana Fuentes Berain, "Hands Across North America," *New York Times*, March 28, 2005
[92] Council on Foreign Relations, "Council Joins Leading Canadians and Mexicans to Launch Independent Task Force on the Future of North America," News release, October 15, 2004; http://www.cfr.org/publication/7454/
[93] Ibid. Council on Foreign Relations
[94] Council on Foreign Relations, *Building a North American Community: Independent Task Force Report No.53*, published by the Council on Foreign Relations, 2005, p.3
[95] Jerome R. Corsi, "Meet Robert Pastor: Father of the North American Union," *Human Events*, July 25, 2006
[96] Robert A. Pastor, "A Modest Proposal To the Trilateral Commission," presented to the Trilateral Commission, Toronto, Canada, November 1-2, 2002, p.4; Also see Jerome R. Corsi, *The Late Great USA: NAFTA, the North American Union, and the Threat of a Coming Merger with Mexico and Canada*, published by Simon & Schuster, 2009
[97] Ibid. Pastor, p.6
[98] Ibid. Pastor, p.7
[99] Robert A. Pastor, "A North American Community Approach to Security," Testimony before the Subcommittee on the Western Hemisphere, U.S. Senate Foreign Relations Committee, June 9, 2005; http://www.cfr.org/publication/8173/north_american_community_approach_to_security.html
[100] Ibid. Pastor
[101] Ibid. Pastor
[102] Council on Foreign Relations, *Building a North American Community: Independent Task Force Report No.53*, published by the Council on Foreign Relations, 2005, p.8
[103] Ibid. Council on Foreign Relations, p.11
[104] Ibid. Council on Foreign Relations, p.18
[105] Ibid. Council on Foreign Relations, p.29
[106] Robert A. Pastor, *Toward a North American Community: Lessons from the Old World for the New*, published by Peterson Institute, p.115
[107] Ibid. Pastor, p.15
[108] Ibid. Pastor, p.39
[109] Ben Steil, "The End of National Currency," *Foreign Affairs*, Vol.86, No.3, May/June 2007
[110] "US and EU agree 'single market,'" *BBC News*, April 30, 2007

CHAPTER 8: *The Bush Administration and Beyond*

[1] Leo Strauss and Kenneth Hart Green, *Jewish Philosophy and the Crisis of Modernity: Essays and Lectures in Modern Jewish Thought*, published by SUNY Press, 1997, p.29 (citation refers to Strauss as Zionist)
[2] James Atlas, "The Nation: Leo-Cons; A Classicist's Legacy: New Empire Builders," *New York Times*, May 4, 2003
[3] Kenneth L. Deutsch and John Albert Murley, *Leo Strauss, the Straussians, and the American regime*, published by Rowman & Littlefield, 1999, p.410
[4] Shadia B. Drury, *Leo Strauss and the American Right*, published by Palgrave Macmillan, 1999, p.231
[5] Ibid. Drury, p.23
[6] George Bush, "Bush and Preventive War," *Air Force Magazine*, Vol.90, No.6, June 2007, p.68
[7] "Rebuilding America's Defenses: Strategy, Forces and Resources For a New Century," published by The Project for the New American Century, September 2000, p.iv
[8] Ibid. Project for the New American Century, p.75, 4
[9] Ibid. Project for the New American Century, p.14
[10] Ibid. Project for the New American Century, p.10
[11] Madeline Albright interview, *60 Minutes*, Columbia Broadcasting System (CBS), May 12, 1996; video from author's personal collection
[12] Ibid. Project for the New American Century, p.51
[13] Gustave Mark Gilbert, *Nuremberg Diary*, published by Da Capo Press, 1995, p.278-279
[14] Ibid. Gilbert, p.278-279
[15] Ray McGovern, "It sounds crazy, but..." *Asia Times*, March 3, 2005
[16] Dick Cheney, remarks to the Veterans of Foreign Wars 103rd National Convention, August 26, 2002; video from author's personal collection
[17] "Transcript of Powell's U.N. presentation," Cable News Network (CNN), February 5, 2003; http://edition.cnn.com/2003/US/02/05/sprj.irq.powell.transcript/index.html
[18] Transcript of Bush's televised address to the nation, "Bush: 'Leave Iraq within 48 Hours,'" CNN, March 18, 2003
[19] Seymour M. Hersh, "Selective Intelligence," *New Yorker*, May 12, 2003
[20] Seymour M. Hersh, "Selective Intelligence," *New Yorker*, May 12, 2003
[21] Neil Mackay, "Revealed: the secret cabal which spun for Blair," *Sunday Herald (Scotland)*, June 8, 2003
[22] "The secret Downing Street memo," *The Times (London)*, May 1, 2005
[23] Elisabeth Bumiller, "Bush and Blair Deny 'Fixed' Iraq Reports," *New York Times*, June 8, 2005
[24] Ibid, Bumiller
[25] "Downing Street and Beyond: Hearing Builds Momentum for Full Investigation," *Democracy Now!*, July 17, 2005; audio from author's personal collection; http://www.democracynow.org/2005/6/17/downing_street_and_beyond_hearing_builds
[26] Ibid. Democracy Now!
[27] Remarks by Ray McGovern at "Downing Street Memo" public conference, Washington DC, June 16, 2005; video from author's personal collection
[28] Ray McGovern, "Proof Bush Fixed The Facts," *TomPaine.com*, May 4, 2005
[29] Don Van Natta Jr., "Bush Was Set on Path to War, British Memo Says," *New York Times*, March 27, 2006
[30] Don Van Natta Jr., "Bush Was Set on Path to War, British Memo Says," *New York Times*, March 27, 2006
[31] Don Van Natta Jr., "Bush Was Set on Path to War, British Memo Says," *New York Times*, March 27, 2006
[32] Ibid. Van Natta Jr.
[33] Peter Spiegel, "Protests, tough questions for Rumsfeld: Former CIA analyst says defense chief lied about Iraq arms," *San Francisco Chronicle*, May 5, 2006; video from author's personal collection
[34] Bob Herbert, "It Just Gets Worse," *New York Times*, July 11, 2005
[35] Oliver Burkeman and Julian Borger, "War critics astonished as US hawk admits invasion was illegal," *London Guardian*, November 20, 2003

36 Brian Knowlton, "U.S. Must Act on 'Murky' Data to Prevent Terror, Wolfowitz Says," *New York Times*, July 27, 2003

37 John Deutch testimony before U.S. House Intelligence Committee, July 24, 2003; http://www.pbs.org/newshour/bb/congress/july-dec03/hearing_07-24.html

38 Tim Russert interview with Paul Wolfowitz, "Meet the Press," National Broadcasting Company (NBC), July 27, 2003; http://www.defenselink.mil/transcripts/transcript.aspx?transcriptid=2909

39 Nat Hentoff, "Don't Ask, Don't Tell," *Village Voice*, January 24, 2006

40 Ibid. Hentoff

41 Charlie Savage, "Miers has backed wide executive role," *Boston Globe*, October 5, 2005

42 Bob Egelko, "Obama lawyers argue to drop Yoo torture suit," *San Francisco Chronicle*, March 7, 2009

43 Dan Eggen, "Bush Authorized Domestic Spying," *Washington Post*, December 16, 2005

44 Ibid. Egelko

45 Geoffrey R. Stone, "Taking Liberties," *Washington Post*, November 5, 2006

46 Carolyn Lochhead and Carla Marinucci, "Freedom And Fear Are At War," *San Francisco Chronicle*, September 21, 2001

47 Dan Ackman, "Bring Us Osama Bin Laden," *Forbes*, September 21, 2001

48 William Safire, "Essay; Seizing Dictatorial Power," *New York Times*, November 15, 2001

49 Gustave Mark Gilbert, *Nuremberg Diary*, published by Da Capo Press, 1995, p.278-279

50 Kelly O'Meara, "Police State," *Insight Magazine*, November 9, 2001

51 Ibid. O'Meara

52 Kim Zetter, "ACLU Chief Assails Patriot Spin," *Wired Magazine*, September 23, 2003

53 Kim Zetter, "ACLU Chief Assails Patriot Spin," *Wired Magazine*, September 23, 2003

54 Caroline Fredrickson and Greg Nojeim, "ACLU Letter to Congress Urging A 'No' Vote On the USA PATRIOT Improvement and Reauthorization Act Conference Report," December 7, 2005; http://www.aclu.org/safefree/general/22394leg20051207.html

55 Michael C. Rupert, *Crossing the Rubicon: The Decline of the American Empire at the End of the Age of Oil*, published by New Society Publishers, 2004, p.484

56 Andrew P. Napolitano, *A Nation of Sheep*, published by Thomas Nelson Inc, 2007, p.141

57 Andrew P. Napolitano, *A Nation of Sheep*, published by Thomas Nelson Inc, 2007, p.146-147

58 "A Dangerous New Order" *New York Times*, October 19, 2006

59 "National yawn as our rights evaporate," *Countdown with Keith Olbermann*, MSNBC, October 18, 2006

60 "Jose Padilla Charged," *NewsHour With Jim Lehrer*, Public Broadcasting Service, November 23, 2005; http://www.pbs.org/newshour/bb/law/july-dec05/padilla_11-23.html

61 Paul Craig Roberts, "A Grave Blow to the Constitution: Padilla Jury Opens up Pandora's Box," *Counter Punch*, August 20, 2007; http://www.counterpunch.org/roberts08202007.html

62 Andrew P. Napolitano, *A Nation of Sheep*, published by Thomas Nelson Inc, 2007, p.77

63 James Bovard, "Working for the Clampdown: What might the president do with his new power to declare martial law?" *American Conservative*, April 23, 2007

64 Ibid. Bovard

65 Lewis Seiler and Dan Hamburg, "Rule by fear or rule by law?" *San Francisco Chronicle*, February 4, 2008

66 Peter DeFazio, remarks to Congress, August 2, 2007; video from author's personal collection

67 Albert Pike, *The Magnum Opus or The Great Work of the Ancient and Accepted Scottish Rite of Freemasonry*, published by Kessinger Publishing, 2004

68 Jeff Ferrell, "Homeland Security Enlists Clergy to Quell Public Unrest if Martial Law Ever Declared," KSLA-12 News (Shreveport, Louisiana), 2007 (day unknown); video from author's personal collection

69 Jeff Ferrell, "Homeland Security Enlists Clergy to Quell Public Unrest if Martial Law Ever Declared," KSLA-12 News (Shreveport, Louisiana), 2007 (day unknown); video from author's personal collection

70 David Nicholls, *Deity and Domination: Images of God and the State in the Nineteenth and Twentieth Centuries*, published by Routledge, 1989, p.25

71 InfraGard website, http://www.infragard.net

72 Matt Rothschild, "The FBI Deputizes Business," *The Progressive*, March 2008

73 Spencer Hsu and Ann Scott Tyson, "Pentagon to Detail Troops to Bolster Domestic Security," *Washington Post*, December 1, 2008

74 Gina Cavallaro, "Brigade homeland tours start Oct. 1," *Army Times*, September 30, 2008

75 Ibid. Cavallaro

76 Ibid. Cavallaro

77 Spencer Hsu and Ann Scott Tyson, "Pentagon to Detail Troops to Bolster Domestic Security," *Washington Post*, December 1, 2008

78 Matthew Rothschild, "The Pentagon Wants Authority to Post Almost 400,000 Military Personnel in US," *The Progressive*, August 12, 2009

79 Jay Ostrich, "Pennsylvania Guard Called Up for G-20 Summit," *American Forces Press Service*, September 22, 2009; also see http://www.defenselink.mil/news/newsarticle.aspx?id=55941

80 Michael Rubinkam, "Civil liberties groups: Police overreacted at G-20," *Associated Press*, September 25, 2009

81 Jerome Sherman, "LRAD lets police have loudest word," *Pittsburgh Post-Gazette*, October 9, 2009

82 Ibid. Rubinkam

83 Richard Wolf, Kathy Kiely, Fredreka Schouten, and John Fritze, "Politics, fear spell doom for bailout," *USA Today*, September 30, 2008

84 Lee Brodie, "See What People Are Saying About...Outrage Over Bailout," CNBC, September 25, 2008

85 Les Blumenthal, "Inslee joins state's three Republicans in voting against bailout package," *McClatchy Newspapers*, September 29, 2008

86 Brad Sherman, remarks to Congress, October 2, 2008; video from author's personal collection

87 James Inhofe, interview on 1170 KFAQ, Tulsa, Oklahoma, November 19, 2008; audio from author's personal collection

88 "Obama urges bailout bill's passage, offers plan," Forbes.com, September 30, 2008

89 Joe Nocera, "Fear and loathing in Congress over bailout," *International Herald Tribune*, October 5, 2008

90 Lewis Seiler and Dan Hamburg, "Will Obama restore constitutional government?" *San Francisco Chronicle*, February 20, 1999

91 Lewis Seiler and Dan Hamburg, "Will Obama restore constitutional government?" *San Francisco Chronicle*, February 20, 1999

PART II: FALSE ENVIRONMENTALISM

[1] Robert Andrews, *The Routledge Dictionary of Quotations*, published by Routledge, 1987, p.224
[2] Library of Congress website, http://www.loc.gov/loc/lcib/9806/danpre.html

CHAPTER 9: Eugenics: The Roots of False Environmentalism

[1] Francis Galton, *Inquiries Into Human Faculty and Its Development*, published by J.M. Dent & Sons, 1911, p.17
[2] Ibid. Galton, p.201
[3] Edwin Black, *War Against The Weak: Eugenics And America's Campaign To Create A Master Race*, published by Thunder's Mouth Press, 2003, p.276
[4] Ibid. Black, p.270
[5] Ibid. Black, p.284, 288, 296
[6] Ibid. Black, p.365
[7] Ibid. Black, p.38
[8] Indiana Supreme Court Legal History Lecture Series, "Three Generations of Imbeciles are Enough: Reflections on 100 Years of Eugenics in Indiana," State of Indiana website, http://www.in.gov/judiciary/citc/cle/eugenics/index.html
[9] United States Supreme Court, *Carrie Buck v. James H. Bell*, 274 U.S. 200 (1927)
[10] Ibid. Black, p.56-57
[11] Elof A. Carlson, *The Unfit: The History of a Bad Idea*, published by CSHL Press, 2001, p.267
[12] "Eugenicists Dread Tainted Aliens," *New York Times*, September 25, 1921
[13] "Eugenicists Dread Tainted Aliens," *New York Times*, September 25, 1921
[14] "Eugenicists Dread Tainted Aliens," *New York Times*, September 25, 1921
[15] "Major Darwin Predicts Civilization's Doom Unless Century Brings Wide Eugenic Reforms," *New York Times*, August 23, 1932
[16] "Birth Control Peril To Race, Says Osborn," *New York Times*, August 23, 1932
[17] Ibid.
[18] Ibid.
[19] Ibid.
[20] Planned Parenthood website, http://www.plannedparenthood.org/about-us/who-we-are/history-and-successes.htm
[21] F. William Engdahl, *Seeds of Destruction: The Hidden Agenda of Genetic Manipulation*, published by Global Research, 2007, p.76
[22] Jonathan Peter Spiro, *Defending the Master Race: Conservation, Eugenics, and the Legacy of Madison Grant*, published by UPNE, 2008, p.194
[23] Margaret Sanger, Esther Katz, Cathy Moran Hajo and Peter Engelman, *The Selected Papers of Margaret Sanger: The Woman Rebel, 1900-1928*, published by University of Illinois Press 2002, p.320
[24] Margaret Sanger, *The Pivot of Civilization*, published by Brentano's, 1922, p.46
[25] Ibid. Sanger, p.ix-xvi
[26] Linda Gordon, *The Moral Property of Women*, published by University of Illinois Press, 2002, p.235
[27] Angela Franks, *Margaret Sanger's Eugenic Legacy: The Control Of Female Fertility*, published by McFarland, 2005, p.43
[28] Ibid Engdahl, p.92
[29] Julian Huxley, *Man Stands Alone*, published by Harper & Brothers, 1941, p.66
[30] Ibid. Huxley, p.53
[31] Julian Huxley, "Eugenics and Society," *Eugenics Review*, Vol.28, No.1, 1936; http://www.eugenicsarchive.org/html/eugenics/index2.html?tag=1823
[32] Ibid Engdahl, p.70
[33] "Porto Ricochet," *Time*, February 15, 1932
[34] Ibid.
[35] Ibid. Black, p.418
[36] Ibid. Black, p.422
[37] Ibid. Black, p.424
[38] Grace Lichtenstein, "Fund Backs Controversial Study of 'Racial Betterment,'" *New York Times*, December 11, 1977
[39] Ibid. Lichtenstein
[40] Ibid Engdahl, p.91-92
[41] C.G. Darwin, *The Next Million Years*, published by Rupert Hart-Davis, 1952, p.114
[42] Ibid. Darwin, p.131
[43] Ibid. Darwin, p.125
[44] Ibid. Darwin, p.132
[45] Ibid. Darwin, p.193
[46] Ibid. Darwin, p.149

CHAPTER 10: Applied Eugenics: The Depopulation Agenda

[1] Francis P. Felice, "Population Growth," *The Compass*, 1974; As cited in Jacqueline Kasun, *The War Against Population*, published by Ignatius Press, 1988, p.37
[2] Peter Bauer and Basil Yamey, "The Third World and the West: An Economic Perspective," chapter from Scott Thompson, *The Third World: Premises of U.S. Policy*, published by Contemporary Studies, 1978, p.108; As cited in Jacqueline Kasun, *The War Against Population*, published by Ignatius Press, 1988, p.22
[3] John Cunningham Wood, *John Maynard Keynes: Critical Assessments*, published by Taylor & Francis, 1983, p.535
[4] Colin Clark, *Population Growth: The Advantages*, published by R. L. Sassone, 1972, p.44
[5] Ibid. Clark, p.48
[6] Roger Revelle "The Resources Available for Agriculture," *Scientific American*, Vol.235, No.3, September 1976, p.177; As cited in Jacqueline Kasun, *The War Against Population*, published by Ignatius Press, 1988, p.35
[7] Roger Revelle, "The World Supply of Agricultural Land," chapter from Julian Simon and Herman Kahn, *The Resourceful Earth: A Response to Global 2000*, published by Basil Blackwell, 1984, p.186; As cited in Jacqueline Kasun, *The War Against Population*, published by Ignatius Press, 1988, p.35
[8] Julian Simon, *The Resourceful Earth: A Response to Global 2000*, published by Basil Blackwell, 1984 (one of many books by Simon extolling the virtues of population growth)

9 Max Singer, "The Population Surprise," *The Atlantic*, August 1999

10 Jonathan Wald, "UN: Population to top 9 billion by 2050," *CNN News*, February 25, 2005

11 "World Population To Grow From 6.5 Billion To 9.1 Billion By 2050," United Nations press release, February 24, 2005

12 Jonathan Wald, "U.N.: Population to top 9 billion by 2050," *CNN.com*, February 25, 2005

13 Brian Clowes, *Kissinger Report 2004: A Retrospective on NSSM-200*, published by Human Life International, 2004, p.19

14 Ibid. Clowes, p.19

15 Ibid. Clowes, p.19

16 Kuotsai Tom Liou, *Handbook of Economic Development*, published by CRC Press, 1998, p.152

17 W. David Hopper, "The Development of Agriculture in Developing Countries," *Scientific American*, Vol.235, No.3, September 1976; As cited in Jacqueline Kasun, *The War Against Population*, published by Ignatius Press, 1988, p.36

18 Rogelio Maduro and Ralf Schauerhammer, *The Holes in the Ozone Scare: The Scientific Evidence that the Sky Isn't Falling*, published by 21st Century Science Associates, 1992, p.260

19 Barry Stewart Clark, *Political Economy: A Comparative Approach*, published by Greenwood Publishing Group, 1998, p.45

20 Thomas Malthus, *An Essay on the Principle of Population: With an Inquiry Into Our Prospects Respecting the Future Removal Or Mitigation of the Evils which it Occasions*, published by Cambridge University Press, 1989, p.115

21 Barry Stewart Clark, *Political Economy: A Comparative Approach*, published by Greenwood Publishing Group, 1998, p.45

22 U.S. Census Bureau, *Global Population Profile: 2002*, published by U.S. Census Bureau, 2002

23 Thomas Malthus, *An Essay on the Principle of Population: Or, A View of Its Past and Present Effects on Human Happiness; with an Inquiry Into Our Prospects Respecting the Future Removal Or Mitigation of the Evils which it Occasions*, published by J. Murray, 1817, p.228

24 Keir B. Sterling, Richard P. Harmond, George A. Cevasco, and Lorne F. Hammond, *Biographical Dictionary of American and Canadian Naturalists and Environmentalists*, published by Greenwood Press, 1997, p.798

25 William Vogt, *Road to Survival*, published by William Sloane Associates, 1948, p.238

26 Ibid. Vogt, p.78

27 "Eat Hearty," *Time*, November 8, 1948

28 Ibid.

29 Ibid.

30 Donald Gibson, *Environmentalism: Ideology and Power*, published by Nova, 2002, p.41

31 "A Study of Mankind's Future," *New York Times*, August 17, 1953

32 "Population Curb Held Key to Peace," *New York Times*, October, 25, 1951

33 Al Gore, *Earth in the Balance: Earth in the balance: Ecology and the Human Spirit*, published by Houghton Mifflin Harcourt, 2000, p.127

34 Earth Policy Institute website, http://www.earth-policy.org/About/Lester_bio.htm

35 Lester Brown, *Man and His Environment: Food*, published by Harper & Row, 1972

36 Negative Population Growth website, http://www.npg.org/projects/malthus/lb_remarks.htm

37 Negative Population Growth website, http://www.npg.org/whatis.html

38 Negative Population Growth website, http://www.npg.org/faq.html#anchor3

39 Negative Population Growth website, http://www.npg.org/whatis.html

40 Negative Population Growth website, http://www.npg.org/projects/malthus/lb_remarks.htm

41 Jacqueline Kasun, *The War Against Population*, published by Ignatius Press, 1988, p.79

42 F. William Engdahl, *Seeds of Destruction: The Hidden Agenda of Genetic Manipulation*, published by Global Research, 2007, p.84

43 Population Council website, http://www.popcouncil.org/about/bodycopy50s.htm

44 Population Council website, http://www.popcouncil.org/about/history.html

45 Population Council website, http://www.popcouncil.org/about/bodycopy50s.htm

46 Ibid. Engdahl, p.86

47 James Reston, "Kennedy Opposes Advocacy By U.S. Of Birth Control." *New York Times*, November 28, 1959

48 Donald Gibson, *Environmentalism: Ideology and Power*, published by Nova, 2002, p.54

49 Ibid. Gibson, p.57

50 John Fitzgerald Kennedy and Allan Nevins, *The Strategy of Peace*, published by Harper, 1960, p.225

51 Ibid. Gibson, p.58

52 James Reston, "Washington; The President Recognizes the Population Problem," *New York Times*, January 8, 1965

53 Bonnie Mass, *Population Target: The Political Economy of Population Control in Latin America*, published by Latin American Working Group, 1976, p.152

54 Richard Nixon, "Special Message to the Congress on Problems of Population Growth, July 18, 1969," Public Papers of the Presidents, No.271, p.521, Office of the Federal Register, National Archives, Washington, DC, 1971

55 Ibid. Nixon

56 Commission on Population Growth and the American Future, *Population and the American Future: The Report of the Commission on Population Growth and the American Future*, published by New American Library, 1972, p.150

57 Richard Nixon, "Statement About the Report of the Commission on Population Growth and the American Future, May 5, 1972," Public Papers of the Presidents, No.142, p. 576, Office of the Federal Register, National Archives, Washington, DC, 1974

58 James Scheuer, "A Disappointing Outcome," *Social Contract Journal*, Vol.2, No.4, Summer 1992

59 Ibid. Engdahl, p.57

60 Ibid. Engdahl, p.57

61 Steven D. Mumford, *The Life & Death of NSSM 200: How the Destruction of Political Will Doomed a U. S. Population Policy*, published by Center for Research on Population and Security, 1996

62 Ibid. Engdahl, p.56

63 Henry Kissinger, *National Security Study Memorandum 200: Implications of Worldwide Population Growth For U.S. Security and Overseas Interests*, December 10, 1974 (declassified July 3, 1989) Part 1, Chapter 3

64 Ibid. Kissinger, Executive Summary, Paragraph 8

65 Ibid. Kissinger, Part 1, Chapter 3

66 Ibid. Kissinger, Executive Summary, Paragraph 10

67 Ibid. Kissinger, Part 2, Section IV.B

68 Ibid. Kissinger, Part 2, Section II.B

69 Ibid. Kissinger, Executive Summary, Paragraph 28

70 Ibid. Kissinger, Executive Summary, Paragraph 29

71 Ibid. Kissinger, Executive Summary, Paragraph 37

72 Ibid. Kissinger, Part 2, Section I.F

73 Ibid. Kissinger, Part 1, Chapter 5

74 Ibid. Kissinger, Part 2, Section I.F

[75] Gerald R. Ford Presidential Library & Museum website, http://www.ford.utexas.edu/LIBRARY/DOCUMENT/NSDMNSSM/nsdm314a.htm
[76] Ibid. Kissinger, Part 2, Section I.B
[77] F. William Engdahl, *Seeds of Destruction: The Hidden Agenda of Genetic Manipulation*, published by Global Research, 2007, p.65
[78] Ibid. Engdahl, p.65
[79] Ibid. Franks, p.171
[80] Duff Gillespie, "Reimert T. Ravenholt, USAID's Population Program Stalwart," Population Reference Bureau; http://www.prb.org/Articles/2000/ReimertTRavenholtUSAIDsPopulationProgramStalwart.aspx
[81] Reimert T. Ravenholt, "Africa's Population-Driven Catastrophe Worsens," June 2000, published at http://www.ravenholt.com/
[82] Ibid. Ravenholt
[83] Brian Clowes, *Kissinger Report 2004: A Retrospective on NSSM-200*, published by Human Life International, 2004, p.4
[84] Ibid. Clowes, p.4
[85] Ibid. Clowes, p.5
[86] Gerald O. Barney, *The Global 2000 Report to the President: Entering the Twenty-First Century*, published by Seven Locks Press, 1980, p.46
[87] Ibid. Barney, p.1
[88] Ibid. Barney, p.2-3
[89] Ibid. Barney, p.3
[90] Ibid. Barney, p.4
[91] Ibid. Barney, p.40-41
[92] Ibid. Barney, p.41
[93] "Spanish judge seeks Kissinger," *CNN News*, April 18, 2002
[94] "Tatchell loses battle for Kissinger's arrest," *BBC News*, April 24, 2002
[95] John Pilger, "Fake faith and epic crimes," *New Statesman*, April 2, 2009
[96] Walter Isaacson, *Kissinger: A Biography*, published by Simon & Schuster, 2005, p.72
[97] Nobel Prize website, http://nobelprize.org/nobel_prizes/peace/laureates/1973/kissinger-bio.html
[98] Bob Woodward and Carl Bernstein, *The Final Days*, published by Simon and Schuster, 2005, p.194
[99] Bob Woodward and Carl Bernstein, *The Final Days*, published by Simon and Schuster, 2005, p.194
[100] Christopher Hitchens, "The Case Against Henry Kissinger: Part Two, Crimes against humanity," *Harper's Magazine*, March 2001, p.61
[101] Christopher Hitchens, *The Trial of Henry Kissinger*, published by Verso, 2002, p.95-96
[102] Ibid. Hitchens, p.99
[103] Stanford University website, http://www.stanford.edu/group/CCB/Staff/Ehrlich.html
[104] Bernard Dixon, *What is Science for?* published by Harper & Row, 1973, p.198
[105] Robert Gottlieb, *Forcing the Spring: The Transformation of the American Environmental Movement*, published by Island Press, 2005, p.330
[106] Paul R. Ehrlich, *The Population Bomb (Revised and Expanded Edition)*, published by Ballantine Books, 1971, p.xii
[107] Paul R. Ehrlich, *The Population Bomb (Revised and Expanded Edition)*, published by Ballantine Books, 1971, p.44
[108] Ibid. Ehrlich, p.157
[109] Ibid. Ehrlich, p.xii
[110] Ibid. Ehrlich, p.130
[111] Ibid. Ehrlich, p.132
[112] Ibid. Ehrlich, p.84
[113] Ibid. Ehrlich, p.82, 151
[114] Angela Franks, *Margaret Sanger's Eugenic Legacy: The Control Of Female Fertility*, published by McFarland, 2005, p.171
[115] Ibid. Franks, p.171
[116] Ibid. Ehrlich, p.84
[117] Ibid. Ehrlich, p.86
[118] Ibid. Ehrlich, p.86
[119] Ibid. Ehrlich, p.88
[120] Stephen R. Fox, *The American Conservation Movement: John Muir and His Legacy*, published by University of Wisconsin Press, 1985, p.313
[121] Vernon H. Heywood, *Global Biodiversity Assessment*, Cambridge University Press, 1996, p.773
[122] Bahgat Elnadi and Adel Rifaat, "Jacques-Yves Cousteau," *UNESCO Courier*, November 1991
[123] Claire Chambers, *The SIECUS Circle*, published by Western Islands, 1977, p.330
[124] William F. Jasper, *Global Tyranny – Step by Step*, published by Western Islands, 1992, p.106
[125] Prince Philip, *Down to Earth: Speeches and Writings of His Royal Highness Prince Philip Duke of Edinburgh on the Relationship of Man with His Environment*, published by Collins, 1988, p.8
[126] Fred Hauptfuhrer, "Vanishing Breeds Worry Prince Philip, but Not as Much as Overpopulation," *People*, Vol.16, No.25, December 21, 1981
[127] Scott Thompson, "The Ehrlichs: Two Genocidal Maniacs Whom Al Gore Loves," *Executive Intelligence Review*, Vol.26, No.27, July 9, 1999
[128] Dipesh Gadher, "'Eco-warrior' Prince Philip attacks big families," *London Times*, May 11, 2008
[129] Jonathan Grant, Stijn Hoorens et al., *Low Fertility and Population Ageing: Causes, Consequences, and Policy Options*, published by RAND Corporation, 2004
[130] Jeff Jacoby, "A world full of good news," *Boston Globe*, May 13, 2007
[131] Paul Watson, "The Beginning of the End for Life as We Know it on Planet Earth? There is a Biocentric Solution," Sea Shepherd website, May 4, 2007 http://www.seashepherd.org/news-and-media/editorial-070504-1.html
[132] Tracy C. Rembert, "Ted Turner: Billionaire, Media Mogul…and Environmentalist," *E Magazine*, Jan/Feb 1999
[133] John Harlow, "Billionaire club in bid to curb overpopulation," *London Times*, May 24, 2009
[134] Ibid. Harlow
[135] Ibid. Harlow
[136] Ibid. Harlow
[137] Paul Armstrong, "Gates: $10B vaccine program could save 8.7M lives," *CNN.com*, January 29, 2010

CHAPTER 11: *The Environmental Movement: Of, By and For Elites*

[1] Mayer N. Zald and John D. McCarthy, *Social Movements in an Organizational Society*, published by Transaction Publishers, 1990, p.18

[2] Ibid. Zald and McCarthy, p.18

[3] Ron Arnold and Alan Gottlieb, *Trashing the Economy: How Runaway Environmentalism is Wrecking America*, published by Free Enterprise Press, 1994, p.74

[4] Foundation Center website, http://foundationcenter.org/about/

[5] Foundation Center website, http://foundationcenter.org/about/

[6] Donald Gibson, *Environmentalism: Ideology and Power*, published by Nova, 2002, p.82

[7] Ibid. Gibson, p.83

[8] Ibid. Gibson, p.86

[9] "Ford Foundation Grants Millions to Spur Training in Ecology," *New York Times*, June 30, 1968

[10] Ibid.

[11] Kathleen Teltsch, "Rockefeller Foundation Starts Ecology Effort," *New York Times*, July 24, 1990

[12] Stephen R. Fox, *The American Conservation Movement: John Muir and His Legacy*, published by University of Wisconsin Press, 1985, p.220

[13] Roy Rosenzweig and Elizabeth Blackmar, *The Park and the People: A History of Central Park*, published by Cornell University Press, 1992, p.418

[14] Robin W. Winks, *Laurence S. Rockefeller: Catalyst for Conservation*, published by Island Press, p.43

[15] Rogelio Maduro and Ralf Schauerhammer, *The Holes in the Ozone Scare: The Scientific Evidence that the Sky Isn't Falling*, published by 21st Century Science Associates, 1992, p.245

[16] Leslie Spencer, Jan Bollwerk and Richard Morais, "The Not So Peaceful World of Greenpeace," *Forbes*, November 11, 1991

[17] Greenpeace International, *Annual Report 2007*, published by Greenpeace International, p.8

[18] Ibid. Greenpeace International, p.7

[19] Greenpeace International website, http://www.greenpeace.org/international/about/our-mission

[20] Center for Consumer Freedom, http://www.activistcash.com/index.cfm

[21] Robert Hunter, *Warriors of the Rainbow: A Chronicle of the Greenpeace Movement*, published by Holt, Rinehart and Winston, 1979

[22] Leslie Spencer, Jan Bollwerk and Richard Morais, "The Not So Peaceful World of Greenpeace," *Forbes*, November 11, 1991

[23] Ibid. Arnold and Gottlieb, p.183

[24] Gerd Leipold (guest) on HardTalk with Stephen Sackur, *BBC*, August 5, 2009

[25] Ron Arnold and Alan Gottlieb, *Trashing the Economy: How Runaway Environmentalism is Wrecking America*, published by Free Enterprise Press, 1994, p.577

[26] Center for Consumer Freedom, http://www.activistcash.com/index.cfm

[27] Center for Consumer Freedom, http://www.activistcash.com/foundation.cfm?did=104

[28] Ibid. Arnold and Gottlieb, p.577

[29] Lester R. Brown, *World Without Borders*, published by Vintage Books, 1972, p.308

[30] Hilary F. French, Erik Hagerman, Megan Ryan and Worldwatch Institute, *After the Earth Summit: The Future of Environmental Governance*, published by Worldwatch Institute, 1992, p.8

[31] Ibid. French, Hagerman, Ryan and Worldwatch Institute, p.8

[32] Ibid. French, Hagerman, Ryan and Worldwatch Institute, p.23

[33] Worldwatch Institute website, http://www.worldwatch.org/node/24

[34] Society of Conservation Biology website, http://www.conbio.org/AboutUs/

[35] Michael E. Soulé, *Conservation Biology: The Science of Scarcity and Diversity*, published by Sinauer Associates, 1986, p.6

[36] World Wide Fund for Nature, *Living Planet Report 2008*, published by WWF, 2008

[37] Ibid.

[38] Wildlands Network website, http://www.twp.org/cms/page1090.cfm

[39] Wildlands Network website, http://www.twp.org/cms/page1129.cfm

[40] "The Wildlands Project," *Wild Earth*, Special Issue, 1992, p.3

[41] John Davis, "The Role of Wild Earth in The Wildlands Project," *Wild Earth*, Special Issue, 1992, p.9

[42] Charles Mann and Mark Plummer, "The High Cost of Biodiversity," *Science*, Vol.260, June 25, 1993

[43] Michael S. Coffman, *Savior of the Earth? The Politics and Religion of the Environmental Movement*, published by Northfield Publishing, 1994, p.138

[44] Charles Mann and Mark Plummer, "The High Cost of Biodiversity," *Science*, Vol.260, June 25, 1993

[45] David Foreman, *Confessions of an Eco-Warrior*, published by Crown Publishing Group, 1991, p.7

[46] Vernon H. Heywood, *Global Biodiversity Assessment*, Cambridge University Press, 1996, p.993

[47] Ibid. Heywood, p.787

[48] Wildlands Network website, http://www.twp.org/cms/page1125.cfm

[49] Wildlands Network website, http://www.twp.org/cms/page1089.cfm

[50] Ibid. Coffman, p.92

[51] David M. Graber, "Mother Nature as a Hothouse Flower," *Los Angeles Times*, October 22, 1989

[52] Jamie Mobley, "Whirlwind of controversy surrounds UT prof," *Seguin Gazette-Enterprise* (Texas), April 4, 2006

[53] Jamie Mobley, "Doomsday: UT prof says death is imminent," *Seguin Gazette-Enterprise* (Texas), April 2, 2006

[54] Jamie Mobley, "Whirlwind of controversy surrounds UT prof," *Seguin Gazette-Enterprise* (Texas), April 4, 2006

[55] Jamie Mobley, "Whirlwind of controversy surrounds UT prof," *Seguin Gazette-Enterprise* (Texas), April 4, 2006

[56] Jamie Mobley, "Whirlwind of controversy surrounds UT prof," *Seguin Gazette-Enterprise* (Texas), April 4, 2006

[57] Jamie Mobley, "Doomsday: UT prof says death is imminent," *Seguin Gazette-Enterprise* (Texas), April 2, 2006

[58] Eric Pianka, *The Vanishing Book of Life on Earth*, published at http://uts.cc.utexas.edu/~varanus/Vanishing.Book.text.pdf

[59] Center for Consumer Freedom, http://www.activistcash.com/organization_financials.cfm/oid/271

[60] Earth First! website, http://www.earthfirst.org/about.htm

[61] Ibid.

[62] Ron Arnold and Alan Gottlieb, *Trashing the Economy: How Runaway Environmentalism is Wrecking America*, published by Free Enterprise Press, 1994, p.562

[63] Jeffrey Jacob, *New Pioneers: The Back-to-the-Land Movement and the Search for a Sustainable Future*, published by Penn State Press, 2006, p.253; Also see Daniel T. Oliver, *Animal Rights: The Inhumane Crusade*, published by Merrill Press, 1999, p.109

[64] Bill Devall, "A Spanner in the Woods: An Interview with David Foreman," *Simply Living* (Manley, NWS, Australia), Vol.2, No.12, 1987, p.40

[65] Robert Locke, "Ecoterrorism and Us," *Front Page Magazine*, November 15, 2001

[66] Miss Ann Thropy (pseudonym for Christopher Manes), "Population and AIDS," *Earth First!* Vol.7, No.5, May 1, 1987

[67] Tom Stoddard, "Oh, What a Wonderful Famine," *Earth First!* Vol.6, No.5, May 1, 1986

68 Tom Stoddard, "Oh, What a Wonderful Famine," *Earth First!* Vol.6, No.5, May 1, 1986
69 Brent L. Smith, *Terrorism in America: Pipe Bombs and Pipe Dreams*, published by State University of New York Press, 1994, p.127
70 Ibid. Smith, p.127
71 Center for Consumer Freedom, http://www.activistcash.com/organization_financials.cfm/oid/194
72 Joshua W. Busby, *Climate Change and National Security: An Agenda for Action*, published by Council on Foreign Relations, Council Special Report No.32, November 2007
73 Richard N. Gardner, "The Hard Road to World Order," *Foreign Affairs*, Vol.52, No.3, April 1974, p.560
74 Ibid. Gardner, p.560-561
75 George F. Kennan, "To Prevent a World Wasteland," *Foreign Affairs*, Vol.48, No.3, April 1970, p.401
76 Ibid. Kennan, p.409
77 Ibid. Kennan, p.408
78 Ibid. Kennan, p.408
79 Ibid. Kennan, p.413
80 Michael Gorbachev, "Gorbachev Planning To Put Soviets On The Ecology Track," *Deseret News* (Salt Lake City), April 22, 1990
81 Flora Lewis, "Gorbachev Turns Green," *New York Times*, August 14, 1991
82 Ibid. Lewis

CHAPTER 12: The WWF: A Case Study in False Environmentalism

1 World Wide Fund for Nature website, http://www.worldwildlife.org/species/index.html
2 Julian Huxley, *UNESCO: Its Purpose And Its Philosophy*, published by Public Affairs Press, 1948, p.45
3 James Bone, "UN chief is forced to stand down over harassment claims," *London Times*, February 21, 2005
4 Sylvia Washington, Heather Goodall and Paul Rosier, *Echoes from the Poisoned Well: Global Memories of Environmental Injustice*, published by Lexington Books, 2006, pg.371
5 Jad Mouawad, "Shell to Pay $15.5 Million to Settle Nigerian Case," *New York Times*, June 8, 2009
6 WWF website, Who We Are, The Sixties, http://www.panda.org/who_we_are/history/sixties/
7 WWF website, Who We Are, Organization, Presidents, http://www.panda.org/who_we_are/organization/presidents/
8 WWF website, Who We Are, Organization, Presidents, http://www.panda.org/who_we_are/organization/presidents/
9 WWF website, Who We Are, Organization, Presidents, http://www.panda.org/who_we_are/organization/presidents/
10 WWF website, Who We Are, Organization, Presidents, http://www.panda.org/who_we_are/organization/presidents/
11 Ralph J. Desmarais, "Tots and Quots (act. 1931–1946)," *Oxford Dictionary of National Biography*, published by Oxford University Press, 2007
12 Ibid. Desmarais
13 Kevin Dowling, "WWF—An Unnatural History part 2," *Noseweek*, No.21, March 1998
14 Ibid. Desmarais
15 Ibid. Desmarais
16 John Symonds, *The King of the Shadow Realm: Aleister Crowley, His Life and Magic*, published by Duckworth, 1989, p.414-415
17 Solly Zuckerman, *From Apes to Warlords*, published by Hamilton, 1978, p.111
18 Kevin Dowling, "WWF—An Unnatural History part 2," *Noseweek*, No.21, March 1998
19 Kevin Dowling, "WWF—An Unnatural History part 2," *Noseweek*, No.21, March 1998
20 Max Nicholson, *The New Environmental Age*, published by Cambridge University Press, 1989, p.105
21 Ibid. Nicholson, p.105
22 Kristina Dronamraju, *If I Am to be Remembered: The Life and Work of Julian Huxley with Selected Correspondence*, published by World Scientific, 1993, p.85
23 Kevin Dowling, "World Wildlife Fund—An Unnatural History," *Noseweek*, No.20, December 1997
24 Ibid. Dowling
25 World Wide Fund for Nature website, http://www.panda.org/who_we_are/history/sixties/
26 Ibid. Dowling
27 Fred Pearce, *Green Warriors: The People and the Politics Behind the Environmental Revolution*, published by Bodley Head, 1991, p.7
28 Kevin Dowling, "WWF—An Unnatural History part 3," *Noseweek*, No.22, May 1998
29 Alfred William Brian Simpson, *In the Highest Degree Odious: Detention without Trial in Wartime Britain*, published by Clarendon Press, 1992, p.111
30 Ibid. Simpson, p.306
31 Kevin Dowling, "WWF—An Unnatural History part 3," *Noseweek*, No.22, May 1998
32 Kevin Dowling, "World Wildlife Fund—An Unnatural History," *Noseweek*, No.20, December 1997
33 Kevin Dowling, "WWF—An Unnatural History part 5," *Noseweek*, No.24, October 1998
34 Kevin Dowling, "WWF—An Unnatural History part 3," *Noseweek*, No.22, May 1998
35 Kevin Dowling, "WWF—An Unnatural History part 3," *Noseweek*, No.22, May 1998
36 Kevin Dowling, "WWF—An Unnatural History part 2," *Noseweek*, No.21, March 1998
37 Kevin Dowling, "WWF—An Unnatural History part 3," *Noseweek*, No.22, May 1998
38 John McCormick, *Reclaiming Paradise*, Published by Indiana University Press, 1991, p.45
39 Ibid. McCormick, p.45
40 Ibid. McCormick p.46
41 Ibid. McCormick, p.41
42 Ibid. McCormick, p.46
43 European Commission website, http://bioval.jrc.ec.europa.eu/PA/
44 James Ramsay Montagu Butler, *Lord Lothian (Philip Kerr) 1882-1940*, published by St. Martin's Press, 1960, p.68
45 See Chapter 1
46 Ibid. Butler, p.332
47 Julian Huxley, "America Revisited. III. The Negro Problem," *The Spectator*, November 29, 1924
48 Ibid. Huxley
49 Max Nicholson, *The Environmental Revolution: A Guide for the New Masters of the World*, published by McGraw-Hill, 1970, p.278
50 "Who Wants To Take Your Food Away?" *Executive Intelligence Review*, Vol.35, No.18, March 2, 2008
51 Ibid.
52 "Who Wants To Take Your Food Away?" *Executive Intelligence Review*, Vol.35, No.18, March 2, 2008
53 Ibid.
54 Ibid.

CHAPTER 13: *The UN: Epicenter of False Environmentalism*

[1] United Nations Environment Programme website, http://www.unep.org/
[2] James Brooke, "The Earth Summit; U.N. Chief Closes Summit With an Appeal for Action," *New York Times*, June 15, 1992; Also see http://www.un.org/geninfo/bp/enviro.html
[3] Ibid. Brooke
[4] Ibid. Brooke
[5] Gerald Piel, "Defusing the 'Population Bomb'; Third World Development," *The Nation*, Vol.258, March 21, 1994
[6] Daniel Sitarz, *Agenda 21: The Earth Summit Strategy to Save Our Planet*, published by EarthPress, 1994, back cover
[7] Ibid. Sitarz, p.2
[8] Ibid. Sitarz, p.1
[9] Ibid. Sitarz, p.6
[10] Ibid. Sitarz, p.6
[11] Ibid. Sitarz, p.7
[12] Ibid. Sitarz, p.9
[13] Ibid. Sitarz, p.44
[14] Ibid. Sitarz, p.45
[15] Ibid. Sitarz, p.6
[16] Dixy Lee Ray, *Environmental Overkill: Whatever Happened To Common Sense?* published by Regnery Gateway, 1993, p.10-11
[17] Earth Policy Institute website, http://www.earth-policy.org/About/Lester_bio.htm
[18] Lester R. Brown (interview), "A transition to a new era?" *Terraviva*, June 3, 1992; As quoted in William F. Jasper, *Global Tyranny – Step by Step*, published by Western Islands, 1992, p.94
[19] Ibid. Brown, p.94
[20] Ibid. Brown, p.94
[21] Lester Brown et al., *State of the World 1991: A Worldwatch Institute Report on Progress Toward a Sustainable Society*, published by Earthscan, 1991, p.3
[22] Dixy Lee Ray, "Science and the Environment," *Religion and Liberty*, Fall Special Edition 1992; http://www.acton.org/publications/randl/rl_interview_52.php
[23] Dixy Lee Ray, *Environmental Overkill: Whatever Happened To Common Sense?* published by Regnery Gateway, 1993, p.10
[24] "The New World Army," *New York Times*. March 6, 1992
[25] Rogelio Maduro and Ralf Schauerhammer, *The Holes in the Ozone Scare: The Scientific Evidence that the Sky Isn't Falling*, published by 21st Century Science Associates, 1992, p.218
[26] See Chapter 15
[27] United Nations, *Report of the World Commission on Environment and Development: Our Common Future*, published by United Nations, June 1987, Chapter 2, paragraph 1; http://www.un-documents.net/wced-ocf.htm
[28] Ibid. United Nations, Introduction, paragraph 29
[29] Ibid. United Nations, Chapter 4, paragraph 23
[30] Ibid. United Nations, Chapter 4, paragraph 68
[31] Ibid. United Nations, Introduction, paragraph 42-43
[32] William F. Jasper, *Global Tyranny – Step by Step*, published by Western Islands, 1992, p.87
[33] Ronald Bailey, "International Man of Mystery: Who is Maurice Strong?" *National Review*, Vol.49, September 1, 1997
[34] Ibid. Bailey
[35] Maurice Strong, "From Stockholm to Rio: A Journey Down a Generation," published by United Nations Conference on Environment and Development, 1992; Also see *United States Congressional Serial Set*, published by U.S. Government Printing Office, 1994, p.95
[36] Commission on Global Governance, *Issues in Global Governance: Papers Written for the Commission on Global Governance*, published by Martinus Nijhoff Publishers, 1995, p.445
[37] Ronald Bailey, "International Man of Mystery: Who is Maurice Strong?" *National Review*, Vol.49, September 1, 1997
[38] Ibid. Commission on Global Governance, p.453
[39] Ibid. Commission on Global Governance, p.458
[40] Ibid. Commission on Global Governance, p.458
[41] Daniel Wood, "The Wizard of the Baca Grande," *West Magazine* (Alberta, Canada), May 1990
[42] Ibid. Wood
[43] Ronald Bailey, "International Man of Mystery: Who is Maurice Strong?" *National Review*, Vol.49, September 1, 1997
[44] Ibid. Bailey
[45] John Origen Herrick, The Life of William Jennings Bryan, published by Kessinger Publishing, 2005, p.184

CHAPTER 14: *The New World Religion*

[1] Earth Charter website, http://earthcharterinaction.org/about_charter.html
[2] Earth Charter website, http://www.earthcharterinaction.org/who.html
[3] Earth Charter website, http://www.earthcharterinaction.org/resources/files/Peoples%20Earth%20Charter.doc
[4] Nathan Gardels, "Environment: Act Globally, Not Nationally," *Los Angeles Times*, May 8, 1997
[5] "Gore wins Nobel Peace Prize for climate campaign," *London Times*, October 12, 2007
[6] Al Gore, *Earth in the Balance: Forging a New Common Purpose*, published by Earthscan, 2007, p.177
[7] Sharon Begley, "The Evolution Of An Eco-Prophet," *Newsweek*, November 9, 2009
[8] Rockefeller Brothers Fund website, http://www.rbf.org/trustees/trustees_show.htm?doc_id=495673
[9] Steven C. Rockefeller, *Spirit and Nature: Why the Environment is a Religious Issue*, published by Beacon Press, 1992, p.1
[10] Ibid. Rockefeller, p.141
[11] Ibid. Rockefeller, p.144
[12] Ibid. Rockefeller, p.143
[13] Mikhail Gorbachev, *The Search For A New Beginning: Developing A New Civilization*, published by HarperCollins, 1995, p.15
[14] Ibid. Gorbachev, p.64
[15] Ibid. Gorbachev, p.13
[16] Michael York, *The Emerging Network: A Sociology of the New Age and Neo-pagan Movements*, published by Rowman & Littlefield, 1995, p.34
[17] Ibid. York, p.34

[18] Al Gore, *Earth in the Balance: Forging a New Common Purpose*, published by Earthscan, 2007, p.263

[19] Dixy Lee Ray, "Science and the Environment," *Religion and Liberty*, Fall Special Edition 1992; http://www.acton.org/publications/randl/rl_interview_52.php

[20] William Irwin Thompson, *Thinking Together At The Edge of History: The Story of the Lindisfarne Association*, 2008, http://www.williamirwinthompson.org/lindisfarne/pilgrimagetolindisfarne.html

[21] William Irwin Thompson website, http://www.williamirwinthompson.org/lindisfarne/history.html

[22] William F. Jasper, *Global Tyranny — Step by Step*, published by Western Islands, 1992, p.144

[23] Ibid. Jasper, p.144

[24] William Irwin Thompson website, http://www.williamirwinthompson.org/lindisfarne/listoffellows.html

[25] Ted Morgan, "Looking for: Epoch B," *New York Times*, February 29, 1976

[26] Debbie Bookchin and Jim Schumacher, *The Virus and the Vaccine: The True Story of a Cancer-Causing Monkey Virus, Contaminated Polio Vaccine, and the Millions of Americans Exposed*, published by St. Martin's Press, 2004, p.318

[27] Douglas Valentine, *The Strength of the Wolf: The Secret History of America's War of Drugs*, published by Verso, 2004, p.132

[28] Ibid. Valentine, p.132

[29] Ibid. Morgan

[30] Ibid. Morgan

[31] William W. Kellogg and Margaret Mead, *The Atmosphere: Endangered and Endangering*, published by U.S. Government Printing Office, 1977, p.73

[32] William Welch Kellogg and Margaret Mead, *The Atmosphere: Endangered and Endangering*, published by U.S. Department of Health, Education, and Welfare, 1977

[33] Stewart Copinger Easton, *Rudolf Steiner: Herald of a New Epoch*, published by Steiner Books, 1980, p.187

[34] Luis S.R. Vas, *The Mind of J. Krishnamurti*, published by Jaico Publishing House, 2000, p.259-260

[35] Lucis Trust website, http://www.lucistrust.org/en

[36] Michael S. Coffman, *Saviors of the Earth? The Politics and Religion of the Environmental Movement*, published by Northfield Publishing, 1994, p.225

[37] Edward James et al., *Notable American Women, 1607-1950: A Biographical Dictionary*, published by Harvard University Press, 1971, p.175

[38] Helena Petrovna Blavatsky, *The Secret Doctrine*, published by Quest Books, 1993, p.323

[39] Ibid. Blavatsky, p.377

[40] Ibid. Blavatsky, p.415

[41] Ibid. Blavatsky, p.198

[42] Helena Petrovna Blavatsky, *The Secret Doctrine: The Synthesis of Science, Religion, and Philosophy*, published by Theosophical Publishing Company, 1893, p.439

[43] Alice Bailey, *The Externalisation of the Hierarchy*, published by Lucis Publishing Company, 1957, p.191

[44] Ibid. Bailey, p.638

[45] Alice Bailey, *Discipleship in the New Age*, published by Lucis Publishing Company, 1944, p.17

[46] Ibid. Bailey, p.18

[47] Ibid. Bailey, p.18

[48] Ibid. Bailey, p.18

[49] Alice Bailey, *The Externalisation of the Hierarchy*, published by Lucis Publishing Company, 1957, p.70

[50] Lucis Trust website, http://www.lucistrust.org/en/service_activities/world_goodwill__1/key_concepts/the_new_group_of_world_servers

[51] Ibid. Bailey, p.76

[52] Ibid. Bailey, p.77

[53] Alice Bailey, *The Rays and the Initiations*, published by Lucis Publishing Company, 1960, p.635

[54] Alice Bailey, *Education in the New Age*, published by Lucis Publishing Company, 1954, p.111-112

[55] Jackson Spielvogel and David Redles, "Hitler's Racial Ideology: Content and Occult Sources," Simon Wiesenthal Center Annual, Vol.3, 1986; http://motlc.wiesenthal.com/site/pp.asp?c=gvKVLcMVIuG&b=395043

[56] Ibid. Spielvogel and Redles

[57] Ibid. Spielvogel and Redles

[58] Ibid. Spielvogel and Redles

[59] Ibid. Spielvogel and Redles

[60] Ibid. Spielvogel and Redles

[61] William Lawrence Shirer, *The Rise and Fall of the Third Reich: A History of Nazi Germany*, published by Simon and Schuster, 1960, p.21

[62] George Lachmann Mosse, *Nazi Culture: Intellectual, Cultural and Social Life in the Third Reich*, published by University of Wisconsin Press, 2003, p.100

[63] Janet Biehl and Peter Staudenmaier, *Ecofascism: Lessons from the German Experience*, published by AK Press, 1995, p.14

[64] Ibid. Biehl and Staudenmaier, p.1

[65] Lucis Trust website, http://www.lucistrust.org/en/service_activities/world_goodwill__1/purposes_objectives; also see United Nations website, Department of Public Information, Non-Governmental Organization Section, http://www.un.org

[66] Lucis Trust website, http://www.lucistrust.org/en

[67] Lucis Trust website, http://www.lucistrust.org/en/service_activities/world_goodwill__1/purposes_objectives

[68] Lucis Trust website, http://www.lucistrust.org/en/service_activities/world_goodwill__1/purposes_objectives; also see United Nations website, Department of Public Information, Non-Governmental Organization Section, http://www.un.org

[69] Lucis Trust website, http://www.lucistrust.org/en/service_activities/world_goodwill__1/purposes_objectives

[70] Lucis Trust website, http://www.lucistrust.org/en/service_activities/world_goodwill__1/purposes_objectives

[71] Lucis Trust website, http://www.lucistrust.org/en/arcane_school/talks_and_articles/the_esoteric_meaning_of_lucifer

[72] Lucis Trust website, http://www.lucistrust.org/en/arcane_school/talks_and_articles/the_esoteric_meaning_of_lucifer

[73] Lucis Trust website, http://www.lucistrust.org/en/publications_store/the_beacon_magazine

[74] Sarah McKechnie, "Descent and Sacrifice, " *The Beacon*, September/October 1989

[75] Lucis Trust website, http://www.lucistrust.org/en/service_activities/the_great_invocation__1/what_is_the_great_invocation

[76] Alice Bailey, *The Externalization of the Hierarchy*, published by Lucis Trust, 1983, p.453

[77] Alice Bailey, *The Reappearance of the Christ*, published by Lucis Trust, 1948

[78] Alice Bailey, *The Externalization of the Hierarchy*, published by Lucis Trust, 1983, p. 503

[79] Alice Bailey, *The Rays and the Initiations*, published by Lucis Publishing Company, 1960, p.233

[80] Lucis Trust website, http://www.lucistrust.org/en/service_activities/world_goodwill__1/purposes_objectives

[81] United Nations website, Department of Public Information, Non-Governmental Organization Section, http://www.un.org; also see Aquarian Age Community website, http://www.aquaac.org/un/medmtgs.html

[82] Aquarian Age Community website, http://www.aquaac.org/about/about.html

[83] Aquarian Age Community website, http://www.aquaac.org/about/about.html

[84] Andrea Peyser, "Climate-cult con is hard to 'bear,'" *New York Post*, November 30, 2009

[85] George Russell, "Document Reveals U.N.'s Goal of Becoming Rule-Maker in Global Environmental Talks," *Fox News*, November 30, 2009; also see Mark Halle, "The UNEP That We Want Reflections on UNEP's future challenges," *Prangins*, Switzerland, September 17, 2007, http://www.foxnews.com/projects/pdf/113009_IISDreport.pdf

[86] Ibid. Russell

[87] Ibid. Russell

[88] Donald Keys, "Spirituality at the United Nations," published by Aquarian Age Community, http://aquaac.org//un/sprtatun.html

[89] Tijn Touber, "The happiest man on earth," *Ode Magazine*, January/February 2006

[90] Ibid.

[91] Ibid.

[92] Robert Muller, *New Genesis: Shaping a Global Spirituality*, published by Doubleday, 1982, p.183

[93] Ibid. Muller, p.183

[94] Ibid. Muller, p.190

[95] Ibid. Muller, p.190

[96] Ibid. Muller, p.164

[97] Ibid. Muller, p.19

[98] Ibid. Muller, p.127

[99] United Nations Online website, http://www.unol.org/rms/rmltr.html

[100] United Nations Online website, http://www.unol.org/rms/ccl.html

[101] Derek Heater, *World Citizenship: Cosmopolitan Thinking and Its Opponents*, published by Continuum International, p.173; Also see UNESCO website, http://unesco.org

[102] World Core Curriculum website, http://www.worldcorecurriculum.org

[103] Ibid.

[104] Ibid.

[105] James Harder, "U.N. Faithful Eye Global Religion," *Insight on the News*, October 2, 2000

[106] Ibid. Harder

[107] Ibid. Harder

[108] "'Global Ethic' Aiming to Supplant Christian Ethic, Warns Official," *Zenit (Rome)*, February 11, 2003

[109] Ibid.

[110] Baha'is of the United States website, http://www.bahai.us/about-bahai

[111] Baha'is of the United States website, http://www.bahai.us/Q-A

[112] Baha'i International Community website, http://info.bahai.org/article-1-6-0-6.html

[113] Baha'i International Community website, http://info.bahai.org/article-1-6-0-6.html

[114] Baha'is of the United States website, http://www.bahai.us/bahai-united-nations

[115] "Toward the New World Order," *Baha'is Magazine*, 1992, p.73; http://www.bahai.com/thebahais/pg73.htm

[116] "National rivalries, hatreds, and intrigues will cease..." *Baha'is Magazine*, 1992, p.75; http://www.bahai.com/thebahais/pg75.htm

[117] Ibid.

[118] "Climate change creates moral issues, says panel," *Baha'i World News Service*, May 9, 2007; http://news.bahai.org/story/530

[119] "Earth Charter final draft issued," *One Country*, Vol.11, No.4, January-March, 2000

[120] Lynn Townsend White, Jr. "The Historical Roots of our Ecological Crisis," *Science*, Vol.155, March 10, 1967, p.1203

[121] Alston Chase, *Playing God in Yellowstone: The Destruction of America's First National Park*, published by Harcourt Brace Jovanovich, 1987, p.299

[122] Michael S. Coffman, *Saviors of the Earth? The Politics and Religion of the Environmental Movement*, published by Northfield Publishing, 1994, p.81

[123] Ibid. Chase, p.310

[124] Paul R. Ehrlich, *The Population Bomb (Revised and Expanded Edition)*, published by Ballantine Books, 1971, p.155

[125] Carroll Quigley, *The Anglo-American Establishment: From Rhodes to Cliveden*, published by GSG & Associates, 1981, p.161

[126] Arnold J. Toynbee, "The Genesis of Pollution," *New York Times*, September 16, 1973

[127] Ibid.

[128] Ibid.

[129] Steven C. Rockefeller, *Spirit and Nature: Why the Environment is a Religious Issue*, published by Beacon Press, 1992, p.147

[130] Vernon H. Heywood, *Global Biodiversity Assessment*, Cambridge University Press, 1996, p.839

[131] Prince Philip interviewed by Alliance of Religions and Conservation, http://www.arcworld.org/news.asp?pageID=1

[132] Paul Slansky and Arleen Sorkin, "Regrets, we've had a few," *Los Angeles Times*, April 28, 2006

[133] Tracy C. Rembert, "Ted Turner: Billionaire, Media Mogul...and Environmentalist," *E Magazine*, Jan/Feb 1999

[134] James Harder, "U.N. Faithful Eye Global Religion," *Insight on the News*, October 2, 2000

[135] "Turner apologizes to Christians," *Atlanta Business Chronicle*, March 9, 2001

[136] "Ted Turner's 10 Voluntary Initiatives," *WSBTV*, April 1, 2008; http://www.wsbtv.com/news/15761660/detail.html

[137] David Spangler, *Revelation: The Birth of A New Age*, published by Rainbow Bridge, 1976, p.7

[138] "Ted Turner apologizes, joins churches' $200M malaria fight," *Associated Press*, April 1, 2008

[139] Martin Palmer and Victoria Finlay, *Faith in Conservation*, published by World Bank Publications, 2003, p.32

[140] Palmer and Finlay, p.32

[141] Russell E. Train, *Politics, Pollution, and Pandas: An Environmental Memoir*, published by Island Press, 2003, p.313

[142] Ibid. Train, p.314

[143] Ibid. Train, p.314

[144] Ibid. Train, p.314

[145] Alliance of Religions and Conservation website, http://www.arcworld.org/about.asp?pageID=4

[146] Prince Philip interviewed by Alliance of Religions and Conservation, http://www.arcworld.org/news.asp?pageID=1

[147] Martin Palmer and Victoria Finlay, *Faith in Conservation*, published by World Bank Publications, 2003, p.xi

[148] Christopher Joyce, "Anglican Leader Brings Climate to the Pulpit," *All Things Considered*, National Public Radio, April 28, 2008

[149] Ibid. Joyce

[150] Lyndsay Moseley and Anna Jane Joyner, *Faith in Action: Communities of Faith Bring Hope for the Planet*, published by Sierra Club, June 2008, p.3

[151] Sharon Begley, "The Evolution Of An Eco-Prophet," *Newsweek*, November 9, 2009

[152] Earth Day Network website, http://earthday.net/node/73

[153] Earth Day Network website, http://earthday.net/node/73

[154] Earth Day Network website, http://earthday.net/node/65

[155] Rockefeller Brothers Fund website, http://www.rbf.org/grantsdatabase/grantsdatabase_show.htm?doc_id=617322

[156] John Vidal and Tom Kington, "Protect God's creation: Vatican issues new green message for world's Catholics," *London Guardian*, April 27, 2007

[157] Ian Fisher, "In First Christmas Homily, Pope Reflects on World Conflicts," *New York Times*, December 25, 2005

[158] "Cardinal Ratzinger Severely Criticizes U.N.'s Proposal For New World Order," *Zenit (Rome)*, September 16, 2000
[159] Ibid.
[160] Stephen Adams and Louise Gray, "Climate change belief given same legal status as religion," *London Telegraph*, November 3, 2009
[161] Stephen Adams and Louise Gray, "Climate change belief given same legal status as religion," *London Telegraph*, November 3, 2009

CHAPTER 15: *The Club of Rome*

[1] Club of Rome website, http://www.clubofrome.org/eng/about/3/
[2] UNESCO website, http://erc.unesco.org/ong/en/directory/ONG_Desc_portal.asp?mode=gn&code=908
[3] Club of Rome website, http://www.clubofrome.org
[4] Donella H. Meadows, Dennis L. Meadows, Jorgen Randers, William W. Behrens III, *The Limits to Growth: A Report to The Club of Rome's Project on the Predicament Of Mankind*, published by Universe Books, 1972, p.23
[5] Ibid, p.191
[6] Ibid, p.196
[7] Ibid, p.159
[8] "The Worst Is Yet to Be?" *Time*, January 24, 1972
[9] Ibid.
[10] Henry Wallich, "More on Growth," *Newsweek*, March 13, 1972
[11] John Maddox, *The Doomsday Syndrome*, published by Macmillan, 1972, p.22
[12] "The Worst Is Yet to Be?" *Time*, January 24, 1972
[13] Ibid.
[14] Edward Goldsmith, *A Blueprint for Survival*, published by Houghton Mifflin 1972, p.18
[15] Herman Daly, *Toward a Steady-state Economy*, published by W.H. Freeman, 1973, p.151
[16] Herman Daly, "Selected Growth Fallacies," *The Social Contract*, Vol.13, No.3, Spring 2003
[17] Ibid.
[18] George J. Church, "Can the World Survive Economic Growth?" *Time*, August 14, 1972
[19] Julian Simon and Herman Kahn, *The Resourceful Earth: A Response to Global 2000*, published by Basil Blackwell, 1984, p.34
[20] "The Worst Is Yet to Be?" *Time*, January 24, 1972
[21] "Club of Rome Revisited," *Time*, April 26, 1976
[22] Leonard Silk, "Scholars Favor Global Growth," *New York Times*, April 13, 1976
[23] Ibid. Silk
[24] Ibid. Silk
[25] Jan Tinbergen, Antony J. Dolman, Jan van Ettinger, Club of Rome, *RIO: Reshaping the International Order: A Report to the Club of Rome*, published by Dutton, 1976
[26] Ibid. Tinbergen et al., p.63
[27] Ibid. Tinbergen et al., p.184
[28] Ibid. Tinbergen et al., p.172
[29] Ibid. Tinbergen et al., p.84
[30] Ibid. Tinbergen et al., p.109
[31] Ibid. Tinbergen et al., p.110
[32] Ibid. Tinbergen et al., p.107
[33] Ibid. Tinbergen et al., p.108
[34] Ibid. Tinbergen et al., p.184
[35] Mihajlo Mesarovi, Eduard Pestel, Club of Rome, *Mankind at the Turning Point: The Second Report to The Club of Rome*, published by Dutton, 1974, p.15
[36] Ibid. Mesarovi et al., p.7
[37] Ibid. Mesarovi et al., p.57
[38] Ibid. Mesarovi et al., p.viii
[39] Ibid. Mesarovi et al., p.111
[40] Ibid. Mesarovi et al., p.147
[41] Ibid. Mesarovi et al., p.147
[42] Ibid. Mesarovi et al., p.69
[43] Ibid. Mesarovi et al., p.143
[44] Ian Robert Dowbiggin, *Keeping America Sane: Psychiatry and Eugenics in the United States and Canada, 1880-1940*, published by Cornell University Press, 2003, p.129
[45] Alan Gregg, "A Medical Aspect of the Population Problem," *Science*, Vol.121, May 13, 1955
[46] Bertrand Russell, *The Impact of Science on Society (1952)*, published by Routledge, 1985, p.63
[47] George J. Church, "Can the World Survive Economic Growth?" *Time*, August 14, 1972
[48] Alexander King and Bertrand Schneider, *The First Global Revolution: A Report by the Council of the Club of Rome*, published by Orient Blackswan, 1991, p.71
[49] Ibid. King and Schneider, p.71
[50] Ibid. King and Schneider, p.75
[51] Ibid. King and Schneider, p.68
[52] Ibid. King and Schneider, p.70
[53] Ibid. King and Schneider, p.70
[54] Ibid. King and Schneider, p.70
[55] Ibid. King and Schneider, p.75
[56] "'Club of Rome' member warns against council amalgamations," Australia Broadcasting Corporation, June 5, 2007
[57] Club of Rome website, http://www.clubofrome.org/eng/people/honorary_members.asp
[58] Council of the European Union official website,
http://www.consilium.europa.eu/cms3_applications/applications/solana/cv.asp?cmsid=246&lang=EN
[59] James Bone, "UN chief is forced to stand down over harassment claims," *London Times*, February 21, 2005
[60] R. Lubbers and J. Koorevaar, "Governance in an era of Globalization: Paper for the Club of Rome Annual Meeting 1999," published by C.O.R., 1999
[61] Ibid. Lubbers and Koorevaar
[62] Tom Buerkle, "Dutch See Humiliation by Washington After Lubbers Withdraws," *International Herald Tribune*, November 11, 1995
[63] Imran H. Khan Sudahazai, "A.Q. Khan—proliferator or scapegoat?" *San Francisco Chronicle*, February 13, 2009

[64] Club of Madrid website, http://www.clubmadrid.org/cmadrid/index.php?id=2

[65] Club of Madrid website, http://www.clubmadrid.org/cmadrid/index.php?id=36

[66] Club of Madrid website, http://www.clubmadrid.org/cmadrid/index.php?id=131

[67] Club of Madrid website, http://www.clubmadrid.org/cmadrid/index.php?id=16

[68] Global Leadership for Climate Action website, http://www.globalclimateaction.com/index.php?option=com_content&task=view&id=33

[69] Ibid.

[70] Ibid.

[71] Ibid.

[72] Club of Budapest website, http://www.clubofbudapest.org/

[73] Ervin Laszlo and David Woolfson, "State of Global Emergency: Draft Declaration," published by Club of Budapest, October 24, 2008, p.9

[74] Ibid. Laszlo and Woolfson, p.2

[75] Ibid. Laszlo and Woolfson, p.2

[76] Ibid. Laszlo and Woolfson, p.2

[77] WorldShift Network website, http://www.worldshiftnetwork.org/action/

[78] Club of Budapest website, http://www.clubofbudapest.org/members.php

[79] World Commission on Global Consciousness and Spirituality website, http://globalspirit.org/pages/mission.php

[80] World Commission on Global Consciousness and Spirituality website, http://globalspirit.org/pages/wc_councils.php

CHAPTER 16: *Manmade Warm-ups for Manmade Warming*

[1] Jay Lehr and Sam Aldrich, "Alar: The Great Apple Scare," *Environment & Climate News*, March, 2007

[2] Steven J. Milloy, *Junk Science Judo*, published by Cato Institute, 2001, p.18

[3] Center for Consumer Freedom's Activistcash.com, http://www.activistcash.com/organization_financials_full.cfm/oid/19

[4] Ibid. Lehr and Aldrich

[5] "How a PR Firm Executed the Alar Scare," *Wall Street Journal*, October 3, 1989

[6] "How a PR Firm Executed the Alar Scare," *Wall Street Journal*, October 3, 1989

[7] "How a PR Firm Executed the Alar Scare," *Wall Street Journal*, October 3, 1989

[8] Robert M. Schoch, *Case Studies in Environmental Science*, published by Jones & Bartlett, 1996, p.10

[9] Michael S. Coffman, *Savior of the Earth? The Politics and Religion of the Environmental Movement*, published by Northfield Publishing, 1994, p.81; Also see Rogelio Maduro and Ralf Schauerhammer, *The Holes in the Ozone Scare: The Scientific Evidence that the Sky Isn't Falling*, published by 21st Century Science Associates, 1992, p.252

[10] J. Gordon Edwards, "DDT: A Case Study in Scientific Fraud," *Journal of American Physicians and Surgeons*, Vol.9, No.3, Fall 2004, p.83

[11] Nobel Prize website, http://nobelprize.org/nobel_prizes/medicine/laureates/1948/muller-bio.html

[12] Ibid. Edwards, p.83

[13] Ibid. Edwards, p.83

[14] Ibid. Edwards, p.83

[15] Ibid. Edwards, p.83

[16] Rachel Carson, *Silent Spring*, published by Houghton Mifflin, 1962, p.6

[17] Paul Ehrlich, "Eco-Catastrophe!" from *Ramparts Magazine's* book *Eco-Catastrophe*, published by Harper & Row, 1970

[18] George Woodwell, Charles Wurster Jr. and Peter Isaacson, "DDT Residues in an East Coast Estuary: A Case of Biological Concentration of a Persistent Insecticide," *Science*, Vol.156, May 12, 1967

[19] Edward A. Laws, *Aquatic Pollution: An Introductory Text*, published by Wiley, 2000, p.318-320

[20] Ibid. Edwards, p.84

[21] Ibid. Edwards, p.84

[22] Ibid. Edwards, p.84

[23] "EDF Takes DDT Appeals to Court" *New York Times*, December 29, 1969

[24] Ibid. Edwards, p.86

[25] Jeff Lehr and Jay Lehr, *Rational Readings on Environmental Concerns*, published by John Wiley and Sons, 1992, p.208

[26] Ibid. Lehr and Lehr

[27] Ibid. Edwards, p.87

[28] Roll Back Malaria website, http://www.rollbackmalaria.org/aboutus.html

[29] Tina Rosenberg, "What the World Needs Now Is DDT," *New York Times*, April 11, 2004

[30] Ibid. Rosenberg

[31] Ibid. Rosenberg

[32] Betsy McKay, "WHO Calls for Spraying Controversial DDT to Fight Malaria," *Wall Street Journal*, September 15, 2006

[33] Marjorie Mazel Hecht, "In Africa, DDT Makes A Comeback To Save Lives," *Executive Intelligence Review*, June 18, 2004

[34] Press Release, "WHO gives indoor use of DDT a clean bill of health for controlling malaria," September 15, 2006

[35] Betsy McKay, "WHO Calls for Spraying Controversial DDT to Fight Malaria," *Wall Street Journal*, September 15, 2006

[36] Donald G. McNeil Jr., "An Iron Fist Joins the Malaria Wars," *New York Times*, June 27, 2006

[37] Ibid. McNeil Jr.

[38] Ibid. McNeil Jr.

[39] Roll Back Malaria Partnership, *The Global Malaria Action Plan*, published by RBM, 2008, p.146

[40] Bill and Melinda Gates Foundation website, http://www.gatesfoundation.org/annual-letter/Pages/2009-malaria-prevention-vaccine.aspx

[41] Greenpeace website, http://www.greenpeace.org/international/campaigns/oceans/pollution; Sierra Club website, http://www.sierraclub.org/toxics/ddt/, EDF website, http://www.edf.org/article.cfm?contentID=5546

[42] Kim Larsen, "Bad Blood," *OnEarth*, Winter 2008

[43] Tina Rosenberg, "What the World Needs Now Is DDT," *New York Times*, April 11, 2004

[44] Ibid. Rosenberg

[45] Ibid. Rosenberg

[46] Tom Bethell, *The Politically Incorrect Guide to Science*, published by Regnery Publishing, 2005, p.76; Also Walter Williams, "Environmental extremism costs lives," *Deseret News (Salt Lake City)*, August 15, 2007

[47] Helen Altonn, "UH professor emeritus put passion into medical study," *Honolulu Star Bulletin*, April 3, 2008

[48] Robert S. Desowitz, *The Malaria Capers: More Tales of Parasites and People, Research and Reality*, published by W.W. Norton, 1991; Also see Steven Milloy, "Eco-Imperialism's Deadly Consequences," *Fox News*, December 5, 2003

PART III: GLOBAL WARMING

[1] Alan Lindsay Mackay, *A Dictionary of Scientific Quotations*, published by CRC Press, 1991, p.242

CHAPTER 17: *United Nations Ringleaders*

[1] Nobel Prize website, http://nobelprize.org/nobel_prizes/peace/laureates/2007/
[2] IPCC website, http://www.ipcc.ch/about/index.htm
[3] IPCC website, http://www.ipcc.ch/about/working-group1.htm
[4] Rajendra Pachauri at the opening ceremony of the United Nations Framework Convention on Climate Change, Conference of the Parties 14, Poznán, Poland, December 1, 2008; http://www.ipcc.ch/press/popup_news2.htm
[5] IPCC, *The Physical Science Basis: Contribution of Working Group I to the 4th Assessment Report of the IPCC*, published by IPCC, 2007
[6] Lawrence Solomon, "Numbers racket," *National Post (Canada)*, November 6, 2009
[7] Tim Ball and Tom Harris, "UN Climate Agency's implication that 2,500 scientist reviewers agree is a deception," *Canada Free Press*, November 20,2007
[8] Tim Ball and Tom Harris, "UN Climate Agency's implication that 2,500 scientist reviewers agree is a deception," *Canada Free Press*, November 20,2007
[9] Larry King Live, "Could Global Warming Kill Us?" *Cable New Network (CNN)*, January 31, 2007
[10] Patrick J. Michaels, *Sound and Fury: The Science and Politics of Global Warming*, published by Cato Institute, 1992, p.25
[11] Ibid. Michaels, p.25
[12] IPCC, *First Assessment Report, Working Group 1, Summary for Policymakers*, published by IPCC, 1990
[13] IPCC, *First Assessment Report, Working Group 1, Chapter 7, Executive Summary*, published by IPCC, 1990
[14] Richard Lindzen, "Global Warming: The Origin and Nature of the Alleged Scientific Consensus," *Regulation*, Vol.15, No.2, Spring 1992
[15] Frederick Seitz, "A Major Deception on Global Warming," *Wall Street Journal*, June 12, 1996
[16] Ibid. Seitz
[17] *Conspiracy Theory with Jesse Ventura*, "Global Warming," TruTV, December 16, 2009
[18] *Conspiracy Theory with Jesse Ventura*, "Global Warming," TruTV, December 16, 2009
[19] Richard Lindzen "The Press Gets It Wrong," *Wall Street Journal*, June 11, 2001
[20] Testimony of Richard S. Lindzen before the U.S. Senate Environment and Public Works Committee, May 2, 2001; http://epw.senate.gov/107th/lin_0502.htm
[21] Ibid. Lindzen
[22] John Christy "No consensus on IPCC's level of ignorance," *BBC News*, November 13, 2007
[23] Ibid. Christy
[24] Ibid. Christy
[25] U.S. Senate Environment and Public Works Committee Minority Staff Report (Inhofe), "More Than 650 International Scientists Dissent Over Man-Made Global Warming Claims Scientists Continue to Debunk 'Consensus' in 2008," published December 11, 2008; www.epw.senate.gov/minority
[26] Madhav L. Khandekar, "Has The IPCC Exaggerated Adverse Impact Of Global Warming On Human Societies?" *Energy & Environment*, Vol.19, No.5, May 2008
[27] Madhav L. Khandekar, "Has The IPCC Exaggerated Adverse Impact Of Global Warming On Human Societies?" *Energy & Environment*, Vol.19, No.5, May 2008
[28] David Derbyshire, "UN report that said Himalayan glaciers would melt within 25 years was all hot air," *Daily Mail*, January 18, 2010
[29] Stanley Changnon, "Shifting Economic Impacts from Weather Extremes in the United States: A Result of Societal Changes, Not Global Warming," *Natural Hazards*, Vol.29, No.2, June 2003
[30] Jonathan Leake, "UN wrongly linked global warming to natural disasters," *London Times*, January 24, 2010
[31] Jonathan Leake, "UN wrongly linked global warming to natural disasters," *London Times*, January 24, 2010
[32] Jonathan Leake, "UN wrongly linked global warming to natural disasters," *London Times*, January 24, 2010
[33] Jonathan Leake, "UN wrongly linked global warming to natural disasters," *London Times*, January 24, 2010
[34] Jonathan Leake, "UN wrongly linked global warming to natural disasters," *London Times*, January 24, 2010
[35] Richard Gray, "UN climate change panel based claims on student dissertation and magazine article," *London Telegraph*, January 30, 2010
[36] Richard Gray, "UN climate change panel based claims on student dissertation and magazine article," *London Telegraph*, January 30, 2010
[37] Jonathan Leake, "Africagate: top British scientist says UN panel is losing credibility," *London Times*, February 7, 2010
[38] Jonathan Leake, "Africagate: top British scientist says UN panel is losing credibility," *London Times*, February 7, 2010
[39] IPCC, Fourth Assessment Report, *Working Group II: Impacts, Adaption and Vulnerability*, Chapter 10.6.2, *The Himalaya Glacier*, 2007
[40] Neeta Lal, "India turns up heat over 'Glaciergate,'" *Asia Times*, January 21, 2010
[41] David Rose, "Glacier scientist: I knew data hadn't been verified," *Daily Mail*, January 24, 2010
[42] Andrew Bolt, "The billion-dollar hoax," *Herald Sun (Melbourne)*, January 27, 2010
[43] Andrew Bolt, "The billion-dollar hoax," *Herald Sun (Melbourne)*, January 27, 2010
[44] David Rose, "Glacier scientist: I knew data hadn't been verified," *Daily Mail*, January 24, 2010
[45] Gaurav Singh and Alex Morales, "Melting Glaciers Scientist Blames Issue on Misquote ," *Bloomberg.com*, January 20, 2010
[46] David Rose, "Glacier scientist: I knew data hadn't been verified," *Daily Mail*, January 24, 2010
[47] Mathew Knight, "U.N. climate chiefs apologize for glacier error," *CNN*, January 20, 2010
[48] Neeta Lal, "India turns up heat over 'Glaciergate,'" *Asia Times*, January 21, 2010
[49] Neeta Lal, "India turns up heat over 'Glaciergate,'" *Asia Times*, January 21, 2010
[50] "Pachauri admits Himalayan blunder," *India Times*, January 24, 2010
[51] "UN climate panel regrets Himalaya glacier data in report," *Agence France-Presse (AFP)*, January 20, 2010
[52] David Rose, "Glacier scientist: I knew data hadn't been verified," *Daily Mail*, January 24, 2010
[53] Ben Webster, "Climate chief was told of false glacier claims before Copenhagen," *London Times*, January 30, 2010
[54] Ben Webster, "Climate chief was told of false glacier claims before Copenhagen," *London Times*, January 30, 2010
[55] Ben Webster, "Climate chief was told of false glacier claims before Copenhagen," *London Times*, January 30, 2010
[56] Ben Webster, "Climate chief was told of false glacier claims before Copenhagen," *London Times*, January 30, 2010
[57] David Schneider, "Storm Watch," *American Scientist*, May-June, 2005
[58] IPCC, *Summary for Policymakers: A Report of Working Group I to the Third Assessment Report of the IPCC*, published by IPCC, 2001
[59] David Schneider, "Storm Watch," *American Scientist*, May-June, 2005
[60] Lawrence Solomon, "The hurricane expert who stood up to UN junk science," *The Financial Post*, February 2, 2007

[61] Ibid. Solomon

[62] "Chris Landsea Leaves IPCC: An Open Letter To The Community" by Chris Landsea, January 17, 2005; http://sciencepolicy.colorado.edu

[63] Dan Vergano, "Global warming stoked '05 hurricanes, study says," *USA Today*, June 25, 2006

[64] IPCC, *The Physical Science Basis: Contribution of Working Group I to the Fourth Assessment Report of the IPCC*, published by IPCC, 2007

[65] Randolph E. Schmid, "Experts say global warming is causing stronger hurricanes," *USA Today*, September 15, 2005

[66] Ibid. Schmid

[67] National Public Radio, *Morning Edition*, "Climate Change Fuels Debate over Hurricane Threat," January 22, 2008

[68] Cain Burdeau, "Top hurricane forecaster calls Al Gore a 'gross alarmist,'" *USA Today-Associated Press*, April 9, 2007

[69] William M. Gray and Philip J. Klotzbach, "Extended Range Forecast of Atlantic Seasonal Hurricane Activity and U.S. Landfall Strike Probability for 2007," Colorado State University, April 3, 2007

[70] Cain Burdeau, "Top hurricane forecaster calls Al Gore a 'gross alarmist,'" *USA Today-Associated Press*, April 9, 2007

[71] Lawrence Solomon, "Bitten by the IPCC" *Financial Post (Canada)*, March 27, 2007

[72] IPCC, *Second Assessment Report, Working Group II, Summary For Policymakers*, published by IPCC, 1995

[73] House of Lords Select Committee on Economic Affairs, Second Report of Session 2005-06, "The Economic of Climate Change," Vol.1, published July 6, 2005

[74] Ibid.

[75] Ibid.

[76] House of Lords Select Committee on Economic Affairs, Second Report of Session 2005-06, "The Economic of Climate Change," Vol.1, published July 6, 2005

[77] Lawrence Solomon, "IPCC too blinkered and corrupt to save," *Financial Post (Canada)*, October 26, 2007

[78] Chris Barton, "It's hype, hysteria and hot air says climate change nay-sayers," *New Zealand Herald*, November 4, 2006

[79] Lawrence Solomon, "IPCC too blinkered and corrupt to save," *Financial Post (Canada)*, October 26, 2007

[80] Robert Novak, "The Russian Didn't Bark," *CNN.com*, October 16, 2003

[81] Yuri Izrael, "Climate Change: Not A Global Threat," *RIA Novosti (Russia)*, June 23, 2005

[82] Freeman J. Dyson, *A Many-Colored Glass*, published by University of Virginia Press, 2007, p.46

[83] Ibid. Dyson

[84] World Federations of Scientists website, http://www.federationofscientists.org/WFSHist.asp

[85] World Federations of Scientists website, http://www.federationofscientists.org/PMPanels/Climate/ClimatePMP.asp

[86] "Global Warming Natural, Says Expert," *Zenit (Rome)*, April 27, 2007; http://www.zenit.org/article-19481?l=english

[87] Ibid.

[88] Lawrence Solomon, "Some restraint in Rome," *National Post (Canada)*, May 11, 2007

[89] Scott Armstrong and Kesten Green, "Your Request for an Analysis of the U.S. Environmental Protection Agency's Advanced Notice of Proposed Rulemaking for Greenhouse Gases," letter to Senator James Inhofe, January 26, 2009; http://climatebet.files.wordpress.com/2009/01/inhofe-epa16.pdf

[90] Mark Morano, "James Hansen's Former NASA Supervisor Declares Himself a Skeptic—Says Hansen 'Embarrassed NASA,'" Inhofe EPW Press Blog, January 27, 2009; http://epw.senate.gov

[91] House of Lords Select Committee on Economic Affairs, "The Economics of Climate Change," published July 6, 2005; http://www.publications.parliament.uk/pa/ld200506/ldselect/ldeconaf/12/12i.pdf

[92] U.S. Congressional Record, "Departments Of Commerce And Justice Science And Related Agencies For Fiscal Year 2006 Conference Report," November 15, 2005; http://www.c-spanarchives.org/congress/?q=node/77531&id=7405145

[93] Press Release, "Inhofe Expresses Concerns Over IPCC's Lack Of Objectivity In Letter To Chairman Pachauri," December 7, 2005; http://epw.senate.gov

[94] Ibid.

[95] Rama Lakshmi, "India challenges Western data linking climate change, Himalayan melt," *Washington Post*, November 22, 2009

[96] "India unveils National Action Plan on Climate Change," *Times of India*, June 30, 2008

[97] Lawrence Solomon, "India rejects climate doom, pursues economic boom," *National Post (Canada)*, July 26, 2008

[98] Michael Abramowitz, "G-8 Conference Tackles Global Warming Treaty," *Washington Post*, July 9, 2008

[99] Ibid. Solomon

[100] "Get off India's back, Pachauri tells developed nations," *Times of India*, July 8, 2008

[101] Karl Ritter, "Climate change dissenters say they are demonized in debate," *USA Today*, December 17, 2007

[102] "Get off India's back, Pachauri tells developed nations," *Times of India*, July 8, 2008

[103] Michael Duffy, "Truly inconvenient truths about climate change being ignored," *Sydney Morning Herald*, November 8, 2008

[104] Ibid. Duffy

[105] David Henderson, "Misplaced Trust: Inconvenient truths about the UN's global warming panel," *Wall Street Journal*, October 11, 2007

[106] Vaclav Klaus, speech to Second International Conference on Climate Change, Marriott Marquis Hotel, New York City, March 8, 2009; http://www.climatescienceinternational.org

[107] "The Tip of the Climategate Iceberg," *Wall Street Journal*, December 8, 2009

[108] Stephen Dinan, "Researcher: NASA hiding climate data," *Washington Times*, December 3, 2009

[109] Stephen Dinan, "Global warming controversy reaches NASA climate data," *Washington Times*, December 3, 2009

[110] John Lott, "What Are Global Warming Supporters Trying to Hide?" *Fox News*, December 4, 2009

[111] Christopher Booker, "This is the worst scientific scandal of our generation," *London Telegraph*, November 28, 2009

[112] Guy Chazan, "U.K. Says University Broke Law on Turning Over Data," *Wall Street Journal*, January 29, 2010

[113] Juliet Eliperin, "In the trenches on climate change, hostility among foes," *Washington Post*, November 22, 2009

[114] Juliet Eliperin, "Hackers steal electronic data from top climate research center," *Washington Post*, November 21, 2009

[115] Juliet Eliperin, "In the trenches on climate change, hostility among foes," *Washington Post*, November 22, 2009

[116] Jim Efstathiou, Jr. and Alex Morales, "U.K. Climate Scientist Steps Down After E-Mail Flap," *Bloomberg.com*, December 2, 2009

[117] Guy Chazan, "U.K. Says University Broke Law on Turning Over Data," *Wall Street Journal*, January 29, 2010

[118] Daniel Compton, "Climategate: What e-mail really means," *Detroit News*, December 23, 2009

[119] Kim Zetter, "Hacked E-Mails Fuel Global Warming Debate," *Wired*, November 20, 2009

[120] Daniel Compton, "Climategate: What e-mail really means," *Detroit News*, December 23, 2009

[121] "UK climate scientist to temporarily step down," *Associated Press*, December 1, 2009

[122] "Q&A: Professor Phil Jones," *BBC News*, February 13, 2010

[123] Jonathan Leake, "Climate change data dumped," *London Times*, November 29, 2009

[124] Jonathan Leake, "Climate change data dumped," *London Times*, November 29, 2009

[125] Thomas R. Karl, "Global warming: What the science tells us," *Washington Post*, December 11, 2009

[126] James Hansen, "The Temperature of Science," Columbia University, December 16, 2009; www.columbia.edu/~jeh1/mailings/.../20091216_TemperatureOfScience.pdf
[127] John Dickerson, "What in the Hell Do They Think Is Causing It?" *Slate.com*, December 8, 2009
[128] "UK Met Office to publish climate records," *CNN News*, December 6, 2009
[129] Wendell Goler, "Obama Ignores 'Climate-Gate' in Revising Copenhagen Plans," *Fox News*, December 5, 2009
[130] "UN body wants probe of climate e-mail row," *BBC News*, December 4, 2009
[131] "UN body wants probe of climate e-mail row," *BBC News*, December 4, 2009
[132] Marlowe Hood, "UN scientists defend 'targeted' colleagues," *Agence France-Presse (AFP)*, December 8, 2009
[133] Brian Winter, "Obama heads to U.N. climate change talks in Copenhagen," *USA Today*, December 18, 2009
[134] Brian Wallace, "U.N. chief weighs in on climate talk expectations," *Los Angeles Times*, December 16, 2009
[135] Brian Wallace, "U.N. chief weighs in on climate talk expectations," *Los Angeles Times*, December 16, 2009
[136] Ban Ki-moon, "We Can Do It," *New York Times*, October 25, 2009
[137] "Obama says disappointment over Copenhagen is valid," *BBC News*, December 23, 2009
[138] Andrew Revkin and John Broder, "A Grudging Accord in Climate Talks," *New York Times*, December 19, 2009
[139] John Vidal, "Copenhagen climate summit in disarray after 'Danish text' leak," *London Guardian*, December 8, 2009
[140] John Vidal, "Copenhagen climate summit in disarray after 'Danish text' leak," *London Guardian*, December 8, 2009
[141] Amy Goodman interview with John Perkins, "Confessions of an Economic Hit Man: How the U.S. Uses Globalization to Cheat Poor Countries Out of Trillions," *Democracy Now*, November 9, 2004
[142] Amy Goodman interview with John Perkins, "Confessions of an Economic Hit Man: How the U.S. Uses Globalization to Cheat Poor Countries Out of Trillions," *Democracy Now*, November 9, 2004
[143] Lucy Komisar, "Interview with Joseph Stiglitz," *The Progressive*, June 2000
[144] Joseph Stiglitz, "Wall Street's Toxic Message," *Vanity Fair*, July 2009
[145] Jeremy van Loon and Sandrine Rastello, "Soros Seeks $100 Billion in IMF Funds for Green Plan," *Bloomberg.com*, December 10, 2009
[146] Floyd Norris, "French court upholds Soros conviction," *New York Times*, March 25, 2005
[147] "Hungary fines Soros firm $2.2 mln for OTP deals," *Reuters*, March 26, 2009
[148] John Vidal, "Copenhagen climate summit in disarray after 'Danish text' leak," *London Guardian*, December 8, 2009
[149] Richard Black, "Copenhagen climate summit negotiations 'suspended,'" *BBC News*, December 14, 2009
[150] Andrew Revkin and John Broder, "A Grudging Accord in Climate Talks," *New York Times*, December 19, 2009
[151] *Copenhagen Accord*, Draft decision CP15, December 18, 2009; see UNFCC website, http://unfccc.int/resource/docs/2009/cop15/eng/l07.pdf
[152] Interview with Christopher Monckton, *The Alex Jones Show*, Genesis Communications Network, December 17, 2009

CHAPTER 18: *Never-Ending Climate Change*

[1] Fred Singer and Dennis Avery, *Unstoppable Global Warming: Every 1,500 Years*, published by Rowman & Littlefield, 2007, p.52
[2] Ibid Singer and Avery, p.51
[3] Robert M. Carter, "Knock, Knock: Where is the Evidence for Dangerous Human-Caused Global Warming?" *Economic Analysis & Policy*, Vol.38, No.2, September 2008
[4] A.N. Salamatin et al., "Ice core age dating and paleothermometer calibration based on isotope and temperature profiles from deep boreholes at Vostok Station (East Antarctica)," *Journal of Geophysical Research*, Vol.103, March 1998, p.8963-8977
[5] U.K. Meteorological Office, Hadley Center 2008; http://www.metoffice.gov.uk/research/hadleycentre/CR_data/Monthly/Hadplot_globe.gif
[6] Robert M. Carter, "Knock, Knock: Where is the Evidence for Dangerous Human-Caused Global Warming?" *Economic Analysis & Policy*, Vol.38, No.2, September 2008
[7] Richard Tkachuck, "The Little Ice Age," *Origins*, Vol.10, 1983; As cited in S. Fred Singer and Dennis T. Avery, *Unstoppable Global Warming: Every 1,500 Years*, published by Rowman & Littlefield, 2007, p.46
[8] Ibid. Tkachuck
[9] "Tree rings challenge history," *BBC News*, September 8, 2000
[10] John E. Oliver, *Climate and Man's Environment*, published by Wiley, 1973, p.365
[11] Ibid Singer and Avery, p.42
[12] Press release, "20th Century Climate Not So Hot," Harvard-Smithsonian Center for Astrophysics, March 31, 2003
[13] Willie Soon and Sallie Baliunas, "Reconstructing Climatic and Environmental Changes of the Past 1,000 Years: A Reappraisal," *Energy & Environment*, Vol.14, No.2-3, March 2003
[14] William Happer, "Global Warming and Climate Change in Perspective: Truths and Myths About Carbon Dioxide, Scientific Consensus, and Climate Models," *Capitalism Magazine*, February 28, 2009
[15] Michael Mann et al., "Global-scale temperature patterns and climate forcing over the past six centuries," *Nature*, Vol.392, April 23, 1998
[16] Al Gore, *An Inconvenient Truth*, directed by Davis Guggenheim, Lawrence Bender Productions, 2006.
[17] Christopher Monckton, "Climate chaos? Don't believe it," *London Telegraph*, November 6, 2006
[18] David Deming, testimony before U.S. Senate Committee on Environment & Public Works, December 6, 2006; http://epw.senate.gov/hearing_statements.cfm?id=266543
[19] Ibid. Deming
[20] "Hockey Stick on Ice," *Wall Street Journal*, February 18, 2005
[21] Steve McIntyre and Ross McKitrick, "Corrections to the Mann et al (1998) Proxy Data Base and Northern Hemispheric Average Temperature Series," *Energy & Environment*, Vol.14, No.6, November 2003
[22] "Hockey Stick on Ice," *Wall Street Journal*, February 18, 2005
[23] J. T. Houghton, Y. Ding, D.J. Griggs, M. Noguer, P. J. van der Linden and D. Xiaosu (Eds.), *Climate Change 2001: The Scientific Basis*, published by Cambridge University Press, 2001
[24] J.T. Houghton, G.J. Jenkins and J.J. Ephraums, *Scientific Assessment of Climate Change: Report of Working Group I*, Cambridge University Press, 1990, p.202
[25] Eric Berger, "Congress told climate on hot streak," *Houston Chronicle*, June 23, 2006
[26] Beth Daley, "National panel supports '98 global warming evidence," *Boston Globe*, June 23, 2006
[27] Ibid. Berger
[28] Andrew Revkin, "Panel Supports a Controversial Report on Global Warming," *New York Times*, June 23, 2006
[29] Beth Daley, "National panel supports '98 global warming evidence," *Boston Globe*, June 23, 2006
[30] Keay Davidson, "It's official: We live in hot times," *San Francisco Chronicle*, June 23, 2006
[31] Ibid. Davidson
[32] Kevin Grandia, "Michael Mann in his own words on the stolen CRU emails," *Huffington Post*, November 26, 2009

[33] Lawrence Solomon, "Statistics needed," *Financial Post (Canada)*, February 2, 2007
[34] H. Sterling Burnett, "When warming's 'hockey stick' breaks," *Washington Post*, August 3, 2006
[35] William J. Broad, "From a Rapt Audience, a Call to Cool the Hype," *New York Times*, March 13, 2007
[36] Mason Inman, "Earth Hotter Now Than in Past 2,000 Years, Study Says," *National Geographic News*, September 2, 2008
[37] T.R. Oke, "City Size and the Urban Heat Island," *Atmospheric Environment*, Vol.7, No.8, August 1973
[38] Eugenia Kalnay and Ming Cai, "Impact of urbanization and land-use change on climate," *Nature*, Vol.423, May 29, 2003
[39] Ibid Singer and Avery, p.viii
[40] Board on Atmospheric Sciences and Climate, *Adequacy of Climate Observing Systems*, published by National Academy Press, 1999, p.1 (executive summary); http://www.nap.edu/catalog/6424.html
[41] Surface Stations website, http://www.surfacestations.org/about.htm
[42] Anthony Watts, *Is the U.S. Surface Temperature Record Reliable?* published by Heartland Institute, 2009
[43] Anthony Watts, *Is the U.S. Surface Temperature Record Reliable?* published by Heartland Institute, 2009
[44] Saeed Ahmed, "From underwater, Maldives sends warning on climate change," *CNN.com*, October 17, 2009
[45] Oren Dorell, "Polar bears caught in a heated eco-debate," *USA Today*, March 9, 2008
[46] "Scientists say polar bears have survived climate change before," *Daily Mail (London)*, February 28, 2010
[47] Joseph Brean, "Gore pays for photo after Canada didn't," *National Post (Canada)*, March 23, 2007
[48] Lewis Smith, "Al Gore's inconvenient judgment," *London Times*, October 11, 2007
[49] Lawrence Solomon "Polar scientists on thin ice," *Financial Post (Canada)*, February 2, 2007
[50] Duncan Wingham et al., "Mass Balance of the Antarctic Ice Sheet," *The Journal of the Royal Society*, Vol.364, No.1844, July 15, 2006, p.1627
[51] Greg Roberts, "Antarctic ice is growing, not melting away," *The Australian*, April 18, 2009
[52] Greg Roberts, "Antarctic ice is growing, not melting away," *The Australian*, April 18, 2009
[53] http://www.esa.int/esaLP/LPcryosat.html
[54] Aalok Mehta, "North Pole May Be Ice-Free for First Time This Summer," *National Geographic*, June 20, 2008
[55] Aalok Mehta, "North Pole May Be Ice-Free for First Time This Summer," *National Geographic*, June 20, 2008
[56] Steven Goddard, "Arctic ice refuses to melt as ordered," *The Register*, August 15, 2008
[57] National Snow and Ice Data Center website (compiled from Arctic Sea Ice Press Announcements) http://nsidc.org/arcticseaicenews/
[58] Press Release, "Urgent action needed as Arctic ice melts," *Greenpeace*, July 15, 2009
[59] Gerd Leipold (guest) on *HardTalk with Stephen Sackur*, BBC, August 5, 2009
[60] NASA press release, "NASA Sees Arctic Ocean Circulation Do an About-Face," November 13, 2007; http://www.jpl.nasa.gov/news/news.cfm?release=2007-131
[61] NASA press release, "NASA Sees Arctic Ocean Circulation Do an About-Face," November 13, 2007; http://www.jpl.nasa.gov/news/news.cfm?release=2007-131
[62] Christopher Booker, "Rise of sea levels is 'the greatest lie ever told'" *London Telegraph*, March 28, 2009
[63] Christopher Booker, "Rise of sea levels is 'the greatest lie ever told'" *London Telegraph*, March 28, 2009
[64] Christopher Booker, "Rise of sea levels is 'the greatest lie ever told'" *London Telegraph*, March 28, 2009
[65] Nils-Axel Mörner, "Why the Maldives aren't sinking," *Spectator*, December 3, 2009
[66] Nils-Axel Mörner, "Why the Maldives aren't sinking," *Spectator*, December 3, 2009

CHAPTER 19: *Deadly Carbon Dioxide*

[1] Al Gore, *Earth in the Balance: Forging a New Common Purpose*, published by Earthscan, 2007, p.93
[2] Juliet Eilperin, "Long Droughts, Rising Seas Predicted Despite Future CO2 Curbs," *Washington Post*, January 27, 2009
[3] Ibid. Gore, p.325
[4] Press Release, "Inhofe Slams New Cap-and-Trade Bill as All 'Economic Pain For No Climate Gain,'" U.S. Senate Committee on Environment and Public Works, October 18, 2007; http://epw.senate.gov
[5] "Everyone in Britain could be given a personal 'carbon allowance,'" *London Telegraph*, November 9, 2009
[6] Christian Schwägerl, "German Climate Adviser: 'Industrialized Nations Are Facing CO2 Insolvency,'" *Der Spiegel*, September 4, 2009
[7] Elisabeth Rosenthal, "To Cut Global Warming, Swedes Study Their Plates," *New York Times*, October 22, 2009
[8] Audra Ang, "Pelosi appeals for China's help on climate change," *Associated Press* via *San Francisco Chronicle*, May 27, 2009
[9] Jim Tankersley, "Obama finalizes 'endangerment finding' on global warming," *Los Angeles Times*, December 7, 2009
[10] Kimberley Strassel, "The EPA's Carbon Bomb Fizzles," *Wall Street Journal*, December 11, 2009
[11] Al Gore, *An Inconvenient Truth*, directed by Davis Guggenheim, Lawrence Bender Productions, 2006.
[12] Christopher Monckton, "Climate chaos? Don't believe it," *London Telegraph*, November 5, 2006
[13] J.R. Petit et al., "Climate and atmospheric history of the past 420,000 years from the Vostok ice core, Antarctica," *Nature*, Vol.399, June 3, 1999
[14] Nicolas Caillon et al., "Timing of Atmospheric CO_2 and Antarctic Temperature Changes Across Termination III," *Science*, Vol.299, March 14, 2003
[15] Laurie David and Cambria Gordon, *The Down-to-earth Guide to Global Warming*, published by Scholastic/Orchard, 2007, p.18
[16] Press Release, "A Fundamental Scientific Error in 'global warming' Book for Children," *Science and Public Policy Institute*, September 13, 2007
[17] Laurie David and Cambria Gordon, *The Down-to-earth Guide to Global Warming*, published by Scholastic/Orchard, 2007, p.18
[18] Urs Siegenthaler et al., "Stable Carbon Cycle–Climate Relationship During the Late Pleistocene," *Science*, Vol.310, No.5752, November 25, 2005
[19] Hubertus Fischer et al., "Ice Core Record of Atmospheric CO_2 Around the Last Three Glacial Terminations," *Science*, Vol.283, No.5408, March 12, 1999
[20] Nicolas Caillon et al., "Timing of Atmospheric CO_2 and Antarctic Temperature Changes Across Termination III," *Science*, Vol.299, No.5613, March 14, 2003
[21] S. Fred Singer and Dennis T. Avery, *Unstoppable Global Warming: Every 1,500 Years*, published by Rowman & Littlefield, 2007, p.37
[22] Lawrence Solomon, "Models Trump Measurements," *National Post (Canada)*, July 7, 2007
[23] Ibid. Solomon
[24] Lawrence Solomon, "Models Trump Measurements," *National Post (Canada)*, July 7, 2007
[25] Ibid. Solomon
[26] S. Fred Singer and Dennis T. Avery, *Unstoppable Global Warming: Every 1,500 Years*, published by Rowman & Littlefield, 2007, p.108
[27] Ibid. Singer and Avery, p.105
[28] Peter Doran et al., "Antarctic Climate Cooling and Terrestrial Ecosystem Response," *Nature*, Vol.415, January 31, 2002
[29] Ibid. Singer and Avery, p.105
[30] Patrick J. Michaels, *Sound and Fury: The Science and Politics of Global Warming*, published by Cato Institute, 1992, p.10
[31] Robert Berner, "The Rise of Plants and Their Effect on Weathering and Atmospheric CO_2," *Science*, Vol.276, No.5312, April 25, 1997
[32] Ibid. Berner

[33] Patrick J. Michaels, *Sound and Fury: The Science and Politics of Global Warming*, published by Cato Institute, 1992, p.10

[34] Ibid. Michaels, p.12

[35] Sherwood Idso, Craig Idso and Keith Idso, "The Specter of Species Extinction," *George C. Marshall Institute*, July 29, 2003

[36] Bruce Kimball, "Carbon Dioxide and Agricultural Yield: An Assemblage and Analysis of 430 Prior Observations," *Agronomy Journal*, Vol.75, September 1983; K.E. Idso and S.B. Idso, "Plant responses to atmospheric CO2 enrichment in the face of environmental constraints: a review of the past 10 years," *Agricultural and Forest Meteorology*, Vol.69, 1994

[37] Sherwood Idso, Craig Idso and Keith Idso, "The Specter of Species Extinction," *George C. Marshall Institute*, July 29, 2003

[38] Robert Uhlig, "Feast and famine in Europe as global warming scorches farms," *Telegraph (London)*, August 21, 2003

[39] Hyrum B. Mayeaux et al., "Yield of Wheat Across a Subambient Carbon Dioxide Gradient," *Global Change Biology*, Vol.3, No.3, June 1997

[40] S. Fred Singer and Dennis T. Avery, *Unstoppable Global Warming: Every 1,500 Years*, published by Rowman & Littlefield, 2007, p.193

CHAPTER 20: Here Comes the Sun King

[1] Lloyd Keigwin, "The Little Ice Age and Medieval Warm Period in the Sargasso Sea," *Science*, Vol.274, November 29, 1996

[2] Richard Kerr, "A Variable Sun Paces Millennial Climate," *Science*, Vol.294, November 16, 2001

[3] Gerard Bond et al., "Persistent Solar Influence on North Atlantic Climate during the Holocene," *Science*, Vol.294, November 16, 2001

[4] William K. Stevens, "Danes Link Sunspot Intensity to Global Temperature Rise," *New York Times*, November 5, 1991

[5] Lawrence Solomon, "Science, not politics," *National Post (Canada)*, April 13, 2007

[6] Sami Solanki et al., "Unusual activity of the Sun during recent decades compared to the previous 11,000 years," *Nature*, Vol.432, October 28, 2004

[7] S. Fred Singer and Dennis T. Avery, *Unstoppable Global Warming: Every 1,500 Years*, published by Rowman & Littlefield, 2007, p.6

[8] Mark Peplow, "Sunspot record reveals Sun's past," *Nature News*, October 27, 2004

[9] Mark Peplow, "Sunspot record reveals Sun's past," *Nature News*, October 27, 2004

[10] Michael Leidig and Roya Nikkhah, "The truth about global warming—it's the Sun that's to blame," *London Telegraph*, July 19, 2004

[11] Jonathan Leake, "Wildlife groups axe Bellamy as global warming 'heretic,'" *London Times*, May 15, 2005

[12] Nigel Calder, "An experiment that hints we are wrong on climate change," *London Times*, February 11, 2007

[13] Richard Black, "Sun and global warming: A cosmic connection?" *BBC News*, November 14, 2007

[14] Lawrence Solomon, "The sun moves climate change," *Financial Post (Canada)*, February 2, 2007

[15] Ibid. Singer and Avery, p.34

[16] Henrik Svensmark, "Cosmoclimatology: a new theory emerges," *Astronomy & Geophysics*, Vol.48, No.1, February 2007

[17] Richard Black, "Sun and global warming: A cosmic connection?" *BBC News*, November 14, 2007

[18] Sid Marris, "Coalition MPs dispute climate finding," *The Australian*, August 13, 2007

[19] Robin McKie, "Cosmic conspiracy revealed: global warming is universal," *Taipei Times*, reprint from *London Observer*, December 16, 2007

[20] Jonathan Rosenblum, "Think Again: Not quite 10 minutes to doomsday," *Jerusalem Post*, March 17, 2009

[21] Howard Spery, "Global warming is minimal," *Denver Post*, February 28, 2009

[22] Alex Jones, *Endgame: Blueprint for Global Enslavement*, Alex Jones Film Productions, 2007

[23] Kate Ravilious, "Mars Melt Hints at Solar, Not Human, Cause for Warming, Scientist Says" *National Geographic News*, February 28, 2007

CHAPTER 21: What Consensus?

[1] Lawrence Solomon, "They call this a consensus?" *National Post (Canada)*, June 2, 2007; Also see Congressional Record, Todd Akin, May 13, 2009, 18:24, http://www.c-spanvideo.org/congress/?q=node/77531&id=8965630

[2] Interview with Al Gore, *MSNBC Today Show with Meredith Vieira*, broadcast November 5, 2007

[3] Richard Lindzen, "Global Warming: The Origin and Nature of the Alleged Scientific Consensus," *Regulation*, Vol.15, No.2, Spring 1992

[4] Lawrence Solomon, "They call this a consensus?" *National Post (Canada)*, June 2, 2007; Also see Congressional Record, Todd Akin, May 13, 2009, 18:24, http://www.c-spanvideo.org/congress/?q=node/77531&id=8965630

[5] Naomi Oreskes, "Beyond The Ivory Tower: The Scientific Consensus on Climate Change," *Science*, Vol.306, No.5702, December 3, 2004

[6] Al Gore, *An Inconvenient Truth*, directed by Davis Guggenheim, Lawrence Bender Productions, 2006.

[7] Robert Matthews, "Leading scientific journals 'are censoring debate on global warming,'" *London Telegraph*, May 1, 2005

[8] Ibid. Matthews

[9] Ibid. Matthews

[10] Ibid. Matthews

[11] Ibid. Matthews

[12] Donald Kennedy, "Climate Change and Climate Science," *Science*, Vol.304, No.5677, June 11, 2004

[13] American Association for the Advancement of Science website, http://www.aaas.org/aboutaaas/

[14] Press release, "AAAS Board Releases New Statement on Climate Change," American Association for the Advancement of Science, February 18, 2007

[15] Edward W. Lempinen, "AAAS News and Notes," *Science*, December 23, 2005, Vol.310, No.5756, p.1917

[16] Anne Ehrlich, Paul Ehrlich and John Holdren, *Ecoscience: Population, Resources, Environment*, published by W.H. Freeman, 1977, p.943

[17] Anne Ehrlich, Paul Ehrlich and John Holdren, *Ecoscience: Population, Resources, Environment*, published by W.H. Freeman, 1977, p.837

[18] Anne Ehrlich, Paul Ehrlich and John Holdren, *Ecoscience: Population, Resources, Environment*, published by W.H. Freeman, 1977, p.786

[19] Anne Ehrlich, Paul Ehrlich and John Holdren, *Ecoscience: Population, Resources, Environment*, published by W.H. Freeman, 1977, p.787-8

[20] John Holdren, *Energy: A Crisis in Power*, published by Sierra Club, 1972, p.29

[21] Jonathan Leake, "Focus: The war on hot air," *London Times*, September 3, 2006

[22] John Holdren "Convincing the climate-change skeptics," *Boston Globe*, August 4, 2008

[23] Richard Lindzen, "Climate Science: Is it currently designed to answer questions?" http://arxiv.org/abs/0809.3762

[24] Amanda Staudt, Nancy Huddleston and Ian Kraucunas, *Understanding and Responding to Climate Change*, published by National Academy of Sciences, May 19, 2008

[25] "Climate Change: An Information Statement of the American Meteorological Society," *Bulletin of the American Meteorological Society*, Vol.88, No.7, July 2007

[26] James Taylor, "Meteorologists Reject U.N.'s Global Warming Claims," *Environment & Climate News*, February 2010

[27] James Taylor, "Meteorologists Reject U.N.'s Global Warming Claims," *Environment & Climate News*, February 2010

[28] The Royal Society, "The Science Of Climate Change," May 17, 2001; http://royalsociety.org/displaypagedoc.asp?id=13619

[29] The National Academies, "Joint science academies' statement: Global response to climate change," June 7, 2005; http://nationalacademies.org/onpi/06072005.pdf

[30] "Joint science academies' statement on growth and responsibility: sustainability, energy efficiency and climate protection," May 16, 2007; http://www.leopoldina-halle.de/energy-climate.pdf

[31] "Joint Science Academies' Statement: Climate Change Adaptation and the Transition to a Low Carbon Society," June 16, 2008; http://nationalacademies.org/includes/climatechangestatement.pdf; "G8+5 Academies' joint statement: Climate change and the transformation of energy technologies for a low carbon future," May 2009

[32] Oregon Petition Project website, http://www.petitionproject.org/

[33] Graham Tibbetts, "Scientists sign petition denying man-made global warming," *London Telegraph*, May 30, 2008

[34] Interview with Al Gore, *MSNBC Today Show with Meredith Vieira*, broadcast November 5, 2007

[35] Union of Concerned Scientists, "U.S. Scientists and Economists' Call for Swift and Deep Cuts in Greenhouse Gas Emissions," May 2008; http://www.ucsusa.org/climateletter

[36] National Academy of Sciences website, http://www.nasonline.org

[37] Union of Concerned Scientists, "U.S. Scientists and Economists' Call for Swift and Deep Cuts in Greenhouse Gas Emissions," May 2008; http://www.ucsusa.org/climateletter

[38] Ben Webster, "Top scientists rally to the defense of the Met Office," *London Times*, December 10, 2009

[39] "Statement from the UK science community," *London Times*, December 9, 2009

[40] Ben Webster, "Top scientists rally to the defense of the Met Office," *London Times*, December 10, 2009

[41] Lawrence Solomon, "32,000 deniers," *Financial Post (Canada)*, May 17, 2008

[42] Greg Roberts, "Skeptics put their case to UN chief," *The Australian*, December 14, 2007

[43] U.S. Senate Environment and Public Works Committee Minority Staff Report (Inhofe), "More Than 650 International Scientists Dissent Over Man-Made Global Warming Claims Scientists Continue to Debunk 'Consensus' in 2008," published December 11, 2008; www.epw.senate.gov/minority

[44] Gordon Jaremko, "Causes of climate change varied: poll," *Edmonton Journal*, March 6, 2008

[45] Thomas Sowell, "Cold water on 'global warming,'" *Washington Times*, March 2, 2008

[46] International Climate Science Coalition website, www.climatescienceinternational.org

[47] Albert Gore, "What Is Wrong With Us?" *Time*, January 2, 1989

[48] "Aliens Cause Global Warming," speech by Michael Crichton at California Institute of Technology, January 17, 2003 http://www.michaelcrichton.com/speech-alienscauseglobalwarming.html

CHAPTER 22: Who is Al Gore?

[1] Al Gore interviewed on Late Edition with Wolf Blitzer, *Cable News Network (CNN)*, March 9, 1999

[2] Katie Hafner and Matthew Lyon, *Where Wizards Stay Up Late: The Origins of the Internet*, published by Simon & Schuster, 1996

[3] John Broder, "Gore's Dual Role: Advocate and Investor," *New York Times*, November 2, 2009

[4] Climate Exchange Plc website, http://www.climateexchangeplc.com/investor-relations/shares-in-issue-top-10-holders

[5] Chicago Climate Exchange website, http://www.chicagoclimatex.com/

[6] Matt Taibbi, "The Great American Bubble Machine," *Rolling Stone*, July 13, 2009

[7] Ed Barnes, "Obama Years Ago Helped Fund Carbon Program He Is Now Pushing Through Congress," *Fox News*, March 25, 2009

[8] Chicago Climate Exchange website, http://www.chicagoclimatex.com/content.jsf?id=67

[9] John Broder, "Gore's Dual Role: Advocate and Investor," *New York Times*, November 2, 2009

[10] Ronald Bailey, "Political Science," *Reason Magazine*, December 1993

[11] Michael Gough, *Politicizing Science: The Alchemy of Policymaking*, published by Hoover Institution Press, 2003

[12] Holman Jenkins, Jr., "Al Gore Leads a Purge," *Wall Street Journal*, May 25, 1993

[13] Ronald Bailey, "Political Science," *Reason Magazine*, December 1993

[14] Al Gore, *Earth in the Balance: Ecology and the Human Spirit*, published by Houghton Mifflin Harcourt, 2000, p.39

[15] James Rodger Fleming, *Historical Perspectives on Climate Change*, published by Oxford University Press US, 1998, p.122

[16] R. Revelle and H. Suess, "Carbon dioxide exchange between atmosphere and ocean and the question of an increase of atmospheric CO2 during the past decades," *Tellus*, Vol.9, 1957

[17] Environmental Pollution Panel of the President's Science Advisory Committee, *Restoring the Quality of Our Environment*, published by Government Printing Office, 1965, p.111-133

[18] Al Gore, *Earth in the Balance*, published by Earthscan, 2007, p.5

[19] Fred Singer, Chauncey Starr, and Roger Revelle, "What To Do About Greenhouse Warming: Look Before You Leap" *Cosmos*, Vol.1, 1991, p.28

[20] Michael Gough, *Politicizing Science: The Alchemy of Policymaking*, published by Hoover Institution Press, 2003, p.288

[21] "The 1992 Campaign; Excerpts From the Debate Among Quayle, Gore and Stockdale," *New York Times*, October 14, 1992

[22] Fred Singer, "Gore's 'global warming mentor,' in his own words" *Environment & Climate News*, January 2000

[23] Lawrence Solomon, "Gore's guru disagreed," *Financial Post (Canada)*, April 28, 2007; also Fred Singer, "Gore's 'global warming mentor,' in his own words" *Environment & Climate News*, January 2000

[24] Fred Singer, "Gore's 'global warming mentor,' in his own words" *Environment & Climate News*, January 2000

[25] George Will, "Al Gore's Green Guilt," *Washington Post*, September 3, 1992; Gregg Easterbrook, "Green Cassandras," *New Republic*, July 6, 1992

[26] Ibid Gough, p.289

[27] Ibid Gough, p.291

[28] Scott Allen "Global Warming At Center Of Libel Suit" *Boston Globe*, December 27, 1993

[29] Ibid Gough, p.291

[30] Jonathan Adler, *Washington Times*, July 27, 1994; As cited in U.S. Congressional Record, *The National Security Implications of Climate Change: Hearing Before the Subcommittee on Investigations and Oversight of the Committee on Science and Technology*, House of Representatives, One Hundred Tenth Congress, First Session, September 27, 2007, published by Government Printing Office, 2008, p.30

[31] Ted Koppel, "Is Environmental Science for Sale?" Transcript from *Nightline*, ABC News, February 24, 1994

[32] Ibid Gough, p.297

[33] Ioana Patringenaru, "Gore's Effort to Spread Message of Renowned Scripps Director Rewarded with Roger Revelle Prize," *UC San Diego News*, March 9, 2009; http://ucsdnews.ucsd.edu/thisweek/2009/03/09_gore.asp

[34] Lewis Smith, "Al Gore's inconvenient judgment," *London Times*, October 11, 2007

[35] "Weather Channel Founder: Sue Al Gore for Fraud," Fox News, March 14, 2008

CHAPTER 23: *The Rhetoric of Alarmists*

[1] News Corporation, Global Warming promotional segment, 2009; video from author's private collection
[2] Ellen Goodman, "No change in political climate," *Boston Globe*, February 9, 2007
[3] James Hansen, Columbia University website, http://www.columbia.edu/~jeh1/mailings/2007/20071121_NMAletters.pdf
[4] Arthur Max, "UN: Ignoring global warming is 'criminally irresponsible,'" *USA Today*, November 12, 2007
[5] Michael Hawthorne, "Blunt answers about risks of global warming," *Chicago Tribune*, August 3, 2008
[6] Piers Akerman, "Stern's report scare-mongering," *Sydney Daily Telegraph*, November 5, 2006
[7] Piers Akerman, "Stern's report scare-mongering," *Sydney Daily Telegraph*, November 5, 2006
[8] Benny Avni, "Mayor Compares Threat of Global Warming to Terrorism," *New York Sun*, February 12, 2008
[9] Ibid. Avni
[10] Tim Wirth, *PBS Frontline*, January 17, 2007; http://www.pbs.org/wgbh/pages/frontline/hotpolitics/interviews/wirth.html
[11] Robbie McKie, "We have only four years left to act on climate change—America has to lead," *The Observer*, January 18, 2009
[12] Robbie McKie, "We have only four years left to act on climate change—America has to lead," *The Observer*, January 18, 2009
[13] Jim Hansen, "The Threat to the Planet," *New York Review of Books*, Vol.53, No.12, July 13, 2006
[14] James Delingpole, "James Hansen: Would you buy a used temperature data set from THIS man?" *London Telegraph Blogs*, January 22, 2010
[15] James Delingpole, "James Hansen: Would you buy a used temperature data set from THIS man?" *London Telegraph Blogs*, January 22, 2010
[16] Mark Morano, "James Hansen's Former NASA Supervisor Declares Himself a Skeptic—Says Hansen 'Embarrassed NASA,'" Inhofe EPW Press Blog, January 27, 2009; http://epw.senate.gov
[17] Stephen Schneider and S.I. Rasool, "Atmospheric Carbon Dioxide and Aerosols: Effects of Large Increases on Global Climate," *Science*, Vol.173, July 9, 1971
[18] Robert Gillette, "No Ice Age Soon, Scientists Say," *Los Angeles Times*, February 18, 1978
[19] Robert Gillette, "No Ice Age Soon, Scientists Say," *Los Angeles Times*, February 18, 1978
[20] Stephen Schneider (guest on Ockham's Razor presented by Robyn Williams), "The global warming debate—Professor Stephen Schneider's response to Professor Don Aitkin," *ABC Radio National*, May 18, 2008
[21] Ross Gelbspan, "Racing to an Environmental Precipice," *Boston Globe*, May 31, 1992
[22] Jonathan Schell, "Our Fragile Earth," *Discover*, October 1989
[23] Terrence Corcoran, "Global Warming: The Real Agenda," *National Post (Canada)*, December 26, 1998
[24] Terence Corcoran, "Global Warming: The Real Agenda," *National Post (Canada)*, December 26, 1998
[25] Dwight Lee, "Eco-hype working against the cause?" *Washington Times*, November 27, 1992; also Fred Kilbourne, "Global Warming: Just a Lot of Hot Air?" *The Actuarial Update*, Vol.26, No.3, March 1998
[26] Fred S. Singer, *Climate Policy From Rio to Kyoto: A Political Issue for 2000 and Beyond*, published by Hoover Institution, Stanford University, 2000
[27] John H. Cushman Jr. and David E. Sanger, "Global Warming; No Simple Fight" *New York Times*, December 1, 1997
[28] United Nations Foundation website, http://www.unfoundation.org/our-solutions/campaigns/global-leadership-for-climate-action/
[29] Charlie Rose Show, "A conversation with Ted Turner," *Public Broadcasting Service (PBS)*, April 1, 2008; also Mike Morris, "Ted Turner: Global warming could lead to cannibalism," *Atlanta Journal-Constitution*, April 3, 2008
[30] Tracey C. Rembert, "Ted Turner: Billionaire, Media Mogul...and Environmentalist," *E Magazine*, January/February 1999
[31] United Nations Foundation website, http://www.unfoundation.org/global-issues/women-and-population/
[32] Charlie Rose Show, "A conversation with Ted Turner," *Public Broadcasting Service (PBS)*, April 1, 2008; also Mike Morris, "Ted Turner: Global warming could lead to cannibalism," *Atlanta Journal-Constitution*, April 3, 2008
[33] Ibid.
[34] Press release, "Ted Turner Announces First-Ever Global Sustainable Tourism Criteria at World Conservation Congress," October 8, 2008; http://www.sustainabletourismcriteria.org
[35] Interview with David Mayer de Rothschild, *The Alex Jones Show*, Genesis Communications Network, July 6, 2007
[36] Peter Griffiths, "Global businesses demand ambitious new climate deal," *Reuters*, September 21, 2009
[37] "Environment and Society: Our Position," BP website, www.bp.com
[38] "Environment and Society: Our Position," BP website, www.bp.com
[39] "The business of climate change—Shell Global Solutions," Shell website, www.shell.com
[40] Jeroen van der Veer, "States should create a climate for change," *Financial Times*, January 23, 2007
[41] Jeroen van der Veer, "States should create a climate for change," *Financial Times*, January 23, 2007
[42] "Energy and Environment: Climate and Emissions: Our Views on Climate Change," Exxon website, www.exxon.com
[43] "Energy and Environment: Climate and Emissions: Our Views on Climate Change," Exxon website, www.exxon.com
[44] Rex Tillerson, "Promoting energy investment and innovation to meet U.S. economic and environmental challenges," speech at Economic Club of Washington, DC, October 1, 2009; http://www.exxonmobil.com/Corporate/news_speeches_20091001_rwt.aspx
[45] Rex Tillerson, "Promoting energy investment and innovation to meet U.S. economic and environmental challenges," speech at Economic Club of Washington, DC, October 1, 2009; http://www.exxonmobil.com/Corporate/news_speeches_20091001_rwt.aspx
[46] United Nations, "Rio Declaration on Environment and Development," *Agenda 21 Earth Summit: United Nations Program of Action from Rio*, published by United Nations, 1992
[47] David Sanger, "Bush Cites Iraqi Threat Posed to U.S. and Allies," *New York Times*, October 7, 2002
[48] Vaclav Klaus, speech to Second International Conference on Climate Change, Marriott Marquis Hotel, New York City, March 8, 2009; http://www.climatescienceinternational.org/

PART IV: LAPDOG MEDIA

[1] Gore Vidal, *The Decline and fall of the American Empire*, published by Odonian Press, 1992, p.40

[2] Thomas Jefferson and Thomas Jefferson Randolph, *Memoir, Correspondence, and Miscellanies, from the Papers of Thomas Jefferson*, published by Gray and Bowen, 1829, p.343

[3] Walter Karp, "All The Congressmen's Men: How Capitol Hill Controls The Press," *Harper's*, July 1989

[4] Noam Chomsky, *Necessary Illusions: Thought Control in Democratic Societies*, published by South End Press, 1989, p.8

[5] Dennis W. Mazzocco, *Networks of Power: Corporate TV's Threat to Democracy*, published by South End Press, 1994, p.viii; Also see Hofstra University website, http://www.hofstra.edu/Academics/soc/AVF/soc_avf_faculty_bios.cfm

[6] Ibid. Mazzocco, p.xiv

[7] Tom Fenton, *Bad News: The Decline of Reporting, the Business of News, and the Danger to Us All*, published by HarperCollins, 2005, p.20

[8] Ibid. Fenton, p.83

[9] Richard Davis and Diana Owen, *New Media and American Politics*, published by Oxford University Press, 1998, p.193

[10] Howard Kurtz, *Hot Air: All Talk, All the Time*, published by Times Books, 1996, p.15-16

[11] Neil Hickey, "Roone At The Top," *Columbia Journalism Review*, May/June 1994; Also see Tom Fenton, *Bad News: The Decline of Reporting, the Business of News, and the Danger to Us All*, published by HarperCollins, 2005, p.58

CHAPTER 24: Ruling-Class Control

[1] Wolfgang Saxon, "F. Lundberg, 92, Author Who Wrote of the Rich," *New York Times*, March 3, 1995

[2] Ibid. Saxon

[3] Ferdinand Lundberg, *America's 60 Families (1937)*, published by Read Books, 2008, p.247

[4] Ibid. Lundberg, p.252

[5] Ibid. Lundberg, p.257

[6] Carroll Quigley, *Tragedy and Hope: A History of the World in Our Time*, published by Macmillan, 1966, p.133

[7] Ibid. Quigley, p.953

[8] Ibid. Lundberg, p.244

[9] Ibid. Lundberg, p.300

[10] "For Press Investigation; Moore Asks Inquiry Into Charges On Preparedness Campaign," *New York Times*, February 14, 1917

[11] "For Press Investigation; Moore Asks Inquiry Into Charges On Preparedness Campaign," *New York Times*, February 14, 1917

[12] "Pacifists House Topic," *Washington Post*, February 16, 1917

[13] "For Press Investigation; Moore Asks Inquiry Into Charges On Preparedness Campaign," *New York Times*, February 14, 1917

[14] Douglas W. Johnson, "An Irresponsible Attack," *New York Times*, February 16, 1917

[15] Douglas W. Johnson, "An Irresponsible Attack," *New York Times*, February 16, 1917

[16] "Pacifists House Topic," *Washington Post*, February 16, 1917

[17] "To J. Hampton Moore," *Washington Post*, February 17, 1917

[18] "Moore Asks An Inquiry, *Washington Post*, February 18, 1917

[19] See Chapter 1

[20] Randolph T. Holhut, *The George Seldes Reader*, published by Barricade Books, 1994, p. 227

[21] "Ickes and Gannett Debate Free Press," *New York Times*, January 13, 1939

[22] "Ickes and Gannett Debate Free Press," *New York Times*, January 13, 1939

[23] "Ickes and Gannett Debate Free Press," *New York Times*, January 13, 1939

[24] Council on Foreign Relations website, http://www.cfr.org/about/corporate/roster.html

[25] *Council on Foreign Relations Annual Report 2008*, published by Council on Foreign Relations

[26] Phyllis Schlafly and Chester Ward, *Kissinger on the Couch*, published by Arlington House Publishers, 1975, p.148

[27] Richard Harwood, "Ruling Class Journalists," *Washington Post*, October 30, 1993

[28] Ibid. Harwood

CHAPTER 25: Covert Government Propaganda

[1] William Schaap, Expert testimony, *King v. Jowers and Other Unknown Co-Conspirators*, Circuit Court of Shelby County, Tennessee for the Thirtieth Judicial District at Memphis, November 30, 1999; Video footage from author's personal collection, also available online; Transcript available from King Center website, http://www.thekingcenter.org/news/trial/Volume9.html

[2] Edward Bernays, *Propaganda*, published by Horace Liveright Publishing, 1928, p.9-10

[3] Ibid. Schaap

[4] Ibid. Schaap

[5] Ibid. Schaap

[6] Ibid. Schaap

[7] Ibid. Schaap

[8] Ibid. Schaap

[9] Ibid. Schaap

[10] Ibid. Schaap

[11] Ibid. Schaap

[12] Erik Hedegaard, "The Last Confessions of E. Howard Hunt," Rolling Stone, April 5, 2007

[13] E. Howard Hunt and Greg Aunapu, *American Spy: My Secret History in the CIA, Watergate, and Beyond*, published by John Wiley and Sons, 2007 p.148-149

[14] "Final Report of the Select Committee to Study Government Operations With Respect to Intelligence Activities," published by U.S. Government Printing Office, 1976, p.192; Also see E. Howard Hunt and Greg Aunapu, American Spy: My Secret History in the CIA, Watergate, and Beyond, published by John Wiley and Sons, 2007 p.152

[15] Ibid.

[16] "The CIA Report The President Doesn't Want You To Read," *Village Voice*, February 16, 1976; Also see Philip Agee, Otis Pike, and United States Congress House Select Committee on Intelligence, *CIA: The Pike Report*, published by Spokesman Books, 1977, p.224

[17] Ibid.

[18] Ibid. Hunt, p.153
[19] Sean Gervasi, "CIA Covert Propaganda Capability," *Covert Action Information Bulletin*, No.7, December 1979/January 1980; As cited in Daniel Brandt, "Journalism and the CIA: The Mighty Wurlitzer," *Newsline*, No.17, April-June 1997
[20] Carl Bernstein, "The CIA and the Media," *Rolling Stone*, October 20, 1977
[21] Ibid. Bernstein
[22] Ibid. Bernstein
[23] *Washington Post* website, http://www.washpostco.com/dir_kg.htm
[24] Stephen L. Vaughn, *Encyclopedia of American Journalism*, published by CRC Press, 2008, p.201; Also see Mark Perry, "The Case Against William Webster," *Regardie's Magazine*, Vol.10, No.5, January 1990
[25] Ibid. Bernstein
[26] John M. Crewdson and Joseph B. Treaster, "The CIA's 3-Decade Effort To Mold the World's Views," *New York Times*, December 25, 1977
[27] Ibid. Crewdson and Treaster
[28] Ibid. Crewdson and Treaster
[29] Ibid. Schaap
[30] "Ron Paul: After 'CIA coup,' agency 'runs military,'" *Raw Story*, January 20, 2010
[31] Ralph W. McGehee, *Deadly Deceits: My 25 Years in the CIA*, published by Sheridan Square Publications, 1983, p.191
[32] "Censorship by the CIA Challenged in Court Suit," *New York Times*, March 29, 1981
[33] Ibid. McGehee, p.194
[34] Ibid. McGehee, p.192
[35] Ibid. McGehee, p.192-193
[36] Ibid. McGehee, p.194
[37] Ibid. McGehee, p.194
[38] Ibid. Hunt, p.153
[39] Philip H. Melanson, *Secrecy Wars: National Security, Privacy, and the Public's Right to Know*, published by Brassey's, 2002
[40] Freedom of Information Act, *Bingham v. U.S. Department of Justice and FBI*, Case No. 1:05-00475; Judicial Watch website, http://www.judicialwatch.org/archive/2006/doubletreeaffadavit.pdf
[41] James Risen and Eric Lichtblau, "Early Test for Obama on Domestic Spying Views," *New York Times*, November 18, 2008
[42] Scott Shane, David Johnston and James Risen, "Secret U.S. Endorsement of Severe Interrogation," *New York Times*, October 4, 2007
[43] U.S. Department of State website, http://www.state.gov/r/pa/ho/time/cwr/17603.htm
[44] U.S. Code, Title 50, Section 413 (a)(2); http://uscode.house.gov/search/criteria.shtml
[45] U.S. Code, Title 50, Section 413 (a)(2)(b); http://uscode.house.gov/search/criteria.shtml
[46] U.S. Code, Title 50, Section 415b (d)(3)(A); http://uscode.house.gov/search/criteria.shtml
[47] U.S. Code, Title 50, Section 415b (e); http://uscode.house.gov/search/criteria.shtml
[48] U.S. Code, Title 50, Section 415b (f); http://uscode.house.gov/search/criteria.shtml
[49] National Security Archive Electronic Briefing Book No. 255, "New Kissinger 'Telcons' Reveal Chile Plotting At Highest Levels Of U.S. Government," Secretary Rogers, September 14, 1970; http://www.gwu.edu/~nsarchiv/NSAEBB/NSAEBB255/index.htm
[50] Ibid.
[51] National Security Archive, "Chile and the United States: Declassified Documents relating to the Military Coup, 1970-1976," CIA, Genesis of Project FUBELT, September 16, 1970; http://www.gwu.edu/~nsarchiv/NSAEBB/NSAEBB8/nsaebb8.htm
[52] National Security Archive, "Chile and the United States: Declassified Documents relating to the Military Coup, 1970-1976," CIA, Operating Guidance Cable on Coup Plotting, October 16, 1970; http://www.gwu.edu/~nsarchiv/NSAEBB/NSAEBB8/nsaebb8.htm
[53] Carl Bernstein, "The CIA and the Media," *Rolling Stone*, October 20, 1977
[54] Stephen Kinzer, *All the Shah's Men: An American Coup and the Roots of Middle East Terror*, published by John Wiley and Sons, 2007, .67
[55] "Challenge of the East," *Time* magazine, January 7, 1952
[56] James Risen, "SECRETS OF HISTORY: The C.I.A. in Iran—A special report," *New York Times*, April 16, 2000
[57] James Risen, "C.I.A. Tried, With Little Success, to Use U.S. Press in Coup," *New York Times*, April 16, 2000
[58] Ibid. Risen
[59] Ibid. Risen
[60] Ibid. Risen
[61] "A Conversation with Michael Hayden," transcript from Council on Foreign Relations address, September 7, 2007; http://www.cfr.org
[62] Barack Obama, "Remarks by the President on a New Beginning," speech at Cairo University, Cairo, Egypt, June 4, 2009
[63] Helene Cooper and David Sanger, "Obama Condemns Iran's Iron Fist Against Protests," *New York Times*, June 23, 2009
[64] Tim Shipman, "Bush sanctions 'black ops' against Iran," *London Telegraph*, May 27, 2007
[65] Tim Shipman, "Bush sanctions 'black ops' against Iran," *London Telegraph*, May 27, 2007
[66] Richard Spencer, "Iran threatens revenge against Britain over bombing," *London Telegraph*, October 19, 2009
[67] Tim Weiner, "Word for Word/The Bay of Pigs," *New York Times*, March 25, 2001
[68] Edward R. Drachman and Alan Shank, *Presidents and Foreign Policy: Countdown to Ten Controversial Decisions*, published by State University of New York Press, 1997, p.87
[69] Ibid. Drachman and Shank
[70] "Protest Against Theft," *Time*, January 25, 1960
[71] Tim Weiner, "Word for Word/The Bay of Pigs," *New York Times*, March 25, 2001
[72] Howard Jones, *The Bay of Pigs*, published by Oxford University Press, 2008, p.19
[73] Gabriel Molina, *Diario de Girón*, published by Editora Política, 1983, p.53; National Security Archive, Bay of Pigs Chronology, http://www.gwu.edu/~nsarchiv/bayofpigs/chron.html
[74] Ibid. Molina, p.67-68
[75] Tim Weiner, "C.I.A. Had Ability to Plant Bay of Pigs News, Document Shows," *New York Times*, March 24, 2001
[76] Ibid. Weimer
[77] National Security Archive, "Documents Reveal CIA'S Dulles Wanted Cuba To Ask For Soviet Bloc Arms In 1959," White House, Memorandum, TOP SECRET, [Arthur Schlesinger, Jr. to President Kennedy], February 11, 1961; http://www.gwu.edu/~nsarchiv/bayofpigs/press2.html
[78] Ibid.
[79] National Security Archive, "Pentagon Proposed Pretexts for Cuba Invasion in 1962," Chairman, Joint Chiefs of Staff, Justification for US Military Intervention in Cuba, March 13, 1962; http://www.gwu.edu/~nsarchiv/news/20010430/
[80] Ibid.
[81] Ibid.
[82] U.S. Code, Title 50, Section 415b (f); http://uscode.house.gov/search/criteria.shtml

[83] Francisco Gil-White, ""Did the National Security Act of 1947 destroy freedom of the press?" *Historical and Investigative Research*, January 3, 2006; http://www.hirhome.com/national-security.htm
[84] Peter Grose, "Uncle Sam's Nazi's," *Washington Post*, April 24, 1988
[85] Ibid. Gil-White
[86] Wolfgang Benz and Thomas Dunlap, *A Concise History of the Third Reich*, published by University of California Press, 2006, p.26
[87] Gerald Caplan, "Republican Party Home to Racists, Fascists," *Toronto Star*, January 2, 1989
[88] Ibid. Caplan
[89] Edward P. Morgan, "Press Betrayed. Purge Due. Please Tell George Bush. And Publisher," *New York Times*, May 31, 1976
[90] Ibid. Morgan
[91] Henry Fowles Pringle, "Politicians and the Press," *Harper's*, April 1928
[92] Henry Fowles Pringle, "Politicians and the Press," *Harper's*, April 1928
[93] Joseph Kraft, "Politics of the Washington Press Corps," *Harper's*, June 1965
[94] Leonard Downie, Jr., *The New Muckrakers*, published by New Republic Book Company, 1976 p.51
[95] Doris Kearns Goodwin, *Lyndon Johnson and the American Dream*, published by Harper & Row, 1976, p.247
[96] Walter Karp, "All The Congressmen's Men: How Capitol Hill Controls The Press," *Harper's*, July 1989
[97] Stephen Hess, *The Washington Reporters*, published by Brookings Institution Press, 1981, p.99
[98] Ibid. Karp
[99] Ibid. Karp
[100] Ibid. Karp
[101] Ibid. Karp
[102] Ibid. Karp
[103] David Halberstam, "The Power and the Profits: Part II," *The Atlantic*, February 1976
[104] Bob Woodward, "Gadhafi Target of Secret U.S. Deception Plan," *Washington Post*, October 2, 1986
[105] Alfonso Chardy, "Secrets Leaked to Harm Nicaragua, Sources Say," *Miami Herald*, October 13, 1986
[106] Ibid. Chardy
[107] Frank Rich, "The White House Stages Its 'Daily Show,'" *New York Times*, February 20, 2005
[108] Greg Toppo, "Education Dept. paid commentator to promote law," *USA Today*, January 7, 2005
[109] Ibid. Toppo
[110] Frank Rich, "The White House Stages Its 'Daily Show,'" *New York Times*, February 20, 2005
[111] David Barstow and Robin Stein, "Under Bush, a New Age of Prepackaged TV News," *New York Times*, March 13, 2005
[112] Ibid. Barstow and Stein
[113] Ibid. Barstow and Stein
[114] Ibid. Barstow and Stein
[115] "Bush on Iran, Baseball, DNA and Judges," *New York Times*, March 17, 2005
[116] David Barstow, "Behind TV Analysts, Pentagon's Hidden Hand," *New York Times*, April 20, 2008
[117] Ibid. Barstow
[118] Ibid. Barstow
[119] Ibid. Barstow
[120] James Dao and Eric Schmitt, "Pentagon Readies Efforts to Sway Sentiment Abroad," *New York Times*, February 19, 2002
[121] Ibid. Dao and Schmitt, "Pentagon Readies Efforts to Sway Sentiment Abroad," *New York Times*, February 19, 2002
[122] James Dao and Eric Schmitt, "A 'Damaged' Information Office Is Declared Closed by Rumsfeld," *New York Times*, February 27, 2002
[123] James Dao and Eric Schmitt, "A 'Damaged' Information Office Is Declared Closed by Rumsfeld," *New York Times*, February 27, 2002
[124] Eric Schmitt, "Pentagon and Bogus News: All Is Denied," *New York Times*, December 5, 2003
[125] Eric Schmitt, "Pentagon and Bogus News: All Is Denied," *New York Times*, December 5, 2003

CHAPTER 26: Big Media and Government Collusion

[1] Tom Fenton, *Bad News: The Decline of Reporting, the Business of News, and the Danger to Us All*, published by HarperCollins, 2005, p.243
[2] Ibid. Fenton, p.131
[3] Ibid. Fenton, p.131
[4] Ibid. Fenton, p.244
[5] Ibid. Fenton, p.128
[6] "Australia to implement mandatory internet censorship," *Herald Sun (Melbourne)*, October 29, 2008
[7] Communications Act of 1934, United States Code, Title 47, Section 315; http://www.fcc.gov/Reports/1934new.pdf
[8] Robert Kane Pappas (Director), *Orwell Rolls in his Grave*, Sky Island Films, 2004
[9] Dennis W. Mazzocco, *Networks of Power: Corporate TV's Threat to Democracy*, published by South End Press, 1994, p.162
[10] Federal Communications Commission website, http://www.fcc.gov/telecom.html
[11] John Nichols and Robert McChesney, "FCC: Public Be Damned," *The Nation*, June 2, 2003
[12] Maurice Hinchey, House Resolution 3302, 109th Congress, 1st Session, July, 14, 2005; http://www.thomas.gov/home/gpoxmlc109/h3302_ih.xml
[13] Ben Bagdikian website, http://benbagdikian.net/Docs/bio.htm
[14] Dennis W. Mazzocco, *Networks of Power: Corporate TV's Threat to Democracy*, published by South End Press, 1994, p.5
[15] Ben H. Bagdikian, *The Media Monopoly*, published by Beacon Press, 1997
[16] Ben H. Bagdikian, *The New Media Monopoly*, published by Beacon Press, 2004, p.3
[17] Ibid. Bagdikian, p.8
[18] Dennis W. Mazzocco, *Networks of Power: Corporate TV's Threat to Democracy*, published by South End Press, 1994, p.141
[19] Ibid. Mazzocco, p.101
[20] Robert McChesney, *Corporate Media and the Threat to Democracy*, published by Seven Stories Press, 1997, p.43
[21] Tom Feran, "Media Access Examined: Bill Moyers Special," *Columbus Dispatch*, June 8, 1999; Also see Otto Lerbinger, *Corporate Public Affairs: Interacting with Interest Groups, Media, and Government*, published by Routledge, 2006, p.368
[22] Ibid. Feran
[23] Otto Lerbinger, *Corporate Public Affairs*, published by Routledge, 2006, p.368
[24] Eric Boehlert, "One big happy channel?" *Salon.com*, June 28, 2001
[25] Frank Ahrens, "Clear Channel Sale to End Era," *Washington Post*, November 17, 2006
[26] Peter DiCola, Kristin Thomson and Future of Music Coalition, *Radio Deregulation: Has It Served Citizens and Musicians?* published by Future of Music Coalition, 2002; http://www.futureofmusic.org/images/FMCradiostudy.pdf

27 Robert Kane Pappas (Director), *Orwell Rolls in his Grave*, Sky Island Films, 2004
28 Ibid. Pappas
29 Common Cause Education Fund, *The Fallout from the Telecommunications Act of 1996: Unintended Consequences and Lessons Learned*, published by Common Cause, 2005
30 "Rewriting the Rules," *NewsHour with Jim Lehrer* on PBS, June 2, 2003; http://www.pbs.org/newshour/bb/media/jan-june03/powell_6-2.html
31 Ibid.
32 Ibid.
33 Ibid.
34 John Nichols and Robert McChesney, "FCC: Public Be Damned," *The Nation*, May 15, 2003
35 Ibid. Nichols and McChesney
36 Ibid. Nichols and McChesney
37 Ibid. Nichols and McChesney
38 Ibid. Nichols and McChesney
39 Ibid. Nichols and McChesney
40 Stephen Labaton, "F.C.C. Media Rule Blocked In House In A 400-21 Vote," *New York Times*, July 24, 2003
41 Ibid. Labaton
42 Ibid. Labaton
43 Amy Goodman, "Court Rejects FCC Attempt to Rewrite Nation's Media Ownership Laws," *Democracy Now!* June 25, 2004
44 "Lawyer: FCC Ordered Media Ownership Study Destroyed," Associated Press via *Fox News*, September 14, 2006
45 Ibid.
46 Ibid.
47 "Chairman Kevin J. Martin Proposes Revision to the Newspaper/Broadcast Cross-Ownership Rule," FCC Press Release, November 13, 2007; http://hraunfoss.fcc.gov/edocs_public/attachmatch/DOC-278113A1.pdf
48 Kevin Martin, "The Daily Show," *New York Times*, November 13, 2007
49 Louise Witt, "FCC commissioner Michael Copps vs. Big Media," *Salon.com*, December 3, 2007
50 "The FCC Vote," *Bill Moyers Journal* on PBS, December 21, 2007
51 "FCC Update," *Bill Moyers Journal* on PBS, November 16, 2007; http://www.pbs.org/moyers/journal/11162007/transcript1.html
52 Ibid.
53 Eric Pryne and Stuart Eskenazi, "Seattle crowd blasts FCC on big media," *Seattle Times*, November 10, 2007
54 Josh Silver, "Seattle Opens Can 'o Whoop Ass on FCC Chairman," *Huffington Post*, November 10, 2007
55 "FCC Update," *Bill Moyers Journal* on PBS, November 16, 2007; http://www.pbs.org/moyers/journal/11162007/transcript1.html
56 Amy Goodman, "'Today's Decision Would Make George Orwell Proud'—FCC Commissioner Michael Copps on the FCC's Vote to Rewrite the Nation's Media Ownership Rules," *Democracy Now!* December 26, 2007; http://www.democracynow.org/2007/12/26/fcc_michael_copps
57 Robert Greenwald (producer and director), *Outfoxed: Rupert Murdoch's War on Journalism*, distributed by Brave New Films, 2004; See transcript at http://www.outfoxed.org/docs/outfoxed_transcript.pdf
58 Ibid. Greenwald
59 Ibid. Greenwald
60 Ibid. Greenwald
61 Ibid. Greenwald
62 Sheldon Rampton and John Stauber, "This Report Brought To You By Monsanto," *The Progressive*, July 1998
63 Paul Kingsnorth, "Monsanto's Bovine Growth Hormones," *The Ecologist*, October 1998
64 Ibid. Rampton and Stauber
65 Jane Akre and Steve Wilson BGH Bulletin website, http://www.foxbghsuit.com/exhibit%20c.htm
66 Ibid. Rampton and Stauber
67 Cythnia Cotts, "Does Fox Slant the News?" *Village Voice*, November 28, 2000
68 Ibid. Rampton and Stauber
69 "Judge: Reporters not liable for lawyer fees," *St. Petersburg Times (Florida)*, August 18, 2004
70 Cythnia Cotts, "Does Fox Slant the News?" *Village Voice*, November 28, 2000
71 *WTVT-TV v. Jane Akre*, Florida Second District Court of Appeals, Case No.2D01-529, opinion filed February 14, 2003; http://www.2dca.org/opinion/February%2014,%202003/2D01-529.pdf
72 Federal Communications Commission, "Complaints About Broadcast Journalism," *Mass Media Bureau Publication 80*, June 1997; http://www.fcc.gov/mb/enf/forms/pub80.html
73 *Red Lion Broadcasting Co. v. FCC*, U.S. Supreme Court, 395 U.S. 367 (1969), opinion filed June 9, 1969
74 Ibid. Greenwald

CHAPTER 27: *Media Duplicity: Case Studies*

1 Oswald Spengler and Edwin Franden Dakin, *Today and Destiny: Vital Excerpts from The Decline of the West of Oswald Spengler*, published by A.A. Knopf, 1940, p.64
2 Mike Clark, "Vote Fraud Scandal Casts a Shadow on Mayoral Election," *Los Angeles Times*, November 13, 1997
3 Gregory Palast, "Florida's 'Disappeared Voters': Disfranchised by the GOP," *The Nation*, January 18, 2001
4 Ibid. Palast
5 Note: Database Technologies later merged with ChoicePoint, Inc.
6 Gregory Palast, "Vanishing Votes," *The Nation*, May 17, 2004
7 Gregory Palast, "Florida's 'Disappeared Voters': Disfranchised by the GOP," *The Nation*, January 18, 2001
8 Ibid. Palast
9 Ibid. Palast
10 Ibid. Palast
11 Gregory Palast, "Silence of the Lambs: The Election Story Never Told," *MediaChannel*, March 12, 2001; http://www.mediachannel.org/views/whistleblower/palast.shtml
12 Robert Capps, "Sex-slave whistle-blowers vindicated," *Salon.com*, August 6, 2002
13 Kelly Patricia O'Meara, "DynCorp disgrace," *Insight on the News*, February 4, 2002
14 Ibid. O'Meara
15 Ibid. O'Meara
16 Ibid. O'Meara

[17] Robert Capps, "Outside the law," *Salon.ccm*, June 26, 2002
[18] Robert Capps, "Crime without punishment," *Salon.com*, June 27, 2002
[19] John Crewdson, "Contractor tries to avert repeat of Bosnia woes; Sex scandal still haunts DynCorp," *Chicago Tribune*, April 19, 2003
[20] Ibid. O'Meara
[21] Ibid. O'Meara
[22] Robert Capps, "Sex-slave whistle-blowers vindicated," *Salon.com*, August 6, 2002
[23] Ibid. O'Meara
[24] Colum Lynch, "U.N. Halted Probe of Officers' Alleged Role in Sex Trafficking," *Washington Post*, December 27, 2001
[25] Ibid. Lynch
[26] Ibid. Lynch
[27] Ibid. Lynch
[28] Ian Traynor, "Nato force 'feeds Kosovo sex trade,'" *London Guardian*, May 7, 2004
[29] Ian Traynor, "Nato force 'feeds Kosovo sex trade,'" *London Guardian*, May 7, 2004
[30] Ibid. Traynor
[31] Wendy McElroy, "Is the U.N. Running Brothels in Bosnia?" *Fox News*, January 22, 2002
[32] Cam Simpson, "Into a war zone, on a deadly road," *Chicago Tribune*, October 10, 2005
[33] Cam Simpson, "Desperate for work, lured into danger," *Chicago Tribune*, October 9, 2005
[34] David Rohde, "The Struggle For Iraq: Foreign Labor; Indians Who Worked in Iraq Complain of Exploitation," *New York Times*, May 7, 2004
[35] Ibid. Rohde
[36] Cam Simpson, "U.S. stalls on human trafficking: Pentagon has yet to ban contractors from using forced labor," *Chicago Tribune*, December 27, 2005
[37] Jarrett Murphy, "Cheney's Halliburton Ties Remain," *Associated Press* via *CBS News*, September 26, 2003
[38] Ibid. Murphy
[39] Ibid. Murphy
[40] Ibid. Murphy
[41] Henry A. Waxman, "Halliburton's Performance Under the Restore Iraqi Oil 2 Contract," Committee on Government Reform, U.S. House of Representatives, March 28, 2006; http://oversight.house.gov/story.asp?ID=1032
[42] Clifford Krauss, "Former KBR Executive Pleads Guilty to Bribery," *New York Times*, September 3, 2008
[43] "Sale of KBR Bolsters Profit at Halliburton," *New York Times*, July 24, 2007
[44] Christine Lagorio, "Halliburton's Dubai Move Sparks Outcry," *Associated Press* via *CBS News*, March 12, 2007
[45] "Halliburton bails out of Iraq, KBR and now America," *HalliburtonWatch.org*, March 12, 2007
[46] Cam Simpson, "U.S. stalls on human trafficking: Pentagon has yet to ban contractors from using forced labor," *Chicago Tribune*, December 27, 2005
[47] CSPAN coverage of FY 2006 Defense Department Budget, House Armed Service Committee, March 11, 2005; Also see http://www.fromthewilderness.com/free/ww3/031505_mckinney_transcript.shtml
[48] A.M. Rosenthal, "On My Mind; The New World Order," *New York Times*, May 5, 1998
[49] Ibid. Rosenthal
[50] Ibid. Rosenthal
[51] Ibid. Rosenthal
[52] Ibid. Rosenthal
[53] Gideon Rachman, "And now for a world government," *Financial Times (London)*, December 8, 2008
[54] Ibid. Rachman
[55] Ibid. Rachman

CONCLUSION

[1] The National Archives and Michael Beschloss, *Our Documents: 100 Milestone Documents from the National Archives*, published by Oxford University Press, 2006, p.222
[2] Andy Wachowski and Lana Wachowski (Directors), *The Matrix*, Groucho II Film Partnership, 1999
[3] http://www.solari.com
[4] http://www.themoneymasters.com
[5] "Rebuilding America's Defenses: Strategy, Forces and Resources For a New Century," published by The Project for the New American Century, September 2000, p.60
[6] Gary Null, Carolyn Dean, Martin Feldman, Debora Rasio and Dorothy Smith, "Death by Medicine," published October 2000
[7] Benjamin Franklin, William Temple Franklin and William Duane, *Memoirs of Benjamin Franklin, Volume 2*, published by Derby and Jackson, 1859, p.99

www.ingramcontent.com/pod-product-compliance
Lightning Source LLC
Chambersburg PA
CBHW081144270326
41930CB00014B/3029